Carlos A. Bana e Costa (Ed.)

Readings in Multiple Criteria Decision Aid

With 124 Figures

Springer-Verlag
Berlin Heidelberg New York
London Paris Tokyo
Hong Kong Barcelona

Carlos A. Bana e Costa
Technical University of Lisbon (IST)
IST-CESUR
Av. Rovisco Pais
1000 Lisbon, Portugal

ISBN-13:978-3-642-75937-6 e-ISBN-13:978-3-642-75935-2
DOI: 10.1007/978-3-642-75935-2

2142/7130-543210

FOREWORD

Multiple Criteria Decision Aid is a field which has seen important developments in the last few years. This is not only illustrated by the increasing number of papers and communications in the scientific journals and Congresses, but also by the activities of several international working groups. In 1983, a first Summer School was organised at Catania (Sicily) to promote multicriteria decision-aid in companies and to encourage specialists to exchange didactic material. The second School was held in 1985 at Namur (Belgium) and I am pleased now to present the selected readings from the "Third International Summer School on Multicriteria Decision Aid: Methods, Applications and Software", which took place in Monte Estoril (Portugal), in 1988.

Such was the quality of the contributions presented by the lecturers during the Summer School that I have decided to take advantage of this opportunity to produce a more carefully prepared and homogeneous book rather than a simple volume of proceedings. All the initial versions of the selected papers were revised and some, although not included in the programme of the School, were written in order to give a more complete overview of the MCDA field.

I believe **"Readings in Multiple Criteria Decision Aid"** has achieved its main goal: to cover the different trends in MCDA research, not only in terms of "methods" but also in discussion of fundamental methodological aspects, basic theoretical concepts and conditions and problems embodying a robust real-world decision aid practice. I am convinced that the papers specially related with this last issue will show the readers that the practice of decision aid is something more complex, elaborate and responsible than a simple direct application of multicriteria "methods" and "software" manipulation.

It is especially important to emphasise "decision-aid" as an human art activity in an epoch where the advent of new powerful computer facilities has given rise to an increasing production of more and more attractive, user-friendly, visual interactive MCDA software instruments. On the one hand, to use some "good" (in strict computational terms) software based on a "bad" (theoretically inconsistent) method is a danger that MCDA practitioners have to avoid. On the other hand, it should not be forgotten that even the "best" method and software always requires human verification of its adequacy to the specific characteristics of each particular decision-making process, and careful interpretation and validation of the robustness of its out-puts. Bearing this in mind, I am sure that the MCDA software described in various papers of this book will be significant tools for **aiding "analysts" in aiding "decision-makers"**.

The International Summer Schools on MCDA are promoted by an International Permanent Committee. Due to our strong friendship links, I have invited Philippe Vincke, current President of this Committee, to share with me the introductory chapter of this book.

ACKNOWLEDGEMENTS

The Third International Summer School in MCDA would never have achieved its good working relationships and enjoyable atmosphere without the active collaboration of all of its one hundred and twenty participants; included the twenty six scientists with lecturing responsibilities and of course the accompanying persons.

A special emphasis must be given to the fundamental aid of Commanders José Cervaens Rodrigues and César Dinis das Neves of the Portuguese Navy and to the enthusiastic collaboration of its Operations Research Centre (CIOA) staff. I sincerely thank the Portuguese Navy for their support in the organisation of this School, namely the facilities provided by the Naval War Institute.

The Portuguese Operations Research Society (APDIO) and the Centre of Urban and Regional Systems (CESUR) of the Technical University of Lisbon greatly help the Organising Committee and the support provided by the following Portuguese institutions was vital for the Summer School:

- Ministério da Indústria e Energia
- Ministério dos Negócios Estrangeiros
- Secretaria de Estado da Ciência e Tecnologia
- Junta Nacional de Investigação Científica e Tecnológica
- Instituto Nacional de Investigação Científica
- Fundação Calouste Gulbenkian
- Fundação Luso-Americana para o Desenvolvimento
- Departamento de Engenharia Civil - Instituto Superior Técnico
- Administração do Porto de Lisboa
- Imprensa Nacional - Casa da Moeda
- Banco de Fomento Nacional
- Caixa Geral de Depósitos
- Companhia de Seguros Bonança
- CIMPOR, Cimentos de Portugal
- Estaleiros Navais de Viana do Castelo
- Olivetti
- Apple
- Jamoral
- MULTISIS
- CISED
- QUADRIFORMA
- Clube Mimosa

Many people were involved in the organisation of the School. Henrique Sequeira and César Dinis das Neves shared responsibility with me on the Organising Committee. The assistance of Graça Evaristo da Silva, Antónia Barroso and Manuela Ferrão, both before and during the School, was fundamental to its success. Wilques Erlacher and Eduardo Vieitas also deserve to be mentioned for their work in typing and preparing this volume.

Finally, I wish to thank my wife and sons for their patience and encouragement.

Lisbon, May 1990 CARLOS A. BANA E COSTA

CONTENTS

CHAPTER II
THE OUTRANKING APPROACH

CHAPTER III
VALUE AND UTILITY THEORY APPROACH

CHAPTER IV
INTERACTIVE MULTIPLE OBJECTIVE PROGRAMMING

CHAPTER V
GROUP DECISION AND NEGOTIATION

CHAPTER VI
THE SCHOOL CASE-STUDY

INTRODUCTORY CHAPTER

MULTIPLE CRITERIA DECISION AID: AN OVERVIEW

Carlos A. Bana e Costa
IST/CESUR - Technical University of Lisbon, PORTUGAL
and
Philippe Vincke
Free University of Brussels, BELGIUM

1. MODELLING DECISION SITUATIONS

What do we mean by "Multiple Criteria Decision Aid (MCDA)"? To answer this question is far from being an easy task. Some papers in this book contain important elements of response.

First of all, why "decision aid" instead of "decision making"? In his first paper in this book Bernard Roy clarifies the main differences between decision aid and decision making attitudes when facing a decision situation. As he states, "the development (of multicriteria approaches) makes us fully aware of the limitations on objectivity encountered in the field of decision aid and, consequently, of the virtual impossibility of providing a truly scientific foundation for an optimal decision. Work carried out under the rubric of MCDM (Multiple Criteria Decision Making) bases its claims to legitimacy on a framework in which these limitations are left aside. MCDA must be seen from a different perspective. Its aim is, above all, to enable us to enhance the degree of conformity and coherence between the evolution of a decision-making process and the value systems and objectives of those involved in this process. (...) For that purpose, concepts, tools and procedures must be conceived to help us make our way in the presence of ambiguity, uncertainty and an abundance of bifurcations." Denis Bouyssou synthesizes these thoughts: "decision-aid consists in trying to provide answers to questions raised by actors involved in a decision process using a clearly specified model".

Second, to say "multiple criteria decision aid" and not only "decision aid" means that we are referring to multiple criteria approaches for decision aiding. Following again Bouyssou, "in a mono-criterion approach the analyst builds a unique criterion capturing all the relevant aspects of the problem. The comparisons (between alternatives) that are deduced from that criterion are to be interpreted as expressing "global preferences", i.e. preferences taking all the relevant points of view into account. In a multiple criteria approach the analyst seeks to build several criteria using several points of view. These points of view represent the different axes along which the various actors of the decision process justify, transform and argue their preferences. The comparisons deduced from each of these criteria should therefore be interpreted as partial preferences, i.e. preferences restricted to the aspects taken into account in the point of view underlying the definition of the criterion".

Another fundamental issue in MCDA is the role played by the analyst in the decision aid process. Anna Ostanello brings a very interesting contribution to the understanding of "how the analyst (or model builder) actually carries out decision aid activities in order to produce some "valid" answers to a client". "Actual decision aiding situations can be very different and made differently complex by the nesting of technical, organisational and political constraints and problems. These factors not only condition the analyst's acting, but they can determine an evolution of the analyst's and client's interaction". By analysing two actual decision aid processes and their results (formal representations, problem formulation, operative tools, models) Anna Ostanello helps us in recognising the kinds of problems, the nature and the structuring of the analyst's activities. She concludes that "the multicriteria approach can be usefully used not only for the specific multicriteria methods - that cannot always be applied - but also as: either a soft tool allowing, for instance, different elements of representation to be arranged into formal schemes, so to create a link between conceptual and formal models, or even a framework, suitable to outline logics of connection between development and other kind activities and thus to control the modelling process evolution within the decision aid process".

Thus, the activity of multicriteria decision aiding can not be only restricted to the resolution of a problem where one has to aggregate <u>given</u> preferences on a <u>given</u> set of potential actions. The identification of the set of actions (and of the fuzziness of its frontier), the construction of the criteria and the preference modelling are fundamental and often difficult aspects of decision-aid. These aspects are considered in the papers of Denis Bouyssou, Jean-Claude Vansnick and Philippe Vincke.

As Bouyssou states, "building criteria is an important and difficult phase of the decision-aid process. (...) A criterion is a model allowing to establish preference relations between alternatives". So, "when building a criterion, the analyst should keep in mind that it is necessary that all the actors of the decision process adhere to the comparisons that will be deduced from that model. This implies a number of important consequences". Between those referred by Bouyssou, we emphasise that "the choice of a particular way to build a criterion must take into account the quality of the "data" used to build it. In particular, the comparisons deduced from the criterion should take into account the elements of uncertainty, imprecision and/or inaccurate determination affecting the data used to build it". Making use of adequate examples, Bouyssou illustrates some problems emerging in the activity of building criteria, such as the preferential independence, the problems of imprecision and uncertainty, or those of "measuring" various kinds of consequences. He presents some standard techniques for construction criteria and he also treats the problem of coherence and dependence between criteria given some remarks concerning the choice of a consistent family of criteria.

The search for a consistent mathematical basis for MCDA it is a fundamental prerequisite for analysts' activities to have a sense.

Basic concepts of Measurement Theory (presented in Jean-Claude Vansnick's paper) and of Preference Modelling (reviewed in Philippe Vincke's paper) must then be an indispensable theoretical background of all analysts, without which they can not seriously exercise decision aid.

Vansnick reveals the "importance of the notion of measurement for decision aid" and he discusses the problem of meaningfulness associated with the different types of scales that can be used to represent preference information. He also emphasises the crucial importance of the property of preference independence in MCDA.

Knowing the implicit assumptions of the various preference models which can be used in decision-aid is a must for the scientist. Operational research, economy, finance, statistical decision theory are usually based on the "traditional model" where a decision "a" is considered as better than another "b" if its value (for a given function) is greater (or smaller) than the value of "b". The underlying preference structure implies the absence of any incomparability, of any indifference or preference threshold, of any "degree" of preference. In order to be able to take these inevitable phenomena into account, some other models have been introduced in the literature these last years and should be more and more often incorporated in decision-aid procedures: they are presented in Vincke's paper with the corresponding underlying assumptions. A very brief introduction is also given to preference modelling under uncertainty, geometrical representation of preferences and adjustement problems.

"Decision Making in Ill-structured Environments and with Multiple Criteria" by Hans-Jürgen Zimmermann closes the chapter having "modelling" as keyword. It illustrates the use of fuzzy sets concepts in decision-aid for modelling "vague information" and it shows "that for this type of uncertainty (vagueness) (...) models and methods exist which are also adequate for MCDM (...)".

Three types of Operational Approaches for MCDA can be distinguished: outranking, value and utility theory and interactive approaches.

2. THE OUTRANKING APPROACH

The outranking approach consists first in building, on the set of actions, a relation (called outranking relation) to represent the solid part of the preferences of the decision-maker; this relation is not necessarily transitive nor complete ("incomparability" is a key outranking concept). The second step is the exploitation of this relation in order to help the decision-maker in his choice, or sorting, or ordering problems.

In "The Outranking approach and the Foundations of ELECTRE Methods", Bernard Roy presents the fundamental aspects of this approach: the basic principles in the construction of the outranking relation, the concepts of concordance and discordance (including the notion of importance of each criterion), the properties of the results obtained from the exploitation of an outranking relation and the main characteristics of the six ELECTRE methods (I, IS, II, III, IV and A). How to use outranking methods for decision-aid, how to choose among ELECTRE methods and how to give numerical values to indifference, preference and veto thresholds and to the coefficients of importance of the criteria constitute the subject of a fundamental section of Roy's paper devoted to practical considerations.

Daniel Vanderpooten argues that "the outranking approach promotes a realistic and prudent preference modelling that tolerates situations of incomparability, intransitivity of preferences and even cyclic preferences". But "the resulting model is sometimes difficult to exploit so as to derive prescriptions. Thus, in his paper Vanderpooten introduces "several representative concepts and techniques for constructing prescriptions adapted to different problem statements":

- select a subset, as restricted as possible, containing the most satisfactory actions (choice),
- assign each action into pre-defined categories (sorting),
- rank the actions by decreasing order of preference (ordering).

The original ideas embodying the outranking approach and the ELECTRE methods developed by Roy inspired the development of many other methods "leaving room for incomparability", such as PROMETHEE and MAPPAC which are described in this book.

Jean-Pierre Brans and Bertrand Mareschal present the basic concepts and the successive steps of PROMETHEE methods, illustrated by a numerical example. They also explain how it is possible to construct a geometrical representation of the results obtained by PROMETHEE and so, a "visualisation" of the problem and of the conflictual character of the criteria; these methods are at the origin of two software: PROMCALC and GAIA. A third software, called BANKADVISER, allows to cope with the industrial evaluation problems using PROMETHEE methods.

Benedetto Matarazzo developed the MAPPAC and PRAGMA methods for discrete choice problems in the presence of quantitative criteria. As shown in his paper, a characteristic of these methods is that the comparisons of actions are carried out with reference to all the pairs of criteria and these partial results are then aggregated. This approach increases the number of calculations but brings a lot of information about partial (i.e. for subfamilies of criteria) preferences of the decision-maker.

Although the "Outweigh Approach" of Carlos Bana e Costa also uses the concept of outranking relation and incomparability, the basic model it uses is an additive value function. That is why this paper is included in the next chapter, precisely devoted to aggregation models of this type.

3. VALUE AND UTILITY THEORY APPROACH

In value or utility operational approaches, the partial preferences modelled by the multiple criteria are aggregated in a unique function measuring the overall (global) preferability of each alternative. The corresponding preference relation is always complete and transitive. Researches on this topic are concerned with mathematical properties implied by such an aggregation, particular forms of the aggregation function and methods to obtain this function. A fundamental notion in this domain is "compensation" - that is why we also include the Analytic Hierarchy Process within this operational approach.

All the papers in this chapter deals with additive functions, surely the most used type of aggregation model.

Rakesh Sarin presents the notion of Conjoint Measurement and necessary and sufficient conditions, a representation theorem and a constructive procedure for additive conjoint measurement.

The Analytic Hierarchy Process (AHP) of Thomas Saaty is presented in this book by Ernest Forman: he resumes the theory of AHP, namely justifying the use of eigenvalues and eigenvectors, and he makes use of an application of AHP in a problem of retail site choice to present the potentialities of Expert Choice, the powerful software based on AHP. A key concept in AHP is "inconsistency" of preferences. As Forman says "the Analytic Hierarchy Process is, in many ways, similar to Multi Attribute Utility Theory. However, unlike MAUT, AHP does not prescribe when or when not to allow for rank reversals. AHP allows the decision makers to decide how much inconsistency is reasonable (...)".

To know the strengths and weaknesses of each particular approach to decision aiding it is essential for the analyst activity. Those characteristics are discussed by Valerie Belton and Stephen Vickers for the simple value function approach.

Their paper describes VISA, a computer program based on that type of aggregation model, which incorporates a hierarchical structure of criteria thus being appropriate to take account of very many criteria. Its main feature is a visual interactive sensitivity analysis on the weighting constants. Belton and Vickers argue that "extensive experience in the use of simple multicriteria models to advise decision makers indicated a need for more sophisticated sensitivity analyses and an effective way of presenting the information gathered from such analyses to the decision maker. Even in problems involving a few criteria the amount of information yielded by a complete multi-dimensional sensitivity analysis can be overwhelming and the analyst is still faced with the problem of extracting what is useful to the decision maker." With VISA it is possible to do this analytically, thus avoiding time consuming ad-hoc processes. VISA makes use of interactive graphics to allow to investigate the effect on the overall scores of any changes on the values of the criteria weights.

In classic aggregation procedures the analyst interrogates the decision-maker to estimate directly the parameters of the model (i.e. weights, trade-offs, indifference values,...). But, as it is well noted by Eric Jacquet-Lagrèze, "in quite many situations it is very difficult for the decision-maker to answer the questions asked by the analyst in order to get the weights. The process involved to assess an overall preference using a decomposed analytical procedure is not natural to him. (...) On the other hand, especially when some alternatives are known to the decision maker, he feels that he can more easily compare these alternatives in a holistic manner". Thus, Jacquet-Lagrèze suggests to ask for holistic judgments or preferences of the decision maker and estimate indirectly the parameters of the model. This is what he calls disaggregation. Arguing that aggregation and disaggregation "are related to each other in a dialectic manner", Jacquet-Lagrèze proposes to use aggregation and disaggregation interactively. This original operational approach was first implemented in PREFCALC, an interactive decision support system described in his paper.

Carlos Bana e Costa remarks also that "those familiar with the practice of decision aid shall recognise that (...) the questioning procedures appropriate to specify the weights of the criteria are intrinsically complex and hard to tackle. This fact by itself can create uncertainty and imprecision and/or may cause the interruption of the process close to initial stages". So, as we have been seeing, the phase of modelling inter-attribute preferences is crucial for the success of MCDA. This topic is also discussed in the paper by Bana e Costa. He remarks that "situations involving poor weighting information are very common in the practice of decision aid. Nevertheless, there exists a significant lack of operational MCDA approaches explicitly devoted to support decision making under those circumstances. As a matter of fact, traditional multicriteria methods only operate with stable and single weights". Also in the context of MAUT, the Outweigh Approach proposed by Bana e Costa is a contribution towards the development of procedures able of directly facing those situations of poor preferential information of three main types: only some conditions relating the weighting constants are known, a rank-order of the weights is known and lower and upper preference bounds for value tradeoffs, or for substitution rates, or for weights are known.

The paper by Janssen, Nijkamp and Rietvelt also deals with ordinal priority information. They discuss indirect and direct methods for assessing the weights, and they present some qualitative multicriteria methods developed in the Netherlands; these methods have in common the fact of being suitable to facing decision problems in which only a rank-order of the criteria in order of importance is known. Furthermore if intra-criteria preference information is also of an ordinal nature the analyst can make use of the REGIME method. This new additive method is concisely presented by Janssen, Nijkamp and Rietvelt. In their paper, they describe also the application of qualitative multicriteria methods to site nuclear plants in the Netherlands.

4. INTERACTIVE MULTIPLE OBJECTIVE PROGRAMMING

We can speak of interactivity in multiple objective programming since 1971, when Benayoun, de Montgolfier, Tergny and Larichev proposed the STEM Method (a reference point procedure), followed one year later by GDF, the Geoffrion-Dyer-Feinberg Procedure (a non-reference point procedure). The field has known a fast development, many new methods or procedures appeared, first being all convergent, after coming progressively closer to the modern concept of interactivity, in which convergence lost its dogmatic character. Ralph Steuer and Lorraine Gardiner discusse in their paper "Interactive Multiple Objective Programming: Concepts, Current Status, and Future Directions" 32 topics that summarize interactive multiple objective programming: "topics 1 to 16 characterize the modest mathematical background necessary for the study of interactive multiple objective programming; topics 17 to 24 describe the current status of the field; topics 25 to 32 indicate future directions."

As remarked by Steuer and Gardiner, "for instruction and research as well as in many small applications, a microcomputer installation of a common computer package would be ideal". The TRIMAP package of João Clímaco and Henggeler Antunes has been recognised as one powerful answer to that aspiration. In their paper, they offer an appealing comparative study of three multiple criteria linear programming interactive methods (STEM, Zionts-Wallenius and Trimap), on basis of an electric power system planning problem. The potentialities of TRIMAP package for teaching multicriteria linear programming are emphasized in a second part of the paper.

One of the most recent and successful Decision Support System for multiobjective linear programming is VIG (Visual Interactive Goal Programming). Two main characteristics of VIG are: the fact that the decision maker does not have to specify the model precisely prior to solving the problem ("evolutionary approach") and the interactive use of computer graphics.

VIG was developed by Pekka Korhonen. In his paper in this book, Korhonen describes the principles of VIG and some real-world applications.

Another micro-computer implementation of a multiobjective programming approach is presented by Teghem, Kunsch, Delhaye and Bourgeois. They illustrate the capabilities of a friendly user-interface with a numerical example of application of the STRANGE method for solving multiobjective programming under uncertainty, a topic which has not been much developed in MCDA literature.

Many are the potential application areas of the interactive procedures. One of them is financial planning. In his paper, Jaap Spronk offers us a state of the art overview of "interactive multifactorial planning": starting with a discussion of some mono-criterion approaches to financial planning, Spronk gives several reasons to use a multiple criteria approach in this field. He also describes "interactive multiple goal programming, which has been especially designed for financial planning but which is also being used in a wide variety of other applications", and he discusses "the multi-factorial approach to financial planning, being a relatively new approach to deal with uncertainties in the projects' cash flows".

5. GROUP DECISION AND NEGOTIATION

Group decison making and negotiation support are more and more studied in the literature and they constitute a field having strong links with multicriteria decision aid. The paper by Tawfik Jelassi, Gregory Kersten and Stanley Zionts provides "an over-view of formal models for group decision making and negotiation with a special focus on those which can be used in developing computer-based support systems." First they distinguish four types of multi-person decision-making situations: individual decision-making in a group setting; hierarchical or bureaucratic decision-making (centralized and decentralized); group decision-making or one-party decision-making; and multi-party decision-making.

Second, they discuss the relationships, similarities between multi-criteria decision making and negotiation. The last part of the paper is devoted to group decision and negotiation support systems, which is a promising open area of research.

Günter Fandel states that game and bargaining approaches can be taken into account as solution methods for group decision problems. Methodology and applications in this context are the subjects of Fandel' s paper about group decision making.

Finally, Gregory Kersten and Wojtek Michalowski show the use of "a rule-based formalism to represent the decision problem, to model interactions with the environment, and to modify problem representation". They show us an example of the potentialities of the application of artificial intelligence concepts to decision support. The expert system shell NEGOPLAN is applied to model and support the negotiations with a hostage-taker.

Our tour of MCDA "world", now finishing, has been a cycle. In fact, when Kersten and Michalowski say that "decision making requires flexibility in modelling and support" and that "it is possible to provide this flexibility with a modelling framework which accomodates changes in problem understanding, and changes in problem perception", we feel that we are again near the modelling preoccupations embodying Anna Ostanello's paper and decision aid attitude in general.

6. THE SCHOOL CASE-STUDY

In the very beginning of the Third International Summer on MCDA the participants were invited for group discussions around a case: "how to rank officers for promotion?". The information given to the groups was exactly the one presented in this book by Carlos Bana e Costa and José Cervaens Rodrigues. The paper by Marc Pirlot describes the ambiance of the working group activities and results found by some of them.

CHAPTER I

MODELLING DECISION SITUATIONS

DECISION-AID
AND
DECISION-MAKING

Bernard Roy

LAMSADE, Université de Paris-Dauphine
Place du Maréchal De Lattre de Tassigny
75775 Paris Cedex 16 - FRANCE

ABSTRACT

The objective of multiple criteria approaches is to help us to make better decisions, but what is the meaning of "better" ? This field has developed considerably in the past twenty years. The development makes us fully aware of the limitations on objectivity encountered in the field of decision aid and, consequently, of the virtual impossibility of providing a truly scientific foundation for an optimal decision. Work carried out under the rubric of MCDM (Multiple Criteria Decision Making) bases its claims to legitimacy on a framework in which these limitations are left aside. Multiple Criteria Decision Aid (MCDA) must be seen from a different perspective. Its aim is, above all, to enable us to enhance the degree of conformity and coherence between the evolution of a decision-making process and the value systems and objectives of those involved in this process. The purpose of decision-aid is, therefore, to help us make our way in the presence of ambiguity, uncertainty and an abundance of bifurcations. We shall analyse multiple criteria concepts, models and procedures from both these viewpoints.

Key-words: Decision analysis, Multiple criteria decision making, Multi-attribute utility theory, Outranking methods.

0. INTRODUCTION

Multiple criteria models are increasingly studied and applied. For understanding what their specific contributions in Operations Research have been, it is necessary, I believe, to distinguish between two complementary fundamental attitudes. They can be characterized by the following key-words: "decision-making" and "decision-aid".

To begin this paper, I wish to recall some characteristics of the traditional mono-criterion approach as it has worked for more than 30 years in OR. Doing this, I will introduce some notations. Afterwards, I will describe what I consider the main features of multiple criteria decision-making: **MCDM**. This will lead me to explain, in the third part of the paper, why some limitations on objectivity have led to a different and complementary attitude, labelled multiple criteria decision-aid: **MCDA**. I will try to summarize what the basic theoretical and methodological preoccupations in MCDA are.

1. THE TRADITIONAL MONO-CRITERION APPROACH

Until the end of the 1960s in OR, decision-making problems were formalized on the three following bases:

(i) A well-defined set A of feasible alternatives a

This set A is supposed to be well-defined. Its definition can take either an **analytic form** or an **enumerative form**. In the first case, the alternative a is defined by means of a vector: $a = (x_1,...,x_m)$, and a well-specified set of constraints leads to define A as a part of $\mathbb{R}^m$. In the second case, A is defined by a list of alternatives without any explicit links to mathematical formulations of constraints.

(ii) A real-valued function g defined on A precisely reflecting the preferences of the decision-maker D

Decisions are studied from a point of view of a unique and well-defined decision-maker D. His preferences are supposed to be correctly taken into account by means of a unique criterion g. The comparison between two alternatives a and a′ is done simply on the basis of the comparison between the numbers g(a) and g(a′):

D prefers a′ over a iff $g(a') > g(a)$,
D is indifferent between a′ and a iff $g(a') = g(a)$.

Within the analytic form of A, we have:

$$g(a) = g(x_1, \ldots, x_m).$$

It is usual to distinguish between the **deterministic case** and the **probabilistic case**. In the first case g(a) is computed without any reference to random variables. On the contrary, in the second case, one or more random variables (denoted below by Y) intervene (by means of some characteristics of probabilistic distributions and, perhaps, Von Neumann-Morgenstern utility functions) in the computation of g(a).

(iii) A well-formulated mathematical problem

It remains now to solve the following problem:

Find a* in A such that $g(a^*) \geq g(a)\ \forall a \in A$.

It is important to note that, in this context, "to find" means "to discover" since the solution, if it exists, is completely determined (in fact created) by the way the problem has been stated. OR preoccupations are precisely oriented towards appropriate statements of such optimisation problems.

2. MAIN FEATURES OF MCDM

2.1. The General Framework

Let us now reconsider the three preceding bases within the general framework of what is usually called MCDM.

(i) A well-defined set A of feasible alternatives a

On this point, nothing changes compared to traditional OR. The analytic form, as well as the enumerative form, remains available to apprehend a set of possible mutually exclusive decisions.

(ii) A model of preferences, well-shaped in D's mind, rationally structured from a set of attributes

A well-defined set of attributes (or consequences) is now explicitly introduced. It is supposed that a well-identified decision-maker D refers to those attributes (and only to them) for founding a judgement with respect to the comparison of two alternatives a' and a. Confronted with such a comparison, D is able to choose (without any ambiguity) one and only one among the following possibilities:

a' P a: a' is strictly preferred to a,
a P a': a is strictly preferred to a',
a' I a: a is indifferent to a'.

Moreover, these preferences are supposed to be completely shaped in D's mind with the following properties:

- the binary relation P (defined on A) is asymmetric and transitive,
- the binary relation I (defined on A) is reflexive, symmetric and transitive.

Consequently, everything occurs as if there were a real-valued function U (defined on A) such that:

a' P a iff U(a') > U(a),
a' I a iff U(a') = U(a).

U is called a value function or a utility function. For defining such a function U, it is again necessary to distinguish the deterministic case and the probabilistic case. In both cases, this definition makes the attributes (or consequences) intervene by means of more or less explicit criteria (see below, § 2.2).

(iii) A well-formulated mathematical problem

As in the mono-criterion approach, the search for a solution consists of the discovery of an optimal alternative a* in A. a* must verify:

$U(a^*) \geq U(a) \; \forall a \in A.$

U works like a unique criterion. The problem is still to discover its optimum.

There is nevertheless something new in MCDM: the criterion U is not a priori explicitly known. When it is not, it intervenes through the answers given by the decision-maker (or by some representative) D to appropriate questions. It is assumed that U, implicitly present in D's mind, dictates the answers. Moreover, it is often assumed that U, as a function of the attributes or criteria, verifies some analytic properties.

2.2. Some Clarifications on the Nature of U

a) The deterministic case

By definition in this case, if we consider a given alternative a, the value taken by each attribute (or consequence) is supposed to be known precisely. Therefore, after a more or less complete sub-aggregation, it is possible to synthesize them by means of n criteria $g_k(a)$, $k \in \{1,\ldots,n\}$. Each k refers to a specified point of view: $g_k(a)$ can be either the value of a given attribute or the result of a sub-aggregation of those connected with the considered point of view. Each criterion reflects partial preferences in the following sense:

if $g_j(a') = g_j(a)\ \forall j \neq k$, then:

- D prefers a′ to a iff $g_k(a') > g_k(a)$;
- D is indifferent between a′ and a iff $g_k(a') = g_k(a)$.

Consequently, we can write:

$$U(a) = V[g_1(a),\ldots,g_n(a)],$$

V being an increasing function of $g_1,\ldots,g_n$.

Within the analytic form of A, $g_k(a) = g_k(x_1,\ldots,x_m)$ and then:

$$U(a) = U(x_1,\ldots,x_m).$$

b) The probabilistic case

By definition, this case corresponds to a type of modelling in which the attributes (not necessarily all but at least one) are viewed as random variables y_k. Such random variables are characterized, for each $a \in A$, by a multivariate probability distribution $\delta^a(y_1,y_2,\ldots)$. U(a) is then a function of such a distribution.

2.3. Some Basic Theoretical Problems in MCDM

The MCDM approach brings out some important theoretical problems widely studied in scientific journals. Here, we will briefly present only three classes of such problems.

Table 1

(1) *General additive form*	$U(a) = \sum_{j=1}^{n} w_j[g_j(a)]$ where $w_j[g_j(a)]$ is a non-decreasing function of $g_j(a)$.
(2) *Weighted sum*	$U(a) = \sum_{j=1}^{n} k_j.g_j(a)$ with $k_j \geq 0$.
(3) *General form of expected utility value*	$U(a) = \sum_{y_1,\ldots,y_n} u(y_1,\ldots,y_n).\delta^a(y_1,\ldots,y_n)$ where $u(y_1,\ldots,y_n)$ is a multi-attribute utility function.
(4) *Utility additive form*	$U(a) = \sum_{j=1}^{n} k_j.g_j(a)$, $k_j \geq 0$ with $\sum_{j=1}^{n} k_j = 1$, $g_j(a) = \sum_{y_j} u_j(y_j).\delta_j^a(y_j)$ where $u_j(y_j)$ is a partial utility function attached to the jth attribute and $\delta_j^a(y_j)$ is the marginal probability distribution of y_j.
(5) *Utility multiplicative form*	$U(a) = \dfrac{\prod_{j=1}^{n} [1+k.k_j.g_j(a)]-1}{k}$, $k_j \geq 0$, $k \neq 0$, $k > -1$ with $\sum_{j=1}^{n} k_j \neq 1$, $g_j(a) = \sum_{y_j} u_j(y_j).\delta_j^a(y_j)$ where $u_j(y_j)$ is a partial utility function attached to the jth attribute and $\delta_j^a(y_j)$ is the marginal probability distribution of y_j.

a) The standard aggregation formula

Diverse standard forms of the utility function U have been proposed, some for purely practical empirical reasons, others as the result of theoretical research. Some of these are very complex. Five are presented in Table 1.

For each one, the problem consists of the explicitation of hypotheses or axioms able to justify the use of such a formula in a given context. The hypotheses and axioms deal mainly with what is in D's mind since U has to reflect exactly what D's preferences are. They also deal with the nature of the information (relative magnitude on an ordinal scale, exact measure on an interval scale, random variables,...) expressed in numerical performance values.

Let us briefly comment on formulas from Table 1. The first and second ones are of course very familiar. It is obvious that the weighted sum is a particular case of the general additive form. They both deal essentially, but not exclusively, with the deterministic case. Stronger independence hypotheses are required to legitimate them.

The formulas (3), (4) and (5) treat the probabilistic case exclusively. Here, the aggregation is based on the concept of expected utility value. For defining the general form of the expected utility value, it is necessary to consider a multi-attribute utility function $u(y_1,...,y_n)$. The axioms able to legitimate this form of aggregation have been widely studied.

Even if the number of attributes is small, it is very difficult in practice to assess directly such a multivariable utility function. This explains the interest for additional hypotheses leading to a decomposition of $u(y_1,...,y_n)$ on the basis of partial utility functions $u_j(y_j)$. Formulas (4) and (5) show the two main forms to which multi-attribute utility theory leads. The way they are written makes criteria g_j, defined as expected utility functions, intervene. The utility additive form does not require the stochastic independence of the n random attributes. On the contrary, this stochastic independence is essential in order to be able to use the multiplicative form as it is expressed in Table 1.

b) The set E of efficient alternatives

A given alternative a is **strongly efficient** in A iff:

$\forall b \in A$ verifying $g_j(b) > g_j(a)$ for some j,
there exists (at least) one criterion g_k such that $g_k(b) < g_k(a)$.

In other words, a is strongly[1] efficient in A (we say also Pareto optimal) if it is impossible to find b in A such that:

$g_j(b) \geq g_j(a)$ $\forall j$ and $g_j(b) > g_j(a)$ for at least one j.

Let E be the subset of all strongly efficient alternatives in A. Among the problems studied we might cite, for example, the following:

- does a given alternative a belong to E?
- how can E be generated?
- how to move in E so as to achieve a good compromise?
- is a given property (connexity, concavity or convexity,...) verified by E?

c) The convergence of some procedures

The procedures involved here are those proposed in the framework of MCDM when U is not explicitely formulated. They all work on the basis of a given organisation of specific questions. It is always supposed that the answers given by the decision-maker (or

----

1. The qualifier "strongly" is used to avoid all confusion with weaker definition of this concept of efficiency.

his representative) are perfectly coherent with the utility function U.

A problem often studied is the following: does the decision automatically lead, in a finite number of steps, to the optimum a*? If not, is it possible to obtain this convergence by assuming that U verifies some additional appropriate properties?

2.4. Two Crucial Preoccupations Within the MCDM Framework

a) The main objective is to describe or to discover something which is viewed as a fixed and ever present entity.

This entity can be an optimum a* in A, the exact analytic form of a utility function U or the true values of coefficients (weights for instance) characterizing the specific role devoted to each criterion in an aggregation formula. It is essential to remember that it is the way the problem is stated which creates such entities.

b) Efforts of researchers are oriented towards concepts, axioms and theorems which are consequently liable to be used for the following purposes:

- to define conditions under which the existence of the entity which must be discovered is guaranteed (the legitimacy of procedures is considered in this perspective);

- to help to dictate the right solution to the decision-maker: if the rationality corresponding to the axioms is accepted by D, he must agree with the solution obtained.

3. FROM MCDM TO MCDA

3.1. Some Limitations on Objectivity

The practice of OR and MCDM has shed light on some fundamental limitations on objectivity. Five major aspects have to be taken into account:

(i) The frontier of A is often fuzzy. Because of this, the borderline between what is and what is not feasible has inevitably a certain amount of arbitrariness. A more crucial limitation on objectivity comes from the fact that this borderline is frequently modified in the light of what is found through the decision process itself.

(ii) In many real world problems, D, as a person truly able to make the decision, does not really exist: usually, several people take part in the decision process and we tend to confuse the one who ratifies the decision with what is called the decision-maker.

(iii) Even when D is not a mythical person, D's preferences very seldom seem well-stated: in and among areas of firm convictions lie hazy zones of uncertainty, half-held belief or, indeed, conflicts and contradictions. We have to admit, therefore, that the study itself contributes to answering questioning, solving conflicts, transforming contradictions and destabilizing certain convictions.

(iv) Data such as the numerical values of performances $g_k(a)$, the analytic forms of distributions such as $\delta_k^a(y_k)$ or $\delta^a(y_1,\ldots,y_n)$ and numerical values of the characteristics of those distributions are, in many cases, imprecise and/or defined in an arbitrary way.

(v) In general, it is impossible to say that a decision is a good one or a bad one by referring only to a mathematical model:

organizational, pedagogical, and cultural aspects of the whole decision process which leads to making a given decision also contribute to the quality and success of this decision.

If we want to avoid leaving aside these five limitations on objectivity, it seems impossible to provide a truly scientific foundation for an optimal decision. Of course, this does not mean that it is without any interest in decision contexts to solve optimization problems.

3.2. Two Crucial Preoccupations Within the Framework of MCDA [1]

a) The main objective is to construct or to create something which, by definition, does not pre-exist. This entity to be constructed or created is viewed as liable to help an actor taking part in the decision process:

- either to shape, and/or to argue, and/or to transform his preferences,
- or to make a decision in conformity with his goals.

b) Efforts of researchers are oriented towards concepts, properties and procedures which are consequently liable to be used for the following purposes:

- to extract from the available information what appears to be really meaningful (in the perspective of what needs to be built);
- to help to shed light on D's behavior by bringing to him arguments able to strengthen or weaken his own convictions.

---

1. It is important to consider this section in relation to 2.4.

3.3. The General Framework of MCDA

In order to give the salient characteristics of this new framework, we shall once again refer to three different bases. It is clear that these bases (described below) have many features in common with the three given in § 2.1 [1].

(i) A not necessarily stable set A of potential actions a

The way to conceive the set A is not so restrictive here as in MCDM. The term **potential action** is used, in the same way as the term alternative, to designate something through which a decision will materialize. The concept of potential action is more general than the one of feasible alternative for two reasons:

- two (or more) potential actions, unlike alternatives, can be put (if necessary) jointly into operation;
- the feasibility of a potential action is not positively imposed: A may contain certain actions for which it is difficult to say if they are feasible or not: either because they are ideal actions which can serve as points of reference, or because the line between what is feasible and what is not is ambiguous.

Moreover, the type of reasoning used in decision-aid does not necessarily imply assuming that A is a stable set. A may evolve during the study just as it often evolves during the decision process.

(ii) Comparisons based on n criteria (or pseudo-criteria) g_k

Criteria g_k are introduced to reflect, possibly with a certain fuzziness, the preferences of one or several actors. The actors in question are either those who intervene in the decision process

---

1. The reader should refer to this paragraph in order to see these differences clearly.

or those on behalf of whom decision-aid is provided. We will denote by F the family of the n criteria considered. Such a family has to be built in reference to what is taken into consideration by actors for shaping, and/or arguing, and/or transforming their preferences.

It is essential to conceive F in such a way that as broad consensus as possible exists on the following point: the comparison between two potential actions a′ and a can be founded and discussed through the comparison of the two **performance vectors:**

$$g(a') = (g_1(a'),\ldots,g_n(a')); \quad g(a) = (g_1(a),\ldots,g_n(a)).$$

For this purpose, each criterion must take into account one or more precise attributes (or consequences) allowing us to clearly apprehend a well-identified point of view (see 2.2 a). The consensus deals with the way of defining, and consequently computing, the performances $g_k(a)$. On the other hand, it does not deal with the importance which should be given to each of these criteria in order to carry out the comparisons, as the value systems on which different actors base their opinions on the topic may not be the same.

For taking into account uncertainty, imprecision and inaccurate determination, probability distributions can still be used. In addition or instead, we may also consider fuzzy numbers and/or different types of thresholds. A very simple manner of introducing thresholds is provided by the concept of pseudo-criterion.

A **pseudo-criterion** g_k is a criterion to which one or two thresholds q_k and p_k have been assigned. The aim is to work with a partial preference model attached to the point of view represented by g_k which is not so rough as the one introduced in 2.2 a). For a′ and a verifying $g_j(a') = g_j(a)$ $\forall j \neq k$, a′ is no longer strictly preferred to a iff $g_k(a') > g_k(a)$. When the positive difference $g_k(a') - g_k(a)$ is sufficiently small (more precisely $\leq q_k$), then a′ and a are considered indifferent.

To have a strict preference, it is necessary that the positive difference $g_k(a') - g_k(a)$ be sufficiently large (more precisely $> p_k$). The case:

$$q_k < g_k(a') - g_k(a) \leq p_k$$

is interpreted as a hesitation between indifference and strict preference. It leads to what we call a **weak preference.** This concept allows us to apprehend the ambiguity inherent in the presence of imprecision, uncertainty, or inaccurate determination.

(iii) An ill-defined mathematical problem

Contrary to MCDM, MCDA does not lead to a well-stated optimisation problem. On the basis of:

- the family F,
- additional inter-criteria information (weights, veto thresholds, substitution rates,...),

the problem is now to elaborate:

- either a mathematical model allowing us to compare potential actions in a comprehensive way, i.e. unlike partial comparisons for which only one criterion comes into play (the $n - 1$ others having the same value),

- or a procedure helping to reflect upon and to progress in the formulation of comprehensive comparisons between potential actions.

Finally, the aid can be provided by referring to three different perspectives corresponding to the three problem statements:

P.α) the selection of a better action, optimum or satisfecum;

P.β) the assignment of each action to an appropriate pre-defined category according to what we want it to become afterwards (for instance acceptance, rejection or delay for additional information);

P.γ) the ranking of those actions which seem to be the most satisfactory according to a total or partial pre-order.

3.4. Some Basic Theoretical and Methodological Preoccupations in MCDA

The following list is of course non-exhaustive. It would be interesting to compare the topics of concern treated below with the problems presented in 2.3.

a) The nature, the generation and the formal definition of potential actions to be considered.

b) The nature and the formal definition of criteria

How can attributes or consequences be formalized? What are the main sources of imprecision, uncertainty and inaccurate determination? In order to take them into account, is it better to introduce probability distributions, fuzzy numbers, or thresholds? Does the personality and the number of actors influence the definition and the nature of some criteria?

c) The conditions to be satisfied by the family F in order to be appropriate to play the role devoted to it.

For this topic, it is important to consider both organizational and cultural aspects carefully.

d) The nature and the quantification of the inter-criteria information which is required to model comprehensive preferences (i.e., taking into account the all n criteria at one time).

Here, a major problem comes from the fact that each criterion has to play a specific role according to its own importance. It is essential to bear in mind that this notion of importance has no absolute meaning. To be meaningful, the importance of a criterion must refer to a given aggregation model.

e) The logic and the properties of the aggregation models by which comprehensive preferences are totally or partially formalized

Among those models, we can isolate those which, as in MCDM, use a synthesizing single criterion. Of the other models, I will mention only those which are founded on one or more binary relations, fuzzy or not; they usually refer either to outranking or to strict preference situations.

f) The logic and the properties of the procedures, interactive or not, by which the final selection, assignment or ranking is made.

Considering the limitations on objectivity presented above (cf. 3.1) and the general orientations of MCDA (cf. 3.2), it is impossible to consider that there exists somewhere the right solution, or assignment, or ranking which could be considered independent from any procedure. So, it becomes impossible to found the validity of a procedure either on a notion of approximation or on a mathematical property of convergence. Therefore, the final solution, or assignment, or ranking is more like a creation than a discovery. That is why the validity of a given procedure depends:

- on the one hand, on some mathematical properties which make it conform to given requirements ;

- on the other hand, on the way it is used and integrated in a decision process.

4. CONCLUSION

These last considerations lead us to examine the following question:

What should the objectives of multicriteria approaches be?

The natural answer is: to help managers to make better decisions. But **what is the meaning of better**?

In this paper, I have tried to make it clear that this meaning depends, in part, on the process by which the decision is made and implemented. This point of view leads us naturally to the three following assertions:

- Our concepts, tools and procedures cannot be conceived on the implicit premise of discovering, with more or less approximation, pre-existing truths which can be universally imposed.

- Nevertheless, methodical decision-aid based upon appropriate concepts and procedures can play a significant and beneficial role in guiding a decision-making process.

- Therefore, solutions obtained by solving well-formulated MCDM problems constitute a fundamental background for MCDA.

Consequently, we consider that the aim of MCDA is, above all, to enable us to enhance the degree of conformity and coherence between:

- the evolution of a decision-making process, and
- the value systems and the objectives of those involved in this process.

For that purpose, concepts, tools and procedures must be conceived to help us make our way in the presence of ambiguity, uncertainty and an abundance of bifurcations.

REFERENCES

Keeney, R.L. and Raiffa, H. (1976), *Decisions With Multiple Objectives: Preferences and Value Tradeoffs*, John Wiley and Sons, New York.

Roy, B. (1985), *Méthodologie Multicritère d'Aide à la Décision*, Economica, Paris.

Schärlig, A. (1985), *Décider sur Plusieurs Critères - Panorama de l'Aide à la Décision Multicritère*, Presses Polytechniques Romandes, Lausanne.

Steuer, R. (1986), *Multiple Criteria Optimization: Theory, Computation and Application*, John Wiley and Sons, New York.

Vincke, Ph. (1989), *L'Aide Multicritère à la Décision*, Editions de l'Université de Bruxelles et Editions Ellipses, Brussels.

Zeleny, M. (1982), *Multiple Criteria Decision Making*, McGraw Hill, New York.

ACTION EVALUATION AND ACTION STRUCTURING: DIFFERENT DECISION AID SITUATIONS REVIEWED THROUGH TWO ACTUAL CASES

Anna Ostanello

Dipartimento di Automatica e Informatica
Politecnico di Torino - Corso Duca degli Abruzzi 24,
Torino, ITALY

1. INTRODUCTION

This paper intends to bring some contributions to refine our understanding of how the *Analyst (or model builder)* actually carries out decision aid activities in order to produce some «valid» answers to a *Client* ; this is normally a subject demanding for support about a problem his/her belonging organisation is involved with (organizational or main process).

The Client's demand (or proposed action (Roy, 1985)) normally concerns a «problem situation», i.e. a situation «perceived by one or more organizational actors as presenting dissatisfaction with the performance of a system or awareness of existence of opportunities» (Landry et al., 1983).

Actual decision aiding situations can be very different and made differently complex by the nesting of technical, organizational and political constraints and problems. These factors not only condition the Analyst's acting, but they can determine an evolution of the Analyst's and Client's interaction (A/C interaction).

The analysis of actual Decision Aid (DA) processes and of their results (formal representations, problem formulation, operative tools, models) can help in recognising the kinds of problems, the nature and the structuring of the Analyst's activities.

For that purpose we shall comparatively discuss two actual cases of Decision Aiding for Clients, members of Public Administration involved in «public» decision processes. (For a specification of the concept of «public decision» see Ostanello et al., 1987).

Different reconstructions of the DA processes could be performed, and that depending on the assumed paradigm, on the research purposes and, not last, on the researcher's spatial and temporal location in the studies process (cf, for instance, Witte, 1972; Moscarola and Siskos, 1983; Nutt, 1984).

For what concerns the presented cases, they have been both conducted by the author and her collaborators (Simoni and Vernoni, in 1978; Norese, in 1982-1983); for each of them, the story and most of the activities had been registered during the process, memorized and verified.

To our reconstruction we shall use a framework which has been derived by an empirical research on organizational DA process (cf Ostanello, 1987; Norese and Ostanello, 1988 and 1989).

One of the main assumptions on which the frame is founded, concerns the necessity of integrating organizational and political dimensions in the process of model building within actual organizational contexts (cf, for instance, Moscarola, 1981; Tomlinson, 1981; Checkland, 1984; Schwarz and Thompson, 1985). That allows us to explain different meanings and functions of the different kinds of activities undertaken by the analyst to reach some valid results, and more precisely: on the one hand, of activities - other than strictly rational - undertaken to reach valid formal results (on that point see, for instance, Landry et al., 1983), and, on the other hand, of «rational» activities to help Client's comprehension on the «problem» structuring or learning of the problem situation, to facilitate actors communication, compromising, negotiation and, thus, to reach valid organizational and political results (cf Moscarola, 1981).

Then the modelling process - a sub-process of the DA process - shows generally a more complex structure than that suggested by the rational paradigm. That is particularly evident when the proposed action is tackled by the *Multicriteria Approach* (cf de Montgolfier and Bertier, 1978).

In the following, we shall firstly summarize the frame model and then we shall use it to characterise the Decision Aid situations of the actual cases. Finally the modelling activities shall be compared.

2. FRAME MODEL

The frame model does not assume a priori phases in the process. It assumes five different Routines - empirically derived - as basic steps, leaving a phase recognition as a result of the process reconstruction (cf also: Witte, 1972; Mintzberg et al., 1976). Some combinations of the Routines, named Development Procedures, - also empirically identified - help in giving a more structured representation of the DA process.

2.1. Routines

The concept of *«routine»* is here somehow equivalent to that of a task sub-process, where different standard activities are developed.

A routine can be essentially characterised by a *main domain* of activity application and by the *nature (or the purpose) of the prevailing activities*. Possible domains of application of the activities may be, for instance: the «actors system» of the organization decision process and / or the decisional system (decisional context); a structured Information Base; a formal and / or informal Information System; the internal or external organization environment (operational context).

Table 1: Main features of Routines

ROUTINES	IDENTIFICATION	STRUCTURING	CONTROL	DEVELOPMENT	COMUNICATION
MAIN DOMAIN	. Main process . Decisional context	. Operational context . Decisional system	(1). Operational context and information (2). Organizational and decisonal context	Information base	. Operational context . Decision system
PREVAILING ACTIVITIES	Inquiry to catch: . C's motivations . the project functions . symptonms of dissatisfaction . nature of the stimula ... (Mintzberg, 1976)	. Exploration of lines of inquiry . Search, activation, explorations of information channels . Development and control of intervention alternat. in operational context . Search: of alternatives to identify criteria, of reference systems, of criteria to identify alternatives,...	. Inquiry . Verification . Comparison . Evaluation . Justification . Negotiation . Persuasion . Cooptation . Defending . Counterattacking,...	Modelling	. Investigation . Dissemination . Explanation . Clarification . Justification . Persuasion . Negotiation
EXPECTED RESULTS	. Identification of the problem situation . Identification of the operational context . "Choice of a particular angle of attack" (Landry, 1983) so to activate: information, relationships, value systems,...	. Operational character of the proposed action . Formal representations	(1). Quality of the information . Stability of the representations . Pertinence of data . Efficiency of the operational tools (cognitive/operational purposes) (2). "Piloting" of the intervention . Inducing local decisions . Identifying following process steps . Identifying following process steps (political and operational purposes)	Formalized representations	Operational and political
VALIDITY DIMENSIONS (Landry et al.,1983	. Conceptual coherence . Operational legitimacy	. Logical validity . Operational validity . Technical legitimacy	. Data validity . Experimental, logical, operational and conceptual validity . Technical and political legitimacy	. Logical . Experimental . Operational	. Different kinds . Techinical and political legitimacy

The routines can be considered both as *«contexts»* (*«main routines»*), having one or more identified objects of action, and as *«supporting routines»*, i.e. as tools to support the action in a main routine.

DA process representations can then be based on five identified routines (see Norese and Ostanello, 1989), which have been named: **Identification (R1)**, **Structuring (R2)**, **Control (R3)**, **Development (R4)** and **Communication (R5)**.

2.2. D-procedures

As can be seen from table 1, modelling activities have been synthesized by the Development routine.

A closer representation of the structuring of the modelling process can be given by some combinations of the routines, called *«D-procedures»*, respectively of: **Re-structuring**, **Evaluation**, **Identification**, **Elaboration**.

Their «shapes» by the routines and some of the characterizing features are summarized in the table 2.

Our reconstruction of the DA processes shows that these procedures are, actually, differently applied in different contexts; their activation increases with the growing of both the technical/operational and organizational/political complexity. Different applications essentially depend on the present possible different states of the proposed action (in the sense given in Roy, 1985), on the possible different states of the Information Base and on the present context of development.

The main operational interest of these procedures is, in our view, of pointing out some differences between modelling moments; particularly, they allow us to make some correlations between formalization activities and «context variables» (Moscarola, 1980).

Table 2: Characterizing features of the D-procedures

DENOMINATION	RE-STRUCTURING r(i)	EVALUATION m(e)	IDENTIFICATION		ELABORATION	
			By search and screen i(ss)	By multicr. approach i(ma)	By design e(d)	By multicr. approach e(ma)
FEATURES						
STATE OF THE ACTIONS	Given and developed	.Given or available in the organization environment .Partialy developed	. Existing but not available . «Read-made» . to be searched for (in the market supply, in the environment) and screened . Not clearly differentiated	. Abstract of verbal type: a) actor's old or new proposals b) partial actor's representations	. Not existing . At least one <u>must</u> be elaborated	
INFORMATION BASE	Available and partially structured	. Essencially formal . Fully or partially or nothing usable	. Both formal and informal, not available in the organization . To be identified, evaluated, structurated	. Essentially informal to be: - searched for - evaluated - integrated - structurated - re-structurated	. Both formal and informal . Identified . To be evaluated structured and elaborated	. Both formal and informal . Identified . To be evaluated integrated structured elaborated
INFORMATION SOURCE	«Unicity»: one or more cooperating sources	One or more sources generally cooperating, with possible coflicting points of view	. Multiple sources both internal and external . Not necessarily cooperating . Conflicting points of view	Multiples sources, mainly internal showing divergences, redundances, conflicts	. One or few coperating sources	Multiples sources, mainly internal, normally cooperating, bringing different elements that can vary from structured data to partial analyses, informal representat., verbal indications,...
STARTING STATE OF THE MODEL	Developed, stable	. Partially developed . Unstable	. Developed . Stable	. Partially developed . Unstable	. Totally undeveloped	. Totally undeveloped
«SHAPE» BY THE ROUTINES	R3 / R4	R4/R2 ⟷ R3/R5	R2 <-> R5 ⟷	R4 <-> R3	R4 <-> R3 ⟷	R4 <-> R3

3. DA PROCESS PROFILES

Different configurations of the DA processes can be identified by this frame (see Norese and Ostanello, 1988). They may be profiled on a two dimensional map, where some *contexts or stages* (cf Nutt, 1984) are correlated with step activities or *steps*, as follows.

The *contexts* are distinct at two levels of activity: individual - of the Analyst - and collective - of the Analyst/Client's interaction.

At the *individual level*, they are assumed coinciding with the main routines of the frame model. At the *collective level*, they are correspondingly named: **Communication and Identification** (C/I), **Communication and Structuring** (C/S), **Communication and Development** (C/D), **Communication and Control** (C/Co).

The *steps* shall be distinguished by the nature of the prevailing activities, respectively *operational/cognitive* and *organizational/political*.

The D-procedures are taken to represent the operational/cognitive activities - summarized by R4 -, whereas the organizational/political activities are summarized by the supporting routines R1, R2, R3, R5; by their definition, these routines develop in collective contexts, i.e. the internal or organizational and external environment (cf table 1).

The structuring of each process is discovered through a *path in the map (profile)* shown in the figure 1.

The *context situations* of a DA process, or of different stages of it, can thus be taken into account by expanding the profile from the first and second orthant (non complex cases) to the third and fourth orthant (complex cases).

Figure 1: Map to profile the DA processes

	STEPS				
CONTEXTS	DEVELOPMENT (R4)	ORGANIZATIONAL/POLITICAL			
	r(i) m(e) i(ss)...	R1	R2	R3	R5
A's					
Identific. Structur. Developm. Control Comunic.	I (no C's involvement)	III (with organization involvement)			
A and C					
C/I C/S C/Co C/Co C/D	II (C's involvement)	IV (with both C's and organization involvement)			

We are now in the condition of presenting two extreme situations of DA process, that we shall discussed on two actual cases.

4. COMPARISON OF TWO ACTUAL CASES

4.1. Presentation of the Cases

The cases refer to DA processes developed after a Client's demand of support about a «public» decision.

The problem situation concerned, in both cases, Public Administrators, respectively for: a territorial decision (case 1) and the modality of funds investments on public facilities (case 2).

Some relevant characteristics of the organization and decisional contexts for the DA processes are summarized in table 3.

Table 3: Organization and decisional context

	CLIENT'S ORGANIZATION	CLIENT'S POSITION	ORGANIZATION PROCESS	PRESENT PHASE	DECISION SYSTEM
CASE 1	Regional Administrat.	Head Service of a Regional Study Centre	Relocation of firms at regional level Strategical	Decision/ Negotiation on the relocation area	. Regional Council . System of firm owners with a strong leader
CASE 2	Municipal Administrat.	Two City Councillors	Investments on public facilities Periodic Decison	Negotiation on the priority of different kinds of facilities	. City Council . Client

The *problem situation* has been presented by each Client, respectively as follows:

Case 1) Need for the Administration of controlling the coherence of the industry relocation process with the Regional Development Plan; the control power is however uncertain for several reasons, one of which is the ownership of a wide area by the Leader of the entrepreneurs.

> Client's Demand: To elaborate multicriteria evaluation elements on the «sectotial» vocation of some given territorial areas to receive localization of specific kinds of industries.

Case 2) Dissatisfaction with the present decision routine, essentially based on inter-personal relationships and heavily conditioned by political criteria. Intention of elaborating some action to change this situation.

> Client's Demand: To produce cognitive elements useful for the elaboration of an investment plan in the Sport Sector.

4.2. Case Typologies and Profiles

The problem situations, as presented by the Clients, and the context variables - principally the available information System - as recognized by the Analyst, made the two DA processes totally different, although an apparent similarity of the demands.

Following, the classification introduced in Norese and Ostanello (1988), the two cases have been identified as belonging to process classes respectively called: *Action Evaluation* (Case 1) and *Action Structuring* (case 2).

Action Evaluation class

The processes of this class are the closest to the «classical» conception of DA process, within the Multicriteria Approach. They are characterized by:

- a high level of the demand structuring, i.e. of the problem formulation;
- a set of alternatives or actions, given or easy to reach and to elaborate;
- the presence of an «unique» information source, owning a structured Information System available through the Client. Unicity of the information source may consist in a multiplicity of cooperating individual sources.

The structuring, on the map, characterizing this class can be summarized as follows:

STEPS - Process steps are essentially «operational/cognitive» (R4); depending on the process, they are distributed on different sequences at both individual and collective levels (I and II orthants of the map).
The state of the proposed action and of the Information System keeps, generally, the activities within r(i), m(e) and i(ma) procedures.

STARTING STAGE - It is normally «Structuring» in a context of Communication (A/C interaction).

CENTRAL STAGE - It is «Development» at the Analyst's level, sometimes reinforced at A/C level.

ENDING STAGE - It is normally «Control» at the A/C level, excepting the cases where «Development» has been conducted at this level.

Different *typologies* may be identified in this class, depending on: a) the level of Client's participation; b) the nature (for instance, concrete or abstract) of the actions; c) the adequacy of the present structuring of the Information Base to the problem situation (in the sense of «problematique» (Roy, 1985)) and to the Analyst's operational approach.

Case 1 typology: The case corresponds to a reduced participation of the Client and to the presence of a rich Information Base - available through the Client - to be structured for both the problem situation and the adopted multicriteria approach.

The process profile is illustrated in the figure 2.

Figure 2: Action Evaluation - profile of the case 1

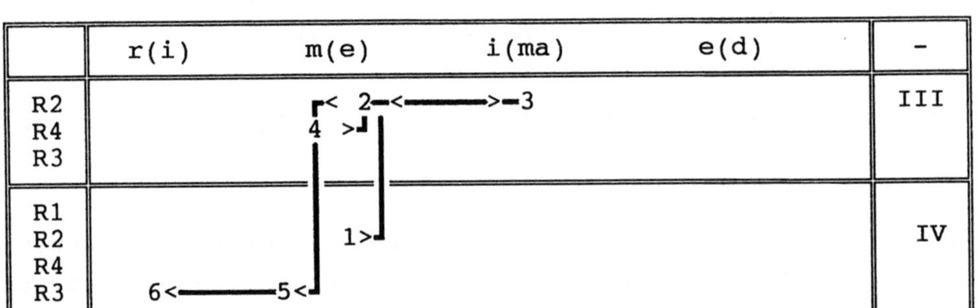

Action Structuring class

The case 2 has been identified as belonging to the *Action Structuring class*. The denomination comes from the core routines of the profiles. Within the set of actual cases of our empirical research, the processes of this class are the most frequent. They are characterized by:

- a low level of structuring of the proposed action; the Client is principally interested in acquiring valid information and legitimated (or legitimating) argumentations on a problem situation surrounded by *uncertainty*; the problem situation may have shown opportunity for action more than critical conditions to C (cf Mintzberg et al., 1976);
- alternatives are not given or those available are estimated not adequate to the present or the perceived situation;
- the available Information System is inadequate, obsolete (mainly the formal one) or incomplete and not usable. Information has to be searched for in the organization or in its environment, verified and elaborated. The *learning function* of the process is here more important than in any other identified class.

The structuring, on the map, characterizing this class is summarized as follows:

STEPS - They are predominantly of «Structuring» in different contexts (I,II,III,IV orthants). Modelling activities develop to: shape «abstract actions of verbal type» and partial collected representations, to identify alternatives and criteria, to formalize criteria and evaluation scales on potential actions, previously searched for and screened. The high level of instability of the representations along the process makes artificial a stage structuring of the map representation. Indeed, as a stable problem representation is achieved, the DA process normally ends up.

STARTING STAGE - It is generally «Structuring» at collective level (II and IV orthants in the map).

CENTRAL STEPS -

a) In cases of «participating Client», they are «Development» activities, mainly conducted at collective level; complexity arises in the case of organization involvement (see figure 3);

b) In cases of «non-participating Client», «Structuring» at individual level is heavily based on modelling activities (R4), as i(ss), i(ma), e(ma), iterating with «Communication» for «Structuring and Control» in the organization environment.

ENDING STAGE - Depending on the Client and the organization involvement, it can be: «Control» (which may indicate either a low C's involvement or interruptions generally because of external events); «Development» (in case of a strict C's participation); or «Identification» (for instance, of new opportunities of action).

Case 2 typology: The case corresponds to a reduced participation of the Client (decision maker) in Development activities; he is however easily reachable for Control and Communication.

The discovery of the absence of a structured Information Base in the organization, contradicted the starting Analyst's assumptions; this was the first result of the process (point 8 in the map). Such problem identification determined process complexity, with heavy organization and environment involvement, and thus the incoming of a multiplicity of different information sources and of a variety of both technical/operational and organizational/political problems.

The profile of the case is illustrated in the figure 3.

Figure 3: Action Structuring - profile of case 2

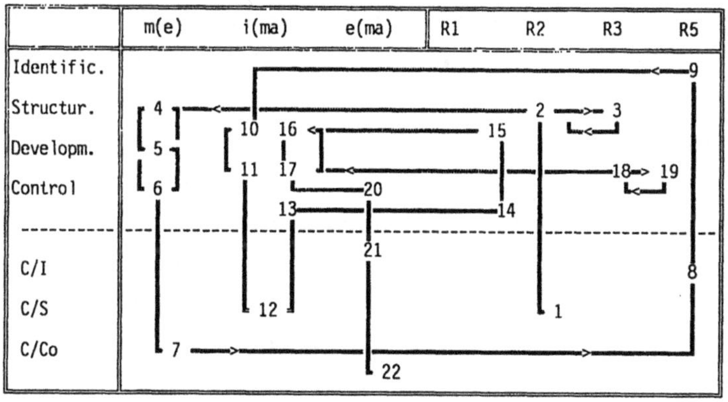

Two different process phases can be recognized from the profile of figure 3: a starting «Action Evaluation phase» (points 1 to 7 of the profile) and then an «Action Structuring phase» with environment involvement (points 10 to 22 of the profile). The two phases are separated by identification of an organizational and political problem, in a Communication context.

4.4. Comparison of the Modelling Activities

As it can be seen in the figures 2 and 3, both cases started as modelling processes by a Multicriteria Approach. Indeed, similar initial assumptions on the available Information Base were deduced from Clients' declarations.

Thus, the differences that can be noticed between the profile of figure 2 and the first phase profile of figure 3, essentially correspond to different distributions of the Clients' organizations: *concentrated*, in the case 1, and *distributed* among different Administration Services, in the case 2.

However the principal factors that, in our view, may have produced different process typologies are the Client's role in the organization process and the project function perceived by the organization environment (both internal and external). (For a discussion on this point, see Norese and Ostanello, 1984).

4.4.1. Case 1

The modelling process fits schematically to the «phase» sequence presented in (Roy, 1975), with cognitive and operational activities. At the n-points of the profile of figure 2, the activities may be described as follows:

1) Structuring stage:

Assessment of the conceptual model with the Client. This allowed the Analyst to delineate the following representation operational scheme:

Actions - They are identified with couples (t,i), with t and i respectively elements of the sets of territorial areas (T) and industrial sectors (I). An initial set of 41 potential zones was reduced to 14 areas, by a structuring analysis where both physical and political factors were activated.

Process actors - Three collective subjects were identified as principal actors, and named: «private» actor, A1 (to represent the Entrepreneurs Group, expressing an homogeneous policy), «public» actor, A2 (to represent both residents in the areas and workers occupied in the relocating firms).

Actors' objectives - For each of the actors, the expressed objectives could be identified:

A1 - to increase of the productive efficiency,

A2 - to re-organize and to re-equilibrate the territory,

A3 - to defend employment places and to avoid negative housing and services consequences.

A relevant part of the operational activities, at the structuring stage, was referring to the acquisition of the necessary information elements (through the Client): formal of official documentation, identification and activation of relationships with the information sources.

2 ⇄ 3 *Consequence analysis:*

The «problem» dimension (Roy, 1975) were deduced iterating from official documentation and verbal communications of the sources.

Compatibly with the available time, a satisfying multidimensional Information Base was then structured, suitable to build representation of identified conflicting logics.

4) Criteria modelling:

Three criteria were elaborated on the base, by applying ELECTRE II to aggregate different factors. «Weighting» problems were solved by both analysing official documents and discussing with competent information sources.

5 ——> 6) Simulation:

Simulations on the actions behaviour were conducted interactively with the Client, after he had verified the model validity (both conceptual and logical). Results of the modelling process are reported in (Ostanello et al., 1978).

4.4.2. Case 2

The starting «Evaluation phase», conducted by a «line of action» similar to that of the Case 1, did not produced any kind of model, due to both organizational/political and conceptual difficulties.

Indeed, the demand of the Client (a political decision maker) of cognitive elements on a organizational decision process - traditionally controlled by public functionaries - rose internal resistances and conflicts (see Norese and Ostanello, 1984). These internal problems pointed out an existing organizational problem concerning the information System.

Conceptual difficulties were principally relating to an insufficient development of a cultural debate on planning of the specific Service Sector and to the spread of knowledge and competences, mostly not available in the organization.

Thus the structuring of a problem representation (points 10-20 of figure 3) was heavily based on «Identification» of information and knowledge sources (point 15) and on «Control», both operational and political, in the organization context (points 18-19).

The evolution of modelling frame is synthesized in table 4 and the cognitive activities are illustrated in table 5.

Results of the modelling process are reported in (Norese and Ostanello, 1985).

A confrontation of the modelling activities of the two processes has been conducted on the map profiles. By it we have tried to point out, particularly, some differences between the «rational» activities that can be, contingently, required to the Analyst searching for validity of the modelling process within organization contexts. These activities can vary from «classical» solving method application, to «softer» activities concerning search, control, integration, elaboration of data and of «problem» representations.

Table 4: evaluation of modelling frame

	EVALUATION PHASE (points 1-7)	IDENTIFICATION (points 10-16)	STRUCTURING PHASE (points 17-21)
OBJECTS	. Facilities . Users	«Sport-Points»	Shared representation of «Sport-Points»
PREVAILING RELATIONS	Demand-Supply	Between Sport activities (practice levels and their organization) and activity spaces (organizational complexity and availability of facilities)	Between representation «bearers» and between representations
LINES OF INQUIRY	. On existing facilities . On users' present and forseen behaviour	. Structuring of the demand . Facilities classification relating to the binomial: (activity level, activity space)	. Modelling of shared types of facility kinds (criteria) . Elaboration of an interpretative model of the users behaviours
EXPECTED RESULTS	. Evaluation of service needs and priority identification	. Elaboration of general qualitative standards suitable to evaluated needs	. Identification of situations having different critical levels . Elaboration of priorities and of modalities of intervention

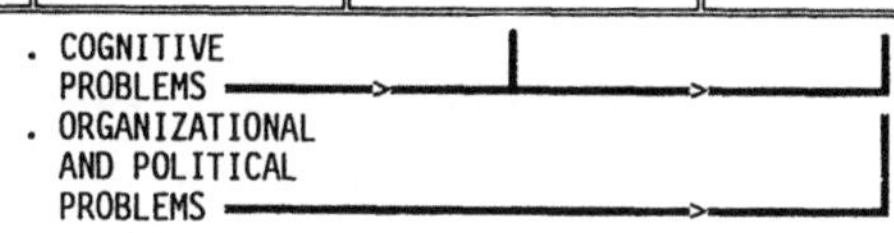

In the paper, we have also underlined that a possible evolving of the complexity of the analyst's work may depend on the Client's role within the organization process and on the finalities of the project, as perceived by the organization actors. These «not rational» factors can determine different structuring of the operational context and then of the DA process itself.

A such observed behaviour may suggest to the Analyst the opportunity of integrating organizational and political dimensions in the process of model building.

Table 5: Structuring of the cognitive activities

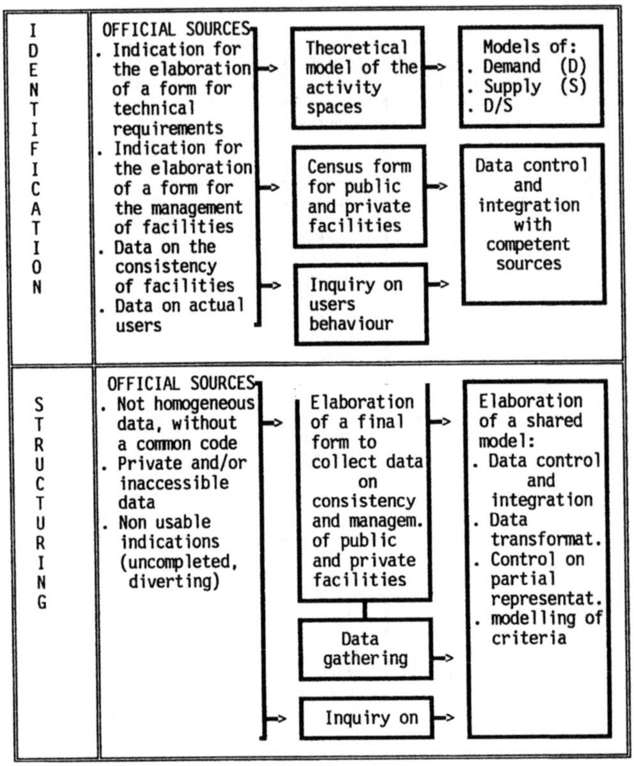

The usefulness of the Multicriteria Approach in building representations, at different levels, is out of question. The Analyst should put a greater attention on the approach, so to be able of deducing softer and flexible tools - beside the «classical» Multicriteria Methods - suitable to support Decision Aid at the increasing of both the technical/operational and political//organizational complexity. Indeed the Multicriteria approach can be usefully used not only for the specific Multicriteria Methods - that cannot always be applied- but also as: either a soft tool allowing, for instance, different elements of representation to be arranged into formal schemes, so to create a link between conceptual and formal models, or even a framework, suitable to outline logics of connection between development and other kind activities and thus to control the modelling process evolution within the DA process.

CONCLUSIVE CONSIDERATIONS

Two different cases of actual Decision Aiding by a Multicriteria Approach have been illustrated and their types represented on an empirically deduced map. The map offers a possibility of identify different complexity levels of the DA process, by refereeing the process to a representation of the context situations.

REFERENCES

Bourgine, P. and Espinasse, E. (1987), "Aide à la décision : une approache constructiviste", in *AFCET, Aide à la Décision dans l'Organisation*, Paris, 47-55.

Checkland, P. (1984), "Rethinking the system approach", in R. Tomlinson and J. Kiss (eds.), *Rethinking the Process of Operational Research and Systems Analysis*, Pergamon Press, Oxford, New York.

Checkland, P. (1985), "The approach to plural rationality through soft systems methodology", in M. Grauer, M. Thompson and A.P. Wierzbicki (eds.), *Plural Rationality and Interactive Decision Processes*, Lecture Notes in Economics and Mathematical Systems, 248, Springer-Verlag, Berlin, 8-61.

Hirsch, G., Jacquet-lagrèze, E., Moscarola, J., Roy, B. (1978), "Définition d'un processus de décision - I: Quelques concepts", *Cahier du LAMSADE 13*, Université Paris Dauphine.

Jelassi, M.T. (1986), "From 'stand-alone' methods to integrated and intelligent DSS", in Y. Sawaragi, K. Inoue and H. Nakayama (eds.), *Toward Interactive and Intelligent Decision Support Systems*, Lecture Notes in Economics and Mathematical Systems, 286, vol.2, Springer-Verlag, Berlin, 90-99.

Landry, M. (1987), "Les rapports entre la complexité et la dimension cognitive de la formulation des problèmes, in *AFCET, Aide à la Décision dans l'Organisation*, Paris, 3-29.

Landry, M., Malouin, J.L., Oral, M. (1983), "Model validation in Operations Research", *European Journal of Operational Research*, 17, 207-220.

Mintzberg,H., Raisinghani, D., Théoret, A. (1976), "The structure of the unstructured decision process", *Administrative Science Quarterly*, 21, 246-276.

de Montgolfier, J., Bertier, P. (1978), *Approach Multicritère des problèmes de Dècision*, Editions Hommes et Techiques, Suresnes.

Moscarola, J. (1980), "Les tendances actuelles des travaux sur les processus de décision dans les organisations", *Enseignement et Gestion*, 15, 7-20.

Moscarola, J. (1981), "Efficiency of decision, how to assist decision-making in the organization", in J. P. Brans (ed.), *Operational Research 81*, North-Holland, 91-103.

Moscarola, J. (1984), "Organizational decision process and ORASA intervention", in R. Tomlinsom and I. Kiss (eds.), *Rethinking the Process of Operational Research and Systems Analysis*, Pergamon Press, Oxford.

Moscarola, J., Siskos, J. (1983), "Analyse à posteriori d'une étude d'aide à lá décision en matière de gestion de reseau de distribuition" in E. Jacquet-Lagrèze and J. Siskos (eds.), *Mèthode de Dècision Multicritère*, Editions Hommes et Techniques, Boulogne-Billancourt, 143-167.

Norese, M.F., Ostanello, A. (1984a), "Un cas de modélisation partagée pour un problème de décision collective", in *EIASM: Actes des 19 èmes Journèes du Groupe EURO sur Aide à la decision multicritère*, Liége, Mars 1984.

Norese, M.F., Ostanello, A. (1984b), "Planning processes and technician interventions: an integrated approach", *Sistemi Urbani*, 2, 247-259.

Norese, M.F., Ostanello, A. (1985), "A multicriteria model an evaluation of supply/demand of sport facilities in metropolitan area", paper presented at TIMS XXVI, Copenhagen, June 1984; *AIRO 1985*, 467-492.

Norese, M.F., Ostanello, A. (1988), "Decision aid process typologies and operational tools", Paper presented at the Annual Meeting of the AIRO, Pisa, October 1989.

Norese, M.F., Ostanello, A. (1989), "Identification and development of alternatives: Introduction to the recognition of process typologies", in A.G. Lockett and G. Islei (eds), *Improving Decision Making in Organisations*, Lecture Notes in Economics and Mathematical Systems, 335, Springer-Verlag, Berlin, 112-123.

Nutt, P. (1984), "Types of organizational decision processes", *Administrative Science Quarterly*, 29, 414-450.

Ostanello, A. (1987), "Comparaison d'approaches pour la définition de poids des critères", Paper presented at 25th Meeting of the EURO Group on MDCM, Bruxelles, March.

Ostanello, A., Mucci, L., Tsoukias, A. (1987), "Processus publics et notion d'espace d'interaction", *Sistemi Urbani*, 2/3.

Ostanello, A., Simoni, P., Veroni, P. (1978), "Multicriteria analysis of industrial localisation: experimentation on the territorial zones of Turin area", *AIRO 1978*, vol. 2, 393-411.

Roy, B. (1975), "Vers une méthodologie générale d'aide à la Décision", Revue Metra, vol. XIV, nº3.

Roy, B. (1985), *Mèthodologie Multicritère d'Aide à la Dècision*, Economica, Paris.

Schneeweiss, Ch. (1987), "On a formalization of the process of quantitative model building", *European Journal of Operational Research*, 29, 24-41.

Schwarz, M.. Thompson, M. (1985), "Beyond the politics of interest", in M. Grauer, M. Thompson and A.P. Wierzbicki (eds.), *Plural Rationality and Interactive Decision Processes*, Lecture Notes in Economics and Mathematical Systems, 248, Springer-Verlag, Berlin, 22-36.

Witte, E. (1972), "Field research on complex decision making - the phase theorem", *International Studies of Management and Organization*, 156-182.

BUILDING CRITERIA: A PREREQUISITE FOR MCDA

Denis Bouyssou

LAMSADE, Université de Paris Dauphine
Place du Maréchal De Lattre de Tassigny
75775 Paris Cedex 16 - FRANCE

1. INTRODUCTION

Following Roy (1985), we will say that decision-aid consists in trying to provide answers to questions raised by actors involved in a decision process using a clearly specified model. In order to do so, the analyst often has to compare "alternatives" (see Vincke, 1989). In an approach using several criteria, the analyst aims at establishing comparisons on the basis of the evaluation of the alternatives according to several criteria. In an approach using a single criterion, the analyst seeks to build a unique criterion taking into account all the relevant aspects of the problem. In either approach, the success of decision-aid crucially depends upon the way in which the unique criterion or the family of criteria have been built. The aim of this paper is to emphasize the importance of this phase by a number of frequently encountered difficulties and some techniques to overcome them.

The paper is organized as follows. We define the notion of criterion in section 2. In section 3, we use an example to show that building criteria is an important and difficult phase of the decision-aid process. Some standard techniques for constructing a criterion are presented in section 4. We conclude, in section 5, with some remarks concerning the choice of a consistent family of criteria.

2. WHAT IS A CRITERION?

a) Definition and remarks.

In what follows, we will call *criterion* a "tool" allowing to compare alternatives according to a particular "significance axis" or a "point of view[1]" (Roy, 1985)).

More precisely, *a criterion is a real-valued function on the set A of alternatives, such that it appears meaningful to compare two alternatives a and b according to a particular point of view on the sole basis of the two numbers g(a) and g(b).*

In a mono-criterion approach, the analyst builds a unique criterion capturing all the relevant aspects of the problem. The comparisons that are deduced from that criterion are to be interpreted as expressing "global preferences", *i.e.* preferences taking all the relevant points of view into account.

In a multiple criteria approach, the analyst seeks to build several criteria using several points of view. These points of view represent the different axes along which the various actors of the decision process justify, transform and argue their preferences. The comparisons deduced from each of these criteria should therefore be interpreted as partial preferences, *i.e.* preferences restricted to the aspects taken into account in the point of view underlying the definition of the criterion. Of course, speaking of partial preference implies the possibility of making *ceteris paribus* comparisons on the aspects that have not been taken into account in the definition of the criterion. This crucial hypotheses is central to MCDA. Its "test" would require the preferences of the actors of the decision process to be highly structured which is rather uncommon in decision-aid contexts.

----

1. Though these two terms are not synonymous (an axis is the operational counterpart of a given point of view (see Roy, 1985)) we will use them interchangeably in this paper.

However there are good reasons to believe that its adoption is not a severe restriction to the ability of MCDA to deal with real-world problems (see Roy and Bouyssou, 1988, chap. 2).

Our definition implies that a criterion is a *model* allowing to establish preference relations between alternatives. The quality of the *construction* of this model is crucial for the quality of decision-aid. An analogy will help to understand the importance of this phase. It is well-known in Statistics that the implementation of sophisticated data analysis methods cannot compensate for the weaknesses of the phase consisting in gathering and preparing the data. The same is true for MCDA: applying sophisticated aggregation procedures is of little use if the criteria have been built in an unconvincing way.

This analogy with Statistics can be further pursued. If there are a number of standard techniques for analyzing various types of statistical data, this is not the case for the phase consisting in gathering and preparing the data. The choice of the statistical variables, the definition of the population, the redaction of a questionnaire, the coding of the data, ... are crucial problems the solution of which depends more on the art and the experience of the statistician than on "Science". In our opinion, the same is true in decision-aid for the construction of the criteria. In this paper, we will thus try more to warn the reader against common difficulties and to present a number of techniques that have proved useful, than to present a well-established methodology that would allow to build criteria in all cases.

b) Some general guidelines for constructing a criterion.

When building a criterion, the analyst should keep in mind that it is necessary that all the actors of the decision process adhere to the comparisons that will be deduced from that model. This implies a number of important consequences.

i- The points of view underlying the definition of the various criteria should be understood and accepted by all the actors of the decision process, even if they disagree on the relative importance that they would like each of them to have in the aggregation model. These points of view should be familiar enough to these actors for them to be willing to discuss and argue on such a basis. Being able to associate to a given point of view a criterion having a clear physical unit may be seen as a great advantage in this respect.

ii- Once a point of view has been defined and accepted, the method allowing to arrive at the evaluation on the criterion for each alternative, should also be understood and accepted by all the actors of the decision process. The search for a simple and transparent model of evaluation should therefore be an important preoccupation of the analyst. Furthermore this method should be as free as possible from elements deeply linked to a particular value system. In fact, the presence of such elements may well lead some actors to question or even reject the validity of the comparisons made on the basis of the criterion.

iii- The choice of a particular way to build a criterion must take into account the quality of the "data" used to build it. In particular, the comparisons deduced from the criterion should take into account the elements of uncertainty, imprecision and/or inaccurate determination affecting the data used to build it.

c) Criteria, consequences and points of view.

The result of the implementation of an alternative can be modelled using a number of consequences or attributes (see Roy, 1985, chap. 8). These consequences are, in general, numerous and concern many aspects: time, money security, quality, image, ... Conceptually it is possible to base the comparison of the alternatives directly in terms of their consequences. However, due to the number of these consequences and to the fact that the evaluation of the alternatives on them often involves many elements of uncertainty, imprecision and inaccurate determination, this is, in general, difficult. A criterion thus appears as a tool allowing to "sum up" a set of evaluation on consequences related to a same point of view so as to be able to establish partial preferences. For instance, in a siting study, the analyst may want to build a criterion "impact on environment", taking into account consequences such as "impact on animal life", "impact on flora", "impact on landscape", ...

Building a criterion implies that one has chosen a point of view along which it seems adequate to establish comparisons. The determination of points of view that are understood and admitted by all actors is an important problem in MCDA. In order to do so, various techniques have been proposed. Roy (1985) considers that these points of view will emerge after a thorough analysis of various class of consequences, taking into account the "culture" of the actors involved. Keeney and Raiffa (1976), Keeney (1988), Saaty (1980), Forman (1990), Belton and Vickers (1990) advocate a "hierarchical" way of building the criteria through the decomposition a unique point of view ("well-being", "social benefits", ...) into sub-points of view that are again decomposed, till the relevant points of view are reached. It is worth noting using a hierarchical approach, one often speaks of "criteria", "sub-criteria", "sub-sub-criteria" depending on the level of the hierarchy. Here, we will restrict the use of the world criteria for the models related to the upper levels of the hierarchy. It is worth noting that the down-up approach of Roy and the hierarchical approach are not exclusive.

Another important problem in an approach using multiple criteria is the choice of a family of criteria. We will turn to this problem in section 5. Let us only mention now that, for reasons related to cognitive limitations of the human mind (see Miller, 1957) and to the necessity of gathering inter-criteria information for the implementation of aggregation procedures, it seems unadvisable to go much further than a dozen of criteria (defined at a high level of the hierarchy if such an approach is used).

d) The discriminating power of a criterion.

How to infer comparisons between alternatives on the basis of a criterion? In the most classical model, it is supposed that[1], for all a, b $\in$ A:

$a \; P_g \; b \Leftrightarrow g(a) > g(b)$ and
$a \; I_g \; b \Leftrightarrow g(a) = g(b)$,
where P_g (resp. I_g) is a binary relation that reads "is strictly preferred to (resp. indifferent to) considering the consequences taken into account in the definition of g". In this type of model called "true-criterion", any difference, as small as it may be, between two evaluations implies a strict preference. As we will see, since the evaluations g(a) and g(b) are often obtained through a model that includes some arbitrariness and on the basis of imprecise, uncertain data, this model may lead to unconvincing comparisons. In fact, it is often reasonable to admit that "small" differences g(a) - g(b) are compatible with an indifference situation between a and b. This leads to the following model of comparison:

$a \; P_g \; b \Leftrightarrow g(a) - g(b) > q$
$a \; I_g \; b \Leftrightarrow |g(a) - g(b)| \leq q$,

where q, the indifference threshold, is the largest difference g(a) - g(b) compatible with an indifference situation. Such a criterion is called a "quasi-criterion". In this model a difference greater than q implies a strict preference even if is very close to q. In order to avoid a sudden change from strict preference to indifference, it is possible to introduce a "buffer zone" in which there is an hesitation between the indifference and strict preference. Denoting this hesitation by a binary relation Q_g, often called "weak preference", we obtain a model with two

----

1. the direction of the inequality being conventional and not restrictive

thresholds, a preference threshold p and an indifference threshold q, called "pseudo-criterion", where[1]:

$$a\ P_g\ b \Leftrightarrow \quad g(a) - g(b) > p,$$
$$a\ Q_g\ b \Leftrightarrow q < g(a) - g(b) \leq p,$$
$$a\ I_g\ b \Leftrightarrow \quad |g(a) - g(b)| \leq q.$$

It is not an easy task to give a value to these two thresholds (see Bouyssou and Roy, 1987). Yet, in many situations, every reasonable non-null value for p and q, leads to a model of preferences that seems more convincing than the one obtained by letting p = q = 0 as done with a true-criterion. However, this model is used in most methods for MCDA[2]. It is true that the elements of arbitrariness inherent to this model can be "corrected" by a sensitivity analysis. However, let us recall from the previous section that the family of criteria will only play its part in the decision-aid model if the comparisons that can be inferred on its basis are not subject to criticisms. The interest of the pseudo-criterion model is then seen more clearly, since the presence of criteria leading to unconvincing comparisons may lead some actors to reject the whole family of criteria. Furthermore, it is worth noticing that the use of criteria with thresholds is of utmost importance if one wants to use afterwards an aggregation procedure based on a noncompensatory logic[3] as this is the case with the ELECTRE methods (Roy, 1990, Vanderpooten, 1990), TACTIC (Vansnick, 1987) or ORESTE (Roubens, 1982, Pastijn and Leysen, 1989).

----

1. These thresholds may vary along the scale of the criterion. When this is the case, the threshold is conventionally attached to the least preferred alternative. We have:

$$a\ P_g\ b \Leftrightarrow g(a) - g(b) > p(g(b)),$$
$$a\ Q_g\ b \Leftrightarrow q(g(b)) < g(a) - g(b) \leq p(g(b)),$$
$$a\ I_g\ b \Leftrightarrow g(a) - g(b) < q(g(b) \text{ and } g(b) - g(a) \leq q(g(a)).$$

2. This has, perhaps, to do with the traditional "culture" of Operational Research.

3. On this notion see Bouyssou and Vansnick (1986).

3. AN EXAMPLE

Suppose that an analyst wishes to build a criterion taking into account the impact in terms of noise of the construction of a new airport on the riparian population in a study aiming at prescribing one or several possible site for the construction of a new airport.

The objective of the construction of this criterion is to associate a figure to each site in such a way that this figure allows, at least, to determine if, from the point of view of the impact on the riparians, a site can be considered as preferable to another site. Though this study is conducted in a complex and conflictual context involving many actors (local and national authorities, air carriers, defense associations, public opinion), we will suppose that there is a consensus to take this point of view into account in order to choose a site.

The construction of this criterion takes place after a preliminary technical study which retained a small number of sites for a thorough analysis. In 1989, the time of the study, it is planned that the final decision will be taken in one or two years and that the construction will start within five years so that the airport could be fully operational around turn of the century.

In order to build such a criterion, the analyst will probably try to estimate first the number of inhabitants that would be, in 1989, affected by the installation of the new airport. Depending on time constraints, she may either count houses on a map and then multiply this number by an average number of inhabitants living in the region in different types of houses or conduct on-site studies. It should be clear that, whichever method is used, the number obtained will be highly imprecise. This imprecision due to the counting operations is however of limited importance compared to the multiple sources of inaccurate determination affecting the construction of this criterion. In particular, any mode of construction implies to take a definite position (explicitly or implicitly) on the following problems.

i- Where to place the boarder between "close" and "far" from the site? Ten years from now, will the airplanes be more or less noisy than they presently are? Will the problem of noise be more important than it presently is?

ii- The nuisance created to the riparians crucially depends on the position of the riparians relative to the runaways and the aerial corridors. At this stage of the study such information are probably not available yet. Should one consider a best guess hypothesis or some kind of average taking into account various possibilities or neglect this problem at that stage?

iii- What is a riparian? In particular, in the counting of the riparians should one give a particular weight (higher or lower?) to the second homes? Is it possible to make the hypothesis that the schools, hospitals, ... presently located near the site will move away from the site or, on the contrary, should these riparians be included in the total with a particular weight?

iv- Is it possible to make the hypothesis that the increase of the number of riparians will be identical on all sites during the period separating the study from the installation of the airport? On the contrary, should one consider that this increase depends on the distance between the projected site and the city center? If this is the case, should one envisage various scenarii for the growth of the population? Should one include in the model the probable consequences of the construction of the airport on the surrounding population (departure of the present riparians, installation of employees of the airport, installation of new firms near the site)? Should one take into account the fact that the construction of the airport will also imply the construction of new roads and railways that also create nuisance?

v- How to take into account the greater or lesser proximity of the riparians to the source of nuisance? Is the classical technique consisting in defining "zones" of noise and in giving a "weight" [1] to each zone satisfactory?

Building the criteria then implies to combine all this information in a formula allowing to compute a figure for each site. Depending on the options taken, one may use a criterion of one of the following types:

$$g(s) = \sum_{i=1}^{n} W_i h_i(s),$$

$$g(s) = \sum_{i=1}^{n} W_i h_i(s)(1+\alpha_i)^d,$$

$$g(s) = \sum_{k=1}^{m_s} \left[\sum_{i=1}^{n} W_i h_i(s)(1+\alpha_i(k))^d \right] P(E_k)$$

where there are n zones surrounding each site w_i being the weight affected to the inhabitants of the ith zone, $h_i(s)$ being the number of people living in the ith zone of site s in 1989 (possibly corrected in order to take into account "privileged" inhabitants), α_i being the annual increase rate of that population for the next d years, $\alpha_i(k)$ being the same rate in scenario E_k having a probability $P(E_k)$, m_s scenarii being considered for the site s. Let us note that one may want to include in these formulas various scenarii concerning the orientation of the runaways or the possible date of installation of the airport.

Even if we complicated this example on purpose and if many problems would not be raised in practice either because of a lack of time or because the various sites are not different in these respects, we hope to have shown that building a criterion may be a long and difficult task.

----

1. Usually inversely proportional to the distance, or the squared distance, between the zone and the site.

This construction implies a large amount of work and during this work a number of crucial options are taken. Furthermore it is often impossible to avoid the introduction of elements of arbitrariness in the definition of a criterion. In our example, this arbitrariness concerns a number of precise problems the solution of which does not seem to depend crucially on a particular value system. Therefore, in spite of this arbitrariness, it is not unlikely to be able to reach a consensus on a definition of this criterion. In our opinion this illustrates one of the major advantages of an approach using several criteria. In a mono-criterion approach, obtaining such a consensus is highly unlikely since elements dependent on a particular value system (*e.g.*, the tradeoffs between transportation time and the amount of nuisance created to the riparians) are inextricably mixed to a large number of other options in a complicated and inevitably opaque model. In our example the construction of the criteria requires a vast amount of data in which imprecision, uncertainty and inaccurate determination are involved. Whatever the final definition of the "impact on riparians" criterion, one has to admit that it will only be an "order of magnitude" of what we were willing to capture [1]. It is crucial to keep this point in mind in the rest of the study if convincing comparisons are to be established. This explains the interest of pseudo-criterion type models. In our example, whatever formula is used at the end, it should be clear that any non-null and not unreasonable value for p and q (for instance, $p(g(s)) = 0.2g(s)$ and $q(g(s)) = 0.1g(s)$) will lead to more realistic comparisons than setting $p = q = 0$.

4. SOME TECHNIQUES FOR BUILDING CRITERIA

As we already mentioned, this section is not a catalogue of standard techniques where one could look for a "solution". More modestly and realistically, we tried to distinguish some simple cases that illustrate some important points. For more details on

----

1. This is not to say that it is possible to give a precise definition of what we were willing to capture.

these various techniques we refer the reader to Roy (1985, chap. 9) and to Bouyssou and Roy (1987).

a) Case of a criterion based on a single consequence.

Let us consider a criterion g that the analyst wants to base on a unique consequence, for instance the cost of a given project.

Suppose first that the analyst considers that it is possible to neglect the elements of imprecision and/or uncertainty affecting the evaluation of the alternatives on that consequence. In such a situation it seems reasonable to build the criterion by letting for all $a \in A$:

$g(a) = c(a)$,
c(a) being the evaluation of alternative a on the consequence.

This technique is simple and, in general, leads to a criterion expressed in a clear physical unit. However, nothing implies that considering four alternatives a, b, c, $d \in A$ such that:

$g(a) - g(b) = g(c) - g(d)$,
one can "conclude" that the "preference difference" between a and b is similar to the preference difference between c and d (the difference between "very poor" and "poor" may well be different from the difference between "average" and "good").

It is easy to show that any monotonic transformation of such a criterion gives rise to a criterion leading to the same comparisons in terms of preference and indifference. It may therefore be interesting, especially if one wishes to use afterwards an aggregation procedure based on the idea of additive utility, to find among the strictly increasing monotonic transformation of g a transformation $\mathcal{X}$ for which it seems legitimate to compare preference differences between alternatives as $\mathcal{X}(g(a)) - \mathcal{X}(g(b))$ and $\mathcal{X}(g(c)) - \mathcal{X}(g(d))$.

Another classical situation concerns the case in which the evaluation of the alternatives on the unique consequence involves a best-guess evaluation c(a), an optimistic evaluation $c^+(a)$ and a pessimistic one $c^-(a)$, a case in which each evaluation is surrounded by an interval of imprecision which is not necessarily symmetric. In such a situation it seems again reasonable to consider that:

$$g(a) = c(a).$$

However it is no more possible to admit that a small difference between g(a) and g(b) implies a strict preference. It seems reasonable to consider in this situation that there is a strict preference for a over b only when $c^-(a) > c^+(b)$, a case in which the two imprecision intervals do not intersect[1]. When c(b) increases, the two intervals intersect. This intersection can be interpreted as an indifference when the best-guess evaluation of each alternative is contained in the imprecision interval of the other alternative[2]. The intermediate situation then corresponds to an hesitation zone that can be interpreted as a weak preference.

It is possible to show (see Roy, 1985, chap. 9) that as long as the differences $c^-(a)-c(a)$ and $c^+(a)-c(a)$ only depend on the value of c(a), this mode of comparison can be modelled using a pseudo-criterion. The reader might want to check that if for all $a \in A$, we have:

$$c^-(a) = c(a) - (\alpha' + \beta' c(a)) \text{ and } c^+(a) = c(a) + (\alpha + \beta c(a)),$$

this mode of comparison defines a pseudo-criterion such that for all $a \in A$,

---

1. Let us recall that the direction of the inequality is conventional.

2. Other conventions are possible. See, *e.g.*, Siskos and Hubert (1983).

$$g(a) = c(a),$$
$$p(g(a)) = [\alpha + \alpha' + (\beta + \beta')g(a)] \; / \; (1 - \beta'),$$
$$q(g(a)) = \text{Min} \; [\alpha + \beta g(a); \; (\alpha' + \beta' \; g(a)) \; / \; (1 - \beta')].$$

Another situation occurs when the evaluation of the alternatives on the consequence is "distributional". In others words, building a criterion amounts to compare distributions on the scale of the consequence. The necessity to consider distributions may come from several sources:

- the evaluation varies in time (problem of "actualization"),
- the evaluation varies in space (evaluation on a consequence of a "linear" project: highway (see Marchet and Siskos, 1979), high voltage line (see Grassin, 1986), distribution of the riparian population between the various zones of noise (in our example),
- the evaluation is uncertain and a probability (or plausibility) distribution is used.

In such a situation, a standard technique, called "point-reduction" in Roy (1985), to build a criterion consists in trying to sum up the distribution by a unique figure. This figure is usually a weighted average. More precisely, in order to build the criterion on the basis of a distribution consisting for an alternative a of a mass $f^a(x)$ associated to each level x of the consequence, a standard technique consists in setting:

$$g(a) = \int u(x) \; f_a(x) \; dx \tag{1}$$

where u is some real-valued function on the set of levels of the consequence.

The most classical technique of point-reduction is "actualization". It consists in summing up a monetary distribution on a time scale. The criterion that is generally used is the Present Value of the distribution:

$$g(a) = \sum_{k=1}^{n} f_a(k)/(1+i)^k,$$

where $f_a(K)$ is the cash flow generated by alternative a in period k, i being an actualization coefficient taking into account the fact that the importance of a flow depends on the period in which it is generated[1]. Letting $u(x) = 1/(1+i)^x$, it is easily seen that actualization is a particular case of (1).

As far as distribution in space are concerned, the procedure we used in our example to define the criterion "impact on the riparians" is obviously a particular case of (1) in which u is defined for each zone by a weight that is a function of the distance between the zone and the airport. Similar techniques have been used in Grassin (1986) and Marchet and Siskos (1979).

When $f_a(x)$ can be interpreted as a probability, formula (1) defines what is usually called an expected utility criterion. When the scale of the consequences is a subset of $\mathbb{R}$, letting $u(x) = x$ in (1) amounts to using the expectation of the distribution as a criterion. Such a criterion would not allow to take into account crucial elements such as risk aversion, the risk of ruin, ... The introduction of the function u, called in that case a von Neumann-Morgenstern utility function, allows to take into account such phenomena. Suppose, for instance, that one of the actors of the decision process declares that she is a "risk adverse" for money, *i.e.* that she has a definite preference for "nonrisky" investments. For instance, she prefers an investment with a certain net present value of[2] 500.000 to an investment that may yield with equal chances either 2 000 000 or 0, even if

---

1. This criterion is often justified by the existence of financial markets on which it is possible to lend and borrow money at an interest rate i. Let us mention that this type of criterion can be justified in the absence of such markets and for a non monetary consequence (see, *e.g.*, Koopmans, 1972).

2. All monetary sums are expressed in ECU (European Currency Unit).

the expected profit of the latter (1 000 000) is far greater than that of the former. In order to build a criterion taking into account such a preference, it suffices to choose u such that:

$$u(500\ 000) > \frac{1}{2}u(0) + \frac{1}{2}u(2\ 000\ 000).$$

There are a number of standard questioning techniques designed to assess a function u compatible with what can be perceived of the preference *vis-à-vis* risk of an actor (see Keeney and Raiffa, 1976, chap. 4). This type of criterion have been extensively studied in litterature. It has many interesting aspects [1] and has been frequently used in real-world situations (see, *e.g.*, Keeney and Nair, 1977).

There is a theory (Expected Utility theory) aiming at justifying the use of such a criterion through an axiomatic analysis of the preferences compatible with formulations of type (1). However, the existence of such a theory does not oblige the analyst to use a formulation of type (1) if she thinks that another type of criterion could lead to more convincing comparisons [2]. Furthermore, the richness of formulation (1) should not lead one to forget that this type of criteria is based on probability distributions that are rarely the only sensible ones and on a function u which has been assessed through a questioning process that might have had an overwhelming influences on the shape of the function (see, *e.g.*, McCord and de Neufville, 1983). Therefore it is often necessary to consider that criteria built in this way are not true-criteria and to use them surrounded by one or two discrimination thresholds.

----

1. For instance, it is easily shown that the shape of u (concavity or convexity) can be interpreted in terms of aversion of preference for risk (see Pratt, 1964).

2. This is all the more true that there are many controversies concerning the normative, descriptive and prescriptive virtues of Expected Utility. For a recent overview of the debates, see, *e.g.*, Munier (1988).

Either because it seems difficult to assess the function u or because the analyst does not wish to sum up a complex set of information by a single figure, it is possible not to use a point-reduction criterion and to sum up the distributional evaluation using several criteria. When the distribution is probabilistic, it is possible, for instance, to use two criteria: a criterion measuring a central tendency (expected value but also median or mode) and a dispersion criterion (variance or standard deviation, semi-variance, interquartile range, probability of ruin, probability of not reaching a specified target, ...), see Colson and Zeleny (1980) or Fishburn (1977).

Another mode of construction is to consider that the source of the distributional evaluation is the existence of various scenarii and to build a criterion per scenario without trying to aggregate them in this phase (see, *e.g.*, Teghem and Kunsch, 1985).

b) Case of a criterion based on several consequences.

Either because of the size of the set of consequences, or because a hierarchical technique of construction is used, or because the actors of the decision process are used to thinking using concepts including several consequences, the analyst often have to build a criterion that takes into account more than one consequence. It is said that such a criterion sub-aggregates a set of consequences. In doing so, it is crucial to keep in mind that the result of this aggregation has to be accepted by all actors and sufficiently transparent and simple in order to be interpreted and easily discussed. That is to say that such a sub-aggregation should only concern a limited number of consequences that are "sufficiently" close from one another in order to keep the model simple and to avoid the introduction of "sensitive" information at this stage.

It is possible to use every aggregation procedure leading to the establishment of a unique criterion in order to build such a criterion. However, taking into account the proximity of the consequences and the necessary transparency of the model, simple methods are, in general, used: lexicographic aggregation, weighted average, sum of ranks, or any ad-hoc combination of these methods.

For instance, Roy et al. (1986), in order to build a criterion taking into account a point of view "level of discomfort" of the users of a subway station [1] chose to aggregate the consequences:

- climatic conditions in the station,
- noise in the station,
- "penibility" of the access to the station (presence of escalators, etc.),
- time lost in going to one platform to another, and
- density of passengers on the trains,

using a three-point scale for each of these consequences:

0 : "no problem has been observed",
1 : "a minor problem exists",
3 : "a major problem exists",

and defining the value of the criterion for a given station on the basis of the sum of the evaluations on the five consequences (after modifying this sum in an ad-hoc way to take into account particular circumstances).

5. CONCLUSION

The construction of criteria is a phase of the decision-aid process that takes place after a preliminary phase consisting in defining the set of alternatives, the problem formulation of the study and the strategy of intervention in the decision process.

---

1. In a study aiming at prescribing a renovation plan for these stations.

These two crucial phases represent, in real-world studies, the major part of the work of the analyst [1].

After the construction of the criteria, an alternative, *i.e.* in general, a complex project that may not be completely specified at the time of the study, will only be taken into account [2] through the vector $(g_1(a), g_2(a), \ldots, g_n(a))$. As we saw in the previous sections, it is essential that each component of this vector be a model that all actors understand and accept. But it is also important that this vector, as a whole, be a faithful representation of the alternative a. These two conditions seem to be a prerequisite for the application of MCDA techniques to be really useful in real-world problems.

We already mentioned two important qualities that a family of criteria should have:

- the "legibility", *i.e.* the family should contain a sufficiently small number of criteria so as to be a discussion basis allowing the analyst to assess inter-criteria information necessary for the implementation of an aggregation procedure,
- the "operationality", *i.e.* the family should be considered by all actors as a sound basis for the continuation of the decision-aid study.

However, as noticed by Roy and Bouyssou (1988, chap. 2), a family of criteria must also possess a number of technical properties, leading to the concept of "consistent family of criteria", to be able to be really useful for decision-aid purposes. In an informal way, we will say that a family of criteria is consistent if it is:

----

1. Sometimes the analyst is only hired for these two phases, see, *e.g.* Grassin (1986).

2. It is rather unusual, once the criteria have been built to go back to the initial data, *i.e.*, the evaluation of the alternatives on the various consequences.

- *exhaustive:* the family should contain every important point of view. In particular this condition implies that if for all the criteria in the family we have $g_i(a) = g_i(b)$, every actor must agree to consider that a and b are indifferent,

- *monotonic:* the partial preferences that are modelled by each criterion have to be consistent with the global preferences expressed on the alternatives. This condition implies that if a is judged to be better than b taking into account all the points of view, the same judgment will hold for an alternative c that is judged at least as good as a on every criterion,

- *minimal:* for obvious reasons this condition implies not to include in the family unnecessary criteria, *i.e.* which suppression will lead to a family still satisfying the first two conditions.

Very often, the search for a legible, operational and consistent family of criteria leads the analyst to reconsider the definition of some criteria, to introduce new ones in the family, to aggregate some of them, etc. Thus the choice of a family of criteria interacts with the construction of the various criteria.

Let us finally mention that other desirable conditions can be imposed on a family of criteria (see Roy and Bouyssou, 1988, chap. 2). For instance, it seems reasonable to be willing to work with a family of criteria in which *ceteris paribus* comparisons on a sub-family of criteria (and not only on a single criterion) are legitimate and in which there are no functional relations between criteria. Without going into these questions in many details, let us only mention that it is not always possible to build a family that would satisfy these two conditions and still being legible and operational. When this is the case, the task of the analyst becomes very difficult in the aggregation phase either because she cannot implement aggregation procedures based on an addition operation (which is the case for nearly all aggregation procedures based on an addition operation (which is the case for nearly all aggregation procedures) or because some actors may accuse her of "double counting" a number of factors.

REFERENCES

Allais, M. (1953), "Le comportement de l'homme rationnel devant le risque, critique des postulats et axiomes de l'école américaine", *Econometrica*, Vol. 21, 503-46.

Belton, V. and Vickers, S. (1990), "Use of a simple multi-attribute value function incorporating visual interactive sensitivity analysis", in this volume.

Bouyssou, D. (1984), "Expected utility theory and decision-aid: a critical survey", in O. Hagen and F. Wenstøp (eds), *Progress in Utility and Risk Theory*, Reidel, Dordrecht, 181-216.

Bouyssou, D. (1988), "Modelling uncertainty, imprecision and inaccute determination using multiple criteria", *Cahier du LAMSADE* nº 88, Université de Paris Dauphine.

Bouyssou, D. and Roy, B. (1987), "La notion de seuils de discrimination en analyse multicritère", *INFOR*, Vol. 25, 302-313.

Bouyssou, D. and Vansnick, J.-C. (1986), "Noncompensatory and generalized noncompensatory preference structures, *Theory and Decision*, Vol. 21, 251-266.

Colson, G. and Zeleny, M. (1980), "Multicriterion concept of risk under incomplete information", *Computers and Operations Research*, Vol.7, 125-143.

Fishburn, P.C. (1970), *Utility Theory for Decision Making*, Wiley, New York.

Fishburn, P.C. (1977), "Mean-Risk analysis with risk associated with below-target returns", *American Economic Review*, Vol. 67, 116-126.

Forman, E.H. (1990), "Multi criteria decision making and the analytic hierarchy process", in this volume.

Grassin, N. (1986), "Constructing criteria "population" for the comparison of different options of high voltage line routes", *EJOR*, Vol. 26, 42-47.

Keeney, R.L. (1988), "Structuring objectives for problems of public interest", *Operations Research*, Vol. 36, 396-405.

Keeney, R.L. and Nair, K. (1977), "Selecting nuclear power plants sites on the Pacific Northwest using decision analysis", in D.E. Bell, R.L. Keeney and H. Raiffa (eds), *Conflicting Objectives in Decision*, Wiley, New York.

Keeney, R.L. and Raiffa, H. (1976), *Decisions with Multiple Objectives - Preferences and Value Tradeoffs*, Wiley, New York.

Koopmans, T.C. (1972), "Representation of preference orderings over time, in C.B. McGuire and R. Radner (eds), *Decision and Organization*, North Holland, Amsterdam.

Marchet, T.C. and Siskos, J. (1979), "Aide à la décision en matière d'environnment: application au choix de tracé autoroutier", *Sistemi Urbani*, Vol. 2, 65-95.

McCord, M. and de Neufville, R. (1983), "Fundamental deficiency of expected utility analysis, in S. French, R. Hartley, L.C. Thomas, and D.J. White (eds), *Multiobjective Decision-Making*, Academic Press, London, 279-305.

Miller, G.A. (1956), "The magical number seven plus or minus two - Some limits on our capacity for processing information", *The Psychological Review*, Vol. 63, 81-97.

Munier, B. (1988), *Risk, Decision and Rationality*, Reidel, Dordrecht.

Von Neumann, J. and Morgenstern, O. (1947), *Theory of Games and Economic Behaviour*, 2nd edition, Princeton University Press, Princeton.

Pastijn, H. and Leysen, J. (1989), "Constructing an outranking relation with ORESTE", *Mathematical and Computer Modelling*, Vol. 12 (10/11), 1255-1268.

Pratt, J.W. (1964), "Risk aversion in the small and in the large", *Econometrica*, Vol. 32, 122-136.

Roubens, M. (1982), "Preference relations on actions and criteria in multi-criteria decision making", *European Journal of Operational Research*, Vol. 10 (1), 51-55.

Roy, B. (1985), *Mèthodologie Multicritère d'Aide à la Dècision*, Economica, Paris.

Roy, B. (1988), "Main sources of inaccurate determination, uncertainty and imprecision in decision models", in B.R. Munier and M.F. Shakun (eds), *Compromise, Negotiation and Group Decision*, Reidel, Dordrecht, 43-62.

Roy, B. (1990), "the outranking approach and the foundations of ELECTRE methods", in this volume.

Roy, B. and Bouyssou, D. (1988), *Aide Multicritère à la Décision: Méthodes et Cas*, book in preparation, Chapter 2 "Famille cohérente de critères, dépendances et conflicts entre critères", *Document du LAMSADE* nº 37, Chapter 3 "Conflits entre critères et procédures élémentaires d'agrégation multicritère, *Document du LAMSADE* nº 41, Chapter 4 "Procédures d'agrégation conduisant à un critère unique de synthèse", *Document du LAMSADE* nº42, Université de Paris-Dauphine.

Roy, B., Présent, M, and Silhol, D. (1986), "A programming method for determining which Paris metro stations should be renovated", *EJOR*, Vol. 24, 318-334.

Saaty, T.L. (1980), *The Analytic Hierarchy Process*, Mcgraw-Hill, New York.

Siskos, J. and Hubert, Ph. (1983), "Multicriteria analysis of the impact of energy alternatives: a survey and a new comparative approach", *EJOR*, Vol. 13, 278-299.

Teghem, J. and Kunsch, P. (1985), "Multi-objective decision making under uncertainty: an example for power system", in Y. Haimes and V. Chankong (eds), *Decision Making with Multiple Objectives*, Springer Verlag, Berlin, 443-456.

Vanderpooten, D. (1990), "The construction of prescriptions in outranking methods", in this volume.

Vincke, Ph. (1989), *L'Aide Multicritère à la Décision*, Editions de l'Université de Bruxelles et Editions Ellipses, Brussels.

Zeleny, M. (1982), *Multiple Criteria Decision Making*, McGraw-Hill, New York.

MEASUREMENT THEORY AND DECISION AID

Jean-Claude Vansnick

University of Mons-Hainaut
Place Warocqué, 17
B - 7000 Mons BELGIUM

1. INTRODUCTION

When facing a decision problem, the first thing to do is to structure this problem, to *create a model* of the problem in order to be possible to apply to this model a systematic treatment. The elaboration of such a systematic treatment *(decision aid method)* depends on the model but also on the information obtainable from the decision-maker. As the decision aid methods are often based on the notion of real number, it is fundamental, given a set X and some information about this set, to know

1) when it is possible to assign to each element x of X a real number $\mu(x)$ so that, by specifying how to use these numbers, they can represent the information given on X *(representation problem)*;

2) what kinds of mathematical manipulations are possible with these numbers, that is what statements can be meaningfully made using these numbers *(problem of meaningfulness)*.

These are the topics of *measurement theory*.

In that framework, the real number $\mu(x)$ is called the *measure* of x and the application μ: $X \to \mathbb{R}$ (where $\mathbb{R}$ is the set of real numbers) is called a *(numerical) scale of measurement*.

We present briefly in the next sections some fundamental elements of measurement theory (essentially on the basis of examples involving the idea of preference) and we show <u>the importance of the notion of measurement for decision aid</u>. Section 2 introduces the notion of binary relation and presents some basic results concerning the representation problem. Section 3 is devoted to the important problem of meaningfulness, which leads to the notion of scale type. This one is studied in section 4. Finally, section 5 deals with the problem of decision aid in reference to measurement theory; it tries to specify the positions of the «American and French Schools of decision aid» facing the problems of measurement.

For those who wish to get further into the field of measurement we highly recommend the book of F.S. Roberts (1979).

2. THE REPRESENTATION PROBLEM: BINARY RELATIONS AND BASIC RESULTS

In decision aid, the notion of *preference* plays a fundamental role. When a decision-maker says that he prefers x to y, he creates a relationship between x and y so that the mathematical notion of *binary relation* clearly appears as perfectly adapted for modelling the idea of preference. Binary relations also form a basis for much of the theory of measurement. Let us recall that a binary relation on a set X is a subset of the Cartesian product X x X, that is, a set of ordered pairs (x, y) such that $x, y \in X$. If X is the set $(x_1, x_2, x_3, x_4, x_5)$ then examples of binary relations on X are given by

$$R_1 = \{(x_1, x_4), (x_3, x_5), (x_4, x_2), (x_2, x_2)\}$$
$$R_2 = \{(x_4, x_3), (x_3, x_4), (x_2, x_5)\}.$$

Let R a binary relation on X.

In order to express that $(x, y) \in R$ (resp. $(x,y) \notin R$), we will write x R y (resp. $x \not R y$). We will say that

R is *asymmetric* iff $\forall x, y \in X$: $[x\ R\ y] \Rightarrow [y\ \not R\ x]$;

R is *complete* iff $\forall x, y \in X$ with $x \neq y$: $x\ R\ y$ or $y\ R\ x$;

R is *transitive* iff $\forall x, y, z \in X$:

$[x\ R\ x \text{ and } y\ R\ z] \Rightarrow [x\ R\ z]$;

R is *negatively transitive* iff $\forall x, y, z \in X$:

$[x\ \not R\ y \text{ and } y\ \not R\ z] \Rightarrow [x\ \not R\ z]$.

Suppose the decision-maker tells us that, regarding the elements of a set $X = \{x_1, x_2, x_3, x_4\}$, the prefers x_1 to x_2, x_2 to x_3, x_1 to x_3 and x_4 to x_3. We can model this information by the binary relation

$$P = \{(x_1, x_2), (x_2, x_3), (x_1, x_3), (x_4, x_3)\}.$$

The graphical representation of P is given in figure 1.

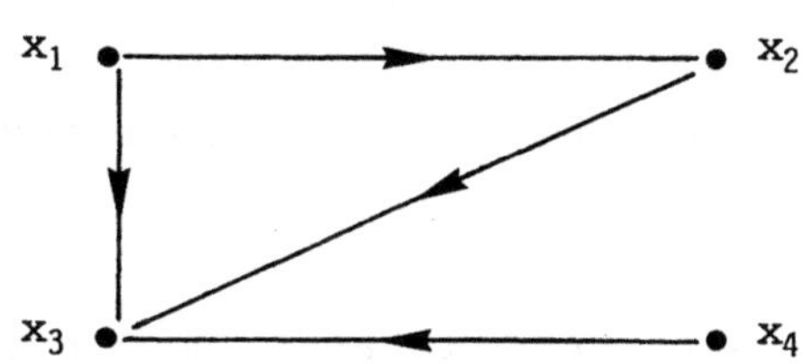

Figure 1

It is possible to represent this relation P through real numbers assigned to the elements of X, using the binary relation > on $\mathbb{R}$? More precisely, is there a function $\mu: X \rightarrow \mathbb{R}$ such that

$$\forall x, y \in X : x\ P\ y \text{ iff } \mu(x) > \mu(y) \text{ ?}$$

Let us observe that the relation P is asymmetric and transitive and yet the answer to the previous question is «no». Indeed, we would must have:

$\mu(x_4) = \mu(x_2)$ because $x_2\ \not P\ x_4$ and $x_4\ \not P\ x_2$

$\mu(x_4) = \mu(x_1)$ because $x_1\ \not P\ x_4$ and $x_4\ \not P\ x_1$

$\mu(x_1) > \mu(x_2)$ because $x_1\ P\ x_2$,

which is impossible.

The following theorem of Cantor concerning the problem stated above is a basic result of measurement theory (for a proof, see Roberts, 1979, th. 3.3, p. 109):

Theorem 1 - Suppose X is a finite or countable set and P is a binary relation on X. Then, there is a real-valued function μ on X satisfying [$\forall x, y \in X$: x P y iff $\mu(x) > \mu(y)$] if only if P is asymmetric and negatively transitive.

If we consider that P is a relation modelling preferences, the property of asymmetry is very natural since, when you prefer x to y, you do not prefer y to x. The property of negative transitivity which requires that, if you do not prefer x to y, and do not prefer y to z, then you do not prefer x to z is less known than the property of transitivity. The following proposition explains the connections between these two properties for an asymmetric relation.

Proposition 1 - Suppose P is an asymmetric binary relation on a set X. Then P is negatively transitive if and only if

1) P is transitive
2) the relation I defined by [$\forall x, y \in X$: x I y iff $x \not P y$ and $y \not P x$] is transitive.

Remark : When P is a preference relation, the relation I can be viewed as a indifference relation.

A binary relation which is asymmetric and negatively transitive is often called *«strict weak order»*. When, in addition, this relation is complete, it is called *«strict simple order»* or *«linear order»*. When P is a strict simple order, P is a strict weak order and, furthermore, [$\forall x, y \in X$ with $x \neq y$, x P y or y P x] so that it immediately results from theorem 1 that:

Theorem 2 - Suppose X is a finite or countable set and P is a binary relation on X. Then, there is a real-valued function μ on X satisfying [$\forall x, y \in X$: x P y iff $\mu(x) > \mu(y)$ and $x \neq y$ iff $\mu(x) \neq \mu(y)$] if and only if P is a strict simple order.

Remarks :

1) When P is asymmetric and complete, it can be proved that P is negatively transitive if and only if P is transitive, so that a strict simple order can be defined as an asymmetric, transitive and complete binary relation.

2) When you can assign to each element x of a set X a real number $\mu(x)$ so that the information given on X can be represented through the relation «strictly greater than» on the real numbers, you can order, thanks to these numbers, the elements of X. When, in addition, [$\mu(x) \neq \mu(y)$ iff $x \neq y$), it means that there is no ex-aequo in the ranking. It therefore results from theorems 1 and 2 that the mathematical notions of strict weak order and strict simple order correspond, respectively, to the common notions of *ranking with possibility of ex-aequo* and of *ranking without ex-aequo*.

Theorems 1 and 2 gives conditions under which an information on X (modelled by a binary relation) can be expressed by real numbers using a specified relation on $\mathbb{R}$. These are examples of representation theorems in measurement theory. Such theorems present axioms under which measurement is possible. Given an information on a set X, if there is a function $\mu: X \rightarrow \mathbb{R}$ so that this information can be represented by the real numbers $\mu(x)$ ($x \in X$) using some specified mathematical concept(s) on $\mathbb{R}$, then we will call μ a *(numerical) scale of measurement*.

Let us point out that the way of using the numbers $\mu(x)$ in order to represent the information considered would normally be included in the definition of the notion of scale of measurement.

Nevertheless, for simplicity, μ (alone) will be called scale of measurement in the next sections.

3. THE PROBLEM OF MEANINGFULNESS

When we have a (numerical) scale of measurement, it is very important to know what kinds of mathematical manipulations are possible with the real numbers *(measures)* given by the scale: it is the problem of meaningfulness.

Suppose $X = \{x_1, x_2, x_3, x_4\}$ and the decision-maker tells us that he prefers x_1 to x_2, x_2 to x_3, x_1 to x_3, x_4 to x_3 and x_1 to x_4. This information can be modelled by the binary relation P on X:

$$P = \{(x_1, x_2), (x_2, x_3), (x_1, x_3), (x_4, x_3), (x_1, x_4)\}.$$

The graphical representation of P is given in figure 2.

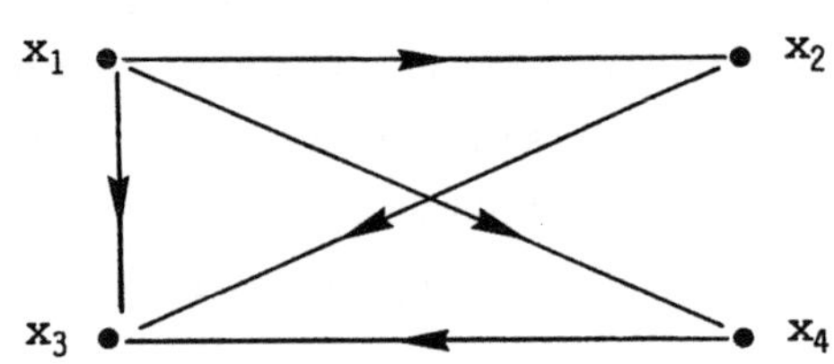

Figure 2

P is asymmetric and negatively transitive so that we know that there is $\mu: X \to \mathbb{R}$ such that $[\forall x, y \in X: x\ P\ y$ iff $\mu(x) > \mu(y)]$.

We can for instance take:

$$\mu: X \to \mathbb{R}: x \to \mu(x) = \#\ \{y \in X \mid x\ P\ y\},$$

which leads to $\mu(x_1) = 3$, $\mu(x_2) = 1$, $\mu(x_3) = 0$ and $\mu(x_4) = 1$.

Let us observe that, with this scale of measurement, we have

$$\mu(x_1) - \mu(x_2) = 2.[\mu(x_4) - \mu(x_3)]$$

and consider the following statement: «the degree of preference for x_1 over x_2 is twice the degree of preference for x_4 over x_3».

Is that statement meaningful ?

The information given by the decision-maker (strict weak order P on X) can also be represented by the scale

$$\mu': X \to \mathbb{R}: x \to \mu'(x) = \begin{cases} 10 & \text{if } x = x_1 \\ 8 & \text{if } x = x_2 \\ 0 & \text{if } x = x_3 \\ 8 & \text{if } x = x_4 \end{cases}$$

since we clearly have

$$\forall x, y \in X : x \text{ P } y \text{ iff } \mu'(x) > \mu'(y).$$

With the scale μ',

$$\mu'(x_1) - \mu'(x_2) \neq 2.[\mu'(x_4) - \mu'(x_3)].$$

This is the reason why we will say that the above statement is not meaningful; it is meaningless.

By definition, a statement involving scales of measurement is meaningful if and only if its truth or falsity remains unchanged when a scale representing the information considered is replaced by any other scale also representing this information.

Note that meaningfulness is not the same as truth. For example, the statement «I weigh more than the Statue of Liberty» is false but meaningful because it is false for all scales of weight (gram, kilogramme, ton, ...).

Meaningfulness however is a very important property.

As Roberts (1985) points out: from the point of view of decision-making, stating an «assertion» which is meaningless seems wrong because, if two different sets of numbers are both legitimate as input for a decision-making problem, and if these sets of numbers lead to different decisions, then there would be chaos.

A fundamental point in measurement theory is that, generally, the scale which can represent the information considered is not unique. Once measurement has been accomplished, it is very important to know what is the *uniqueness* of the resulting scale in order to can study what kinds of manipulations are possible with the real numbers (measures) obtained. The uniqueness problem is usually studied by considering *admissible transformations of scale*. We now introduce this notion.

Suppose X is a set on which we have some information and $\mu: X \to \mathbb{R}$ is a scale of measurement representing this information. Let us consider a function $\tau: \mu(x) \to \mathbb{R}$ that maps the range of μ, the set $\mu(X) = \{\mu(x)) \mid x \in X\}$, into the set of real numbers $\mathbb{R}$. The composition $\tau \circ \mu$ is a function defined on X with values in $\mathbb{R}$

$$\tau \circ \mu: X \to \mathbb{R} : x \to (\tau \circ \mu)(x) = \tau(\mu(x)).$$

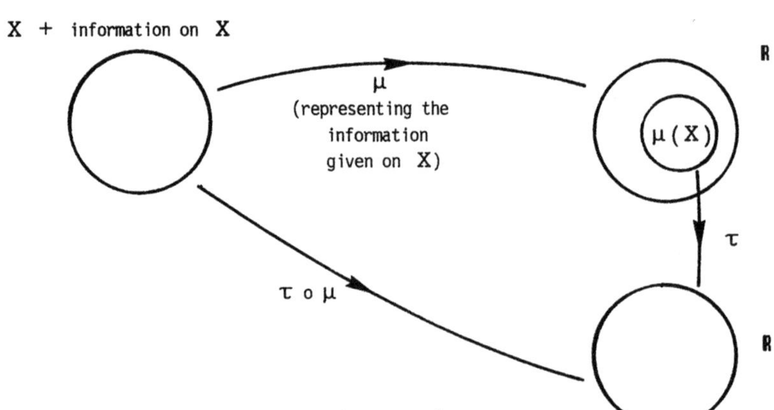

Figure 3

When $\tau \circ \mu$ also represents the information given on X, τ is called an *admissible transformation* of the scale μ.

Example:

As previously, let $X = \{x_1, x_2, x_3, x_4\}$, P the preference relation on X presented in figure 2 and $\mu: X \to \mathbb{R}$ the scale of measurement defined by $\mu(x_1) = 3$, $\mu(x_2) = 1$, $\mu(x_3) = 0$ and $\mu(x_4) = 1$. Consider the function

$$\tau: \mu(X) \to \mathbb{R} : r \to 2^r.$$

We have

$$\tau \circ \mu : X \to \mathbb{R} : x \to (\tau \circ \mu)(x) = \begin{cases} 2^3 = 8 & \text{if } x = x_1 \\ 2^1 = 2 & \text{if } x = x_2 \\ 2^0 = 1 & \text{if } x = x_3 \\ 2^1 = 2 & \text{if } x = x_4 \end{cases}$$

and it immediately appears that

$$\forall x, y \in X : x \, P \, y \text{ iff } (\tau \circ \mu)(x) > (\tau \circ \mu)(y)$$

that is $\tau \circ \mu$ also represents the information given on X. Consequently,

$$\tau : \mu(x) \to \mathbb{R} : r \to 2^r$$

is an admissible transformation of the scale μ.

In several important cases (but not always), the set of all admissible transformations of a scale μ generates, by considering

$$\{\tau \circ \mu \mid \tau \text{ is an admissible transformation of } \mu\},$$

the set of all scales of measurement representing the same information as μ. The scale μ is then called *regular*. In other words, a scale μ is regular if and only if for each scale μ' representing the same information as μ, there is a function

$\tau: \mu(X) \to \mathbb{R}$ such that $\mu' = \tau \circ \mu$.

For a regular representation, the problem of meaningfulness is quite simple since we can say (see Roberts, 1979, p. 59) that «a statement involving (numerical) scales is meaningful if and only if its truth or falsity is unchanged under admissible transformations of all scales in question».

Moreover, for such a representation, the class of admissible transformations defines how unique each scale is, and can be used to define the scale type. For decision aid, three types of scale are particularly important. We study them in the next section.

4. ORDINAL, INTERVAL AND RATIO SCALES

By definition, a scale of measurement $\mu: X \to \mathbb{R}$ is

* an *ordinal scale* when the admissible transformations of μ are all the functions $\tau: \mu(x) \to \mathbb{R}$ such that

  $\forall r_1, r_2 \in \mu(X) : r_1 > r_2$ iff $\tau(r_1) > \tau(r_2)$.

  (In other words, μ is an ordinal scale if and only if it is unique up to a strictly monotone increasing transformation).

* an *interval scale* when the admissible transformations of μ are all the functions $\tau: \mu(x) \to \mathbb{R}$ of the form

  $\tau(r) = \alpha.r + \beta$ with $\alpha, \beta \in \mathbb{R}$ and $\alpha > 0$.

  (In other words, μ is an interval scale if and only if it is unique up to a positive linear transformation).

* a *ratio scale* when the admissible transformations of μ are all the functions $\tau: \mu(x) \to \mathbb{R}$ of the form

  $\tau(r) = \alpha.r$ with $\alpha \in \mathbb{R}$ and $\alpha > 0$.
  (In others words, μ is a ratio scale if and only if it is unique up to a similarity transformation).

Obviously, the less the admissible transformations of a scale, the more the meaningful statements involving that scale. From this point of view, it is more interesting to have a ratio scale than an interval scale and an interval scale than an ordinal scale.

Example:

Suppose $X = \{x_1, x_2, x_3, x_4, x_5\}$ and $\mu: X \to \mathbb{R}$ is a scale of measurement defined by $\mu(x_i) = i$ for each $i \in \{1,2,3,4,5\}$. Consider the following statement: «the average rating of subset $\{x_3, x_4\}$ is higher than the average rating of subset $\{x_1, x_5\}$».

This statement is meaningless if μ is an ordinal scale but it is meaningful if μ is an interval scale or a ratio scale.

Proof:

Case 1) μ is an ordinal scale:

- The statement is true with the scale μ defined above since

$$\left.\begin{aligned} \frac{\mu(x_3) + \mu(x_4)}{2} &= (3 + 4) / 2 = 7/2 \\ \frac{\mu(x_1) + \mu(x_5)}{2} &= (1 + 5) / 2 = 6/2 \end{aligned}\right\} \quad \text{and } 7/2 > 6/2$$

- The statement is false with the scale $\tau \circ \mu$ where τ is the admissible transformation $\tau: \mu(X) \to \mathbb{R} : r \to \tau(r) = 2^r$; indeed,

$$\left.\begin{aligned} \frac{\tau(\mu(x_3)) + \tau(\mu(x_4))}{2} &= (2^3 + 2^4) / 2 = 12 \\ \frac{\tau(\mu(x_1)) + \tau(\mu(x_5))}{2} &= (2^1 + 2^5) / 2 = 17 \end{aligned}\right\} \quad \text{and } 12 < 17.$$

Case 2) μ is an interval scale:

- The statement is true with the scale μ given (see above).
- The statement remains true for each scale $\tau \circ \mu$ where τ is a function $\tau: \mu(X) \to \mathbb{R}$ of the form $\tau(r) = \alpha.r + \beta$ with $\alpha, \beta \in \mathbb{R}$ and $\alpha > 0$; indeed

$$\frac{\tau(\mu(x_3)) + \tau(\mu(x_4))}{2} = \frac{\alpha.\mu(x_3) + \beta + \alpha.\mu(x_4) + \beta}{2} =$$

$$= \frac{\alpha.(3 + 4) + 2.\beta}{2} = 7/2.\alpha + \beta$$

$$\frac{\tau(\mu(x_1)) + \tau(\mu(x_5))}{2} = \frac{\alpha.\mu(x_1) + \beta + \alpha.\mu(x_5) + \beta}{2} =$$

$$= \frac{\alpha.(1 + 6) + 2.\beta}{2} = 6/2.\alpha + \beta$$

and $7/2.\alpha + \beta > 6/2.\alpha + \beta$ since $\alpha > 0$.

If we now consider the statement «the rating of x_4 is twice the rating of x_2», it is easy to verify that this statement is meaningful if the scale μ defined above is a ratio scale but that it is meaningless if μ is an interval scale or an ordinal scale.

The problem of the determination of the type of a scale of measurement is called uniqueness problem. The following result solves this problem for measurement studied in theorems 1 and 2.

<u>Theorem 3</u> Suppose P is a binary relation on X and $\mu: X \to \mathbb{R}$ a function satisfying $[\forall x, y \in X : x \, P \, y \text{ iff } \mu(x) > \mu(y)]$. Then, this representation is regular and μ is an ordinal scale.

For a proof of this theorem, we refer to (Roberts, 1979, th. 3.2., p. 108).

It results from theorems 1 and 3 that, if the decision-maker gives a ranking of the elements of a set X (a strict weak order P on X), he gives an information which allows to determine an ordinal scale on X.

Remark - This ordinal scale is called:

- a value function, representing the decision-maker's preference structure or reflecting the decision-maker's preferences, by Keeney and Raiffa (1976, pp. 80 and 81);
- an ordinal utility function by Fishburn (1976, p. 1104).

In order to obtain an interval scale on X, we necessarily have to get supplementary information from the decision-maker about his preferences.

As an interval scale is characterized by admissible transformations of the form $\tau(r) = \alpha.r + \beta$ $(\alpha > 0)$, we must have, for determining a scale of this type, an information such that, when the two degrees of freedom corresponding to the two parameters α and β are used (for instance by selecting two elements x and y of X with x P y and by imposing that $\mu(y) = 0$ and $\mu(x) = 1$), the measure of each element of X can be uniquely determined.

<u>Example</u>:

If $X = \{x_1, x_2, x_3\}$ and if the only information given by the decision-maker is x_1 P x_2, x_2 P x_3 and x_1 P x_3 (which corresponds to a ranking of the elements of X), it is impossible to obtain an interval scale on X because if we arbitrarily fix $\mu(x_3) = 0$ and $\mu(x_2) = 1$ (use of our two degrees of freedom), the value of $\mu(x_1)$ is not uniquely determined (we can take for $\mu(x_1)$ any real number greater than 1).

We could obtain an interval scale if, for example, the decision-maker would give the supplementary information: «my preference for x_1 over x_2 is twice my preference for x_2 over x_3». Indeed, this information would allow to write:

$$\mu(x_1) - \mu(x_2) = 2.(\mu(x_2) - \mu(x_3)),$$

from which it results that $\mu(x_1) = 3$.

Let us mention that the idea of supplementary information that we have just presented above corresponds, from a mathematical point of view, to the necessity of introducing, along with the usual condition

$$\forall x, y \in X : x \; P \; y \text{ iff } \mu(x) > \mu(y),$$

an additional condition in order to restrict the class of admissible transformations of μ to positive linear functions. This additional condition can assume different forms in connection with different axiom systems for interval scales. The study of these axiom systems is very interesting but rather difficult. We refer for that point to Krantz et al. (1971), Fishburn (1976), Roberts (1979) and Vansnick (1987).

Remarks :

1) Following the axiomatic approach, the resulting interval scale is called differently. For example, when the basis of the axiom system is the notion of intensity of preference, the resulting interval scale is called *measurable value function*; the interval scale corresponding to the introduction of simple probability measures on X is called *utility function* or *von Neumann - Morgenstern utility function* (see Bouyssou and Vansnick, 1988, for a study of the relationships between measurable value functions and utility functions).

2) In the framework of preferences, another terminology for interval scale is *cardinal utility function* (Fishburn, 1976, p. 1105) or *cardinal function* (Dyer and Sarin, 1979, p. 811).

Now, we say a few words about the notion of ratio scale. Such a scale is characterized by admissible transformations of the form $\tau(r) = \alpha.r$ ($\alpha > 0$). For each $\alpha > 0$, such a transformation has one fixed point: $r = 0$. This shows that it is only conceivable to obtain a ratio scale when the measurement concerns a notion for which there is a natural zero. In physics, mass defines a ratio scale as null mass is a natural zero and the only degree of freedom is the choice of the unit (gram, kilogramme, ton, ...). In decision aid, the notion of difference of preference can lead to a ratio scale on $\{(x, y) \in X \times X \mid x P y)$ because speaking of null difference of preference makes sense. It is more difficult to conceive the existence of a ratio scale on X (in the framework of decision aid) but we think that it is possible to obtain such a scale from the idea of algebraic *«attractiveness»* of the elements of X (negative attractiveness = repulsiveness).

5. THE AMERICAN AND FRENCH SCHOOLS OF DECISION AID FACING THE PROBLEMS OF MEASUREMENT

There are two important schools of thought regarding decision aid: the *«American school»* and the *«French school»*. Both use, as basis for the application of theirs methods, the same decision model: the *model A.A.E. (Alternatives, Attributes, Evaluators)*. Other models exist for structuring a decision problem (for instance a hierarchic or network structure, model used in the *«Analytic Hierarchy Process»* of T. Saaty (1980), presented in this book by E. Forman (1990)) but we will focus here on the model A.A.E.. As indicated by its name, this model includes three important parts:

1) a set *A* representing the set of all feasible alternatives,

2) a set $\{X_1, X_2, \ldots, X_m\}$ of attributes, an attribute X_i being a set of at least two elements expressing the different levels of some underlying dimension;

Remark - The levels of an attribute can be real numbers or not. For example, if the underlying dimension is «profit», an usual attribute is $X = \{(x.(1\$) \mid x \in \mathbb{R}\}$. But, if the underlying dimension is «environmental effects from hazardous chemical spills», it is difficult to define the levels of the corresponding attribute by numbers; we refer to Keeney and Raiffa (1976, p. 427) for a possible definition, in common parlance, of these levels.
An attribute is called *quantitative* when its levels are real numbers (expressed in a specified unit) and *qualitative* otherwise.

3) a set $(\mathcal{E}_1, \mathcal{E}_2, \ldots, \mathcal{E}_m)$ of evaluators, an evaluator $\mathcal{E}_i$ being a function

$$\mathcal{E}_i : A \rightarrow \mathcal{P}(X_i) - \phi$$

associating to each feasible alternative *a* a non-empty set of X_i called the evaluation of *a* relative to the dimension i.

To be useful, the set of attributes must verify some properties (see Keeney and Raiffa, 1976 and Roy, 1985). One of the most important is the property of *preference independence*. Roughly, this property says that it must be possible to order, according to the preferences of the decision-maker, the elements of each attribute X_i independently of the other attributes. A classical example of two sets which are not preferentially independent is the following one: X_1 = (white wine, red wine), X_2 = (meat, fish). Indeed, most of people prefers white wine to red wine with fish but red wine to white wine with meat so that it doesn't exist a ranking of the elements of X_1 independently of those of X_2.

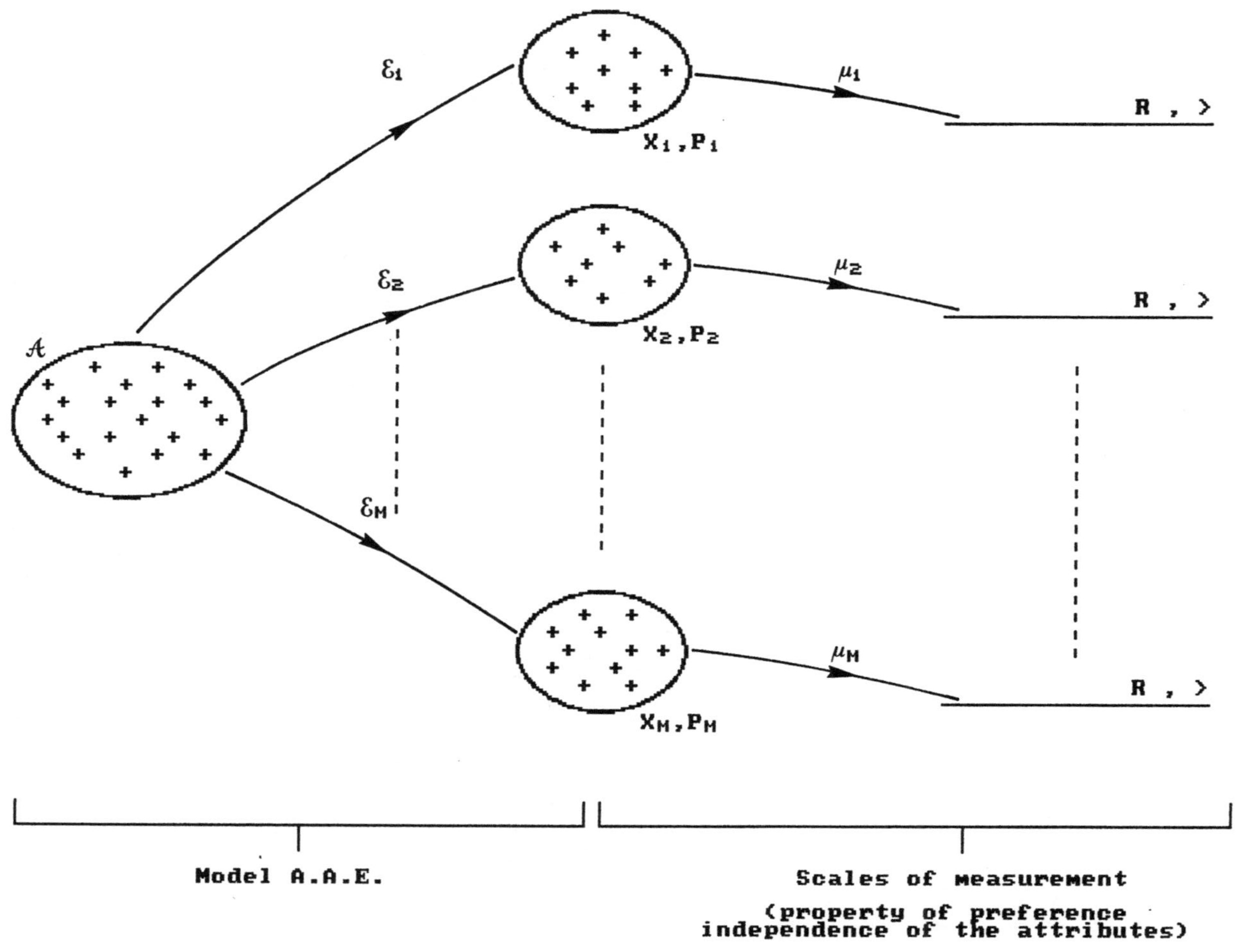

$\mathcal{A}$
$\mathcal{E}_1$
$\mathcal{E}_2$
$\mathcal{E}_M$
X_1, P_1
X_2, P_2
X_M, P_M
μ_1
μ_2
μ_M
R , >
R , >
R , >
Model A.A.E.
Scales of measurement
(property of preference independence of the attributes)

FIGURE 4

The property of preference independence is very important for a set of attributes because it allows to determine on each attribute X_i a strict weak order P_i modelling the preferences of the decision-maker concerning the elements of X_i and consequently (see section 2) to define, for each $i \in \{1,2,...,m\}$ a scale of measurement $\mu_i: X_i \to \mathbb{R}$ such that

$$\forall x_i, y_i \in x_i: x_i \ P_i \ y_i \text{ iff } \mu_i(x_i) > \mu_i(y_i).$$

These scales are ordinal scales (see theorem 3).

Remark - When X is a quantitative attribute, the relation P modelling the decision-maker's preferences is often «greater than (>)» or «less than (<)». When P = >, we can take as function $\mu: X \to \mathbb{R}$ such that $[\forall x, y \in X : x \ P \ y \text{ iff } \mu(x) > \mu(y)]$ the function $\mu(x) = x$. Similarly, when P = <, the function $\mu(x) = -x$ is a scale of measurement on X. We want to emphasize that, in spite of their particular definition, these two scales are only ordinal scales because they represent a preferential information which is only ordinal. Another trap to avoid is presented through the following example: Let X = {v, w, x, y, z} a qualitative attribute and P the linear order modelling the following ranking:

$$z \to y \to x \to w \to v.$$

A special status cannot be given to the scale

	z	y	x	w	v
$\mu(.)$ =	4	3	2	1	0

The scale

$\mu'(.)$ =	12347	923	615	18	-209

can also be used. They are ordinal scales.

We know (see section 4) that it is not easy to work with ordinal scales and that the most interesting mathematical manipulations are only meaningful with interval or ratio scales.

Faced with this situation, two basic attitudes are possible. Either you tackle the difficult problem of working with a poor information, or you tackle the difficult problem of getting additional information in order to obtain an interval scale or a ratio scale and consequently to can use a more sophisticated mathematical treatment.

Of course, the decision aid methods developed in accordance with the first attitude are very different of those developed in accordance with the second one (see Vansnick, 1988, for a study of that point) but they do not oppose because they do not apply to the same cases. We would rather say that they are complementary because they are perfectly adapted to different practical situations; indeed, it is sometimes very difficult to obtain an information leading to an interval scale but sometimes it is quite possible.

The *«ELECTRE Methods»* of B. Roy (leader of the French school of decision aid) (see Roy, 1990) fall into the category of *«poor preferential information»*; on the other hand, the American school works with interval scales.

The decision aid methods based on interval scales lead to associate to each feasible alternative a real number, which determines a strict weak order on A. The Roy's methods, starting with a poorer information, do not try do obtain such a result; an important idea of B. Roy is to allow the method to arrive to the conclusion that two alternatives are incomparable taking into account the available information.

Both approaches are very interesting; together, they provide the analyst with powerful methods allowing to tackle any decision problem structured according to the model A.A.E.

REFERENCES

Bouyssou, D. and Vansnick, J.-C. (1988), "A note on the relationships between utility and value functions", in B. Munier (ed.), *Risk, Decision and Rationality*, Reidel, Dordrecht, 103-114.

Dyer, J.S. and SARIN, R.K. (1979), "Measurable multiattribute value functions", *Operations Research*, 27, 4, 810-822.

Fishburn, P.C. (1976), "Cardinal utility: an interpretive essay", *Rivista Internazionale di Scienze Economiche e Commerciali*, 13, 1102-1113.

Forman, E.H. (1990), "Multicriteria decision making and the analytic hierarchy process", in this volume.

Keeney, R.L. and Raiffa, H. (1976), *Decisions with Multiple Objectives: Preferences and Value Tradeoffs*, John Wiley & Sons, New York.

Krantz, D.H., Luce, R.D., Suppes, P. and Tversky, A. (1971), *Foundations of Measurement. Vol. 1. Additive and Polynomial Representations*, Academic Press, New York.

Roberts, F.S. (1979), *Measurement Theory with Applications to Decision-making, Utility and the Social Sciences*, Addison-Wesley, London.

Roberts, F.S. (1985), "Applications of the theory of meaningfulness to psychology", *Journal of Mathematical Psychology*, 29, 311-332.

Roy, B. (1985), *Mèthodologie Multicritère d'Aide à la Dècision*, Economica, Paris.

Roy, B. (1990), "The outranking approach and the foundations of the ELECTRE methods", in this volume.

Saaty, T.L. (1980), *The Analytic Hierarchy Process*, Mc Graw-Hill, New York.

Vansnick, J.C. (1987), "Intensity of preference", in Y. Sawaragi, K. Inoue and H. Nakayama (eds.), *Toward Interactive and Intelligent Decision Support Systems, Vol. 2*, Lecture Notes in Econ. and Math. Systems, 286, Springer-Verlag, 220-229.

Vansnick, J.C. (1988), "Principes et Applications des Méthodes Multicritères", Université de Mons-Hainaut.

BASIC CONCEPTS OF PREFERENCE MODELLING

Philippe Vincke

Université Libre de Bruxelles
BELGIUM

1. INTRODUCTION

Preferences play a fundamental role both at the individual and at the collective levels. For this reason, *preference modelling* is an important step in decision-aid, economy, sociology, psychology, operations research,...

We present here the basic concepts which are used in most of the works devoted to this field.

2. PREFERENCE STRUCTURE

We assume for the moment that, comparing two actions a and b, the decision-maker will have one of the three following attitudes:

. preference for one action,
. indifference between both actions,
. incomparability of both actions.

We denote

. a P b if a is preferred to b,
. b P a if b is preferred to a,
. a I b if a and b are indifferent,
. a ? b if a and b are incomparable.

The *preference* (P), *indifference* (I) and *incomparability* (?) *relations* are the sets of couples (a, b) such that, respectively, a P b, a I b and a ? b. These relations are used in most of the works devoted to preference modelling.

It is reasonable to consider that these relations verify the following properties: $\forall$ a, b $\in$ A (set of actions):

$$\begin{cases} a \; P \; b \Rightarrow b \; \not P \; a & : P \text{ is asymmetric,} \\ a \; I \; a & : I \text{ is reflexive,} \\ a \; I \; b \Rightarrow b \; I \; a & : I \text{ is symmetric,} \\ a \; \not ? \; a & : ? \text{ is irreflexive,} \\ a \; ? \; b \Rightarrow b \; ? \; a & : ? \text{ is symmetric.} \end{cases}$$

<u>Definition</u>: the three relations {P, I ,?} constitute a *preference structure* in A if they verify the here above properties and if, given two elements a and b in A, one and only one of the following situations holds: a P b, b P a, a I b, a ? b

3. CHARACTERISTIC RELATION OF A PREFERENCE STRUCTURE

Every preference structure can be completely characterized by the relation S such that:

a S b iff a P b or a I b ($S = P \cup I$).

We have then

a P b iff a S b and b $\not S$ a,
a I b iff a S b and b S a,
a ? b iff a $\not S$ b and b $\not S$ a.

4. GRAPHICAL REPRESENTATION OF A PREFERENCE STRUCTURE

Here are our graphical conventions to represent the relations P, I and ?.

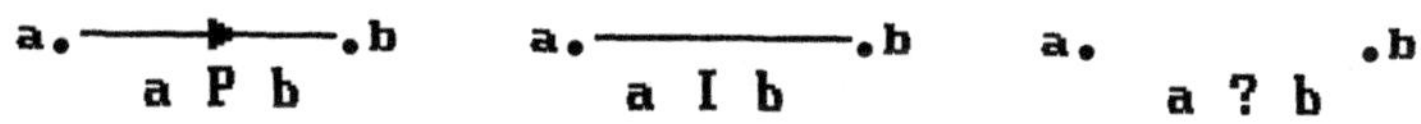

5. THE TRADITIONAL PREFERENCE STRUCTURE

A usual approach consists in replacing a decision problem by the optimization of a function g defined on A: it is the case in operations research, in actuarial sciences and in most of economy models. This approach implies that the decision-makers's preferences verify the following model (for the case where g has to be maximized): ∀ a, b ∈ A:

$$\begin{cases} a \ P \ b \text{ iff } g(a) > g(b), \\ a \ I \ b \text{ iff } g(a) = g(b). \end{cases}$$ **«traditional model»**

It is easy to verify that the implied preference structure has to satisfy the following conditions:

∀ a, b, c ∈ A:

$$\begin{cases} a \not? \ b & : \text{no incomparability,} \\ a \ P \ b \text{ and } b \ P \ c \Rightarrow a \ P \ c & : P \text{ is transitive,} \\ a \ I \ b \text{ and } b \ I \ c \Rightarrow a \ I \ c & : I \text{ is transitive.} \end{cases}$$

When A is finite or enumerable, these conditions are also sufficient for the existence of a function g verifying the «traditional model». When A is infinite, it is necessary to add topological conditions which are generally satisfied in the applications.

The characteristic relation S associated to the «traditional model» is such that, ∀ a, b, c ∈ A:

$$\begin{cases} a \ S \ b \text{ or } b \ S \ a \text{ (not exclusive)} & : S \text{ is complete,} \\ a \ S \ b \text{ and } b \ S \ c \Rightarrow a \ S \ c & : S \text{ is transitive.} \end{cases}$$

Such a relation is called a *complete preorder*: it corresponds to the situation where the elements of A can be ranked from the «best» to the «worse» with eventual ex aequo,

$$a \, S \, b \text{ iff } g(a) \geq g(b).$$

If there is no ex aequo, S is a *complete order*.

<u>Definition</u>: a preference structure is a *complete preorder structure* if it can be represented by the «traditional model»; it is a *complete order structure* if, moreover, the relation I is limited to the couples of the form (a, a).

Remarks:

1) In a complete preorder structure, I is an *equivalence relation* (reflexive, symmetric and transitive) and P is a «weak order» (asymmetric and negatively transitive: $a \not P b$ and $b \not P c \Rightarrow a \not P c$); moreover, knowing P is sufficient to know the whole structure.

2) In a complete order structure P is a *strict complete order*.

6. THE INTRODUCTION OF AN INDIFFERENCE THRESHOLD

The transitivity of indifference, implied by the traditional model, is incompatible with the existence of a sensitivity threshold under which the decision-maker does not feel any difference between two elements or refuses to accept a preference for one of the elements.

This fact was already pointed out by H. Poincaré (1935, p. 69) and, before him, by some Greek philosophers, but it is D. Luce (1956) who introduced this fundamental remark in preference modelling and gave the following example: let T_i be a tea cup with i milligrams of sugar; somebody who has to compare different tea cups will not perceive a difference of 1 milligram

(T_i I T_{i+1}, $\forall$ i) but will have a preference when the difference is very large (T_N P T_0 or T_0 P T_N, when N is large enough): this is in contradiction with the transitivity of indifference.

The introduction of a positive indifference threshold q leads to the following model: $\forall$ a, b $\in$ A:

$$\begin{cases} a \text{ P } b \text{ iff } g(a) > g(b) + q, \\ a \text{ I } b \text{ iff } |g(a) - g(b)| \leq q \end{cases}$$ **«threshold model»**

It is easy to verify that the implied preference structure has to satisfy the following conditions: $\forall$ a, b, c, d $\in$ A:

$$\begin{cases} a \text{ ? } b : \text{no incomparability}, \\ a \text{ P } b,\ b \text{ I } c,\ c \text{ P } d \Rightarrow a \text{ P } d, \\ a \text{ P } b,\ b \text{ P } c,\ a \text{ I } d \Rightarrow d \text{ P } c. \end{cases}$$

When A is finite or enumerable, these conditions are also sufficient for the existence of a function g and a positive threshold q verifying the «threshold model».

When A is infinite, it is necessary to add topological conditions which are generally satisfied in the applications.

The characteristic relation S associated to the threshold model is such that, $\forall$ a, b, c, d $\in$ A:

$$\begin{cases} a \text{ S } b \text{ or } b \text{ S } a \text{ (not exclusive)} & : \text{S is complete}, \\ a \text{ S } b \text{ and } c \text{ S } d \Rightarrow a \text{ S } d \text{ or } c \text{ S } b & : \text{Ferrers' property}, \\ a \text{ S } b \text{ and } b \text{ S } c \Rightarrow a \text{ S } d \text{ or } d \text{ S } c & : \text{S is semitransitive}. \end{cases}$$

Definition: a preference structure is a *semiorder structure* if it can be represented by the «threshold model».

Remarks :

1) In such a structure, P is still transitive.

2) To every semiorder structure, one can associate a complete preorder structure as follows:

$$\begin{cases} a\ P'\ b \text{ iff } g(a) > g\ (b), \\ a\ I'\ b \text{ iff } g(a) = g\ (b). \end{cases}$$

It can be proved that

$$a\ P'\ b \text{ iff } \exists c\text{: } a\ P\ c \text{ and } c\ I\ b \\ \text{or } a\ I\ c \text{ and } c\ P\ b \\ \text{or } a\ P\ c \text{ and } c\ P\ b,$$

I′ being the complementary relation.

3) Graphically, a semiorder structure is characterized by the fact that the following configurations are forbidden (for every choice for the diagonals):

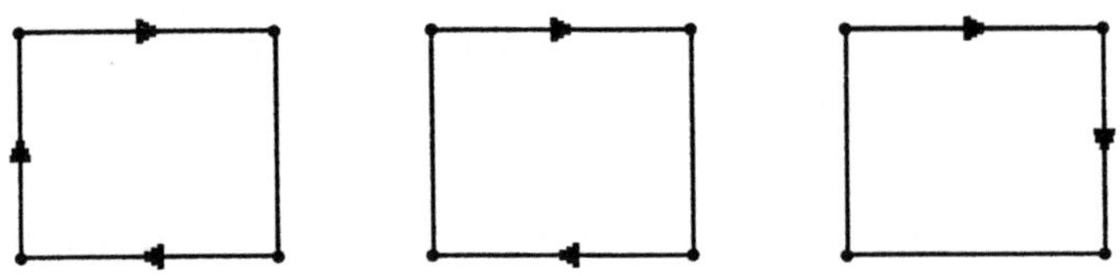

4) The semiorder structure has a very particular matrix representation, leading to the interesting concept of «minimal representation» (Pirlot, 1990) which generalizes the notion of rank usually associated to the complete order.

7. CASE WHERE THE INDIFFERENCE THRESHOLD IS VARIABLE

The previous model only considers a constant threshold. However, in many applications, the indifference threshold may vary (a difference of 1.000$ does not have the same effect for amounts around 10.000$ or around 1.000.000$). It is often useful to introduce a variable indifference threshold such that, $\forall a, b \in A$:

$$\begin{cases} a\ P\ b \text{ iff } g(a) > g(b) + q(g(b)), \\ a\ I\ b \text{ iff } g(a) \le g(b) + q(g(b)), \\ \qquad\qquad g(b) \le g(a) + q(g(a)). \end{cases}$$

«variable threshold model»

Two cases can arise:

(i)

If the function q satisfies the following «coherence condition»: $\forall a, b \in A$:

$$g(a) > g(b) \Rightarrow g(a) + q(g(a)) \ge g(b) + q(g(b)),$$

then we have again a semiorder structure and it is always possible, by a transformation of g and q, to obtain a constant threshold model.

An important example of this situation is the case where the threshold is a percentage of the considered value:

$$q(g(b)) = \alpha.g(b) \quad (\alpha > 0).$$

(ii)

If the function q does not verify the coherence condition then the preference structure implied by the variable threshold model has to satisfy the following conditions: $\forall a, b, c, d \in A$:

$$\begin{cases} a\ ?\ b : \text{no incomparability}, \\ a\ P\ b,\ b\ I\ c,\ c\ P\ d \Rightarrow a\ P\ d. \end{cases}$$

When A is finite or enumerable, these conditions are also sufficient for the existence of two functions g and q verifying the «variable threshold model».

The characteristic relation S associated to the variable threshold model is a complete relation with Ferrers' property (see the previous section).

Definition: a preference structure is an *interval order structure* if it can be represented by the «variable threshold model».

Remarks :

1) P is still transitive.

2) To every interval order structure, one can associate two complete preorder structures as follows:

$$\begin{cases} a\ P'\ b \text{ iff } g(a) > g(b), \\ a\ I'\ b \text{ iff } g(a) = g(b), \end{cases}$$

and

$$\begin{cases} a\ P''\ b \text{ iff } g(a) + q(g(a)) > g(b) + q(g(b)), \\ a\ I''\ b \text{ iff } g(a) + q(g(a)) = g(b) + q(g(b)). \end{cases}$$

These complete preorder structures are connected to the rankings which are obtained by the association, to each action, of the number of actions preferred to it and the number of actions less good than it (see Roubens and Vincke, 1985).

3) Graphically, an interval order structure is characterized by the fact that the following configurations are forbidden (for every choice for the diagonals).

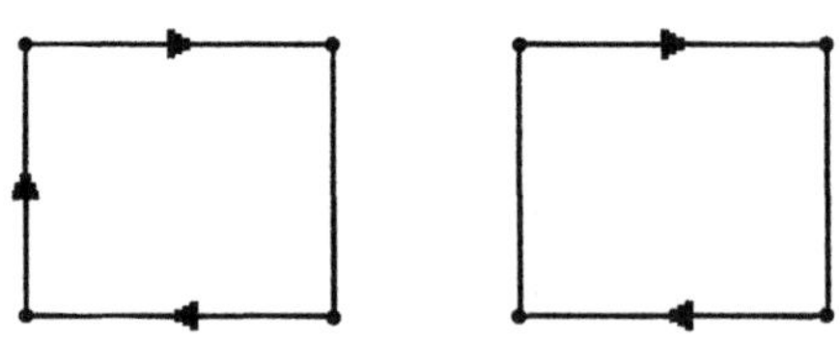

4) The interval order structure also has a particular matrix representation (see Roubens and Vincke, 1985).

5) The interval order structure has strong connections with the concepts of Guttmann's scales, biorders, interval and comparability graphs (see Doignon et al., 1986, Fishburn, 1985, Golumbic, 1980, Roubens and Vincke, 1985).

8. THE COMPARISON OF INTERVALS

The imprecision of the data and the complexity of the decisions often imply a great difficulty for representing the evaluation of an action by a precise number. The evaluation of action a can in this case be an interval (x_a, y_a). Essentially two cases were studied in the literature for the comparison of intervals:

(i)

If the preference only arises when the intervals are disjoint, i.e. if

$$\begin{cases} a \text{ P } b \text{ iff } x_a > y_b, \\ a \text{ I } b \text{ iff } [x_a, y_a] \cap [x_b, y_b] \neq \phi \end{cases}$$

then, putting, $\forall a \in A$:

$$\begin{cases} x_a = g(a), \\ y_a = g(a) + q(g(a)), \end{cases}$$

we obtain the variable threshold model and the preference structure is an interval order structure. We see that this structure naturally appears in the comparison of intervals (hence its name). In the particular case where the intervals have the same length, we have a semiorder structure (constant threshold).

(ii)

In the case where

$$\begin{cases} a \text{ P } b \text{ iff } x_a > x_b \text{ and } y_a > y_b, \\ a \text{ I } b \text{ iff one interval is included in the other,} \end{cases}$$

the relation P is a partial order of dimension 2 and I is the complementary relation. A partial order of dimension 2 is, by definition, a transitive relation which can be seen as the intersection of two strict complete orders. In other words, it is the preference relation obtained when, comparing actions following two points of view, one is able to rank them from the best to the worst for each point of view and one globally prefers a to b if a is before b in both rankings. In the case where intervals are compared, the two rankings are given by the right-end points and the left-end points of the intervals.

9. THE INTRODUCTION OF AN INDIFFERENCE AND A PREFERENCE THRESHOLDS

The practical use of the previous models implies the assessment of the indifference threshold. Experience shows that, usually, there is no precise value giving the limit between indifference and preference but there exists an intermediary region where the decision-maker hesitates between indifference and preference or gives different answers, depending of the way he is questioned. This remark has led to the introduction of a preference model with two distinct thresholds: an indifference threshold under which there is a sure indifference and a preference threshold above which there is a sure preference:

$$\begin{cases} a \text{ P } b \text{ iff } & g(a) > g(b) + p(g(b)), \\ a \text{ Q } b \text{ iff } & g(b) + p(g(b)) \geq g(a) > g(b) + q(g(b)), \\ a \text{ I } b \text{ iff } & g(b) + q(g(b)) \geq g(a), \\ & g(a) + q(g(a)) \geq g(b). \end{cases}$$

«Two threshold model»

The relation Q, called «weak preference» by B. Roy, represents an hesitation between indifference and preference, and not an intensity of preference.

The properties of relations P, Q, I implied by this model depend on the eventual «coherence conditions» imposed to the thresholds (see Vincke, 1980 and 1988).

Let us mention the *pseudo-order structure* (Roy and Vincke, 1984 and 1987) where the following conditions are imposed:

$$g(a) > g(b) \iff g(a) + q(g(a)) > g(b) + q(g(b))$$
$$\iff g(a) + p(g(a)) > g(b) + p(g(b)).$$

Let us remark that another (P, Q, I) structure naturally appears when, comparing intervals as in section 8, one concludes

$$\begin{cases} a \text{ P } b \text{ iff } x_a > y_b, \\ a \text{ Q } b \text{ iff } y_a > y_b > x_a > x_b, \\ a \text{ I } b \text{ iff one interval is included in the other.} \end{cases}$$

The characterization of this structure is still an open problem.

10. PREFERENCE MODELS WITH INCOMPARABILITY

All the previous models imply that the incomparability relation is empty. However, incomparability often appears when the decision-maker does not want to or can not (lack of information) compare two actions. It also naturally appears when different points of view have to be aggregated.

Even if an action is chosen at the end of the decision process, it may be useful, during the decision process, to represent the absence of comparison of certain actions. On the other hand, to conclude that two actions are incomparable, given the present information, is also a decision-aid, as pointing out the aspects of the problem which have to be elucidated.

Preference models with incomparability have not been much studied in the literature: we present here the concepts of partial order and partial preorder and mention the concepts of partial semiorder and partial interval order. The notion of outranking relation (presented in Roy, 1990) also lies in this category.

10.1. The Partial Order Structure

Definition: the partial order structure is characterized by the fact that, $\forall a, b, c \in A$:

$$\begin{cases} a \neq b \Rightarrow a \not I b \text{ (no ex-aequo)}, \\ a\ P\ b,\ b\ P\ c \Rightarrow a\ P\ c : P \text{ is transitive.} \end{cases}$$

In this case, there exists a function g such that:

$$a\ P\ b \Rightarrow g(a) > g(b),$$

all the elements of A having different numerical values.

The characteristic relation S associated to this model is such that, $\forall a, b, c \in A$:

$$\begin{cases} a\ S\ a & : S \text{ is reflexive,} \\ a\ S\ b \text{ and } b\ S\ a \Rightarrow a = b & : S \text{ is antisymmetric,} \\ a\ S\ b,\ b\ S\ c \Rightarrow a\ S\ c & : S \text{ is transitive;} \end{cases}$$

such a relation is called a *partial order*.

Given a partial order structure, it is always possible to replace the incomparabilities by preferences in order to obtain a complete order structure, and that in at least two manners. This result is at the origin of the concept of dimension of a partial order (see Golumbic, 1980). Given several complete order structures, one obtains a partial order structure by taking the preferences which are common to all the complete orders (as in the definition of dominance in a multicriteria problem).

10.2. The Partial Preorder Structure

Definition: the partial preorder structure is characterized by the fact that, $\forall a, b, c \in A$:

$$\begin{cases} a \; P \; b, \; b \; P \; c \Rightarrow a \; P \; c : P \text{ is transitive,} \\ a \; I \; b, \; b \; I \; c \Rightarrow a \; I \; c : I \text{ is transitive,} \\ a \; P \; b, \; b \; I \; c \Rightarrow a \; P \; c, \\ a \; I \; b, \; b \; P \; c \Rightarrow a \; P \; c. \end{cases}$$

In this case, there exists a function g such that

$$\begin{cases} a \; P \; b \Rightarrow g(a) > g(b), \\ a \; I \; b \Rightarrow g(a) = g(b). \end{cases}$$

The characteristic relation S associated to this model is a *partial preorder* (reflexive and transitive relation).

Given a partial preorder structure, it is always possible to replace the incomparabilities by preferences or indifferences in order to obtain a complete preorder structure. Given several complete preorder structures, one obtain a partial preorder structure by taking the preferences and the indifferences which are common to all the complete preorders.

10.3. The Partial Semiorder and Interval Order Structures

The two previous structures are incompatible with the existence of an indifference threshold. Definitions have been proposed for the so called partial semiorder and partial interval order, taking simultaneously into account incomparability and indifference thresholds (see Roubens and Vincke, 1984a and 1985, Roy, 1985 and Doignon, 1988).

11. VALUED PREFERENCE STRUCTURES

The previous models assume that the decision-maker does not make any distinction between «more or less strong preferences». If such a distinction has to be made, one can associate, to each couple (a, b) of elements of A, a value v(a, b) representing the «strength» or the «degree» of preference of a over b. This is also the case when the comparison between a and b is obtained by a voting procedure where some answers are in favour of a and others in favour of b.

Here are some particular properties that a valued preference structure may have:

$$\begin{cases} v(a, b) > 0 \Rightarrow v(b, a) = 0, \\ v(a, b) + v(b, a) = 1, \\ v(a, b) + v(b, a) < 1, \\ v(a, b) \geq \max_c \min [v(a, c), v(c,b)], \\ \max [v(a, b), v(c, d)] \geq \min [v(a, d), v(c, b)], \\ \max [v(a, b), v(b, c)] \geq \min [v(a, d), v(d, c)], \\ v(a, c) = v(a, b) + v(b, c). \end{cases}$$

The first property is a kind of antisymmetry: every positive preference of a over b implies a null preference of b over a; the second gives a probabilistic relation and can be obtained in a voting procedure; the third can be the result of a voting procedure with eventual abstentions; the fourth is a way of generalizing transitivity to valued relations; the fifth is a way of generalizing Ferrers' property to valued relation; with the sixth property, it gives a «valued semiorder». The last property, the additivity, leads to the notion of preference intensity and to numerical representation models where the differences g(a)-g(b) have also a meaning (see the next section).

A lot of works are concerned with valued preference relations: the interested reader may consult Fishburn (1970b and 1973), Doignon et al. (1986), Roubens and Vincke (1983, 1984b, 1985).

12. THE COMPARISON OF PREFERENCES DIFFERENCES

It may happen that the decision-maker can compare preferences differences between actions: for example, «the preference of a over b is stronger than the preference of c over d». This situation may lead to the definition of a preference structure on A x A, completing the preference structure on A. The simultaneous study of these structures is not so easy. We only mention here the «traditional additive model» where the two structures are represented by a unique function g such that, $\forall a, b, c, d \in A$:

$$\begin{cases} a\ P\ b \text{ iff } g(a) > g(b), \\ a\ I\ b \text{ iff } g(a) = g(b), \\ [a,b] \succ [c,d] \text{ iff } g(a) - g(b) > g(c) - g(d), \\ [a,b] \sim [c,d] \text{ iff } g(a) - g(b) = g(c) - g(d), \end{cases}$$

where $[a,b] \succ [c,d]$ means «the preference of a over b is stronger than that of c over d» and $[a,b] \sim [c,d]$ means they are equivalent. Clearly, such a model implies a lot of assumptions on both preference structures: the interested reader may consult Fishburn (1970a), Krantz et al. (1971), Roberts (1979) and Roy (1985).

13. PREFERENCE MODELLING UNDER UNCERTAINTY OR RISK

The main approach to the problem of preference modelling under uncertainty is the expected utility theory of Von Neumann-Morgenstern (1967): given an action a whose consequences are x with a probability p and y with a probability (1-p), the «value» of a is:

$$g(a) = p\ U(x) + (1-p)\ U(y),$$

where U(x) and U(y) are the «values» of x and y. This means that the comparison of probability distributions is reduced to the comparison of numbers: this attitude sometimes leads to criticizable conclusions (see Allais, 1953). Other concepts have been introduced in this field (see Fishburn, 1982, Jaffray et al. 1982, Bouyssou, 1983).

14. GEOMETRICAL REPRESENTATION OF PREFERENCES

The development of computer science has led to new techniques for the treatment of data, in particular to geometrical representations. Some of these techniques are concerned with preferences data (see, for example, Batteau et al., 1981).

15. ADJUSTMENT PROBLEMS

When an individual compares actions two by two, the obtained preference structure is not necessarily one of the particular structures defined in the previous sections. It may be interesting, in this case, to determine the minimal modifications which give one of these particular structures. The problem is then to adjust an observed preference structure by a model or to determine a model «at minimum distance» of a given preference structure. For this kind of problems, see for example Ribeill (1973), Vincke (1978), Barthelemy and Monjardet (1981).

16. REFERENCES

Allais, M. (1953), "Le comportement de l'homme rationnel devant le risque: critique des postulats et axiomes de l'Ecole Américaine", *Econometrica*, 21, 503-546.

Barthelemy, J.P. and Monjardet, B. (1981), "The median procedure in cluster analysis and social science theory", *Mathematical Social Sciences*, 1, 235-267.

Batteau, P. Jacquet-Lagrèze, E., Monjardet, B. (eds.) (1981), *Analyse et Agrègation des Prèfèrences dans les Sciences Sociales, Eonomiques et de Gestion*, Economica.

Bouyssou, D. (1983), "Decision-aid and utility theory: a critical survey", *Cahier du Lamsade 52*, Université Paris-Dauphine.

Doignon, J.P., Monjardet, B., Roubens, M. and Vincke, Ph. (1986), "Biorder families, valued relations and preference modelling", *Journal of Mathematical Psychology*, 30,(4), 435-480.

Doignon, J.P. (1988), "Partial structures of preference in non-conventional preference relation in decision-making", in J. Kacprzyk and M. Roubens (eds), *Non-Conventional Preference Relations in Decision Making*, Springer Verlay 301, 22-35.

Fishburn, P.C. (1970a), *Utility Theory for Decision-Making*, Wiley, New York.

Fishburn, P.C. (1970b), "Utility theory with inexact preferences and degrees of preference", *Synthese*, 21, 204-222.

Fishburn, P.C. (1973), "Binary choice probabilities: on the varieties of stochastic transitivity", *Journal of Mathematical Psychology*, 10, 327-352.

Fishburn, P.C. (1982), *The Foundations of Expected Utility*, Reidel, Dordrecht, Holland.

Fishburn, P.C. (1985), *Interval Orders and Interval Graphs*, Wiley, New York.

Golumbic, M. (1980), *Algorithmic Graph Theory and Perfect Graphs*, Academic Press, New York.

Jaffray, J.Y. Cohen, M. (1982), "Experimental results on decision making under uncertainty", *Methods of O.R.*, 44, 275-289.

Krantz, D., Luce, D., Suppes, P., Tversky, A. (1971), *Foundations of Measurement, Vol. 1, Additive and Polynomial Representations*, Academic Press, New York.

Luce, D. (1956), "Semiorders and a theory of utility discrimination", *Econometrica*, 24, 178-191.

Pirlot, M. (1990), "Minimal representation of a semiorder", *Theory and Decision*, 28 (2), 109-141.

Poincare, H. (1935), *La Valeur de la Sciences*, Flammarion, Paris.

Ribeill, G. (1973), "Equilibre, équivalence, ordre et préordre à distance minimum d'un graphe complet", *Mathèmatiques et Sciences Humaines*, 43, 71-106.

Roberts, F. (1979), *Measurement Theory with Applications to Decision-Making, Utility and Social Sciences*, Addison-Wesley, London.

Roubens, M. and Vincke, Ph. (1983), "Linear fuzzy graphs", *Fuzzy Sets and systems*, 10, 79-86.

Roubens, M. and Vincke, Ph. (1984a), "A definition of partial interval orders", in E. Degreef and J. Van Buggenhaut (eds.), *Trends in Mathematical Psychology*, North-Holland, Amsterdam, 309-316.

Roubens, M. and Vincke, Ph. (1985), *Preference Modelling*, Springer-Verlag, Berlin.

Roy, B. and Vincke, Ph. (1984), "Relational systems of preference with one or more pseudo-criteria: some new concepts and results", *Management Science*, 30, (11), 1323-1335.

Roy, B. (1985), *Mèthodologie Multicritère d'Aide à la Dècision*, Economica, Paris.

Roy, B. and Vincke, Ph. (1987), "Pseudo-orders: definition, properties and numerical representation", *Mathematical and Social Sciences*, 14, 2, 263-274.

Roy, B. (1990), "The outranking approach and the foundations of the ELECTRE methods", in this volume.

Vincke, Ph. (1978), "Ordres et préordres totaux à distance minimum d'un quasi-ordre", *Cahiers du Center d'Estudes de Recherche Opèrationelle*, 20, 3, 451-462.

Vincke, Ph. (1980), "Vrais, quasi, pseudo et précritères dans un ensemble fini: propriétés et algorithmes", *Cahier du Lamsade 27*, Université Paris-Dauphine.

Vincke, Ph. (1988), "(P, Q, I) - preference structures", in J. Kacprzyk and M. Roubens (eds.), *Non-conventional Preference Relations in Decision Making*, Springer-Verlag, Berlin, 72-81.

Von Neumann, J. and Morgenstern, O. (1967), *Theory of Games and Economic Behaviour*, Princeton University Press, 3rd edition.

Note: this paper is an English version of chapter 2 of:

Vincke, Ph. (1989), *L'Aide Multicritère à la Dècision*, Éditions de l'Université de Bruxelles and Éditions Ellipses - Paris.

DECISION MAKING IN ILL-STRUCTURED ENVIRONMENTS AND WITH MULTIPLE CRITERIA

Hans-Jürgen Zimmermann

R.W.T.H. Aachen, Institute of Technology,
Templergraben 55
5100 Aachen (F. R. G.)

ABSTRACT

In Multi Criteria Decision Making one is generally concerned with decisions under certainty, i. e. decisions for which the "state" is assumed to be known with certainty. Multi Criteria Decision Making under risk or uncertainty would imply the superimposition of the problem structures of classical MCDM and that of single criteria decision making under risk, i. e., for instance, the combinations of goal programming with stochastic programming. This would, obviously, become very involved mathematically! In this paper we are <u>not</u> concerned with uncertainties (probabilities) of the Kolmogroroff type but rather with uncertainties as they are considered in the theory of fuzzy sets, possibility theory and the like. It will be shown that for this type of uncertainty (vagueness) which is assumed to be more relevant for MCDM, models and methods exist, which are also adequate for MCDM and which are computationally still feasable.

1. DECISION AND UNCERTAINTY

In decision making under risk (stochastic programming, Baysian decision analysis) it is generally assumed that uncertainty is due to a lack of information about prevailing states and that this uncertainty concerns only the occurance of the states and not the definition of states, results, or criteria themselves.

The type of uncertainty (non-dichotomy) we are considering here is not due to a lack of information - on the contrary it may be due to an abundance of information (Zimmermann, 1987) - and it does not (only) concern the occurance but primarily the definition of states, criteria and weights themselves. The following examples may serve to clarify this difference:

The statement: "The probability of loosing the game is .6", is obviously one which can easily and properly be modelled by probability theory. The "event" (loosing the game) is crisply defined and the degree to which the occurence is going to happen (probability) is numerically defined. Let us consider the statement: "The probability of meeting pleasant company is .7". Here the event "pleasant company" is no longer uniquely defined. If we say "the chances of enjoying ourselves are good", not even the "probability" is adequately defined. If, eventually, we consider statements as they often occur in expert systems, text-books, newspapers or daily conversation, such as, "Expert A believes that in 1990 we will have a dry summer but empirical evidence indicates that it will be rather wet", then neither the events nor the probabilities are defined in a way which lends itself to be modelled by probability theory.

Decisions in environments in which states and probabilities of their occurance are stated and defined crisply will be called "decisions under risk". Decisions of the latter kind, which contain components which are intrinsically vague shall be called "decisions in fuzzy environments" or briefly "fuzzy decisions". Main tools to model multi criteria decisions in fuzzy environments will be the theory of fuzzy sets, possibility theory and the theory of evidence. Abundant literature for these areas is available. For the sake of space economy we will, therefore, renounce a detailed description of these theories here but rather point to available references wherever appropriate.

2. MULTI - ATTRIBUTE DECISIONS WITH VAGUE INFORMATION

2.1. Problem Definition

In order to facilitate the comprehension of fuzzy models in this area we shall first define the classical (crisp) multi-attribute decision model:

Definition. Let $X = \{x_i\}$, $i=1,\ldots,n$ be a (finite) set of decision alternatives and $G = \{g_j\}$, $j=1,\ldots,m$ a (finite) set of goals, attributes, or criteria, according to which the desirability of an alternative is to be judged. The aim of MADM is to determine the optimal alternative x° with the highest degree of desirability with respect to all relevant goals g_j.

In MADM two distinct families of approaches can be distinguished: Aggregation approaches and order-focused approaches. They can be sketched as follows:

Aggregation approaches: Two kinds of aggregation procedures are being used:

Direct aggregation procedures which generally consist of two stages:

1. The aggregation of the judgements with respect to goals and per decision alternative.
2. The rank ordering of the decision alternatives according to the aggregated judgements.

In crisp MADM models it is usually assumed that the final judgements of the alternatives are expresses as real numbers. In this case the second stage does not pose any particular problems and suggested algorithms concentrate on the first stage. Fuzzy models are sometimes justified by the argument that the goals g_j or their attainment by the alternatives x_i, respectively, cannot be defined or judged crisply but only as fuzzy sets. In this case the final judgments are also represented by fuzzy sets which have to be ordered to determine the optimal alternative. Then the second stage is, of course, by far not trivial.

Hierarchical aggregation procedures: In crisply defined situations this approach concentrates on the first stage. It establishes "consistent" weights for the different criteria and then aggregates them hierarchically (see Saaty, 1980).

Order-focussed approach (outranking): This approach focusses attention on the fact that in MADM problems one tries to establish preference orderings of alternatives. If the order relations of alternatives induced by different individual criteria are not consistent with each other or if they do not lead to sufficiently well structured dominance relations then outranking procedures are used to arrive from pairwise dominance statements to the best overall dominance relation which is obtainable. The procedures can roughly be sketched as follows:

a. By pairwise comparison of alternatives i and j with respect to criterion k the relative preferability of alternative i over j, $-1 < d^k < 1$, is determined.

b. A concordance relation $c_{ij} \in [0,1]$ is established by aggregating the relative preferabilities and setting $c_{ij} = 1$ if

$$\sum_K d^k > \sum_K d^k$$

and

$$c_{ij} = 0 \text{ else.}$$

c. A discordance relation $b_{ij} \in [0,1]$ is established. $c_{ij} = 1$ if alternative j dominates alternative i with respect to anyone criterion more than a threshold b. $b_{ij} = 1$ has a veto function against dominance of i over j!

d. The aggregation of the concordance relation yields the final dominance relation. Several types of aggregation have been suggested (Siskos et al., 1984).

2.2. MADM-Models for Fuzzy Situations

2.2.1. Aggregation approaches. Let r_{ij} be the (preferability) ratings of alternative i with respect to criterion j and w_j subjective weeights which express the relative importance of the criteria to the decision maker. In crisp MADM-models a frequently used and nonsophisticated way to arrive at overall ratings of alternatives, R_i, is

$$R_i = \sum_j w_j r_{ij} \tag{1}$$

Generally the R_i are real numbers according to which the alternatives can easily be ranked.

The r_{ij} as well as the w_j, however, will in many cases more appropriately be modelled by fuzzy numbers. This has the following consequences:

In <u>step 1</u>: The aggregation procedure of the single criteria ratings will have to be modified.

In <u>step 2</u>: R_i will no longer be real numbers but fuzzy sets which have to be ranked.

In the following some approaches to handle fuzziness in MADM-aggregation models shall be described exemplarily:

A. <u>Hierarchical Aggregation</u>

1. **Hierarchical aggregation** using crisp weights (Yager, 1978)

Essentially Yager assumes a finite set of alternative actions $X = \{x_i\}$ and a finite set of goals (attributes) $G = \{\tilde{g}_j\}$, $j = 1, \ldots, m$. The $\tilde{g}_j$, $\tilde{g}_j = \{x_i, \mu_{\tilde{G}_j}(x_i)\}$ are fuzzy sets the degrees of membership of which represent the normalized degrees of attainment of goal j by alternative x_i. The fuzzy set decision, $\tilde{D}$, is then the intersection of all fuzzy goals, i.e. $\mu_{\tilde{D}}(x_i) = \min_{j=1}^{m} \mu_{\tilde{g}_j}(x_i)$, $i=1,\ldots,n$,

and the maximizing decision is defined to be the x_i' for which

$$\mu_{\tilde{D}}(x_i) = \max \min \mu_{\tilde{g}_j}(x_i). \tag{2}$$

Yager now allows for different importance of the goals and expresses this by exponential weighting of the membership functions of the goals. If w_j are the weights of the goals the weighted membership functions, $\mu_{\tilde{g}}$, are

$$\mu_{\tilde{g}_j}(x_i) = (\mu_{\tilde{g}_j}(x_i))^{w_j}. \tag{3}$$

For the determination of the w_j Yager suggests the use of Saaty's method, i.e. the determination of the reciprocal matrix by pairwise comparison of the goals with respect to their relative importance. The components of the eigenvector of this $m \times m$ matrix whose total is m are then used as weights.

The rational behind using the weights as exponents to express the importance of a goal can be found in the definition of the modifier "very" (Zimmermann, 1985). There the modifier "very" was defined as the squaring operation. Thus the higher the importance of a goal the larger should be the exponent of its representing fuzzy set, at least for normalized fuzzy sets and when using the min-operator for the intersection of the fuzzy goals.

The measure of ranking the decision alternatives is obviously the $\mu_{\tilde{D}}(x_i)$. For the ranking of fuzzy sets in the unit interval Yager suggests another criterion, which bases on properties of the supports of the fuzzy sets rather than on the degree of membership. This is particularly applicable when ranking different (fuzzy) degrees of truth and similar linguistic variables.

Other models along the same line have been suggested by Laarhoven and Pedrycz (1983). They concentrate on step 1 and suggest Saaty's AHP modified for triangular fuzzy numbers.

A different hierarchical aggregation procedure was suggested by Zimmermann and Zysno (1983). This was not intended for solving MADM-problems but it can also be used for step 1. The aggregation of the individual fuzzy criteria is achieved by an appropriate parametrized operation which was tested empirically before. The result of this aggregation procedure is either a fuzzy set or a number (degree of membership) in [0,1] which can both be used for ranking the alternatives in step 2.

B. Direct (non-hierarchical) Aggregation Models

1. **Rating and Ranking Multiple-Aspect Alternatives Using Fuzzy Sets**

This has already become a "classical" in fuzzy MADM-models. We shall, therefore, describe it in some more detail: The authors refer to Kahn's probabilistic model. He considers the alternatives a_i, i=1,..., n and the criteria c_j, j=1,...,m. r_{ij} denotes the rating of alternative i with respect to criteria j. The "weight" of criteria j, its relative importance, is called w_j. The ranking of the alternatives is performed according to their rank

$$R_i = \frac{\sum_{j=1}^{n} w_j r_{ij}}{\sum_j w_j} \tag{4}$$

The optimal alternative is that for which R_i is maximal.

In Kahne's model the w_j and hence the R_i are assumed stochastic variables and the optimal alternative is determined by using Monte Carlo simulation.

Baas and Kwakernaak (1977) suggest the following fuzzy version of above model:

Let again $X = \{x_i\}$, i=1,...,n be the set of alternatives and $G = \{\tilde{g}_j\}$, j=1,...,m the set of goals. r_{ij} is the fuzzy rating of alternative i with respect to goal j and $w_j \in \mathbb{R}$ is the weight (importance) of goal j.

It is assumed that the rating of alternative i with respect to goal j is fuzzy and is represented by the membership function $\mu_{\tilde{W}}(w_j)$. All fuzzy sets are assumed to be normalized (i.e. have finite supports and take on the value 1 at least once!).

Step 1: (Determination of ratings for alternatives)

The evaluation of an alternative x_i is assumed to be a fuzzy set which is computed on the basis of the r_{ij} and w_j as follows:

Consider a function $g(z)$: $\mathbb{R}^{2n} \to \mathbb{R}$ defined by

$$g(z) = \frac{\sum_{j=1}^{n} w_j r_{ij}}{\sum_j w_j} \tag{5}$$

with $z = (w_1, \ldots w_n, r_{i1}, \ldots, r_{in})$.

On the product space $\mathbb{R}^{2n}$ a membership function $\mu_{\tilde{z}_i}$ is defined as $\mu_{\tilde{z}_i}(z) = \min \{\min \{\mu_{\tilde{w}_j}(w_j)), \min (\mu_{\tilde{r}_{ik}}(r_{ik}))\}$.

Through the function g the fuzzy set $Z_i = \{(z, \mu_{\tilde{z}_i})\}$ induces a fuzzy set $\tilde{R}_i = \{(r, \mu_{\tilde{R}_i})\}$ with the membership function

$$\mu_{\tilde{R}_i}(r) = \sup_{z:\, g(z)=r} \mu_{\tilde{z}_i}(z) \qquad r \in R \tag{6}$$

$\mu_{\tilde{R}}(r)$ is the final rating of alternative x_i on the basis of which the "ranking ordering" is performed in step 2.

Step 2: Ranking

For the final ranking of the x_i Baas and Kwakernaak start from the observation that if the x_i had received crisp rating, $r_i = \tilde{R}_i$, then a reasonable procedure would be to select the x_i that have received the highest rating, i.e. to determine the set of preferred alternatives as $\{x_i \in I \mid r_i > r_j, \forall j \in I\}$, $I = \{1, \ldots, m\}$.

Since here the final ratings are fuzzy the problem is somewhat more complicated. The authors suggest in their model two different fuzzy sets in addition to $\tilde{R}_i$ which supply different kinds of information about the preferability of an alternative.

a) They first determine the conditional set $(\tilde{I} \mid \tilde{R})$ with the characteristic function

$$\mu_{(\tilde{I} \mid \tilde{R})}(x_i, r_1, \ldots, r_m) = \begin{cases} 1 \text{ if } r_i \geq r_j, \forall j \in I \\ 0 \text{ else.} \end{cases} \tag{7}$$

This "membership function" expresses that a given alternative x_i belongs to the preferred set only if
$r_i > r_j \ \forall j \in I$.
The final fuzzy ratings $\tilde{R}$ define on $\mathbb{R}^m$ a fuzzy set $\tilde{R} = (r_1, \ldots, r_m)$ with the membership function

$$\mu_{\tilde{R}}(r_1, \ldots, r_m) = \min_{i=1,\ldots,m} \mu_{\tilde{R}_i}(r_i) \tag{8}$$

This fuzzy set together with the conditional fuzzy set (7) induce a fuzzy set $\tilde{I} = \{(x_i, \mu_{\tilde{I}}(x_i))\}$ with the membership function

$$\mu_{\tilde{I}}(x_i) = \sup_{r_1,\ldots,r_m} (\min \{\mu_{\tilde{R}}(x_i), \mu_{(\tilde{I} \mid \tilde{R})}(x_i)\}) \tag{9}$$

which can be interpreted as the degree to which alternative x_i is the best alternative. If there is a unique i for which the supremum of (9) is attained, $\mu_{\tilde{I}}(x_i) = 1$, then alternative x_i dominates crisply all other alternatives.

In addition to the above information Baas and Kwakernaak determine a fuzzy set $\tilde{P}_i$ which expresses the degree of preferability of alternative x_i over all others.

Summarizing:

1. $\mu_{\tilde{R}_i}(r)$ is the fuzzy rating of x_i
2. $\mu_{\tilde{I}}(x_i)$ is the degree to which x_i is best alternative, and
3. $\mu_{\tilde{P}_i}(p)$ is the degree of preferability of x_i over all other alternatives.

2. Extensions and Modifications of the BK-Approach

Some authors have criticized the BK-approach for not being sensitive enough, i.e. for not differentiating enough between alternatives with similar ratings.

Jain (1977) concentrates on step 2 and suggests a special concept of a "Maximizing set":

Let $S = \bigcup_{i=1}^{r} (\tilde{R}_i)$ (union of all supports of ratings).

Then the maximizing set $\tilde{M}$ is defined as:

$$\tilde{M} = \{(r_{ij}, \mu_{\tilde{M}}(r))\} \qquad (10)$$

with

$$\mu_{\tilde{M}}(r) = \left[\frac{r}{r_{max}} \right]^n \qquad (11)$$

and

$$r_{max} = \sup_{r \in S} r.$$

For each alternative we determine

$$\tilde{R}_i^{o}(r_{ij}) = \{(r_{ij}, \mu_{\tilde{R}_i^{o}}(r_{ij}))\}$$

with

$$\mu_{\tilde{R}_i^{o}}(r_{ij}) = \min(\mu_{\tilde{M}}, \mu_{\tilde{R}_i}).$$

The ranking of alternatives is then performed according to $\mu_{\tilde{R}_i^{o}}(r_{ij})$. Figure 1 illustrates this procedure.

As shown by figure 2, this approach does not either differentiate always between different ratings.

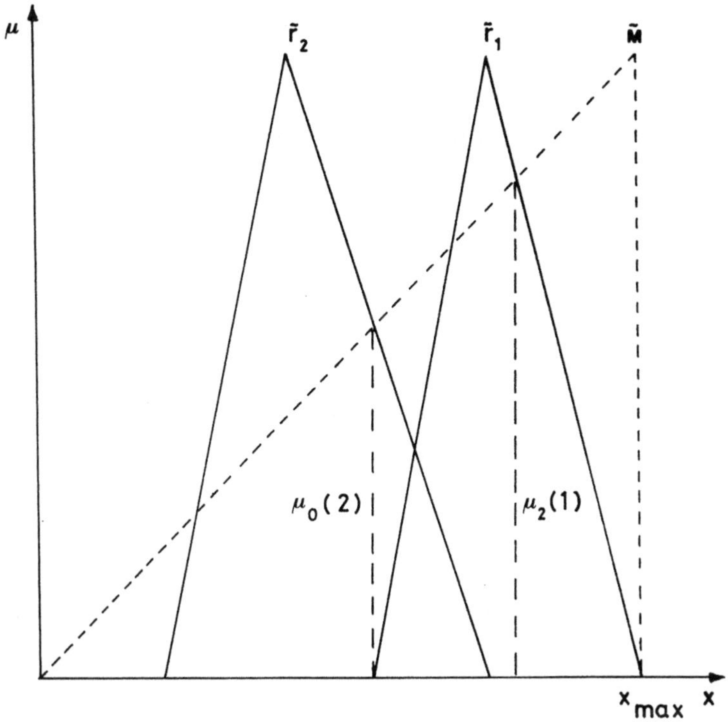

Figure 1

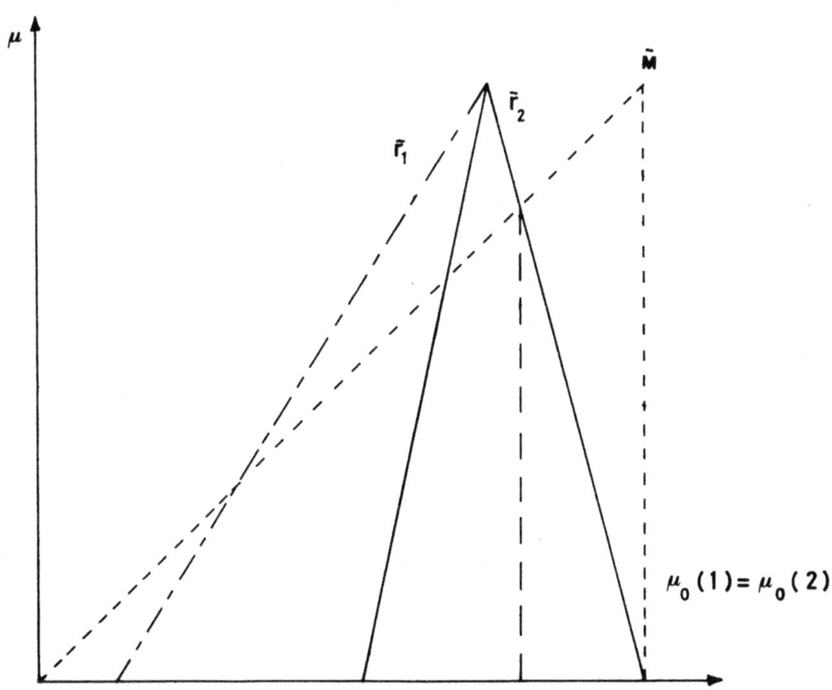

Figure 2

Baldwin and Guild (1979) also tried to increase the sensitivity of the BK-approach. They replace the (crisp) set $(\tilde{I}\,|\tilde{R})$ of Baas and Kwakwernaak by a fuzzy set, use fuzzy relations to rank the fuzzy ratings obtained in step 1 and supply numerically efficient procedures for ratings represented by triangular fuzzy numbers.

Chen (1985) eventually supplements Jain's "maximizing set" by a "minimizing set" in order to increase discrimination in step 2.

Tong and Bonissone (1984) argue quite differently than abovementioned authors: "Previous attempts to solve fuzzy decision problems have produced mumerical rankings of the alternatives. We believe that this is misguided since in situation where fuzzy sets are a suitable way of representing uncertainty the final choice must itself be fuzzy. It is certainly not appropriate to give the decision some artificial precision - solutions should be linguistic rather than numerical".

The authors are only concerned with step 2 of MADM, i.e. the ranking of (normal convex) fuzzy sets, $\tilde{r}_i$, which represent ratings of decision alternatives, x_i, and include already the aggregation of the merits of an alternative with respect to different criteria (goals).

The "ranking" of the final ratings, $\tilde{r}_i$, i=1,...,n, of the alternatives, x_i, i=1,...,n is achieved in 2 steps:

First a fuzzy preference set, $\tilde{R}(x_i)$, is determined which corresponds essentially to Baas and Kwakernaak P_i (6). In a second step a linguistic approximation and truth qualification of the final decision proposal is generated. We shall consider these steps in turn:

Step 1: (determination of fuzzy decision sets)
Dominance sets $\tilde{d}i$ are defined by the following membership functions:

$$\mu_{\tilde{d}_i}(r_i, r_j) = \max_{j=1,\dots,n} \text{Min}\ (\mu_{\leq r}(r_i),\ \mu(r_j)) \quad i=1,\dots,n$$

with

$$\mu_{\leq x_i}(r) = \begin{cases} 1 & \text{for } r_j \leq r^* \\ \mu_i(r_i) & \text{for } r_j \geq r^*, \end{cases} \tag{12}$$

where r^* is the leftmost (lowest) value of r for which $\mu_i(x_i) = 1$.

The rational behind (11) and (12) is to get a clear indication on which alternatives are non-dominated or preferred. (12) does not consider the shapes of the r_i but only the location of the "peaks" of the ratings. In the following figure, for instance, r_3 and r_2 would "dominate" $\tilde{r}_1$ to the same degree, $d(r_2, r_1) = .5$.

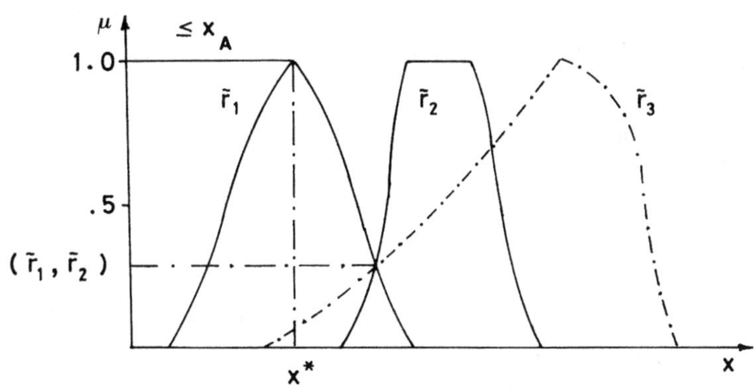

Figure 3

On the basis of the $\tilde{d}_i$ a dominance relation, $\tilde{R}_d(r_i, r_j)$ can be constructed, indicating the dominance of all x_i over xj. This relation is reflexive but neither symmetrical nor min-max transitive and is therefore not on ordering of the x_i! It does not either take into consideration the shapes of the membership functions $\mu_{\tilde{r}_i}(r)$, but it already offers very useful information.

To differentiate further between non-dominated alternatives Tong and Bonissone define a vector $v(x_i) = \min \tilde{R}_d(x_i, xj)$, the components of which indicate the degree to which an alternative dominates all others (on the basis of (12)!). To include also the shapes of the membership functions of r_i they define the following difference function:

$$g_k(r_i) = r_k - \frac{\sum_{\substack{i=1 \\ i \neq k}}^{n} v(x_i) \cdot r_i}{\sum_{\substack{i=1 \\ i \neq k}}^{n} v(x_i)} \tag{13}$$

The index k in (13) corresponds to a position in $v(x_i)$ for which $v(k) = 1$. If in (13) r_i is replaced by $\tilde{r}_i$ the result is a fuzzy set which the authors call $Z_{\tilde{k}}(r_i)$. The membership function of this fuzzy set can be determined via the extension principle as

$$\mu_{\tilde{z}_k}(u) = \max_{r_1,\ldots,r_n} \min_{i=1}^{n} \mu_{\tilde{r}_i}(x_i) \tag{14}$$

such that $g_k(.) = u$.

These fuzzy sets (one for each non-dominated alternative) are now offered as a decision aid, expressing for the non-dominated alternatives the degree of preference over other alternatives.

<u>Step 2</u>: (Truth qualification and linguistic approximation)
The rational behind this second step is the belief that human decision makers understand better a linguistic statement characterizing the decision sets than a (numerical) membership function. The form of the linguistic expressions searched for by the authors is:

"It is very true that x_k is marginally preferred to all other alternatives".

This statement obviously constains three important pieces of information:

- the crisp decision x_k
- the intensity of preference (marginally)
- the degree of truth of the statement (truth qualification).

Formally the statement has the structure:

"x_k is P over all other alternatives in τ".

To translate the decision set Z_k meaningfully in the above-mentioned way the authors interpret P as a fuzzy set on the same universe of discourse as Z_k and as a term of the linguistic variable "truth". They then try to find terms in the term sets of the linguistic variables "Preference" (P) and "Truth" which approximate the unlabelled Z_k as well as possible and then use the linguistic label of these approximating terms to express Z_k in natural language. As a tool for the linguistic approximation they suggest pattern recognition techniques and a context free grammar such as suggested by Zadeh (1977).

Dubois and Prade (1984) suggest a third approach: They do not even try to determine a unique order of the alternatives in step 2. They rather present to the decision maker four measures of dominance and leave it to him to decide on a final ranking. These four measures or indices they suggest are:

1. <u>Possibility of dominance</u>

$$PD(r) = \text{Poss}\ (r_i \geq r_j)$$

$$PD(r_i) = \sup_{\substack{r_i, r_j \\ r_i \geq r_j}} \min[\mu_{\tilde{r}_i}(r_i),\ \mu_{\tilde{r}_j}(r_j)] \tag{15}$$

2. <u>Possibility of strict dominance</u>

$$PSD(r_i) = \text{Poss}\ (r_i > r_j)$$

$$= \sup_{r_i} \inf_{r_j \geq r_i} \min[\mu_{\tilde{r}_i}(r_i),\ 1-\mu_{\tilde{r}_j}(r_j)] \tag{16}$$

3. <u>Necessity of dominance</u>

$$ND(r_j) = Nec\ (r_i > r_j) = \inf_{r_i} \sup_{r_j \le r_i} \max[1-\mu^{\tilde{r}}_i(r_i),\ \mu^{\tilde{r}}_j(r_j)] \quad (17)$$

4. <u>Necessity of strict dominance</u>

$$NDS(r_i) = Nec\ (r_i > r_j) = 1 - \sup_{r_i \le r_j} \min[\mu^{\tilde{r}}_i(r_i),\ 1-\mu^{\tilde{r}}_j(r_j)] \quad (18)$$

2.2.2. Fuzzy order-focussed approach (fuzzy outranking)

Outranking methods were already discusses. It was assumed that crisp dominance judgements could or are being made. In fuzzy outranking this concept is generalized in the sense that concordance, discordance and dominance are considered as matters of degree. The general procedure can be depicted as follows: (see Flowchart in figure 5).

Partial outranking relations which are used to determine the fuzzy set of non-dominated alternatives with the membership function $\mu^{ND}(a)$ have the following structure:

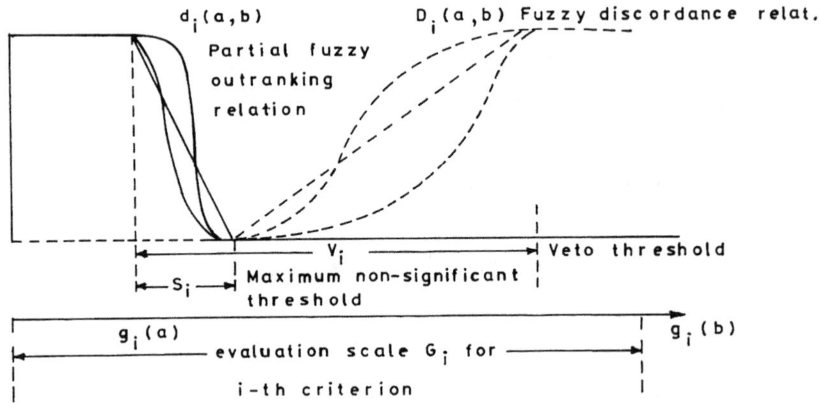

Figure 4

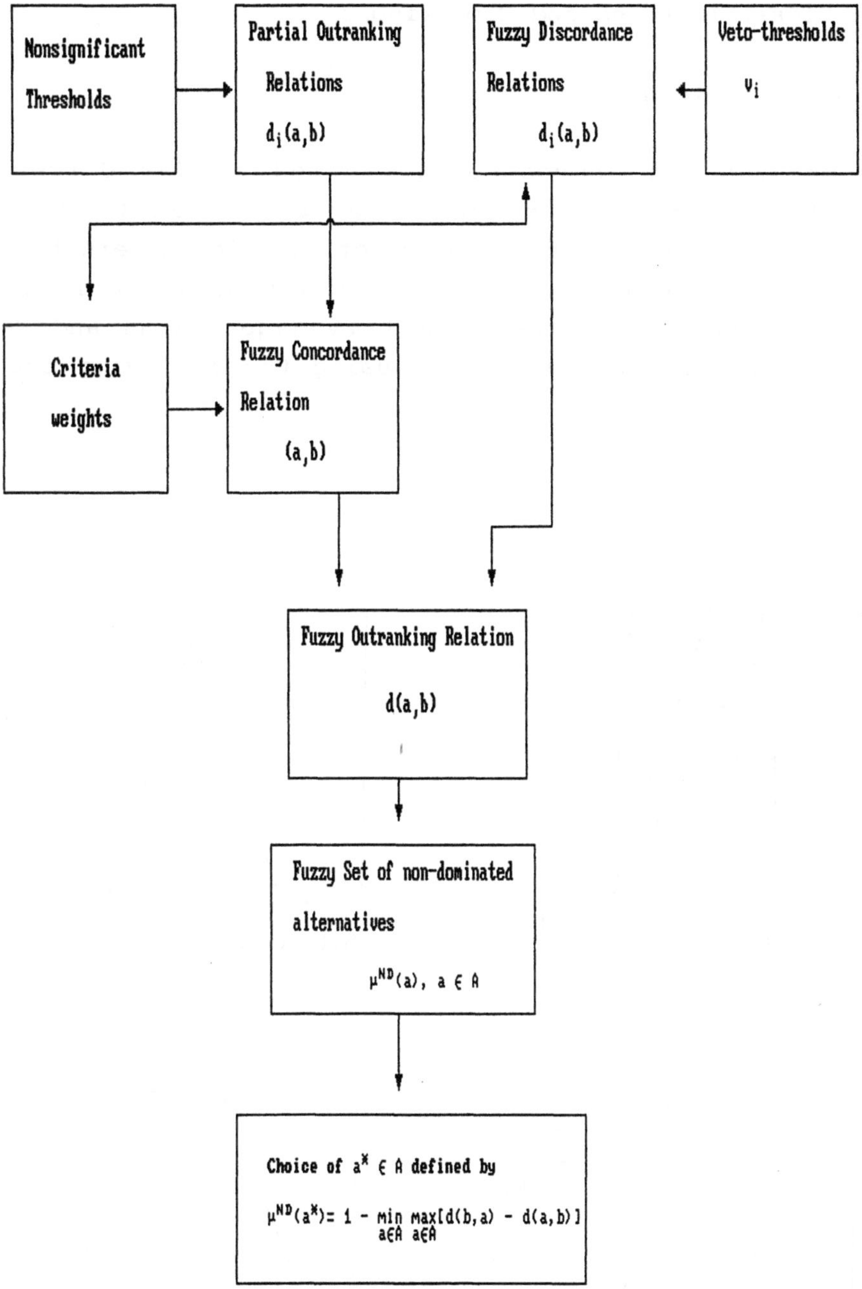
Nonsignificant Thresholds
Partial Outranking Relations $d_i(a,b)$
Fuzzy Discordance Relations $d_i(a,b)$
Veto-thresholds v_i
Criteria weights
Fuzzy Concordance Relation (a,b)
Fuzzy Outranking Relation d(a,b)
Fuzzy Set of non-dominated alternatives $\mu^{ND}(a),\ a \in A$
Choice of $a^* \in A$ defined by
$\mu^{ND}(a^*) = 1 - \min_{a \in A} \max_{a \in A} [d(b,a) - d(a,b)]$

FIGURE 5

2.3. Classifications and Comparisons

2.3.1. Comparisons

The methods and approaches presented so far differ with respect to a number of aspects. They also lead to different results when applied to the same set of decision alternatives. We shall focus our attention on the methods presents in section 2.2.1. We shall compare the methods by Yager, Baas and Kwakernaak, Jain, and Baldwin and Guild using the following set of 6 situations:

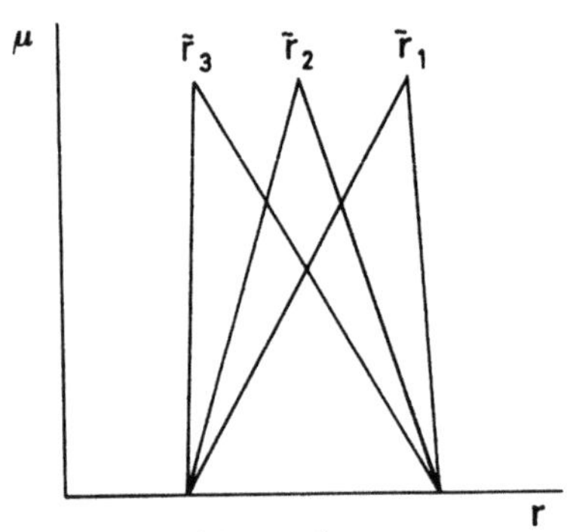

Situation 1

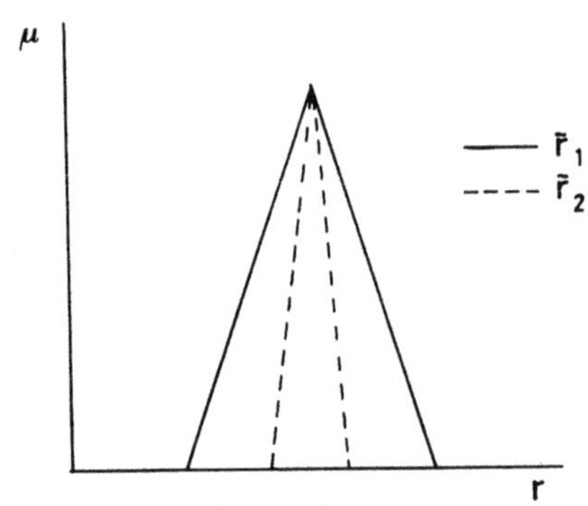

Situation 2

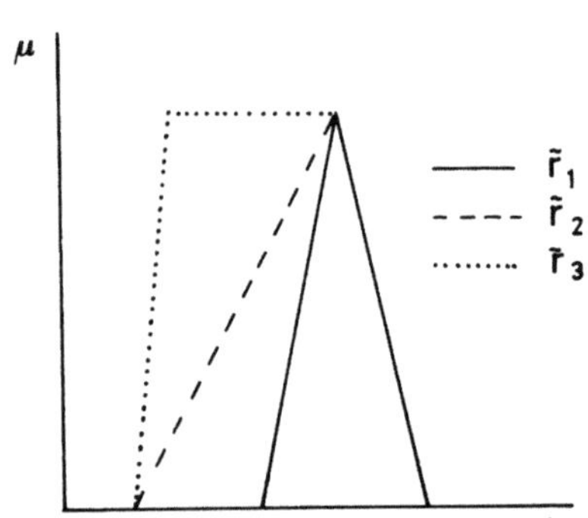

Situation 3

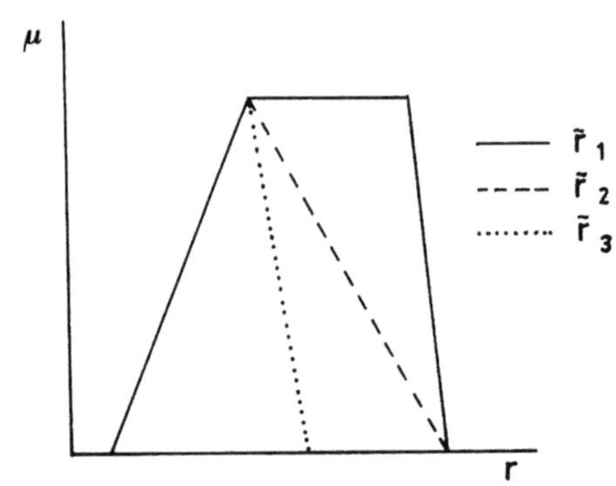

Situation 4

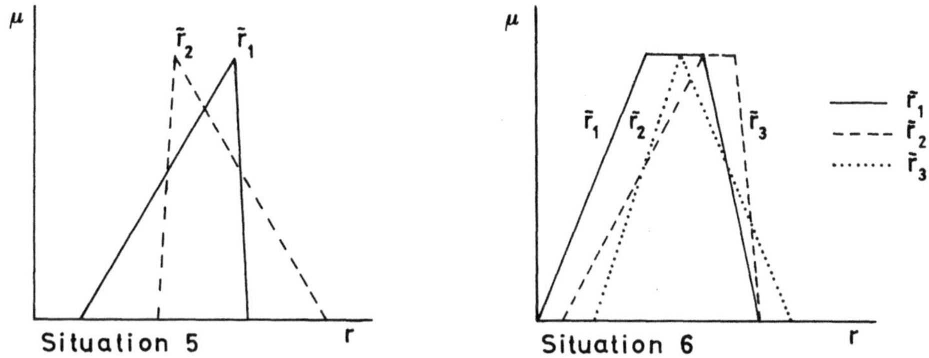

Figure 6

Rather than given the numerical values of the resulting measures, which are not quite comparable, we list the resulting orders of the alternatives in the following table:

Situation	Yager	Baas & Kwakern.	Jain	Baldwin & Guild
1	1 > 2 > 3	1 > 2 > 3	1 > 2 > 3	1 > 2 > 3
2	1 ~ 2	1 ~ 2	1 > 2	1 ~ 2
3	1 > 2 > 3	1 ~ 2 ~ 3	1 > 2 > 3	1 > 2 ~ 3
4	1 > 2 > 3	1 ~ 2 ~ 3	1 > 2 > 3	1 > 2 > 3
5	2 > 1	1 > 2	1 > 2	2 > 1
6	3 > 2 > 1	1 ~ 3 ~ 2	3 > 2 > 1	2 > 3 > 1

The results of Dubois and Prade's are shown referring to the following situations:

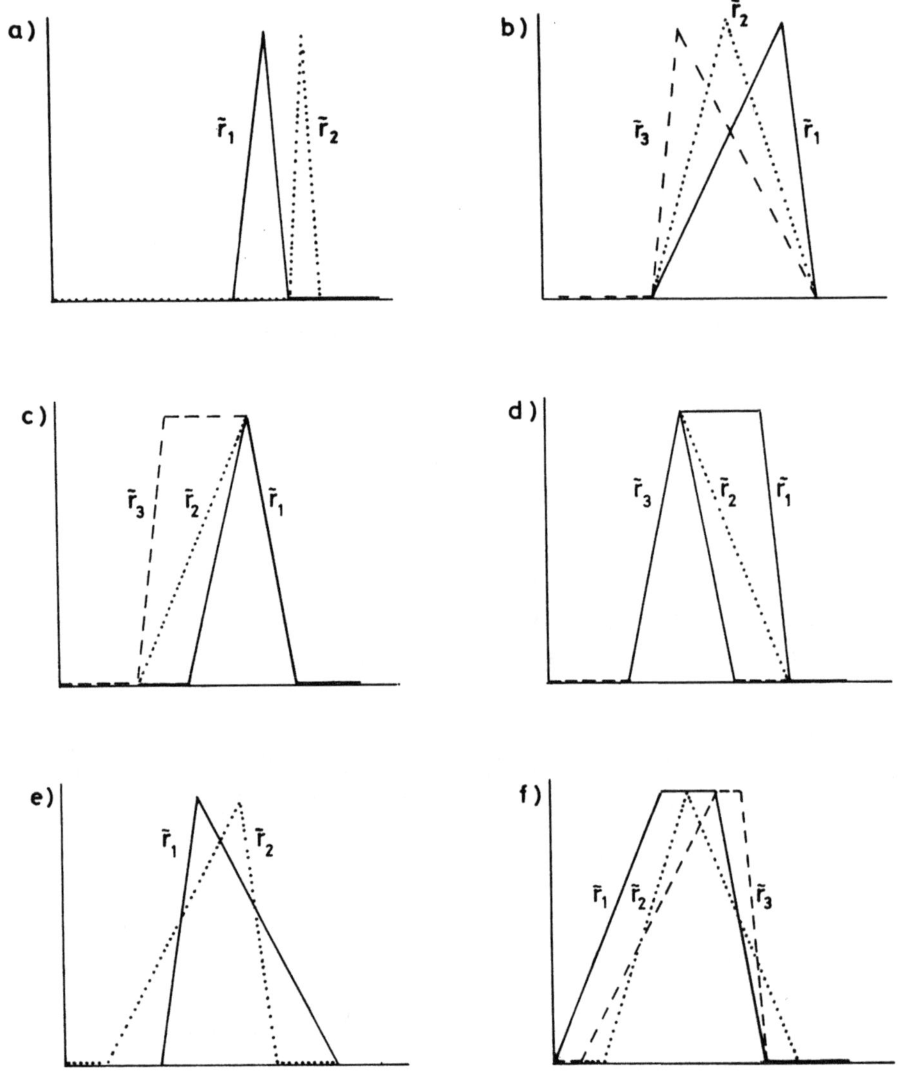

Figure 7

Case	a		b			c			d			e		f		
PD	0	1	1	.74	.60	1	1	1	1	1	1	.84	1	1	.88	1
PSD			.74	.23	.16	.5	.5	.5	.8	.2	0	.54	.46	.3	.4	.6
ND	0	1	.63	.38	.18	.67	.35	0	.5	.5	.5	.54	.46	.3	.5	.5
NSD	0	1	.26	0	0	0	0	0	0	0	0	0	.16	0	0	0

Further details can be found in the paper by Bortolan and Degani (1985).

Looking at the results of applying DP's method one can probably say the following:

Case a is clear cut: $\tilde{r}_2$ dominates obviously $\tilde{r}_i$. The same holds for case b. In c) and d) a slight dominance of $\tilde{r}_i$ over $\tilde{r}_2$ shows a slight possibility of dominance but not if strict dominance and vice versa for necessity of dominance. (The same contradiction can be observed for this case when applying the other methods).

The most difficult case seems to be case f. A possible, though probably not very helpful for the decision maker, interpretation could be: none of the ratings dominates strictly because all three supports overlap. $\tilde{r}_2$ and $\tilde{r}_3$ dominate $\tilde{r}_1$ on the left (ND) while $\tilde{r}_3$ is better than $\tilde{r}_1$ and $\tilde{r}_2$ on the right (PSD). $\tilde{r}_1$ and $\tilde{r}_3$ can both be better than $\tilde{r}_2$ on the right (PD).

It is of interest to compare the rankings of the other methods for this case:

Yager: $\tilde{r}_3 \succ \tilde{r}_2 \succ \tilde{r}_1$; Jain : $\tilde{r}_3 \succ \tilde{r}_2 \succ \tilde{r}_1$;

BK : $\tilde{r}_1 \sim \tilde{r}_3 \sim \tilde{r}_1$ and

Baldwin and Guild: $\tilde{r}_2 \succ \tilde{r}_3 \succ \tilde{r}_1$.

Obviously none of the orders coincides with any other!

Summarizing it can be said, that all the methods mentioned lead to almost the same results if the problems have clear cut solutions. If this is not the case the results contradict each other to different degrees. DP's method seems to allow the most detailed interpretations. Of course, the methods also differ with respect to their computational effort and to their assumptions.

2.3.2. Classification of Methods

The question arises: "Which is the best MADM-method so far?" This question can certainly not be answered because the answer will depend largely on the situation for which the method is to be used, in subjective evaluations and on other factors. Some help in selecting a suitable method might, however, be provided by a classification of the methods according to a number of criteria.

A first classification could be done by looking at the three aspects scope, process and focus as shown in the next table:

Criterion

Scope	Step 1	Step 2	Steps 1 and 2
Process	Simultaneous	Hierarchical	Interactive
Focus	Aggregation	Distance	Order-relation

This classification is rather mechanistic. It considers the method from a technical point of view. More appropriate would probably be a multidimensional classification taking into consideration more aspects of the different approaches. The following figure sketches some possible dimensions.

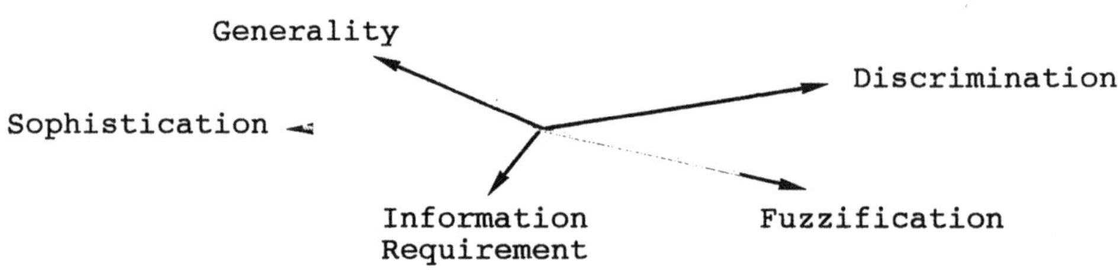

By "<u>generally</u>" we mean the degree of general applicability of the method, i.e. are special types of fuzzy sets assumed or can the ratings have arbitrary forms. Is the method restricted to special operators or can they be adopted to the context etc.

<u>Discrimination</u> refers to the capability of a method to differentiate between alternatives the ratings of which differ only slightly from each other. As mentioned above authors have different views on whether a method should be very discriminatory or rather "stable".

<u>Fuzzification</u>: Obviously different components of the MADM-problem can be represented by fuzzy sets. One extreme would be to only consider the relative weights of the criteria as fuzzy sets. Another extreme could be to consider the ratings, the weights and the alternatives as fuzzy.

<u>Information Requirements</u>: The more standardized the input data the less information has to be processed but the rougher might be the model of the real problem! If, for instance, only triangular fuzzy numbers are allowed each of them can be characterized by three real numbers. If arbitrary fuzzy sets are used much more information has to be provided and processed. The amount of information to be processed would increase even further if Type-2-fuzzy sets are used.

<u>Sophistification</u> refers to the mathematical tools which are being used in steps 1 and/or 2.

An evaluation of the methods according to the above-mentioned criteria would obviously be rather subjective. The following table, therefore, rather describes than evaluates the methods mentioned above.

	Phase	Weights	Criteria	Aggreg.(Phase I)	Crit. f. ranking	Solution
Yager I	I + II	fuzzy	fuzzy sets (norm)	weighted sum	max + min (Phase II)	crisp
Baas & Kwaker-naak	I + II	fuzzy	fuzzy sets	max - min	preference sets	crisp
Laarhiven & Pedrycz	I	fuzzy triangular number	crisp	hierarchical aggregation	-----	fuzzy ratings
Jain	II	----	fuzzy	----	maximizing set	crisp
Baldwin & Guild	II	----	fuzzy (special)	----	relative preference	fuzzy
Chen	II	----	fuzzy	----	maximizing and minimizing set	crisp
Dubois & Prade	II	----	fuzzy	----	multiple	vectorial

Table: Comparison of Fuzzy MADM-Methods

3. MULTI-OBJECTIVE MATHEMATICAL PROGRAMMING IN FUZZY ENVIRONMENTS

In 1970 Bellman and Zadeh (1970) defined a decision in a fuzzy environment as the "Confluence of Goals and Constraints", i.e. as the appropriate aggregation (intersection) of all the fuzzy sets representing either constraints or objectives.

DEFINITION 1. Let $\mu_{\tilde{C}_i}(x)$, $i=1,\ldots,m$, $x \in X$, be membership functions of constraints defining the decision space, and $\mu_{\tilde{G}_i}(x)$, $j=1,\ldots,n$, $x \in X$ the membership functions of objective (utility) functions or goals. A decision is then defined by its membership function

$$\mu_{\tilde{D}}(x) = \mu_{\tilde{C}_i}(x) * \mu_{\tilde{G}_j}(x),\quad i=1,\ldots,m,\quad j=1,\ldots,n \tag{19}$$

where * denotes an appropriate, possibly context-dependent, "aggregator" (connective).

Let M be the set of points $x \in X$ for which $\mu_{\tilde{D}}(x)$ attains its maximum, if it exists. Then M is called the maximizing decision. If $\mu_{\tilde{D}}(x)$ has a unique maximum at x_M, then the maximizing decision is a uniquely defined crisp decision which can be interpreted as the action which belongs to all fuzzy sets representing either constraints or goals with the highest possible degree of membership (which might be quite low).

We shall first consider the standard LP model

$$\begin{aligned} \text{Max}\quad & z = cx \\ \text{s.th.}\quad & Ax \le b \\ & x \ge 0 \end{aligned} \tag{20}$$

in which an aspiration level, z_0, has been established for the objective function and in which the constraints and the objective function only have to be satisfied in an fuzzy, approximate, way:

Find x such that

$$c^T x \lesssim z_0$$
$$Ax \lesssim b, \qquad (21)$$
$$x \geq 0.$$

Here $\lesssim$ denotes the fuzzified version of $\leq$ and has the linguistic interpretation "essentially smaller than or equal". The objective function in (20) might have to be written as a minimizing goal in order to consider z as an upper bound.

We see that (20) is symmetric with respect to objective function and constraints, and we now make that even more obvious by substituing

$$\begin{pmatrix} c \\ A \end{pmatrix} = B \text{ and } \begin{pmatrix} z \\ b \end{pmatrix} = d.$$

Then (20) becomes

Find x such that

$$Bx \lesssim d, \qquad (22)$$
$$x \geq 0.$$

Each of the m+1 rows of (21) will now be represented by a fuzzy set, the membership functions of which are $\mu_i(x)$, which can be interpreted as the degree to which x fulfills (satisfies) the fuzzy unequality $(Bx)_i \leq d_i$, $i=1,\ldots,m+1$, where $(Bx)_i$ denotes the ith row of (21). Following Definition 1, the membership function of the fuzzy set "decision" of the model (21) is

$$\mu_{\tilde{D}}(x) = \min_{i=1}^{m+1} \mu_i(x) \qquad (23)$$

Assuming that the decision maker is interested not in a fuzzy set but in a crisp "optimal" solution x°, we could suggest to him the "maximizing solution" to (22) which is the solution to the possibly programming problem

$$\max_{x\geq 0} \min_{i=1}^{m+1} \mu_i(x) = \mu_M(x^\circ) \tag{24}$$

We now have to specify the membership functions $\mu_i(x)$. They should be 0 if the constraints (including objective function) are strongly violated, 1 if they are very well satisfied (i.e. satisfied in the crisp sense); and they should increase monotonically from 0 to 1, i.e.

$$\mu_i(x) \begin{cases} = 1 & \text{if } (Bx)_i \leq d_i \\ \in [0,1] & \text{if } d_i \leq (Bx)_i \leq d_i+p_i, \quad i=1,\ldots,m+1 \\ = 0 & \text{if } (Bx)_i > d_i+p_i \end{cases} \tag{25}$$

Using the simplest type of membership function, we assume them to be linearly increasing over the tolerance interval $[d_i,d_i+p_i]$

$$\mu_i(x) = \begin{cases} 1 & \text{if } (Bx)_i \leq d_i \\ 1 - \dfrac{(Bx)_i-d_i}{p_i} & \text{if } d_i \leq (Bx)_i \leq d_i+p_i, \quad i=1,\ldots,m+1 \\ 0 & \text{if } (Bx)_i > d_i+p_i \end{cases} \tag{26}$$

The p_i are subjectively chosen constants of admissible violations of the constraints and the objective function. Substituing (25) into (23) yields, after some rearrangements,

$$\max_{x\geq 0} \min_{i=1}^{m+1} \left(1 - \frac{(Bx)_i-d_i}{p_i}\right) \tag{27}$$

Introducing one new variable, x, which corresponds essentially to $\mu_D(x)$ in (23), we arrive at

$$\begin{aligned} &\text{Maximize} && \lambda \\ &\text{such that} && \lambda p_i+(Bx)_i \leq d_i+p_i, \quad i=1,\ldots,m+1, \\ &&& 0 \leq \lambda \leq 1, \\ &&& x \geq 0. \end{aligned} \tag{28}$$

If the optimal solution to (28) is the vector (λ°, x°), then x° is the maximizing solution (23) to the model (20), assuming membership functions as specified in (26).

The reader should realize that this maximizing solution can be found by solving one standard (crisp) LP with only one more variable and one more constraint than the model (21). This makes this approach computationally very efficient.

So far the objective function as well as all constraints were considered fuzzy. If some of the constraints are crisp, $Dx \le b$, then these constraints can easily be added to the formulation (28). Thus (28) would, for instance, become

$$\begin{aligned} &\text{Maximize} && \lambda \\ &\text{such that} && \lambda p_i+(Bx)_i \le d_i+p_i, \; i=1,\ldots, m+1, \\ &&& Dx \le b \\ &&& \lambda \le 1 \\ &&& x, \lambda \ge 0. \end{aligned} \tag{29}$$

So far we have only considered one objective function. Due to the symmetry of the model more calibrated objective functions can, however, be added without difficulties.

Assuming that the decision maker has upper and lower bounds $\overline{c}_i$ and $\underline{c}_i$ for his aspirations, such that he does not accept solutions for which $c_i x$ is smaller than $\underline{c}_i$ and that he is fully satisfied whenever $c_i x$ is equal or larger than $\overline{c}_i$, we can express his objectives by fuzzy sets whose membership functions increase monotonically from 0 at $\underline{c}_i$ to 1 at $\overline{c}_i$. Denoting these fuzzy sets as fuzzy constraints C(x). (21) can be written as

$$\begin{aligned} &\text{Find } x \\ &\text{such that} && Cx \lesssim \underline{c} \\ &&& Ax \le b \\ &&& x \ge 0. \end{aligned} \tag{30}$$

Now we can apply directly model (29). The only problem that remains is how to establish the $\underline{c}_i$ and $\overline{c}_i$. We can either assume that the decision maker establishes apiration levels for himself, or we can compute those as a function of the solution space. For the latter approach it was suggested to use the individual optima as upper bounds and "least justifiable" solutions as lower bounds.

If the solution to (30) is a singleton, then this is also the desired optimal solution. Werners (1984) has pointed to the fact, however, that the final simplex tableau of (30) might indicate multiple optimal solutions (dual degeneracy). In this case not all of the "optimal solutions" might be efficient solutions to the vector maximum problem, i.e. it might be possible to improve these nonefficient solutions by increasing λ_0 further by using slack in the constraints which were not binding with respect to the first optimal solution.

Let us now consider the case in which fuzzy constraints have to be taken into consideration. The problem is then of the type

$$\begin{array}{lrl} \text{"Maximize"} & z & = Cx \\ \text{such that} & Ax & \lesssim b \\ & Dx & \leq b' \\ & x & \geq 0. \end{array} \tag{31}$$

where C is a K x n matrix, z is a vector with k components, and all the other elements are of the usual dimension. Essentially we could use the approach described above. Two additional complications, however, arise.

(1) In classical (crisp) vector maximum theory "efficiency" of solutions has been defined under the assumption that solutions can only be split into feasible and nonfeasible solutions and that only the former are considered as candidates for efficient solutions. In (31), however, solutions do not only differ with respect to the associated values of the objective function, but also with respect to their degree of "feasibility". Above definition therefore has to be extended. Werners (1984) offers the following notion of fuzzy efficiency:

DEFINITION. Let $f_j\colon x \to \mathbb{R}^1$, $j=1,\ldots,k$, be objective functions of the problem (31), and $\mu_i \in [0,1]$, $i=1,\ldots,m$ the membership functions representing the fuzzy constraints of (31). A solution $x \in X$ is called fuzzy efficient if there is no $x' \in X$ such that

$f_j(x') \geq f_j(x)$ for all $j=1,\ldots,k$

and

$\mu_i(x') \geq \mu_i(x)$ for all $i=1, \ldots,m$,

and

$f_{j_0}(x') > f_{j_0}(x)$ for at least one $j_0 \in \{1,\ldots,k\}$

or

$\mu_{i_0}(x') > \mu_{i_0}(x)$ for at least one $i_0 \in \{1,\ldots,m\}$.

(2) There are no uniquely defined individual optima which can be used as upper bounds to calibrate the fuzzy sets representing the objective functions. The "minimal justifiable" solution is not available either. We can, however, employ the approach used to derive model (29) from above-mentioned contribution. The extension to multiple objectives is straightforward.

Some modifications have to be made concerning the lower bounds when calibrating the fuzzy goal set: We still consider as a lower bound of aspiration, for which the degree of membership is equal to 0, the least functional value a decision maker can obtain if the individual optima, x_i°, x_i^1, of the other objective functions are chosen. Since, however, theses optima are no longer uniquely defined, the respective degrees of belonging to the solution space also have to be taken into account. $\mu_l(x)$, $l=1,\ldots,k$, is therefore considered to be zero if either

$$f_l(x) \leq \min_{l'=1}^{k} f_l(x_{l'}^1) = f_l' \qquad (32)$$

or

$$f_1(x) \le \min_{1'=1} f_i(x^o_{1'}) = f'_o$$

Then $f_{o1}(x) = \min (f_1, f_{1'})$. The crisp model equivalent to (30) is then

$$\begin{aligned} &\text{Maximize} \quad \lambda \\ &\text{such that} \quad \lambda(f'_1 - f'_o) - Cx \le -f'_o \\ &\qquad p + Ax \le b + p \\ &\qquad Dx \le b' \\ &\qquad \lambda,\ x \ge 0. \end{aligned} \tag{33}$$

Here $f'_o = (f_{o1}, \ldots, f_{ok})^T$, $f'_1 = (f_{11}, \ldots, f_{1k})^T$, C is the k x n matrix defined in (30), and all other coefficients are defined as before, f_{11} are the fuzzy individual optimal values.

For the sake of generality it should be pointed out that nonlinear membership function and other operators than the "min" can also be used - the resulting equivalent model which has to be solved may, however, become nonlinear and more difficult to solve. (Zimmermann, 1987)

REFERENCES

Baas, M.S. and Kwakernaak, H. (1977), "Rating and ranking of multiple-aspect alternatives using fuzzy sets", *Automatica*, 13, 47-58.

Baldwin, J.F. and Guild, N.C.F. (1979), "Comparison of fuzzy sets on the same decision space", *Fuzzy sets and Systems*, 2, 213-232.

Bellman, R.E. and Zadeh, L.A. (1970), "Decision-making in a fuzzy environment", *Mgt. Sc.*, 17, B141-164.

Bortolan, G. and Degani, R. (1985), "A review of some methods for ranking fuzzy subsets", *Fuzzy Sets and Systems*, 15, 1-20.

Buckley, J.J. (1985), "Ranking alternatives using fuzzy numbers", *Fuzzy Sets and Systems*, 15, 21-32.

Charnes, A. and Cooper, W.W. (1961), *Management Models and Industrial Applications of Linear Programming*, Wiley, New York, 1961.

Chen, S.H. (1985), "Ranking fuzzy numbers with maximizing set and minimizing set", *Fuzzy Sets and Systems*, 17, 113-130.

Dubois, D. and Prade, H. (1984), "Criteria aggregation and ranking of alternatives in the framework of fuzzy set theory", in H.-J. Zimmermann et al. (1984), 209-240.

Hannan, E.L. (1981), "Linear programming with multiple fuzzy goals", *Fuzzy Sets and Systems*, 6, 235-248.

Hwang, Ch.-L. and Yoon ,K. (1981), *Multiple Attribute Decision Making: Methods and Applications*, Springer-Verlag, Berlin.

Hwang, Ch.-L., Masud, A.S. (1979), *Multiple Objective Decision Making: Methods and Applications*, Springer-Verlag, Berlin.

Jain, R. (1977), "Procedure for multi-aspect decision making using fuzzy sets", *Int. Journal Systems Science*, 8, 1-7.

Kahne, H.W. and Tucker, A.W. (1951), "Nonlinear Programming", in J. Neyman (ed.), *Proceedings of the Second Berkeley Symposium on Mathematical Statistics and Probability*.

Van Laarhoven, P.J.M. and Pedrycz, W. (1983), "A fuzzy extension of Saaty's priority theory", *Fuzzy Sets and Systems*, 11, 229-241.

Leberling, H. (1981), "On finding compromise solutions in multicriteria problems using the fuzzy min-operator", *Fuzzy Sets and Systems*, 6, 105-118.

Roy, B. (1986), "Partial Preference Analysis and Decision Aid: The Fuzzy Outranking Relation Concept", *SEMA*, Paris.

Rubin, P.A. and Narasimhan, R. (1984), "Fuzzy goal programming with rested priorities", *Fuzzy Sets and Systems*, 14, 115-130.

Saaty, Th.L. (1978), "Exploring the interface between hierarchies, multiple objectives and fuzzy sets", *Fuzzy Sets and Systems*, 1, 57-68.

Saaty, Th.L. (1980), *The Analytic Hierarchy Process*, McGraw-Hill, New York.

Siskos, J., Lochard, J. and Lombard, J. (1984), "A multicriteria decision making methodology under fuzziness: appreciation to the evaluation of radiological protection in nuclear power plants", in H.-J. Zimmermann et al. (1984), 261-284.

Tong, R.M. and Bonissone, P.P. (1984), "Linguistic solutions to fuzzy decision problems", in H.-J. Zimmermann et al. (1984), 323-334.

Werners, B. (1984), *Interaktive Entscheidungsunterstützung durch ein flexibles mathematisches Programmierungssystem, München.*

Yager, R.R. (1978), "Fuzzy decision making including unequal objectives", *Fuzzy Sets and Systems*, 1, 87-95.

Zadeh, L.A. (1977), " Linguistic characterization of preference relations as a basis for choice in social systems", Memo UCB/ERL M77/24, Berkeley.

Zimmermann, H.-J. (1976), "Description and optimization of fuzzy systems, *Internat. J. Gen. Systems*, 2, 209-215.

Zimmermann, H.-J. (1978), "Fuzzy programming and linear programming with several objective functions", *Fuzzy Sets and Systems*, 1, 45-55.

Zimmermann, H.-J. and Zysno, P. (1983), "Decisions and evaluations by hierarchical aggregation of information", *Fuzzy Sets and Systems*, 10, 243-266.

Zimmermann, H.-J., Zadeh, L.A. and Gaines, B.R. (eds.) (1984), *Fuzzy Sets and Decision Analysis*, North-Holland, New York.

Zimmermann, H.-J. (1985), *Fuzzy Set Theory and Its Applications*, Kluwer-Nijhoff Publishing, Boston.

Zimmermann, H.-J. (1987), *Fuzzy Sets, Decision Making, and Expert Systems*, Kluwer-Nijhoff Publishing, Boston.

CHAPTER II

THE OUTRANKING APPROACH

THE OUTRANKING APPROACH
AND
THE FOUNDATIONS OF ELECTRE METHODS

Bernard Roy

LAMSADE, Université de Paris-Dauphine
Place du Maréchal De Lattre de Tassigny
75775 Paris Cedex 16 - FRANCE

The concept of outranking relations was born of difficulties encountered with diverse concrete problems (see Abgueguen (1971), Bétolaud and Février (1973), Buffet et al. (1967), Charpentier and Jacquet-Lagrèze (1976), Laffy (1966)). Since then, numerous applications of the concept have been developed. Among the most recent ones, we can mention: Barda et al. (1989), Clímaco et al. (1988), Martel and Nadeau (1988), Maystre and Simos (1987), Parent and Schnäbele (1988), Rabeharisoa (1988), Renard (1986), Roy et al. (1986), Slowinski and Treichel (1988). Many others will be found in Jacquet-Lagrèze and Siskos (1983), de Montgolfier and Bertier (1978), Roy (1985), Schärlig (1985).

In the first part of the present paper, we shall begin by describing the main features of real-world problems for which the outranking approach is appropriate. We shall then present the concept of outranking relations. The second part of the paper is devoted to basic ideas and concepts used for building outranking relations. The definition of such outranking relations is given for the main ELECTRE methods in part 3. The final part of the paper is devoted to some practical considerations.

1. INTRODUCTION TO THE OUTRANKING APPROACH

1.1. Preliminary Notations and Definitions

For understanding what the outranking approach is and what kind of real-world problems it refers to, it is necessary to specify what is supposed to be given initially.

a) A set A of **potential actions** (see Roy (1990)) (or alternatives) is considered. Such actions are not necessarily exclusive, i.e. they can be put into operation jointly.

b) A consistent **family F of n criteria** g_j has been defined (see Bouyssou (1990)). This means that preferences of actors involved in the decision process are formed, argued and transformed by reference to points of view adequately reflected by criteria of F.

$g_j(a)$ is called the jth **performance** of a. It is not restrictive to suppose that:

- $g_j(a)$ is a real number (even if it reflects a qualitative assessment);
- $\forall a' \in A$ and $a \in A$, $g_j(a') \geq g_j(a) \Rightarrow$ a' is at least as good as a if we consider only the point of view reflected by the jth criterion.

c) Let us emphasize on a given criterion, for instance the kth. The imprecision, and/or the uncertainty, and/or the inaccurate determination of performances (see Roy (1988)) may lead some actor to judge:

- a' indifferent to a when $g_j(a') = g_j(a)$ $\forall j \neq k$ even if $g_k(a') \neq g_k(a)$;
- a' strictly preferred to a when $g_j(a') = g_j(a)$ $\forall j \neq k$ only if the difference $g_k(a') - g_k(a)$ is sufficiently significant.

These two possibilities underline the fact that maps are not territories: the **vector of performances** $\underline{g}(a) = [g_1(a),...,g_n(a)]$ is like a map of that territory which is the action a. We want to compare actions, i.e. territories, not maps. These comparisons are made by means of maps, and we have to avoid working as if maps do not differ from territories.

d) Let us consider now, at the comprehensive level, the comparison of a' and a on the basis of $\underline{g}(a')$ and $\underline{g}(a)$. The actors involved in the decision process may not all have exactly the same judgement. To give meaning to a comprehensive model of preferences, we will refer to a particular actor D called the **decision-maker**. This one may be viewed either as a real person for whom or in the name of whom decision-aid is provided, or as a mythical person whose preferences can be used to enlighten the decision-aid problem. The comprehensive model of preferences in question does not, consequently, pretend to be an accurate description of well-stated preferences thought to be firmly fixed in the mind of a clearly identified decision-maker D. If D is a mythical, inaccessible or vaguely defined decision-maker, the model is only a system of preferences with which it is possible to work in order to bring forward elements of a response to certain questions.

Under these conditions, the comprehensive model of preferences should allow us to take into account hesitations between two of the three following cases:

a' I a: a' indifferent to a,
a' P a: a' strictly preferred to a,
a P a': a strictly preferred to a'.

According to the types of hesitations, we will speak of:

- **weak preference**[1] (relation Q):
  a' Q a if the hesitation is between a' I a and a' P a (being sure that not a P a'),
  a Q a' is the hesitation is between a' I a and a P a' (being sure that not a' P a);

- **incomparability** (relation R): a' R a if the hesitation deals with a' P a and a P a' (at least).

More precisely, the hesitations mentioned above may come from:

- the existence in D's mind (if D is a real person) of zones of uncertainty, half-held belief or conflicts and contradictions;
- the vaguely defined quality of the decision-maker;
- the fact that the scientist who built the model ignores, in part, how D compares a' and a;
- the imprecision, uncertainty, inaccurate determination of the maps $\underline{g}(a')$ and $\underline{g}(a)$ by means of which a' and a are compared.

e) Approaches of the AHP[2] and MAUT[3] type base the comprehensive model of preferences on the explicitation of a value function or a utility function V(a):

$$V(a) = V[g_1(a), \ldots, g_n(a)]$$

aggregating the n criteria in such a way that, in D's mind:

a' I a iff $V(a') = V(a)$,
a' P a iff $V(a') > V(a)$.

---

1. For more details on this concept, see Roy (1985, ch. 7) or Roy and Vincke (1987).

2. Analytic Hierarchy Process (see Saaty (1980)).

3. Multi-Attribute Utility Theory (see Keeney and Raiffa (1976)).

The assumptions made above do not seem readily compatible with such a way of modelling. This is one (but not the only) reason which leads us to formulate the outranking concept[1].

1.2. Outranking Concept: Level of Preferences Restricted to the g_j Criterion

To each criterion g_j, it is possible to associate a **restricted outranking relation** S_j. By definition, S_j is a binary relation: $a' \, S_j \, a$ holds if the values of the performances $g_j(a')$ and $g_j(a)$ give a sufficiently strong argument for considering the following statement as being true in a D's model of preferences:

«a', with respect to the jth criterion only, is at least as good as a».

Let us point out that «at least as good as» must be considered as «not worse than».

Let us consider for instance the case in which an **indifference threshold** q_j is associated with g_j. By definition, q_j is a real positive number such that:

$a' \, I_j \, a$ iff $|g_j(a') - g_j(a)| \leq q_j$.

In this case:

$a' \, S_j \, a$ iff $g_j(a') \geq g_j(a) - q_j$.

This formula can easily be generalized so as to take into account thresholds which are not constant (for example which vary with $g_j(a)$).

----

1. See Roy (1990).

1.3. Outranking Concept: Level of Comprehensive Preferences

Taking into account the whole family of criteria, it is possible to define a **comprehensive outranking relation** S. By definition, S is a binary relation: a' S a holds if the values of the performances entering into $\underline{g}(a')$ and $\underline{g}(a)$ give a sufficiently strong argument for considering the following statement as being true in a D's model of preferences:

«a', with respect to the n criteria, is at least as good as a» (here again, «at least as good as» is synonymous with «not worse than»).

For illustrating this concept, let us consider the following numerical example defined in Table 1. Quite obviously, the assertions b S a, a S b and c S d hold. Apart from very strange cases (in which the last criterion would have an enormous importance), the assertion c S a holds. But, in the absence of an aggregating rule and additional information, none of the following assertions would seem to be accepted: c S b, b S c, d S b, b S d.

This example shows that, for defining a comprehensive outranking relation S on A, it is necessary to formulate a set of appropriate conditions which, when they are satisfied, can be viewed as sufficiently strong arguments for justifying the assertion a' S a.

A \ F	g_1	g_2	g_3	g_4	g_5
a	50	50	50	50	50
b	55	46	48	54	55
c	90	90	90	45	42
d	90	90	90	10	10

$\forall j \in F$: $q_j = 5$

and

$0 \le g_j(a) \le 100$

Table 1: Numerical example

1.4. Fundamental Properties of Outranking Relations

Let us consider the n restricted outranking relations S_j linked to the n criteria of F and a binary relation S aggregating the same criteria. So that S can be viewed as a comprehensive outranking relation (in conformity with the meaning given above to this concept), it seems natural to us to demand that S verifies the following properties:

a) S is reflexive ($a\,S\,a\ \forall a \in A$) and, for $a' \neq a$, the four configurations shown in Figure 1 are meaningful; more precisely, the meaning of each configuration matches what is written in Figure 1.

Let us remark that:

- a' S a and not a S a' cannot be interpreted, without some precautions, as «a' is strictly preferred to a»;
- S is not necessarily a transitive binary relation.

FIGURE 1: Meaning of the four possible configurations in the comparison of two potential actions a and a'

Graphic convention: a'.——→——.a iff a' S a

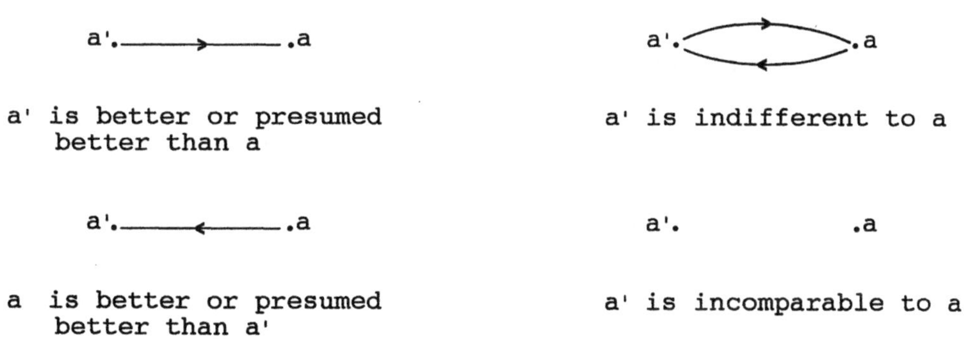

b) Let us denote by Δ_F the dominance relation defined by:

$a \, \Delta_F \, b$ iff $g_j(a) \geq g_j(b) \; \forall j \in F$.

S has to verify, whatever the potential actions a, a', b and b' are:

$a' \, S \, a$ and $a \, \Delta_F \, b \Rightarrow a' \, S \, b$,
$b' \, \Delta_F \, a'$ and $a' \, S \, a \Rightarrow b' \, S \, a$.

Consequently, we have: $a \, \Delta_F \, b \Rightarrow a \, S \, b$.

c) Finally, it is natural to demand:

if $g_j(a') = g_j(a) \; \forall j \neq k$, then $a' \, S \, a$ iff $a' \, S_k \, a$.

This condition can be reinforced as follows:

if $a' \, I_j \, a \; \forall j \neq k$, then $a' \, S \, a$ iff $a' \, S_k \, a$.

From this last property combined with the previous one, we can deduce (for more details, see Roy and Bouyssou (1987a)):

$a' \, S_j \, a \; \forall j \in F \Rightarrow a' \, S \, a$.

2. BASIC CONCEPTS FOR BUILDING OUTRANKING RELATIONS

2.1. General Considerations

The formal expression and even the nature of the conditions which must be satisfied to validate the assertion a' S a can be influenced by many factors. The most important ones seem to be:

- the degree of significance of criteria taken into account in F;
- the nature of basic concepts used: concordance, discordance, substitution rate, intensity of preference, ...;
- the nature of the inter-criteria information required (for a full definition of this term, see Roy and Bouyssou (1987b));

- the strength of the arguments required: the strongest we can imagine is a' dominates a but the outranking concept is an interesting one only because weak arguments can be sufficient; that is why S is usually a notably richer binary relation than Δ_F.

The variety of possible options relevant to each of the four preceding factors explains why there is not a single «best» way of formulating the conditions to be satisfied to validate a' S a. Henceforth, in this paper, we shall present only the way in which these conditions are formulated, within the framework of ELECTRE methods, in order to build the outranking relations upon which they are based. This is why we shall study only those cases in which:

- the degree of significance of each criterion g_j is reflected by means of two thresholds[1] q_j and p_j in conformity with the model of the pseudo-criterion presented in 2.2;
- the basic concepts are those of concordance and discordance (see 2.2 below);
- the inter-criteria information is synthetized in, at most, two kinds of data: for each criterion g_j, its importance coefficient k_j and its veto threshold v_j.

We would like to point out that, by slightly modifying the framework defined by the preceding options but maintaining the central role given to the concept of concordance, Brans et al., (1984) and Brans and Mareschal (1990) on the one hand, and Vansnick (1986), on the other hand, have proposed comprehensive preference models based on binary relations which are not necessarily transitive, leaving room for incomparability; however, although their approach was inspired by the outranking approach

----

1. In this paper, so as to simplify the way formulas are written, they will be given with constant thresholds. They can easily be generalized to thresholds $q_j[g_j(a)]$ and $p_j[g_j(a)]$ varying with $g_j(a)$ (see Roy and Bouyssou (1989)).

and shares many points in common with it, the binary relations they refer to are not, strictly speaking, outranking relations (for more details on this point, see Roy and Bouyssou (1989)).

Let us come back to the last of the four variety factors mentioned at the beginning of 2.1. Of course, it is difficult to fix a minimum degree of strength so that the assertion a' S a is seen as valid if and only if the strength of the arguments which justify it are at least equal to this minimum. For this reason, two types of modelling are envisaged:

TYPE 1: a set of r (r ≥ 1) outranking relations is introduced so as to modelize D's preferences:

$S_1 \subset S_2 \subset \ldots \subset S_r$,

the increasing of the index from 1 through r corresponds to a decreasing strength of the arguments required for validating a' S a.

TYPE 2: instead of one or more crisp outranking relations for modelling D's preferences, it is a fuzzy outranking relation which is introduced; this means that we associate with each ordered pair (a', a) a real number $\sigma(a', a)$ $(0 \leq \sigma(a', a) \leq 1)$ characterizing the degree of strength of the arguments and allowing us (on the basis of the two vectors $\underline{g}(a')$, $\underline{g}(a)$ and additional inter-criteria information) to validate the assertion a' S a. $\sigma(a', a)$ is called the **credibility index** of the outranking a' S a.

The preference model is of type 1 in ELECTRE I, II and IV but of type 2 in ELECTRE III and A. If in practice these two types of modelling differ significantly, from a theoretical point of view, they are nearly equivalent: the only difference between them comes from the fact that, in the second type, the size r of the sequence of crisp outranking relations is not pre-determined.

2.2. Concordance and Discordance Concepts

a) Concordant criterion

By definition, the jth **criterion is in concordance** with the assertion $a' \, S \, a$ if and only if $a' \, S_j \, a$.

According to the definition of S_j (see 1.2) and to the introduction of an indifference threshold q_j, the jth criterion is in concordance with the assertion $a' \, S \, a$ iff:

$$g_j(a') \geq g_j(a) - q_j.$$

The subset of all criteria of F which are in concordance with the assertion $a' \, S \, a$ is called the **concordant coalition** (with this assertion). It is denoted by $C(a' \, S \, a)$.

b) Discordant criterion

By definition, the jth **criterion is in discordance** with the assertion $a' \, S \, a$ if and only if $a \, P_j \, a'$.

According to the pseudo-criterion model[1]:

$$a \, P_j \, a' \text{ iff } g_j(a) > g_j(a') + p_j.$$

This means that the strict preference of a over a' restricted to the jth criterion is significantly established only when the difference $g_j(a) - g_j(a')$ is sufficiently large considering the imprecision, uncertainty and inaccurate determination of the performances.

The subset of all criteria of F which are in discordance with the assertion $a' \, S \, a$ is called the **discordant coalition** (with this assertion). It is denoted by $C(a \, P \, a')$ since it can also be viewed as the concordant coalition with the assertion $a \, P \, a'$.

----

1. For more details, see Roy (1985) or Roy and Vincke (1987).

c) Consequences

From the above definitions, we have:

$C(a' \, S \, a) \cap C(a \, P \, a') = \emptyset$ and $C(a' \, S \, a) \cup C(a \, P \, a') \subset F$.

Let us emphasize that we can have:

$C(a' \, S \, a) \cup C(a \, P \, a') \neq F$.

This inequality holds if and only if there exists at least one criterion which is neither concordant nor discordant with the assertion a' S a. This case appears iff:

$g_j(a) - p_j \leq g_j(a') < g_j(a) - q_j \qquad (p_j > q_j)$.

The subset of F defined by the criteria satisfying this last condition will be denoted by C(a Q a'). Consequently, we have:

$j \in C(a \, Q \, a')$ iff $a \, Q_j \, a'$,

the binary relation Q_j modelizes the weak preference situation (see 1.1 d)) restricted to the jth criterion.

In practice, C(a Q a') is empty for a large number of ordered pairs of potential actions.

Finally, to each ordered pair (a', a), a partition of F into three subsets is associated:

C(a' S a), C(a Q a'), C(a P a').

It is on the basis of this partition that the assertion a' S a is (or is not) validated in ELECTRE methods. Before specifying the conditions of validation, it is important to remark that the partition defined above is robust in the sense given to this term by Roberts (1979). This means that the partition remains invariant when any criterion g_j is replaced by:

$g'_j = \mathcal{X}(g_j)$ with $\mathcal{X}$ monotonous increasing function.

This supposes, of course, that the initial thresholds q_j and p_j would be replaced by new thresholds (not necessarily constant) having the same meaning but taking into account the fact that the way the jth performance is defined has been modified.

2.3. Concordance and Discordance Indices

a) The notion of importance of each criterion

For validating a comprehensive outranking relation S, it is necessary to take into account the fact that the role which has to be devoted to each criterion in the aggregation procedure is not necessarily the same. In other words, we need to characterize what is usually referred to as «the greater of lesser importance» given to each criterion of F. When the aggregation procedure leads to a weighted sum (as in AHP for instance), this notion of importance is taken into account by means of constant substitution rates (currently called weights) assigned to each criterion. In more sophisticated models (MAUT for instance), those substitution rates can vary with performances. In both cases, the coefficients (tradeoffs) so defined are not intrinsic, i.e. they do not depend only on the axis of significance of the criterion g_j to which they refer. They also depend on the way g_j is defined: if g_j is replaced by $\mathcal{X}(g_j)$ ($\mathcal{X}$ monotonous increasing function), then each weight (or any substitution rate) has to be modified.

In ELECTRE methods, the importance of the jth criterion is taken into account by means of, at most, two characteristics:

- its **importance coefficient** k_j (≥ 0) which is intrinsic: the k_j's only intervene in the definition of the concordance degree (see b) below); moreover, these coefficients do not exist in ELECTRE IV;

- its **veto threshold** v_j ($\geq p_j$) [1]: v_j only intervenes in the definition of the discordance degree of criterion g_j (see c) below).

b) The concordance index

By definition, the **concordance index** c(a', a) characterizes the strength of the positive arguments able to validate the assertion a' S a. The strongest among them come from the criteria of c(a' S a) since they are all in favor of the assertion considered. They contribute one part to c(a', a): c_1(a', a). Some weaker positive arguments can also come from criteria of c(a Q a') since such criteria reflect a hesitation between the two following possibilities: a' I_j a (which is in favor of a' S a) and a P_j a' (which is not in favor of a' S a). They contribute a second part to c(a', a): c_2(a', a). Consequently:

$$c(a', a) = c_1(a', a) + c_2(a', a). \tag{2.1}$$

By definition:

$$c_1(a', a) = \frac{1}{k} \sum_{j \in C(a'Sa)} k_j \quad \text{with } k = \sum_{j \in F} k_j, \tag{2.2}$$

$$c_2(a', a) = \frac{1}{k} \sum_{j \in C(aQa')} \varphi_j . k_j \quad \text{with } \varphi_j = \frac{p_j + g_j(a') - g_j(a)}{p_j - q_j} \tag{2.3}$$

---

1. In this paper, so as to simplify the way formulas are written, they will all be given with constant thresholds. They can easily be generalized to thresholds $v_j[g_j(a)]$ varying with $g_j(a)$ (see Roy and Bouyssou (1989)).

Let us now explain how these formulas are justified.

According to the definition, it is natural to set:

$0 \leq c(a', a) \leq 1,$
$c(a', a) = 0$ if $C(a\,P\,a') = F,$
$c(a', a) = 1$ if $C(a'\,S\,a) = F.$

The ratio k_j/k reflects, by definition, the relative strength (in F) assigned to each g_j when this criterion is concordant with a' S a. In other words, k_j can be viewed as the number of representatives supporting the point of view synthesized by the jth criterion in a voting procedure. If $j \in C(a'\,S\,a)$, the whole strength k_j (or all the k_j representatives) contributes to C(a', a). On the contrary, this contribution is null if $j \in C(a\,P\,a')$. If $j \in C(a\,Q\,a')$, it is only a fraction φ_j of k_j which contributes to C(a', a). This fraction φ_j must evidently increase from 0 to 1 when $g_j(a')$ increases from $g_j(a) - p_j$ to $g_j(a) - q_j$. In other words, the more hesitation occurs in favor of a' I_j a, the more «the number $\varphi_j.k_j$ of voters» who defend the assertion a' S a increases. It might seem arbitrary to modelize this growth with a linear formula. Any other formula would be just as arbitrary and would not offer the same simplicity.

c) Veto effect and discordance index

Let us now consider the effect, on the validation of a' S a, of any discordant criterion. Obviously, such a criterion is against the assertion in question but the strength of this opposition can be more or less compatible with the acceptance of the assertion. For reflecting the capacity given to the jth criterion when it is discordant, for rejecting the assertion a' S a without any help of other criteria, a **veto threshold** v_j is defined as follows:

$g_j(a) - g_j(a') > v_j$ is incompatible with the assertion a' S a whatever the other performances are, i.e. even if $c(a', a) = 1 - k_j/k$.

We can also admit that this veto effect can occur for a difference $g_j(a) - g_j(a')$ smaller than v_j when $c(a', a) < 1 - k_j/k$. This leads to reinforce the veto effect all the more as $c(a', a)$ decreases.

The veto effect defined above works on the principle of all or nothing. Consequently, it is appropriate for defining one or more crisp outranking relations. If we consider now the second type of modelling introduced in 2.1 based on a fuzzy outranking relation, it is useful to modulate from 0 to 1 the strength of the opposition to a' S a, according to the position of the difference $g_j(a) - g_j(a')$ on the interval $[p_j, v_j]$. This explains the definition of the **discordant index** (of criterion g_j):

$$d_j(a', a) = \begin{cases} 0 \text{ if } g_j(a) - g_j(a') \le p_j \\ \dfrac{g_j(a) - g_j(a') - p_j}{v_j - p_j} \text{ if } p_j < g_j(a) - g_j(a') \le v_j \\ 1 \text{ if } g_j(a) - g_j(a') > v_j. \end{cases} \quad (2.4)$$

Here again, it is in the interest of simplicity that a linear formula has been chosen.

Let us end this paragraph by emphasizing that, in a certain sense, the criterion g_j is all the more important when v_j is close to p_j. Yet this way of tackling the notion of a criterion's importance (through a veto effect) is fundamentally different from that which prevails when our reasoning is based on positive arguments in the context of concordance (importance coefficient k_j). It is clear that the two criteria rankings according to (i) decreasing values of $v_j - p_j$ and (ii) increasing values of k_j are not unrelated. Despite this, without reflecting any incoherence whatsoever, they may be significantly different.

3. DEFINITION OF OUTRANKING RELATIONS IN ELECTRE METHODS

3.1. ELECTRE IS

In ELECTRE IS, the assertion a' S a is considered valid iff the two following conditions are satisfied:

- $c(a', a) \geq s, \; 1/2 < s \leq s^* \text{ with } s^* = 1 - \dfrac{1}{k} \min_{j \in F} k_j,$ (3.1)
- $g_j(a') + v_j \geq g_j(a) + q_j . w(s, c)$ (3.2)
  (more rigorously, in the above formula, q_j should be replaced by $\min \{q_j, v_j - p_j\}$)

$$\text{with } w(s, c) = \frac{1 - c(a', a) - k_j/k}{1 - s - k_j/k} .$$

The first condition (concordance condition) simply expresses the fact that the value of the concordance index c(a', a) (see (2.1), (2.2) and (2.3)) must be high enough to validate a' S a: s is a parameter called the **concordance level**. This means that a sufficiently high majority of criteria has to be in favor of the assertion; that is why $s > 1/2$. Furthermore, it is easy to prove that if we give, to the concordance level, a value $s > s^*$, then:

- (3.1) is satisfied iff C(a' S a) = F,
- when (3.1) is satisfied, so is (3.2).

Consequently, to set $s > s^*$ means that we want to impose a' S_j a $\forall j \in F$ for validating a' S a. The outranking relation so defined generalizes (for $q_j \neq 0$) the dominance relation (see 1.4 b)). Requiring such unanimity constitutes an extreme case which generally has no interest in practice. In practical applications it seems natural to have the s parameter varying between 3/5 and 4/5.

The condition (3.2) simply expresses the fact that the veto effect does not occur for any criterion. The coefficient w(s, c) is used to modelize the reinforcement of the veto effect introduced in 2.3 c) above.

$w(s, 1 - k_j/k) = 0$: no reinforcement,
$w(s, s) = 1$: maximum reinforcement (magnitude q_j).

3.2. ELECTRE III

The outranking relation in ELECTRE III is a fuzzy binary relation (see 2.1, second type). The credibility index $\sigma(a', a)$ which defines it makes the concordance index $c(a', a)$ intervene again. Moreover it brings in the discordance indices $d_j(a', a)$ (see 2.4) for those discordant criteria verifying $d_j(a', a) > c(a', a)$. In the absence of such discordant criteria, $\sigma(a', a) = c(a', a)$. This credibility value is reduced in the presence of one or more discordant criteria when $d_j(a', a) > c(a', a)$. This reduction is all the greater as $d_j(a', a)$ approaches 1. In conformity with the veto effect, $\sigma(a', a) = 0$ if $d_j(a', a) = 1$ for at least one criterion. More precisely, we have:

$$\sigma(a', a) = c(a', a) \,.\, \prod_{j \in D_c(a', a)} \frac{1 - d_j(a', a)}{1 - c(a', a)} \qquad (3.3)$$

with $D_c(a', a) = \{j \;/\; j \in F,\ d_j(a', a) > c(a', a)\}$.

(The explanations which justify this formula are presented in Roy and Bouyssou (1989); in the interest of brevity, we will not develop them again here).

3.3. Other ELECTRE Methods

Up to now, we have not mentioned either ELECTRE I or ELECTRE II. These methods have been supplanted by ELECTRE IS and ELECTRE III respectively. They are, nevertheless, still interesting from both a pedagogical and historical standpoint.

Let us remember that ELECTRE I (see Roy (1968)) was the first decision-aid method using the concept of outranking relation.

The idea of modulating the credibility of the outranking insertion was introduced in ELECTRE II (see Roy and Bertier (1973)) where two models of preferences are taken into account: the first one being relatively poor but strongly justified and the second one richer but less defensible.

ELECTRE IV is a method in which no k_j is introduced. This does not mean that each criterion has exactly the same «weight». ELECTRE IV is appropriate for cases in which we are not willing or able to introduce information on the specific role (i.e. importance) devoted to each criterion in the aggregation procedure. A sequence of nested outranking relations is introduced:

$$S^1 \subset S^2 \subset S^3 \subset S^4 \subset S^5.$$

Each S^i is defined by referring to concordance and discordance concepts (for an exhaustive definition of these five binary relations, see Roy and Bouyssou (1989)). An application of this method to a ranking problem of suburban line extension projects is presented in Roy and Hugonnard (1982).

Let us mention finally a new ELECTRE method (ELECTRE A, A for Assignment) which has been built to solve some specific problems in the banking sector. It is now used routinely but is not publishable for reasons of confidentiality. The general orientation is that indicated in Moscarola and Roy (1977) and Roy (1981).

4. SOME PRACTICAL CONSIDERATIONS

4.1. How to Use Outranking Models for Decision-Aid

Let us consider a comprehensive model of preferences defined on A. Let us suppose first that this model is nothing more than a single criterion g(a). The way to use it for decision-aid is quite obvious whatever the problem statement considered.

Table 2: Main characteristics of ELECTRE methods

ELECTRE methods	I	IS	II	III	IV	A
Possibility for taking into account indifference and/or preference thresholds	no	yes	no	yes	yes	yes
Necessity of a quantification of the relative importance of criteria	yes	yes	yes	yes	no	yes
Number and nature of outranking relations (1)	1	1	2	1 fuzzy	5	1 fuzzy
Problem statement	α	α	γ	γ	γ	β
Final results	a kernel	a kernel with consistency and connected indices	a partial pre-order	a partial pre-order	a partial pre-order	an assignment to predefined categories

(1) All outranking relations are based on concordance and discordance concepts; except for boxes containing «fuzzy», the figures refer to non-fuzzy relations.

As we have shown (see Roy (1985)), three basic problem statements P.α, P.β, P.γ must be distinguished. Briefly, we can characterize each of them by saying that decision-aid is envisaged according to the following perspective:

with P.α: isolate the smallest subset $A_0 \subset A$ liable to justify the elimination of all actions belonging to $A \setminus A_0$;

with P.β: assign each action to an appropriate pre-defined category according to what we want it to becomes afterwards;

with P.γ: build a partial (or complete) pre-order as rich as possible on a subset A_0 of those among the actions of A which seem to be the most satisfactory.

Let us suppose now that the comprehensive model of preferences is an outranking relation S (crisp or fuzzy) or a sequence of nested outranking relations. Contrary to the preceding case, the way to proceed in order to (α) isolate A_0, (β) assign each action to a predefined category, (γ) build a partial (or complete) pre-order (according to the problem statement chosen) is not obvious. This topic is discussed in the present book by D. Vanderpooten (1990). Let us emphasize here the fact that each ELECTRE method combines:

(i) a given problem statement P.α, P.β or P.γ,
(ii) a way of defining a comprehensive model of preference (see section 3 above).

4.2. How to Choose Among ELECTRE Methods

Before answering such a question, we invite the reader to consider table 2. This table summarizes the main characteristics by which ELECTRE methods can be differentiated. For selecting the most appropriate for a given decision-aid context, we suggest proceeding as follows. Consider first the problem statement chosen (see 4.1 above), then:

if P.α: two ELECTRE methods, ELECTRE I and ELECTRE IS, can be envisaged. ELECTRE I should be selected only if it is truly essential to work with a very simple method and it is realistic to have $p_j = q_j = 0 \ \forall j \in F$;

if P.β: there is presently no choice;

if P.γ: three methods, ELECTRE II, III and IV are in competition; ELECTRE II should be selected only if simplicity is required and $p_j = q_j = 0 \ \forall j \in F$ is realistic; ELECTRE IV is convenient only if there exists a good reason for refusing the introduction of importance coefficients k_j.

4.3. How to Give Numerical Values to Thresholds and Importance Coefficients

Let us remember that the indifference and preference thresholds q_j and p_j have been introduced (see 1.2 above) so as to be able to interpret correctly differences between performances. The simplest way for giving a numerical value to such thresholds consists in coming back to their definition (see Roy (1985, ch. 9)) and in analyzing the main sources of imprecision, uncertainty and inaccurate determination. For more details and a presentation of some more sophisticated techniques, see Bouyssou and Roy (1987) as well as Roy et al. (1986) for an illustration based on a concrete example, which has the advantage of using very diverse processes, given the variety of criteria in question.

In many cases, it is difficult and perhaps arbitrary to fix a precise numerical value for some of the q_j's and/or p_j's. We then can try to insert them between plausible minimum and maximum values. Let us emphasize the fact that:

(i) It is often easy to give to those thresholds a value different from 0 and less arbitrary than the value 0,

(ii) and it is not easier to try to take into account the different sources of imprecision, uncertainty and inaccurate determination (see Roy (1988)) by means of probabilistic distributions (as in MAUT): choosing the form of the distribution, or giving a numerical value to its diverse characteristics (mean value, standard deviation, minimum, maximum, ...) comprises an amount of arbitrariness as considerable as that involved in threshold evaluation.

We are confronted with similar difficulties for characterizing the specificity of the role devoted to each criterion by means of the importance coefficient k_j and the veto threshold v_j. Let us again underline that the k_j's are intrinsic, i.e. they do not depend on the nature of the scale chosen for evaluating performances. This intrinsic characteristic facilitates our examination [1] of the values we can appropriately attribute to these coefficients in order to reflect the relative importance a given decision-maker will assign to different criteria bearing in mind that his ideas concerning this are often rather vague. To do this, we have developed a questioning technique which is illustrated in detail in Roy et al. (1986).

We would like to emphasize that this technique is not designed to «estimate» the value of each k_j «with maximum precision». Indeed we consider that the very idea of estimating is without any basis at all here. The concepts of estimation and approximation refer, of course, to a quantifiable entity whose «real value» exists somewhere. Yet this «somewhere» can only be found in the mind of «someone», namely the decision-maker. We mentioned above (see 1.1 d)) that the latter is often difficult to identify because he is more or less mythical and when he is not, he is frequently not very accessible. When he is accessible, the idea

----

1. In the AHP and MAUT methods, the non-intrinsic character of the weights and substitution rates only further complicates the same examination.

he holds of each criterion's importance is, in most cases, neither formalised nor quantified. The role that each criterion could and should play in designing comprehensive preferences is not something factual which can be observed: it is, in the main, a reflection of a system of values, but also of more fragile opinions, which too detailed a discussion will disturb. This is why, in the questioning technique referred to above, we proceed by a comparison of actions which can be differentiated only through two or three of their performances, asking only qualitative questions. The questions may be asked simultaneously to several actors in order to set forth clearly the areas of consensus and of irreconcilable differences.

The result is a **domain** of values for k_j which are acceptable to a group of actors. This domain, by means of the non-restrictive hypothesis $k_1 \leq k_2 \leq \ldots \leq k_n$, is then re-written in the following form:

$$m_1(k_1) \leq k_2 \leq M_1(k_1)$$
$$m_2(k_1, k_2) \leq k_3 \leq M_2(k_1, k_2)$$
$$\ldots$$
$$m_{n-1}(k_1, \ldots, k_{n-1}) \leq k_n \leq M_{n-1}(k_1, \ldots, k_{n-1}).$$

It is in no way restrictive to put $k_1 = 1$. We can then easily explore the domain of validity for k_j's and deduce from it (when, for certain k_j's the variation interest $[m_{j-1}, M_{j-1}]$ is large) a small number of contrasting sets.

Finally, the numerical value of each veto threshold v_j should be discussed on the basis of its definition (see 2.3 c)). It is often more enlightening to base our reasoning on the v_j/p_j ratio rather than on v_j alone. It is especially important to compare the way in which the criteria are ranked according to the decreasing values of this ratio to the way in which they are ranked according to increasing values of k_j. When there are differences, it is important for them to be based on clear explanations (for example: the desire to make a criterion which is not too important, nonetheless bring the veto into play for a small difference in performance). Here again, if it is difficult to give a precise value to a v_j threshold, we can assign it a variation interval.

It emerges from the preceding that it is not always possible, for the thresholds or for the importance coefficients, to match each one to a well-defined numerical value. Each time we assign a non-negligible amplitude to one or another of these parameters, it is important to explore the effect of this lack of determination. To do so, we can start by applying the ELECTRE method selecting, adopting for each threshold and importance coefficient the value which corresponds to the middle of the interval. Then we must take different combinations of extreme values into consideration. We can thus study the robustness of our conclusions in relation to incompressible margins of arbitrariness which this lack of determination intervals reflect. Examples of such analyses of robustness can be found in Roy and Hugonnard (1982), Renard (1986), Roy et al. (1986).

5. CONCLUSION

It seems important to us to draw attention to the fact that the difficulties mentioned above involve the assigning of numerical values essential to characterizing comprehensive preference models. Such difficulties are in no way specific to multicriteria aggregation procedures of the ELECTRE type. Difficulties of the same kind may be encountered in one way or another in all forms of modelling. We believe that they are inherent in all decision problems. It is important for the approach we adopt not to hide these difficulties into shadow but rather to highlight them (see Roy (1990) and Bouyssou (1988)). The analysis of robustness should, therefore, play a central role in developing a prescription, whatever the type of modelling adopted. One of the advantages of the ELECTRE methods is precisely that they make such robustness analysis particularly easy.

REFERENCES

Abgueguen, R. (1971): *La Sèlection des Supports de Presse*, Robert Laffon.

Barba, H., Dupuy, J. and Lencioni, P. (1989): "Multicriteria location of thermal power plants", *European Journal of Operational Research* (to appear).

Bétolaud, Y. and Février, R. (1973): "Conservation des forêts suburbaines et passage des autoroutes - l'exemple du tracé de l'autoroute A86-A87", *Revue Forestière Française*, 179-200.

Bouyssou, D. (1988): "Modelling inaccurate determination, uncertainty and imprecision using multiple criteria", Université de Paris-Dauphine, *Cahier du LAMSADE* n° 88, 22 p.

Bouyssou, D. (1990): "Building criteria: a prerequisite for MCDA", in this volume.

Bouyssou, D. and Roy, B. (1987): "La notion de seuils de discrimination en analyse multicritère", *INFOR*, vol. 25, nº 4, 302-313.

Brans, J.P. and Mareschal, B. (1990): "The PROMETHEE Methods for MCDM: The PROMCALC, GAIA and BANKADVISOR Software", in this volume.

Brans, J.P., Mareschal, B. and Vincke, Ph. (1984): "PROMETHEE: A new family of outranking methods in multicriteria analysis", in J.P. Brans (ed.), *Operational Research '84*, Elsevier Science Publishers B.V. (North-Holland), 408-421.

Buffet, P., Grémy, J.P., Marc, M. and Sussmann, B. (1967): "Peut-on choisir en tenant compte de critères multiples? Une méthode (ELECTRE) et trois applications", *Revue METRA*, Vol. 6, nº 2, 283-316.

Charpentier, A.R. and Jacquet-Lagrèze, E. (1976): "La promotion de l'électricité dans l'industrie et l'utilisation de méthodes multicritères", in H. Thiriez and S. Zionts (eds), *Multiple Criteria Decision Making*, Lecture Notes in Economics and Mathematical Systems 130, Springer-Verlag, Berlin, 364-377.

Clímaco, J., Martins, A. and Traça-Almeida, A. (1988): "On a multicriteria based approach for energy planning", Communication to the *EURO IX-TIMS XXVIII Congress*, Paris, July 6-8.

Jacquet-Lagrèze, E. and Siskos, J. (1983): *Mèthode de Dècision Multicritère*, Monographies de l'AFCET, Division Gestion Informatisation Decision, Editions Hommes et Techniques, Paris.

Keeney, R.L. and Raiffa, H. (1976): *Decisions with Multiple Objectives, Preferences and Value Tradeoffs*, John Wiley and Sons, New York.

Laffy, R. (1966): "La méthode MARSAN pour la recherche de produits nouveaux", Communication to the *ESOMAR Congress*, Copenhague, September.

Martel, J.M. and Nadeau, R. (1988): "Revealed preference modeling with ELECTRE II, an interactive approach", Communication to the *EURO IX-TIMS XXVIII Congress*, Paris, July 6-8.

Maystre, L.Y. and Simos J. (1987): "Comment pourrait-on gérer les déchets solides urbains de Genève?", in *Actes du Colloque AFCET sur le Dèveloppement des Sciences et Pratiques de l'Organisation*, Thème 1987, "L'Aide à la Décision dans l'Organisation", Paris, 253-258.

de Montgolfier, J. and Bertier, P. (1978): *Approche Multicritère des Problèmes de Dècision*, Editions Hommes et Techniques, Paris.

Moscarola, J. and Roy, B. (1977): "Procédure automatique d'examen de dossiers fondée sur une segmentation trichotomique en présence de critères multiples", *RAIRO Recherche Opèrationnelle*, vol. 11, n° 2, 145-173.

Parent, E. and Schnäbele, P. (1988): "Le choix d'un aménagement aquacole - Exemple d'utilisation de la méthode ELECTRE III et comparaison avec d'autres méthodes multicritères d'aide à la décision", Université de Paris-Dauphine, *Document du LAMSADE* n° 47, 68 p.

Rabeharisoa, V. (1988): "Une application de l'analyse multicritère pour l'évaluation d'alternatives: technologies propres et curatives de la pollution industrielle", Université de Paris-Dauphine, *Cahier du LAMSADE* n° 85, 37 p.

Renard, F. (1986): "Utilisation d'ELECTRE dans l'analyse des réponses à un appel d'offres, Le cas de la machine de tri paquets à la Direction Générale des Postes", Université de Paris-Dauphine, *Cahier du LAMSADE* n° 73, 36 p.

Roberts, F.S. (1979): *Measurement theory with Applications to Decision-making, Utility and the Social Sciences*, Addison-Wesley, London.

Roy, B. (1968): "Classement et choix en présence de points de vue multiples (la méthode ELECTRE)", *RIRO*, n° 8, 57-75.

Roy, B. (1981): "A multicriteria analysis for trichotomic segmentation problems", in P. Nijkamp and J. Spronk (eds), *Multiple Criteria Analysis, Operational Methods*, Gower Press, Aldershot, 245-257.

Roy, B. (1985): *Méthodologie Multicritère d'Aide à la Décision*, Economica, Paris.

Roy, B. (1988): "Main sources of inaccurate determination, uncertainty and imprecision in decision models", in B.R. Munier and M.F. Shakun (eds), *Compromise, Negotiation and Group Decision*, D. Reidel Publishing Company, Dordrecht, 43-62.

Roy, B. (1990): "From decision-making to decision-aid", in this volume.

Roy, B. and Bertier, P. (1973): "La méthode ELECTRE II - Une application au média-planning", in M. Ross (ed), *OR '72*, North-Holland Publishing Company, Amsterdam, 291-302.

Roy, B. and Bouyssou, D. (1987a): "Famille de critères, Problème de cohérence et de dépendance", Université de Paris-Dauphine, *Document du LAMSADE* n° 37, 73 p.

Roy, B. and Bouyssou, D. (1987b): "Conflits entre critères et pro cédures élémentaires d'agrégation multicritère", Université de Paris-Dauphine, *Document du LAMSADE* n° 41, 75 p.

Roy, B. and Bouyssou, D. (1989): "Procédures d'agrégation multicritères non fondées sur un critère unique de synthèse", Université de Paris-Dauphine, *Document du LAMSADE* (to appear).

Roy, B. and Hugonnard, J.C. (1982): "Ranking of suburban line extension projects on the Paris metro system by a multicriteria method", *Transportation Research*, Vol.16A, 4, 301-312.

Roy, B., Présent, M. and Silhol, D. (1986): "A programming method for determining which Paris metro stations should be renovated", *European Journal of Operational Research*, 24, 318-334.

Roy, B. and Vincke, Ph. (1987): "Pseudo-orders: Definition, properties and numerical representation", *Mathematical Social Sciences*, 14, 263-274.

Saaty, T.L. (1980): *The Analytic Hierarchy Process*, McGraw-Hill, New York.

Schärlig, A. (1985): *Décider sur Plusieurs Critères - Panorama de l'Aide à la Décision Multicritère*, Presses Polytechniques Romandes, Lausanne.

Slowinski, R. and Treichel, W (1988): "MCDM methodology for regional water supply system programming", Communication to the *EURO IX-TIMS XXVIII Congress*, Paris, July 6-8.

Vansnick, J.C. (1986): "On the problem of weights in MCDM (the noncompensatory approach), *European Journal of Operational Research*, 24, 288-294.

THE CONSTRUCTION OF PRESCRIPTIONS IN OUTRANKING METHODS

Daniel Vanderpooten

LAMSADE, Université de Paris Dauphine
Place du Maréchal De Lattre de Tassigny
75775 Paris Cedex 16 - FRANCE

1. INTRODUCTION

Multiple Criteria aggregation methods are designed to construct a prescription (a solution) from a set of alternatives in accordance with the preferences of a decision maker (DM) or a group of DMs. In many approaches, the prescription is immediately derived from the aggregation process. However, when the aggregation process is based on the outranking approach, additional treatments are required to construct the prescription. More precisely, it is customary to distinguish two basic stages in outranking methods:

- construction of one (or several) outranking relation(s) for modelling the DM's preferences,
- exploitation of this (or these) outranking relation(s) in order to construct the prescription according to a specific problem statement.

After presenting some introductory concepts about crisp and fuzzy outranking relations and their respective representations through graphs (section 2), we distinguish and illustrate three basic types of prescriptions related to three ways of stating a decision problem (section 3). The reasons why outranking methods do not allow a direct construction of prescriptions are then discussed (section 4). The relationships between crisp and fuzzy outranking relations are briefly discussed so as to simplify our presentation (section 5). The three following sections are devoted to concepts and procedures used to exploit outranking relations so as to derive each type of prescription.

We focus our attention on Electre methods in which most of these concepts were introduced. Precise details of implementation are omitted; the interested reader should consult the references indicated. Concluding remarks are presented in the last section.

2. INTRODUCTORY CONCEPTS

2.1. Outranking Relations

We recall in the first part of this section some basic concepts concerning outranking relations and the way they are established. For more details about these topics, the reader should consult Roy (1990) in this book.

We consider in the following:
- a finite set A of potential actions (or alternatives),
- a consistent family of n criteria g_j (j=1,...,n) (see Bouyssou (1990)).

An outranking relation S is a binary relation such that, considering potential actions a and b, a S b holds if it is reasonable to accept, from the DM's point of view, that «a is at least as good as b». It should be clear from this definition that a situation of indifference between a and b (denoted as a I b) will be represented by a S b and b S a.

The construction of an outranking relation S on A should be conceived as a way of modelling the part of the DM's preferences which can be established with sufficiently strong arguments. In other words, S may be seen as an extension of the dominance relation Δ defined by: a Δ b iff $g_j(a) \geq g_j(b)$ (j=1,...,n).

Arguments to construct S are expressed through conditions reflecting natural basic principles. More precisely, the assertion a S b is accepted if, when comparing a with b, the two following conditions hold:

- a *concordance condition* which ensures that a majority of criteria are concordant with a S b (majority principle),
- a *non discordance condition* which ensures that none of the discordant criteria strongly refutes a S b (respect of minorities principle).

The precise implementation of these conditions in the case of the ELECTRE methods is described in Roy (1990) where some fundamental properties of outranking relations are also presented. For our purpose, it is particularly important to notice that S is not necessarily transitive, which means that the acceptance of a S b and b S c does not entail by itself a S c. Actually a S c will be accepted if and only if concordance and non discordance conditions, regarding the ordered pair (a, c), are verified. Such intransitivities are frequently observed in practice (see *e.g.* Tversky (1969)). It is then important to allow them when modelling preferences. Another significant property is that S is usually not complete. When modelling preferences using outranking relations, if no argument can be found (*i.e.* if the above conditions are not satisfied) to support a S b or b S a, we may only conclude that «a is incomparable with b» (denoted as a R b). Introducing this situation of incomparability is surely more adequate, from the modelling viewpoint, than enforcing situations of preference or indifference. These properties are certainly among the major distinctive features of outranking methods. They will be shown of great importance for the exploitation of outranking relations.

It is clear from the description of the concordance and non discordance conditions that they may be expressed in a more or less demanding way, which results in more or less strongly established outranking relations. It is then interesting to observe the evolution of the outranking relation when different levels of requirement are specified. Two options are available:

- make these levels explicit. This results in the construction of r ($r \geq 1$) nested outranking relations

$$S_1 \subset S_2 \subset \ldots \subset S_r$$

corresponding to weaker and weaker levels of requirement regarding the assertion «a is at least as good as b» (and to richer and richer relations);

- keep these levels implicit. This results in the construction of a fuzzy outranking relation through a credibility index σ, indicating for each ordered pair (a, b) the level of credibility $\sigma(a, b)$ ($0 \leq \sigma(a, b) \leq 1$) of the assertion «a is at least as good as b».

2.2. Graph Representations of Outranking Relations

It is well-known that any binary relation may be given an equivalent mathematical representation through a directed graph *i.e.* a graphical figure consisting of points - called *vertices* - and oriented lines - called *arcs* - joining some of these points. A graph G is then fully described by the doublet (X, U), where X denotes the set of vertices and U the set of arcs.

The way of building graph representations from crisp (unfuzzy) outranking relations and fuzzy outranking relations are now respectively introduced.

- In the case of a sequence of r crisp outranking relations, each relation S_i is associated with a graph

$$G_i = (A, U_i) \qquad (i=1,\ldots,r)$$

where $(a, b) \in U_i$ if and only if $a\ S_i\ b$.

$G_i = (A, U_i)$ is also conveniently denoted by $G_i = (A, S_i)$. Loops, which represent obvious self-outrankings, are omitted.

Figure 2.1 : Graph representation of a crisp outranking relation S_i

$A = \{a_1, a_2, a_3, a_4\}$ $a_1 \; S_i \; a_2, \; a_2 \; S_i \; a_3, \; a_3 \; S_i \; a_2$

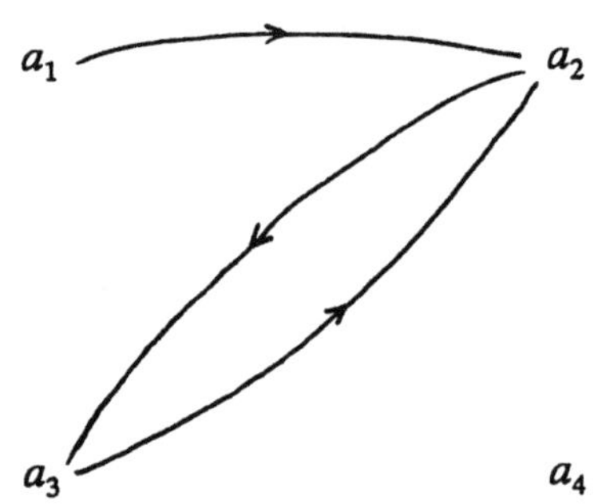

- In the case of one fuzzy outranking relation σ, a complete and symmetric graph G_σ is defined on A (G_σ = (A, A x A)), each arc (a, b) being valued by σ(a, b).

Here again, loops, whose valuations are equal to one, are omitted.

Figure 2.2 : Graph representation of a fuzzy outranking relation

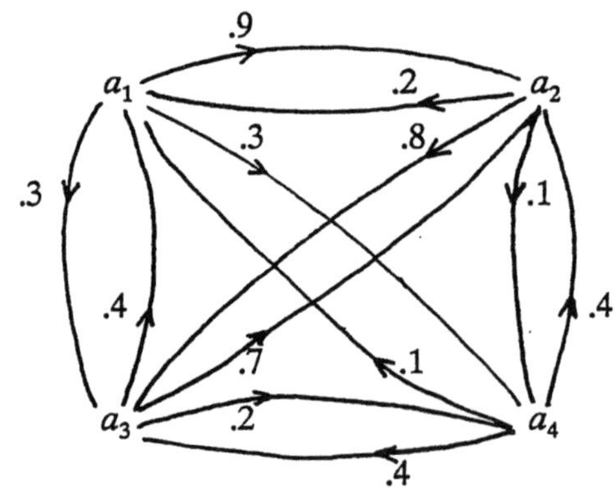

The advantage of such representations through graphs is that they provide a comprehensive and appealing visualization of the preference model. In that respect, considering a crisp relation (possibly derived from a fuzzy relation - see section 5), it is particularly interesting to examine the structure of the corresponding graph (connexity, existence of circuits,...). Moreover, the recourse to graph representations allows to borrow some useful concepts and results from Graph Theory so as to make use of outranking relations in order to construct prescriptions.

3. THREE BASIC TYPES OF PRESCRIPTIONS

The aim of a MCDA study is to provide the DM with guidelines with respect to his/her decision problem. Such guidelines often include a **prescription,** *i.e.* a proposal concerning the decision to be made. This prescription depends on the nature of the decision problem and on the way it is stated by the analyst. We shall distinguish with Roy (1981) three basic problem statements:

- **Choice problem statement (P.α):**
  select a subset, as restricted as possible, containing the most satisfactory actions.

This is the most common formulation which usually consists in stating the problem in terms of the «best choice». Ideally, this choice should be reduced to one action. However, because of the imprecision of the data, the conflicts between criteria, the existence of several relevant value systems,..., it is often more adequate to present the DM with a few actions which represent possible variants of this «best choice» (the selection may be refined afterwards using additional information or further analyses). This problem statement is clearly adapted to selection problems (selection of a location for a plant, selection of candidates for a unique post,...). Formally, a prescription is a subset $A^0 \subset A$.

- **Sorting problem statement (P.β):**

assign each action into pre-defined categories.

This is the relevant formulation when the decision problem consists in examining the *intrinsic* value of each action so as to propose an appropriate recommendation among a set of possible recommendations specified in advance. Each recommendation may be associated with a category. The decision problem is then viewed as sorting potential actions into categories which are defined according to pre-established norms. In many cases, categories are ordered and correspond to recommendations such like «surely accepted», «acceptable», «rejected». Because of the imprecision of the data, the conflicts between criteria, the existence of several relevant value systems,..., it is often more adequate to define additional categories corresponding to hesitations between several categories. Typical problems suited for such a formulation are diagnosis problems where actions are considered one by one (decision for granting credits, medical diagnosis,...). The sorting procedure must be designed so as to assign each action to one and only one category. Formally, a prescription is a partition defined on A.

- **Ordering problem statement (P.γ):**

rank each action by decreasing order of preference.

This is the usual formulation when potential actions are to be differentiated according to their relative interest. Ideally, the ranking which is looked for should be complete. However, because of the imprecision of the data, the conflicts between criteria, the existence of several relevant value systems,..., it is often more adequate to present the DM with a partial ranking which reflects some of the irreducible incomparabilities. Problems where a specified number of actions must be retained are typical ordering problems (selection of candidates for n similar posts, selection of several projects within the limits of a specified budget,...). It should be noticed that, in practice, this ranking may be required only for a subset of the most interesting actions. Formally, a prescription is a partial preorder, *i.e.* a transitive relation, defined on A (or on a subset of A).

4. THE CONSTRUCTION OF PRESCRIPTIONS

We briefly investigate in this section the way prescriptions are constructed in the various multiple criteria approaches. This discussion allows a better understanding of the specific problems in the case of outranking methods.

Considering the way prescriptions are constructed, it is interesting to distinguish:

- approaches which present the DM with «potential prescriptions» progressively adjusted by taking into account preference information supplied in an interactive way,
- approaches which aim at modelling the DM's preferences and exploit the resulting model to derive a prescription.

In the first type of approach, referred to as the interactive approach, no formal model of preferences is built. Prescriptions are directly constructed on the basis of preference information supplied by the DM. From a technical viewpoint, interactive procedures may then be seen as procedures for constructing prescriptions (Vanderpooten (1989)). The concepts and techniques used (based on the optimization of scalarizing functions) are very different from those which will be introduced. Moreover, interactive procedures are mainly oriented towards the choice problem statement.

The second type of approach includes methods modelling preferences through a value function (Multi-Attribute Utility Theory (see *e.g.* Keeney and Raiffa (1976), Analytic Hierarchy Process (Saaty (1980))) and methods modelling preferences through a system of binary relations (Outranking methods). In both cases, the prescription results from the exploitation of the preference model. Insofar as this model is primarily designed to represent the DM's preferences, there is no reason for it to be naturally adapted to derive in a simple way a partition on A, a preorder on A or even a subset of the most satisfactory actions.

Actually, such reasons do exist if the preference model is a value function V ($V : A \to \mathbb{R}$). Let us examine these reasons and the way prescriptions are derived from V. Beforehand, the reader should notice that constructing V and assigning a real value V(a) to each action a amounts to imposing transitivity and completeness, *i.e.* assuming a strong consistency, on the DM's preference structure. We indicated in section 2.1 that imposing these properties may not be relevant when modelling preferences. However, this allows a direct exploitation of V in order to derive prescriptions:

- **P.α case:** select a^* such that $V(a^*) = \max_{a \in A} V(a)$ (or a subset of actions whose values are close to $V(a^*)$).
- **P.β case:** define categories by specifying limits which partition the range of variation of V into intervals and assign each candidate action a to the category whose corresponding interval contains V(a).
- **P.γ case:** construct the complete preorder on A induced by V.

It is a different matter if the model consists of outranking relations. The conditions used to establish outranking relations (concordance and non discordance conditions) are oriented towards a realistic and prudent modelling of preferences. The resulting model, even if more reliable, is generally poorer than V. It is then often only possible to produce «partial prescriptions», *i.e.* a «best choice» consisting of more than one action (P.α), a partition where the categories corresponding to hesitations are nonempty (P.β), a noncomplete preorder (P.γ). Moreover, because of possible intransitivities, it may even be difficult to derive such partial prescriptions. It should be clear that, in this case, some modifications must be performed on the original relations so as to obtain the required prescription (which is quite obvious in the P.γ case where the required result is a transitive relation). The main point consists then in finding acceptable ways of handling these intransitivities without losing too much on the reliability of the original modelling.

5. EXPLOITATION OF CRISP AND FUZZY OUTRANKING RELATIONS

Two types of modelling were introduced in section 2.1: either through a sequence of crisp and nested outranking relations or through one fuzzy outranking relation. Exploitation procedures and concepts should then be adapted to each case. However, as pointed out by Roy (1990), these types of modelling are nearly equivalent from a theoretical viewpoint, the only difference being that r levels are pre-specified in the first type. Consequently, adaptations amount to converting a fuzzy relation to crisp relations or vice-versa (taking some minor precautions).

More precisely, it is possible to derive a sequence of r crisp and nested outranking relations S_i (i=1,...,r) from a fuzzy outranking relation σ by choosing r values $v_1 > v_2 > \ldots > v_r$ ($v_i \in [0, 1]$) and accepting a S_i b if $\sigma(a, b) \geq v_i$. For instance, the crisp outranking relation presented in Fig. 2.1 may be seen as a relation S_i derived from the fuzzy outranking relation presented in Fig. 2.2 by imposing $v_i = 0.7$. The choice of the values v_i may be performed using a formula like $v_i = (r - i)/r$ (i=1,...,r). We should notice that this process may involve discriminations between similar situations (*i.e.* situations where we would get a S_i b and not(c S_i d) whereas the difference $\sigma(a, b) - \sigma(c, d)$ is not significant). Considering this restriction, it is technically possible to apply an exploitation procedure designed for the crisp case to the fuzzy case.

Conversely, it is possible to derive a fuzzy outranking relation σ' from a sequence of r crisp and nested outranking relations S'_i (i=1,...,r) by defining r+1 values $v'_1 \geq v'_2 \geq \ldots \geq v'_r \geq v'_{r+1}$ ($v'_i \in [0, 1]$ and usually $v'_{r+1} = 0$) such that:

- $\sigma(a, b) = v'_{r+1}$ if there is no i such that a S'_i b
- $\sigma(a, b) = v'_i$ if i is the lowest index such that a S'_i b.

Any order preserving formula may be used to define v'_i (*e.g.* $v'_i = (r - i + 1)/r$ $(i=1,\dots,r+1)$). However, the resulting fuzzy relation should be exploited in an ordinal way. Considering this restriction, it is technically possible to apply an exploitation procedure designed for the fuzzy case to the crisp case.

In the following, concepts and procedures used to exploit outranking relations are presented as they were originally introduced in the literature. Their transposition from the crisp case to the fuzzy case or vice versa, when it is of interest, may be achieved easily using the above conversion techniques.

6. EXPLOITATION OF OUTRANKING RELATIONS IN THE P.α CASE

Considering this problem statement, the exploitation procedure must be designed so as to produce a subset A^0 which verifies the two following general requirements :

- A^0 contains the most satisfactory actions, (r.6.1)
- $|A^0|$ is as small as possible. (r.6.2)

We first deal with the case where the preference model only consists of one crisp outranking relation S. Other cases are introduced afterwards. Before presenting any concept or procedure, we must state requirement (r.6.1) in a more precise and operational way. Many operational restatements are possible resulting in various concepts and procedures. We shall restrict our attention to a restatement consistent with the outranking approach in the sense that it emphasizes a prudent attitude. This means that, when characterizing A^0, we must be able, at least, to justify the elimination of any other alternative. We may then restate (r.6.1) as follows:

- $\forall\ b \in A \setminus A^0,\ \exists\ a \in A^0$ such that $a\ S\ b$ (r'.6.1)

Any subset satisfying (r'.6.1) may be seen as a subset which, as a whole, dominates - or more exactly outranks - the set A. In Graph terms, such a subset is referred to as a *dominating subset* (condition (r'.6.1) is also called *external stability*).

6.1. Exploitation of One Crisp Outranking Relation: The Transitive Case

Let us show that a very simple exploitation procedure may be designed when S is transitive. In this case, A can be partitioned in a unique way into equivalence classes C_i which regroup mutually indifferent actions. Considering the graph G, each equivalence class corresponds to a complete and symmetric subgraph called an *indifference clique*. By reducing each indifference clique to a vertex and joining the resulting vertices according to the original arcs of U, we obtain a *reduced* graph H = (C, V) where C is the set of equivalence classes and $(C_i, C_j) \in V$ iff $(a_i, a_j) \in U$ for some $a_i \in C_i$ and some $a_j \in C_j$ (see example in Fig. 6.1). Actually, because of the transitivity of S, the existence of an arc (C_i, C_j) in H is fully justified by the existence in G of the arcs (a_i, a_j) for *any* $a_i \in C_i$ and *any* $a_j \in C_j$. This clearly shows that H is an exact representation of the preference model and may be used as a basis for deriving the prescription. Since H has no circuit, it is possible to distinguish vertices without any predecessor. The corresponding classes certainly contain the most satisfactory actions. Indeed, since H is transitive, the prescription consisting of such actions satisfies (r'.6.1). Moreover, none of these actions is outranked by an eliminated action. However, in order to satisfy (r.6.2) in a strict sense (while still verifying (r'.6.1)), we may define A^0 by arbitrarily selecting one action from each equivalence class without any predecessor. In our example (see Fig. 6.1), we have *indifferently* $A^0 = \{a_1, a_6\}$ or $A^0 = \{a_4, a_6\}$.

Figure 6.1: The transitive case original graph and reduced graph

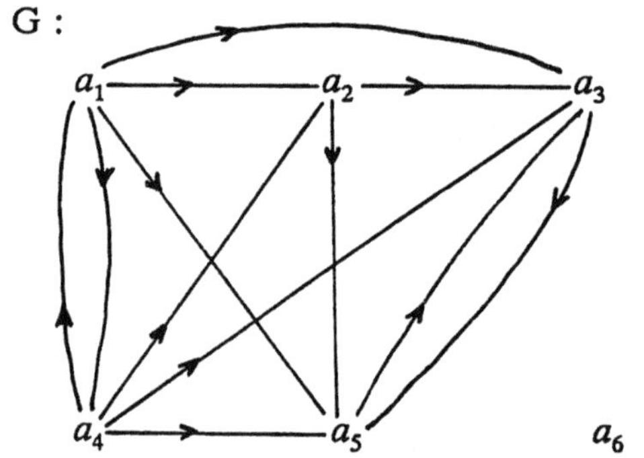

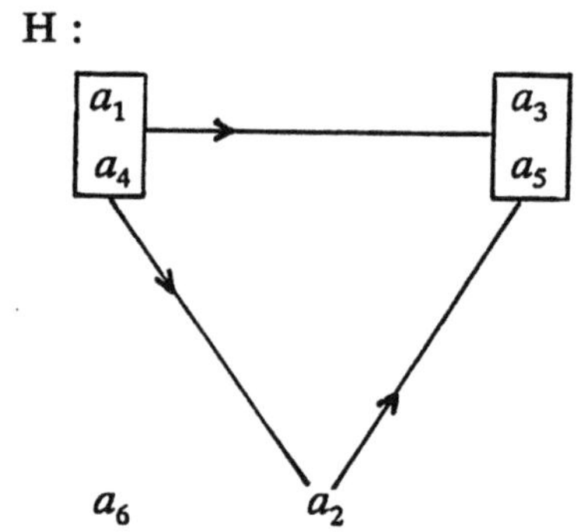

6.2. Exploitation of One Crisp Outranking Relation: The General Case

In the usual case where S is not transitive, the above exploitation procedure is not always possible (once the indifference cliques are reduced, H may still contain circuits) and, when it is possible, the resulting prescription may not satisfy (r'.6.1). An interesting approach consists in directly defining a subset which satisfies (r'.6.1), *i.e.* a dominating (or S-dominating) subset. We should notice that such a subset always exists (A is always a S-dominating subset of A). Thus, what we are looking for is a S-dominating subset which is the smallest in a sense. We are then interested in a *minimal S-dominating subset*, *i.e.* a S-dominating subset such that none of its proper subsets is a S-dominating subset. In example 1 (Fig. 6.2), even if $\{a_1, a_2, a_3, a_4\}$, $\{a_1, a_2, a_3\}$ and $\{a_1, a_2, a_4\}$ are S-dominating subsets, the only minimal S-dominating subset is $\{a_1, a_2\}$.

The concept of minimal S-dominating subset clearly satisfies (r'.6.1) and (r.6.2). Moreover, the prescription A^0 obtained in the transitive case is always a minimal S-dominating subset. Before concluding that we identified the appropriate concept, the reader should consider example 2 (Fig. 6.3) and try to find the minimal S-dominating subset.

Figure 6.2: Example 1

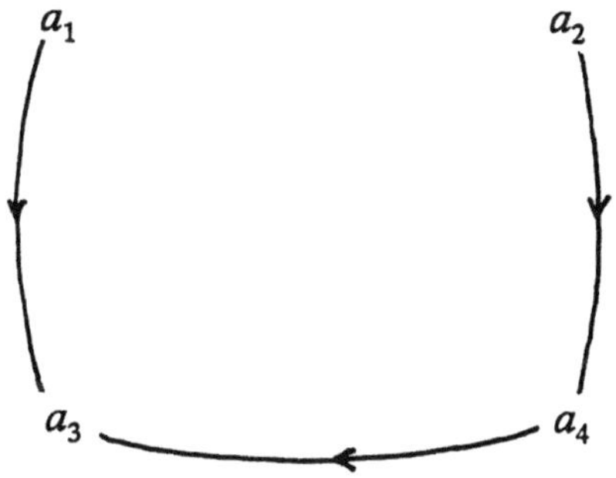

Figure 6.3: Example 2

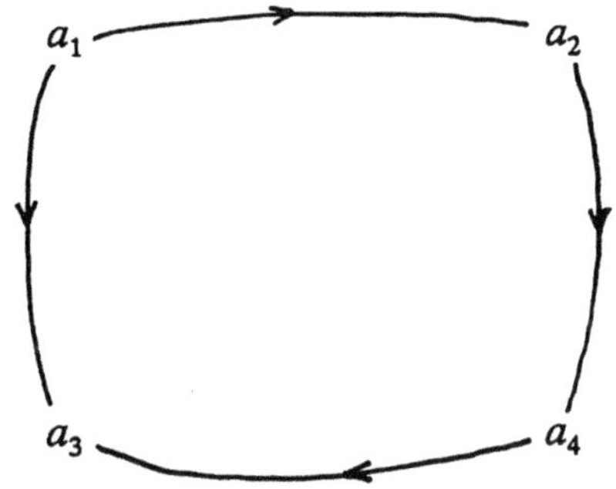

In this case, there exists two minimal S-dominating subsets $\{a_1, a_2\}$ and $\{a_1, a_4\}$. Unlike in the transitive case, where there may exist several, but indifferent, prescriptions, we cannot indifferently propose one of the minimal S-dominating subset. In order to handle this non-uniqueness problem, we may impose additional desirable requirements. Considering again example 2, it may result difficult to justify $\{a_1, a_2\}$ as the prescription insofar as a_1 S a_2 (we might then be induced to propose $\{a_1\}$ but this is not a S-dominating subset). Thus, we may restrict our attention to minimal S-dominating subsets whose actions are pairwise-incomparable. Formally, this requirement is stated as:

- $\forall$ a, a' $\in A^0$ we have not (a S a') and not(a' S a) (r.6.3)

In Graph terms, a subset verifying (r.6.3) is referred to as an *independent subset* (condition (r.6.3) is also called *internal stability*). A subset which is both dominating and independent (verifying (r'.6.1) and (r.6.3)) is called a *kernel*. Because of internal stability, a kernel is also a *minimal* dominating subset (verifying (r.6.2)). The use of the kernel concept was originally introduced in ELECTRE I (Roy (1968)).

The kernel concept clearly improves the exploitation of non-transitive relations (in example 2 (Fig. 6.3) it produces a unique prescription $\{a_1, a_4\}$). In some cases, however, basic difficulties remain since G may admit no or several kernels (see Fig. 6.4).

Figure 6.4: Examples of graphs admitting no or several kernels

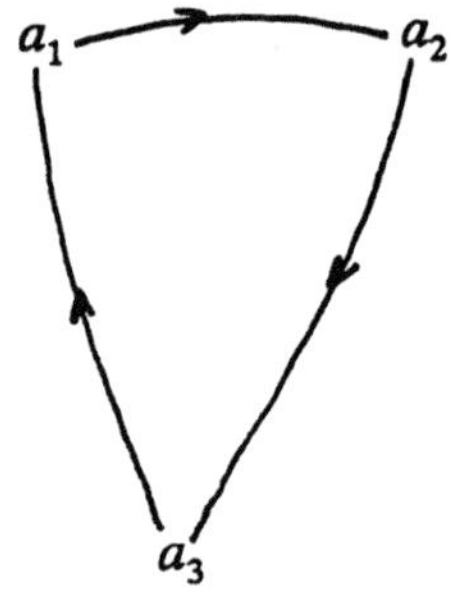

No kernel

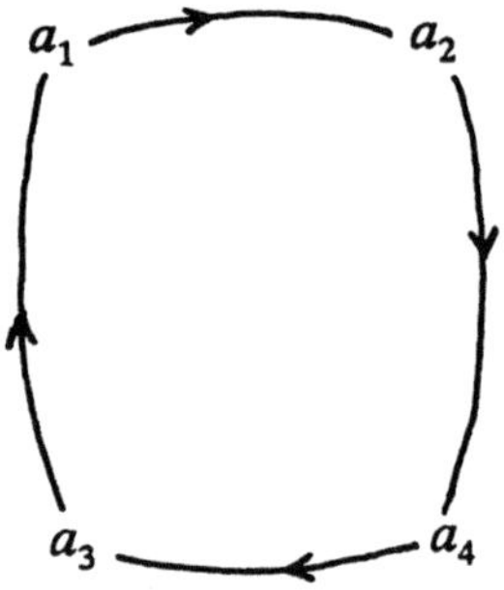

Two kernels : $\{a_1, a_4\}$ and $\{a_2, a_3\}$

Let us examine more closely this type of difficulties. Graph Theory provides us with an important result through the following theorem : *if G has no circuit, then there exists a unique kernel*. This theorem gives a direct answer in some cases. Above all, it allows to identify the type of intransitivities which causes problems: *the existence of circuits*. The reader should be aware that this is an irreducible difficulty (he/she might be convinced by considering examples in Fig. 6.4 and trying to find out a subset corresponding to an appropriate prescription). As indicated at the end of section 4, we are then obliged to perform some - reasonable - modifications on the original modelling when faced with circuits.

The existence of circuits may be handled by considering their impact in the transitive case. If S is transitive, the actions which belong to a same circuit form an indifference clique. We have shown that the reduction of G and its exploitation are then fully justified. The basic idea, in the general case, consists in judging whether or not actions which belong to a same circuit may be considered as mutually indifferent. This judgment may be based on indices reflecting the density of the corresponding subgraph and the credibility of missing arcs (intuitively, the more dense the subgraph, the more relevant the approximation to an indifference clique - see Roy (1989) for more details).

Circuits are then treated as follows:

Considering in G actions which belong to a maximal circuit (*i.e.* a circuit which is not strictly contained in any other circuit) and which do not form an indifference clique:

- if these actions may be considered as mutually indifferent, complement the corresponding subgraph to form an indifference clique;
- otherwise, suppress the circuit by eliminating one of the arcs which represent the less credible outrankings.

By iteratively applying the above procedure on G = (A, U), we obtain a new graph $G' = (A, U')$ such that any circuit in G' forms an indifference clique. It is then possible to partition A into equivalence classes C_i and to derive from G' a reduced graph $H' = (C, V')$ (see section 6.1). However, unlike in the transitive case, arcs of H' may not be fully justified : $(C_i, C_j) \in V'$ implies the existence of at least one arc $(a_i, a_j) \in U'$ (for some $a_i \in C_i$ and some $a_j \in C_j$), but all possible arcs (a_i, a_j) (for ***any*** $a_i \in C_i$ and ***any*** $a_j \in C_j$) do not necessarily exist in G'. Notice however that there is no «contradictory» arc of the (a_j, a_i) type in G'. Any arc $(C_i, C_j) \in V'$ which is insufficiently justified (considering the credibility of missing arcs (a_i, a_j) - see Roy (1989)), is suppressed. We obtain a graph $H'' = (C, V'')$ (with $V' \subset V''$) not necessarily transitive but without any circuit. H'' may be taken as an approximate representation of the preference model. The prescription is then constructed by determining the kernel of H'' and selecting one representative action from each retained equivalence class.

This exploitation procedure is used in Electre IS. Implementation details will be found in Roy and Skalka (1984) and Roy (1989).

6.3. Exploitation of Several Crisp and Nested Relations or of One Fuzzy Relation

Suppose now that the preference model consists of r (r > 1) crisp and nested outranking relations $S_1 \subset S_2 \subset \ldots \subset S_r$ (possibly derived from a fuzzy relation - see section 5). The exploitation procedure consists in first exploiting relation S_1 using the basic procedure described in the previous section. Let K_1 be the prescription (kernel) obtained by applying the basic procedure to $G_1' = (A, S_1)$. If K_1 is restricted enough, we may take it as the final prescription A_0. Otherwise, we may try to analyse K_1 so as to get a refined prescription. Because K_1 is an independent subset of A considering S_1, we cannot use again information from S_1 to refine K_1 (the graph (K_1, S_1) contains no arc). If we wish to refine K_1, we must accept to take more risks by using S_2. Then, we can get a new prescription $K_2 \subset K_1$ by applying the basic procedure to the graph $G_2' = (K_1, S_2)$. This process is repeated until a sufficiently restricted prescription is obtained.

This process which consists in progressively refining the prescription by successively considering relations $S_1,\ldots,S_k$ will generally result in a prescription distinct from that obtained by directly applying the basic procedure to the graph (A, S_k). The first approach seems preferable since it allows a better control on the successive eliminations and does not mix different levels of credibility.

7. EXPLOITATION OF OUTRANKING RELATIONS IN THE P.β CASE

Considering this problem statement, the prescription which is looked for is a partition on A into k pre-defined categories $A^1,\ldots,A^k$. Unlike in the P.α and P.γ cases, procedures for constructing prescriptions are not based on the exploitation of a preference model defined on A: there is no comparison between actions of A, but an evaluation of each action from its intrinsic value. The evaluation process consists in testing each candidate action a so as to assign it in one pre-defined category.

This test will be performed by comparing a with some reference actions which represent the limits of each category.

The procedure for constructing prescriptions must then include two steps:

-definition of the categories,

-construction of a sorting procedure.

7.1 Definition of the Categories

When evaluating a candidate action a so as to decide if it belongs to category A^i, it is natural to compare a with limits which represent:

- the lowest levels compatible with the assignment into A^i,
- the highest levels compatible with the assignment into A^i.

We illustrate the process of defining these levels in a simplified case where the basic recommendations are «accepted», «rejected» (*e.g.* the case of credits granting). So as to determine conditions of acceptance, we may ask the DM to specify on each criterion g_j the lowest level l_j compatible with this recommendation. The value l_j must be specified by simultaneously taking into account the values given to the other criteria, so as to define a consistent profile of levels. In other words, the fictitious (but possibly real) action b such that $g_j(b) = l_j$ (j= 1,...,n) should be considered by the DM as a borderline case for which the recommendation «accepted» remains valid. Different types of actions - *i.e.* actions whose profiles over the criteria are very distinct - may give rise to the same recommendation (a credit file may be accepted either because of the large financial capacities of the applicant or because of the innovative aspects of the project or ...). We are then led to define a reference set B of fictitious actions corresponding to different representative borderline cases of acceptance. We have then:

$$- \exists\, b \in B \;/\; g_j(a) \geq g_j(b) \quad (j=1,\ldots,n) \Rightarrow a \in A^+ \qquad (7.1)$$

where A^+ represents the category of «accepted» actions.

Similarly, we can define a reference set C of fictitious actions corresponding to different representative borderline cases of rejection, such that:

- $\exists\, c \in C \;/\; g_j(c) \geq g_j(a) \quad (j=1,\ldots,n) \Rightarrow a \in A^-$ (7.2)

where A^- represents the category of «rejected» actions.

In some cases, there is a one-to-one correspondance between elements of B and elements of C, each ordered pair (b_i, c_i) representing profiles designed to test the same type of actions (see Fig. 7.1).

Figure 7.1 : Definition of categories

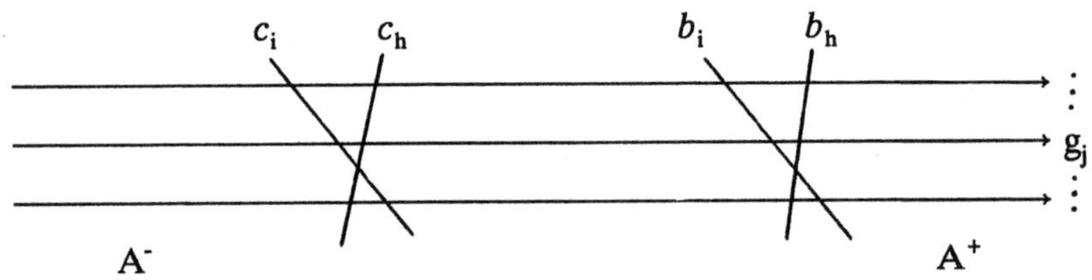

The different profiles must be related by consistency conditions so as to define categories which are distinctive enough. For instance, natural minimal requirements are:

- $\forall\, b \in B,\ \forall\, c \in C,\quad \text{not}(c \;\Delta\; b),$
- $\forall\, b \in B,\ \forall\, b' \in B \setminus \{b\},\quad \text{not}(b' \;\Delta\; b),$
- $\forall\, c \in C,\ \forall\, c' \in C \setminus \{c\},\quad \text{not}(c' \;\Delta\; c).$

An often reasonable requirement is:

- $\forall\, c \in C,\ \exists\, b \in B \;/\; b \;\Delta\; c.$

(above all when profiles are in one-to-one correspondance, in which case we can impose $b_i \;\Delta\; c_i$).

For some candidate actions, we may hesitate between acceptation and rejection. We are then led to define a third category $A^?$ corresponding to a recommendation such like «further information is required». For instance, we may admit that:

- $\exists\, b \in B$ and $\exists\, c \in C \;/\; g_j(c) < g_j(a) < g_j(b) \quad (j=1,\ldots,n) \Rightarrow a \in A^?$ (7.3)

The reader should notice that (7.1), (7.2) and (7.3) are mutually exclusive because of the above minimal requirements. Thus, (7.1-3) could be used for sorting the candidate actions. However, the resulting procedure would not be exhaustive (except for n=1) in the sense that some candidate actions may verify none of these conditions. Moreover, because of the imprecision inherent in the definition of the criteria and in the specification of the reference profiles, these conditions should be modulated. For instance, a may be assigned to A^+ even if some of the inequalities in (7.1) are slightly violated. The use of outranking relations allows to handle both of these difficulties.

7.2. Sorting procedures based on outranking relations

Since the prescription which is looked for is a partition on A, the sorting procedure must verify:

- any action must be assigned to *one and only one* category.

We briefly introduce two types of sorting procedures based on outranking relations:

- procedures using a decision tree (Moscarola and Roy (1977)),
- filtering procedures (Yu (1989)).

Consider again the case with three categories (A^+: «accepted», $A^?$: «further information is required», A^-: «rejected»). We may relax (7.1) as follows:

- $\exists\, b \in B$ / a S_i b (or $\sigma(a, b) \geq v_i$) $\Rightarrow a \in A^+$.

However, specially when the above outranking relation is not well established, we may hesitate to assign a to A^+ if there also exists $b \in B$ such that not(a S_i b) and b S_i a.

On the other hand, we may admit that:

- $\forall\, b \in B$, not(a S_i b) $\Rightarrow a \notin A^+$

Using these considerations, we are in a position to construct a decision tree which guides the examination of different cases. Such a procedure based on a fuzzy outranking relation is presented in Fig. 7.2.

Figure 7.2: Example of a sorting procedure using a decision tree

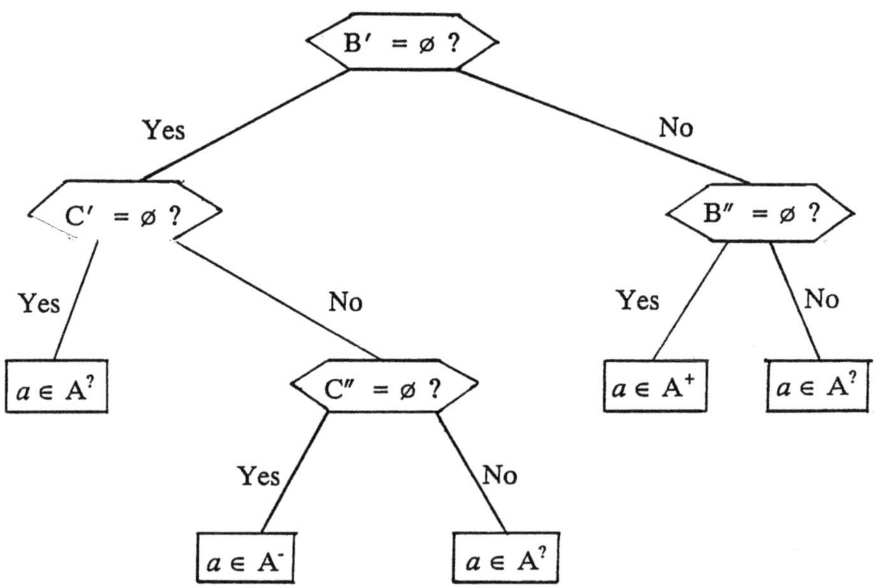

where:

- $B' = \{b \in B \;/\; \sigma(a, b) \geq t\}$, $B'' = \{b \in B \setminus B' \;/\; \sigma(b, a) \geq t'\}$,
- $C' = \{c \in C \;/\; \sigma(c, a) \geq v\}$, $C'' = \{c \in C \setminus C' \;/\; \sigma(a, c) \geq v'\}$.

The decision tree must be defined in connection with the decision problem under consideration. In particular, the definition of the thresholds (in our case t, t', v, v') must be performed by taking into account the consequences of erroneous assignments (*i.e.* the consequences of accepting an action which should have been rejected or conversely, as well as the disadvantages (delay, cost...) involved by the search for additional information).

Unlike procedures based on a decision tree, filtering procedures are particularly adapted to the case of k ordered categories. Such procedures provide a standard treatment, avoiding the construction of specific decision trees. In our presentation, each category A^h is defined by a unique low profile b^{h-1} and a unique high profile b^h (h=1,...,k) (see Fig. 7.3) where b^0 and b^k are introduced, for commodity, as fictitious profiles such that:

- $\forall$ $a \in A$ we have a S b^0, not(b^0 S a), b^k S a and not(a S b^k).

Figure 7.3: Definition of the k categories

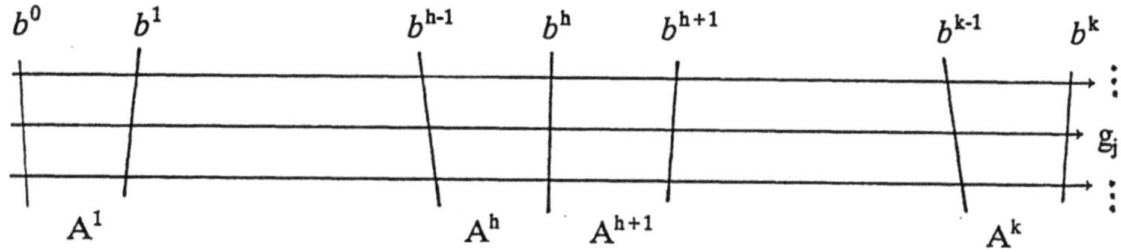

More precisely, we shall assume that b^{h-1} is the lowest profile compatible with the assignment to A^h, whereas b^h is the lowest profile not compatible with the assignment to A^h. We have then:

- a I $b^{h-1} \Rightarrow a \in A^h$ (h=1,...,k). (7.4)

Insofar as profiles delimit ordered categories, it is natural to define them such that:

- a S $b^h \Rightarrow$ a S b^{h-i} (h=0,...,k-1) (i=0,...,h) (7.5)
- b^h S a $\Rightarrow$ b^{h+i} S a (h=0,...,k) (i=0,...,k-h) (7.6)

A simple sufficient condition to satisfy (7.5) and (7.6) is $b^h \Delta b^{h-1}$ (h=1,...,k).

Moreover, in order to avoid ambiguous situations where we could have simultaneously a I b^h and a I b^{h-1} (which would provoke an hesitation as to the assignment of a - see (7.4)), profiles should be conceived so as to verify:

- a S $b^h \Rightarrow$ not(b^{h-1} S a) (h=1,...,k-1) (7.7)

This indicates that categories must be distinctive enough, which may be obtained by imposing lower bounds on the differences $g_j(b^h) - g_j(b^{h-1})$ (such bounds are related to the indifference, preference and/or veto thresholds).

We now introduce two types of filtering procedures:

On the one hand, if we identify h such that a S b^h and not(a S b^{h+1}) then, considering (7.5), we are induced to assign a to A^{h+1}. This results in a procedure which consists in successively testing the assertion a S b^h in a descending order (from h = k-1 down to h = 0).

On the other hand, if we identify h such that b^h S a and not(b^{h-1} S a) then, considering (7.6), we are induced to assign a either to A^{h+1} if we have also a S b^h or to A^h if we have not(a S b^h). This results in a procedure which consists in successively testing the assertion b^h S a in an ascending order (from h = 0 to h = k).

It is interesting to confront the results of each filtering procedure. Suppose that we get $a \in A^{h'}$ from the descending procedure and $a \in A^{h''}$ from the ascending procedure. First, the reader might check that $h' \leq h''$ (using (7.5) and (7.7)). Two cases are then to be distinguished:

- $h' = h''$: we may reasonably decide to assign a to $A^{h'}$,
- $h' < h''$: we have not(a S b^i) (i=h',..., h",..., k) from the descending procedure and not(b^i S a) (i=0,..., h',..., h"-1) from the ascending procedure, which implies that a is incomparable with profiles b^i (i=h',..., h"-1). This clearly reveals an hesitation between categories $A^{h'}, \ldots, A^{h''}$. We may then indicate this hesitation to the DM by assigning a to an additional category and/or use a new relation $S' \supset S$ which still verifies (7.7) (adapted to S') so as to try to reduce the amplitude of this hesitation (the reader might notice that, since S' satisfies (7.7), the previous assignments resulting from the first case remain unchanged).

Filtering procedures may be extended using:

- several reference profiles (regrouped into sets B^h) designed to delimit the categories,
- specific outranking relations S^h designed to test assertions a S^h b^h and b^h S^h a ($b^h \in B^h$) for each category A^h, which is of particular interest if we want to modulate the conditions of assignment (*e.g.* because the importance of the criteria may vary according to the category).

8. EXPLOITATION OF OUTRANKING RELATIONS IN THE P.γ CASE

Considering this problem statement, the exploitation procedure must be designed so as to rank each action by decreasing order of preference. In other words, the procedure must produce a prescription corresponding to a non necessarily complete preorder defined on A.

We first deal with the case where the preference model only consists of one crisp outranking relation S. The case with several crisp relations or one fuzzy relation is introduced afterwards.

8.1. Exploitation of One Crisp Outranking Relation

Different approaches may be envisaged so as to derive a transitive relation from one crisp outranking relation S. Some representative elementary procedures are presented. The way such procedures are used to construct the prescription is then discussed.

A first approach consists in determining a transitive relation S^* which minimizes a distance to S. The corresponding mathematical problem is stated as:

- find S^* such that $d(S^*, S) = \min_{P \in \mathcal{P}} d(P, S)$ (8.1)

where $\mathcal{P}$ is a subset of the set of partial preorders defined on A.

Such a subset may result from constraints on the type of prescription (*e.g.* when the prescription is required to be complete) and/or from specific restrictions which characterize the *a priori* admissible preorders.

The distance usually considered is the *symmetric difference*:

$$d(P, S) = |\{(a, b) \in A \times A \ / \ (a\ P\ b \text{ and not}(a\ S\ b)) \text{ or } (a\ S\ b \text{ and not}(a\ P\ b))\}|$$

Problem (8.1) is actually a very difficult optimization problem and its solution may be non-unique, which considerably restricts the usefulness of this approach.

A second approach consists in determining the *transitive closure* of S, *i.e.* the minimal transitive relation $\hat{S}$ which contains S. Equivalently, considering G = (A, U), the transitive closure $\hat{G} = (A, \hat{U})$ is such that $\hat{U} = U \cup U'$ where U' contains the aminimal number of arcs to make $\hat{G}$ transitive.

$\hat{S}$(or $\hat{G}$) is unique and its construction is easy, which makes this approach interesting. It should be noticed however that actions which belong to a same circuit give rise to an indifference clique and therefore cannot be discriminated (which is particularly embarrassing when there exists a circuit which passes through all the actions).

A third approach consists then in making use of discriminating information so as to produce the prescription. Considering an action a to be evaluated with respect to a subset $B \subset A$, a first discriminating index is the number of actions in B outranked by a. The *power* $p_B(a)$ is defined as:

- $p_B(a) = |\{b \in B \ / \ a\ S\ b\}|$

A second discriminating index is the number of actions in B which outrank a. The *weakness* $w_B(a)$ is defined as:

- $w_B(a) = |\{b \in B \ / \ b\ S\ a\}|$

A third index, derived from the two previous ones, gives an indication of the relative position of a in B. The *qualification* $q_B(a)$ is defined as:

- $q_B(a) = p_B(a) - w_B(a)$

These indices may be taken into account in different ways so as to derive a complete preorder. Considering the qualification index, three elementary procedures may be constructed:

- a direct procedure which produces the preorder induced by q_A (*i.e.* a procedure which ranks actions of A according to their respective qualifications);

- a procedure by iterated choice in a descending order (*i.e.* a procedure which selects actions in B = A whose qualification q_B is maximum as the first class C_1, then actions in $B = A \setminus C_1$ whose qualification q_B is maximum as C_2,... until the last class C_s is identified);

- a procedure by iterated choice in an ascending order (*i.e.* a procedure which selects actions in B = A whose qualification q_B is minimum as the last class D_1, then actions in $B = A \setminus D_1$ whose qualification q_B is minimum as D_2,... until the first class D_t is identified).

Similarly, we can use the power and weakness indices, which results in six other elementary procedures. The reader could check that, if S is a complete preorder, each of these procedures restitutes S. However, as soon as S is not transitive and/or not complete, the resulting preorders may differ.

The construction of the prescription may be performed by selecting one of the above elementary procedures as the exploitation procedure. Since each procedure promotes a specific viewpoint, it seems more appropriate to use a «median» procedure such like the direct procedure based on the qualification index.

However, since the selection of a unique procedure necessarily involves arbitrariness, it is preferable to found the prescription from several procedures. The preorders Z_i resulting from each elementary procedure are confronted as follows:

- $a\ Z_i\ b$ for all $i \Rightarrow a\ Z\ b$

where Z represents the final preorder, *i.e.* the prescription.

This confrontation is illustrated in Fig. 8.1, where transitivity arcs are omitted for Z_1, Z_2 and Z so as to simplify the representations.

Figure 8.1 : Intersection of preorders

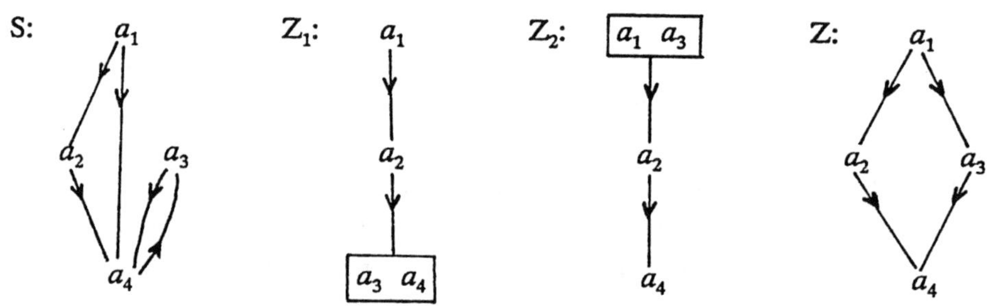

This *intersection* of several (possibly complete) preorders will usually result in a partial preorder which indicates the most established results. Within this approach, it is particularly interesting to confront procedures which promote complementary viewpoints:

- the direct procedure based on the power index and the direct procedure based on the weakness index,
- a procedure by iterated choice in a descending order and a procedure by iterated choice in an ascending order (both of them using the same index),
- the transitive closure and the direct procedure based on the qualification index,...

Our example in Fig. 8.1 illustrates the intersection of complete preorders derived from S using procedures by iterated choice based on the qualification index (Z_1: descending preorder, Z_2: ascending preorder). The reader should notice the way a_3 is treated by each elementary procedure (see Z_1 and Z_2) and the resulting hesitation indicated in the prescription Z.

8.2. Exploitation of Several Crisp and Nested Relations or of One Fuzzy Relation

The previous concepts and procedures may be extended in many ways. We briefly investigate some possible extensions which gave rise to exploitation procedures proposed in the literature.

Suppose that the preference model consists of r (r > 1) crisp and nested outranking relations $S_1 \subset S_2 \subset \ldots \subset S_r$. Like in the P.α case (see section 6.3), we may resort to a process consisting in progressively refining the prescription by successively considering relations S_1, $S_2, \ldots,$ S_r.

The exploitation procedure used in ELECTRE III (Roy (1978)) and ELECTRE IV (Roy *et al.* (1982)) is a typical example of implementation of such a refining process. The general scheme of the procedure is as follows:

- construction of a complete preorder Z_1,
- construction of a complete preorder Z_2,
- construction of the partial preorder $Z = Z_1 \cap Z_2$ as the prescription.

The constructions of Z_1 and Z_2 are respectively performed through a *descending distillation* procedure and an *ascending distillation* procedure which are extensions of procedures by iterated choice based on the qualification index. The current iteration of a descending distillation procedure is as follows: Selection of the class C_h (classes $C_1, \ldots, C_{h-1}$ beeing already selected and $A \setminus (C_1, \ldots, C_{h-1}) \neq \phi$):

(0) $K_0 = A \setminus (C_1 \cup \ldots \cup C_{h-1})$, $i \leftarrow 1$.

(1) Using S_i, construct K_i as the subset of actions from K_{i-1} whose qualification is maximum.

(2) If $|K_i| = 1$ or $i = r$ then $C_h = K_i$ and STOP else $i \leftarrow i+1$ and go to (1).

The progressive use of relations $S_1, \ldots, S_r$ is performed in (1) when further refinements are possible.

The exploitation procedure used in ELECTRE IV strictly amounts to the above one (with $r = 4$). ELECTRE III may be seen as an adaptation of this procedure to the fuzzy case. The crisp relations S_i ($i=1,\ldots,r$) are derived from the fuzzy relation σ as in section 5 with a minor modification:

- $a\ S_i\ b$ if $\sigma(a, b) \geq v_i$ and $\sigma(a, b) > \sigma(b, a) + s[\sigma(a, b)]$.

This modification involving a threshold s aims at eliminating unjustified discriminations which might unduly perturb the qualification index (by overestimating $q_B(a)$). Other implementation details are given in Skalka *et al.* (1983).

Suppose now that the preference model consists of one fuzzy outranking relation σ. Because of the equivalence between the crisp and fuzzy cases, we may adopt the previous approaches as in Electre III. It is also possible to extend the concepts to the fuzzy case and use the procedures presented in section 8.1.

For instance, the power, weakness and qualification indices may be extended as follows:

- $p'_B(a) = \sum_{b \in B} \sigma(a, b)$,
- $w'_B(a) = \sum_{b \in B} \sigma(b, a)$,
- $q'_B(a) = p'_B(a) - w'_B(a)$.

From a rigorous viewpoint, since the summation operator is involved, such exten-sions are meaningful only if we may assume relevant cardinal properties on σ.

The exploitation procedures used in Promethee (Brans *et al.* (1984) and Brans and Mareschal (1990)) and Mappac (Matarazzo (1986) and (1990)) methods implement this type of extensions.

9. CONCLUSION

The outranking approach promotes a realistic and prudent preference modelling that tolerates situations of incomparability, intransitivity of preferences and even cyclic preferences. The resulting model is sometimes difficult to exploit so as to derive prescriptions. We introduced several representative concepts and techniques for constructing prescriptions adapted to different problem statements.

Actually, many exploitation procedures were proposed in the literature, above all in the P.α and P.γ cases. The reader should not be surprised at obtaining different prescriptions when using different procedures. Indeed, exploitation procedures favour different conceptions regarding the definition of a «best» choice, assignment or ranking. It should be clear that the existence of alternative conceptions - none of which is prevailing - stems from the existence of irreducible difficulties (intransitivities, circuits,...).

When selecting (or designing) an exploitation procedure, it is possible to get an intuitive understanding of its behaviour by examining its underlying concepts and testing it on simple representative examples. A more formalized approach consists in imposing some desirable requirements on the prescription (such like (r'.6.1), (r.6.2) and (r.6.3) in the P.α case - see section 6) and trying to derive a procedure which satifies these requirements. Such an axiomatization of exploitation procedures does not suppress the above-mentioned irreducible difficulties: simultaneously imposing all classical desirable requirements usually result in impossibility theorems (see the well-known Arrow's theorem (Arrow (1963)). However, this approach is attractive insofar as it provides guarantees as to the properties of any prescription derived from an axiomatized procedure. Some interesting results were recently proposed in that direction with respect to each problem statement (P.α: Roy (1989), P.β: Yu (1989), P.γ: Perny (1989)).

REFERENCES

Arrow, K.J. (1963), *Social Choice and Individual Values*, Wiley, New-York.

Bouyssou, D. (1990), "Building Criteria: a Pre-requisite for MCDA", in this volume.

Brans, J.P., Mareschal, B. and Vincke, Ph. (1984), "PROMETHEE: a new Family of Outranking Methods in Multicriteria Analysis", in J.P. Brans (ed.), *Operational Research'84*, North-Holland.

Brans, J.P. and Mareschal, B. (1990), "The PROMETHEE Methods for MCDM: The PROMCALC, GAIA and BANKADVISOR Software", in this volume.

Keeney, R.L. and Raiffa, H. (1976), *Decisions with Multiple Objectives: Preferences and Value Tradeoffs*, Wiley, New York.

Matarazzo, B. (1986), "Multicriterion analysis by means of pairwise actions and criterion comparisons (MAPPAC)", *Applied Mathematics and Computation*, 18, 2, 119-141.

Matarazzo, B. (1990), "A Pairwise Criterion Comparison Approach: The MAPPAC and PRAGMA Methods", in this volume.

Moscarola, J. and Roy, B. (1977), "Procédure automatique d'examen de dossiers fondée sur une segmentation trichotomique en présence de critères multiples", *RAIRO*, 11, 2, 145-173.

Perny, P. (1989), "Quelques problèmes concernant les procédures de rangement", Communication presented at the 29th meeting of the Euro Working Group on Multicriteria Decision Aid, March 9-10, Dijon, France.

Roy, B. (1968), "Classement et choix en présence de points de vue multiples (la méthode ELECTRE)", *Revue d'Informatique et de Recherche Opèrationnelle*, 8, 57-75.

Roy, B. (1978), "ELECTRE III: Un algorithme de classements fondé sur une représentation floue des préférences en présence de critères multiples", *Cahiers du Centre d'Etudes de Recherche Opèrationnelle*, 20, 1, 3-24.

Roy, B. (1981), "The Optimisation Problem Formulation: Criticism and Overstepping", *The Journal of the Operational Research Society*, 32, 6, 427-436.

Roy, B. (1989), "About some Requirements to found a Selection of Alternatives on Pairwise Comparisons", Communication presented at EURO IX - TIMS XXVIII conference, July 6 - 8, Paris, France.

Roy, B. (1990), "The Outranking Approache and Foundations of Electre Methods, in this volume.

Roy, B., and Hugonnard, J.C. (1982), " Ranking of suburban line extension projects on the Paris metro system by a multicriteria method", *Transportation Research*, 16A, 4, 301-312.

Roy, B. and Skalka, J.M. (1984), ELECTRE IS - Aspects méthodologiques et guide d'utilisation, *Document du LAMSADE* n°30.

Saaty, T.L. (1980), *The Analytic Hierarchy Process*, Mc Graw Hill.

Skalka, J.M., Bouyssou, D. and Bernabeu, Y.A. (1983), "ELECTRE III et IV - Aspects méthodologiques et guide d'utilisation", *Document du LAMSADE* n°25.

Tversky, A. (1969), "Intransitivity of Preferences", *Psychological Review*, 76, 31-48.

Vanderpooten, D. (1989), "The Interactive Approach in MCDA: a Technical Framework and some Basic Conceptions", *Mathematical and Computer Modelling* (special issue on MCDM), 12, 10/11, 1213-1220.

Yu, W. (1989), "La recherche récente sur la problématique du tri (P.β)", Communication presented at the 29th meeting of the Euro Working Group on Multicriteria Decision Aid, March 9-10, Dijon, France.

THE PROMETHEE METHODS FOR MCDM; THE PROMCALC, GAIA AND BANKADVISER SOFTWARE

Jean Pierre Brans and Bertrand Mareschal

V.U.B. and U.L.B. Universities of Brussels, BELGIUM

CONTENTS:

1. THE PROMETHEE METHODS [1]

1.1. Basic Data

The PROMETHEE Methods are particularly appropriate to treat multicriteria problems of the following type:

$$\text{Max } \{f_1(x), f_2(x), \ldots, f_j(x), \ldots, f_k(x) \mid x \in A\} \tag{1.1}$$

for which A is a finite set of possible alternatives and $f_j(x)$, $j = 1,2,\ldots,k$ a set of k evaluation criteria.

The basic data therefore consist in the following evaluation table.

	$f_1(\cdot)$	$f_2(\cdot)$	...	$f_j(\cdot)$	...	$f_k(\cdot)$
a_1	$f_1(a_1)$	$f_2(a_1)$	...	$f_j(a_1)$	...	$f_k(a_1)$
a_2	$f_1(a_2)$	$f_2(a_2)$	...	$f_j(a_2)$	...	$f_k(a_2)$
..	...	...	...	...	...	...
a_i	$f_1(a_i)$	$f_2(a_i)$	...	$f_j(a_i)$	...	$f_k(a_i)$
..	...	...	...	...	...	...
a_n	$f_1(a_n)$	$f_2(a_n)$	...	$f_j(a_n)$	...	$f_k(a_n)$

In this paper the PROMETHEE I and II Methods are given. PROMETHEE I provides a partial ranking on A, PROMETHEE II a complete one. Both are outranking methods.

---

1. PROMETHEE: Preference Ranking Organisation METHod of Enrichment Evaluations.

1.2. Requisites for Appropriate Multicriteria Methods

The purpose of outranking methods is to enrich the dominance relation. This enrichment has to take place carefully. Let us first consider five small examples on which some basic requisites for a proper enrichment could be formulated. For each example two possible alternatives a and b, and two criteria $f_1(.)$ and $f_2(.)$ are considered. Both criteria have to be maximised.

	EX. I		EX. II		EX. III		EX. IV		EX. V	
	f_1	f_2	f_1	f_2	f_1	f_2	f_1	f_2	f_1	f_2
a	100	100	100	20	100	99	100	99	100	100
b	30	20	30	100	20	100	99	100	99	99

<u>EXAMPLE I</u>: Only a is efficient. a is fairly dominating b on each criterion. a should be recommended to the decision-maker. The notion of efficiency is in agreement with the decision to take.

<u>EXAMPLE II</u>: Both a and b are efficient. No alternative is dominating the other. a is fairly better on the first criterion and b on the second. No mathematical theory could tell which decision is better. a and b are incomparable. It is up to the decision maker to decide if he prefers something better on $f_1(.)$ or something better on $f_2(.)$. Again the notion of efficiency is appropriate: a and b are efficient while there is real hesitation between them.

<u>EXAMPLE III</u>: Again a and b are efficient. But in this case they are not incomparable. a could be preferred to b because a is fairly better than b on the first criterion and nearly equivalent to b on the second. In this case the efficiency theory is wrong: a and b are efficient while only a should be recommended.

<u>EXAMPLE IV</u>: Again a and b are efficient. But they are not incomparable like in Example II. In this case they could be considered as indifferent.

EXAMPLE V: Only a is efficient. It is better than b on both criteria. However the advantage of a on both criteria is negligible so that a and b could be considered as indifferent. Such a conclusion can be extremely helpful, especially when a Real World situation has been modelled. It is possible that some hidden criterion on which b should be better than a has not been modelled. By considering a and b as indifferent, b is not excluded from the decision process.

It is obvious important deviations can be observed between the efficiency theory and an appropriate dominance relation. The efficiency theory is a mathematical notion which often fails for treating multicriteria problems. Other requisites should be formulated for appropriate dominance. The six following ones seem to be crucial.

REQUISITE 1: The **amplitudes of the deviations** between the criteria values should be considered. This information is available, but not used for efficiency. Its importance is obvious in the Examples I to V.

REQUISITE 2: The **scaling effects** should be completely eliminated. Suppose in Example III the first criterion is expressing a number of jobs and the second billions of dollars. It is clear 80 jobs is negligible to 1 billion dollars. Therefore b should strongly be preferred. It is the opposite conclusion compared with the previous one. It is due to the scaling effects. Their importance is considerable and should therefore be totally withdrawn.

REQUISITE 3: **Incomparability** (R) should not be excluded in case of pairwise comparisons. An appropriate method, using only sure information should propose a partial preorder on A. (a P b, b P A, a I b or a R b). When a method is providing systematically a total preorder (complete ranking) it means that more disputable information has been used!

REQUISITE 4: A multicriteria problem is not a mathematically well stated problem. Usually there is no solution optimising all the criteria. Therefore an appropriate method should be "simple" (understandable by the decision-maker). It may not be a "black box" providing a solution, otherwise the decision maker should accept anything without understanding why!

REQUISITE 5: An appropriate method should include no technical parameters having no economical significance. Otherwise these parameters should have again a black box effect.

REQUISITE 6: Each appropriate method should allow the conflictual analysis of the criteria. It is indeed important for the decision-maker to have the opportunity to appreciate the criteria expressing the same, independent or opposite preferences.

The PROMETHEE Methods, completed by the GAIA visual modelling technique, widely take these six requisites into account.

1.3. Principles of the PROMETHEE METHODS

The PROMETHEE Methods request 3 steps:

STEP 1: Enrichment of the preference structure

By considering associated generalised criteria, the amplitudes of the deviations between criteria values will be taken into account. This step is crucial. It can easily be understood by the decision-maker. All the additional parameters to be defined have an economical significance. The scaling effects are totally withdrawn.

STEP 2: Enrichment of the dominance relation

A fuzzy outranking graph is built up. The arcs express how each alternative is dominating the others. Two arcs, one in each direction, are considered for each pair of alternatives.

STEP 3: Exploitation for decision Aid

PROMETHEE I provides a partial relation on A (including possible incomparabilities). PROMETHEE II provides a complete ranking. It looks more efficient, but the information used is more disputable.

1.4. STEP 1: Generalised Criteria

A generalised criterion will be associated to each criterion. Let us therefore consider first one particular criterion f(.). Suppose it has to be maximised:

$$f(a) \qquad f: A \to \mathbb{R} \text{ (to be maximised)} \tag{1.2}$$

Pairwise comparisons between the alternatives of A provide the following natural preference structure: $\forall a,b \in A$:

$$\begin{cases} f(a) > f(b) \Leftrightarrow a\,P\,b \\ f(a) = f(b) \Leftrightarrow a\,I\,b \end{cases} \tag{1.3}$$

which defines the dominance relation. This structure is extremely poor, especially for multicriteria problems, because it does not take the amplitude of deviations ($d = f(a)-f(b)$) into account.

Let us therefore consider a preference function P(a,b) giving the intensity of preference of a over b in function of the deviation d. Let us suppose this intensity defined between 0 and 1 so that:

$$0 \le P(a,b) \le 1 \tag{1.4}$$

$$\begin{cases} P(a,b)=0 \text{ if } d\le 0 \ (f(a)\le f(b)). \text{ No preference, indifference} \\ P(a,b)\approx 0 \text{ if } d>0 \ (f(a)>f(b)). \text{ Weak preference} \\ P(a,b)\approx 1 \text{ if } d>>0 \ (f(a)>>f(b)). \text{ Strong preference} \\ P(a,b)=1 \text{ if } d>>>0 \ (f(a)>>>f(b)). \text{ Strict preference} \end{cases} \tag{1.5}$$

It is clear this preference function will have the following shape:

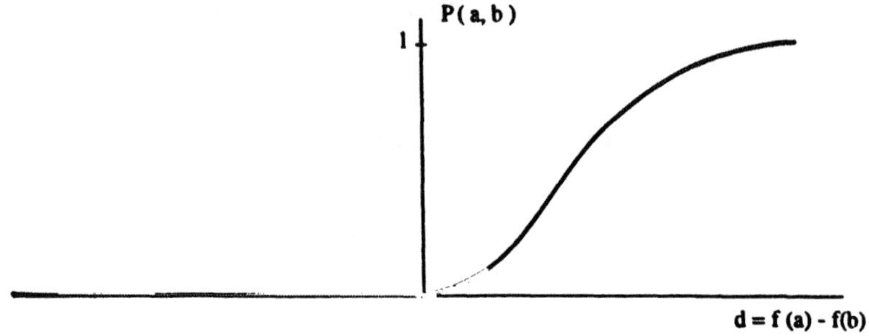

This function is monotonically increasing. The higher d, the higher the intensity of preference of a over b.

For the sake of presentation let us consider H(d) so that:

$$H(d) = \begin{cases} P(a,b) & d \geq 0, \\ P(b,a) & d \leq 0. \end{cases} \tag{1.6}$$

It is obvious H(d) has the following shape:

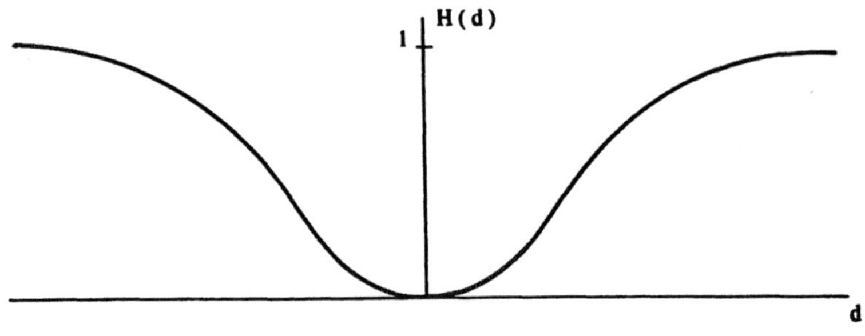

The generalised criterion then consists by definition in the pair:

$$\{f(.),P(a,b)\} \text{ or alternatively } \{f(.),H(d)\}. \tag{1.7}$$

The PROMETHEE Methods request a generalised criterion for each criterion $f_j(.)$, j=1,2,...,k. This is a critical task. In order to facilitate it a set of six possible generalised criteria is proposed to the decision-maker. The effective choice is made interactively by the decision-maker and the analyst according to their feeling of the intensities of preference. In each case 0, 1 or 2 parameters, each having a real economical significance, have to be fixed:

q is a threshold defining an indifference area,
p is a threshold defining a strict preference area,
s is a parameter the value of which lies between p and q.

The following overview table provides the shape of the six possible types of generalised criteria. This table provides assistance to the decision-maker for the selection. The six types are not exhaustive. There is no objection to consider other shapes. For instance, level criteria with more than 2 levels, non symmetrical shapes with variable thresholds depending on the alternatives, on f(a), on f(b), ... However, as far as we know, this overview table has been satisfactory for all the practical applications treated by PROMETHEE.

<u>TYPE I</u>: <u>USUAL CRITERION</u>: There is strict preference as soon as a non zero deviation is observed. It is again the classical (I,P) situation. This shape is less selected, it is totally insensitive to different values of d.

<u>TYPE II</u>: <u>U-SHAPE CRITERION</u>: An indifference area is considered. Deviations smaller than q imply no preference. Only one parameter q (indifference threshold) has to be fixed.

<u>TYPE III</u>: <u>V SHAPE CRITERION</u>: As soon as a non zero deviation is observed, an intensity of preference, increasing linearly with d, is considered. Above a deviation p the preference is strict. Only one parameter p (strict preference threshold) has to be fixed.

Type of generalised criterion — ***H(d) preference function***

Type I : Usual criterion.

$$H(d) \begin{cases} 0 & d = 0 \\ 1 & |d| > 0 \end{cases}$$

No parameter to fix.

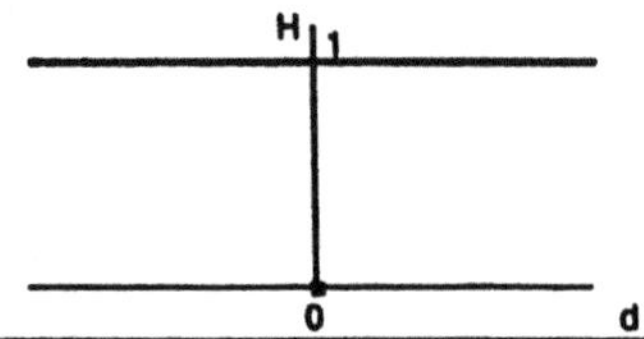

Type II : U-shape criterion.

$$H(d) \begin{cases} 0 & |d| \le q \\ 1 & |d| > q \end{cases}$$

q to fix.

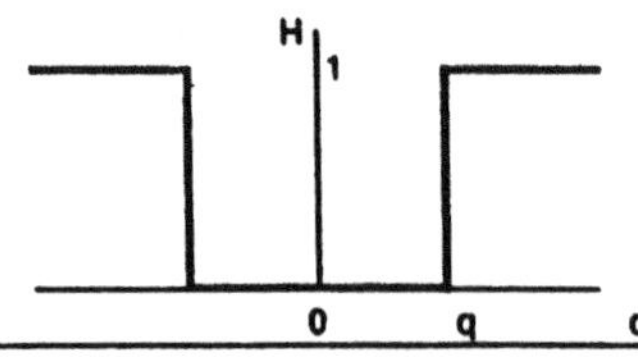

Type III : V-shape criterion.

$$H(d) \begin{cases} |d|/p & |d| \le p \\ 1 & |d| > p \end{cases}$$

p to fix.

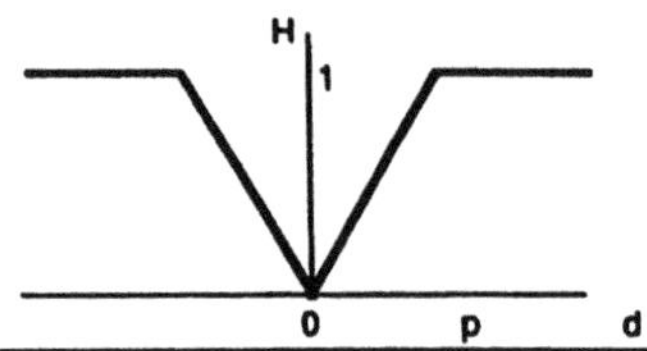

Type IV : Level criterion.

$$H(d) \begin{cases} 0 & |d| \le q \\ 1/2 & q < |d| \le p \\ 1 & |d| > p \end{cases}$$

q and p to fix.

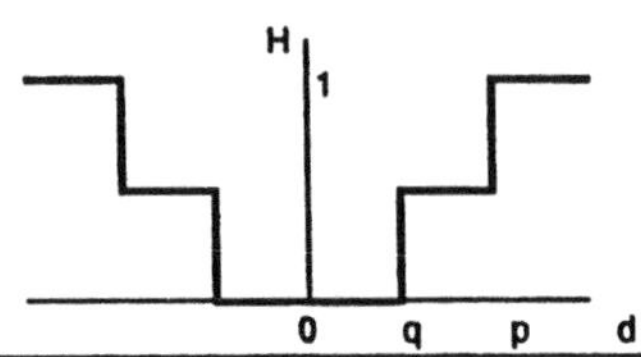

Type V : V-shape with indiff. area.

$$H(d) \begin{cases} 0 & |d| \le q \\ (|d|-q)/(p-q) & q < |d| \le p \\ 1 & |d| > p \end{cases}$$

q and p to fix.

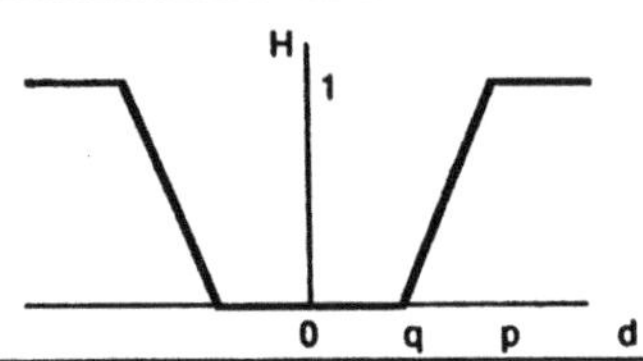

Type VI : Gaussian criterion.

$$H(d) = 1 - \exp(-d^2/2s^2)$$

s to fix.

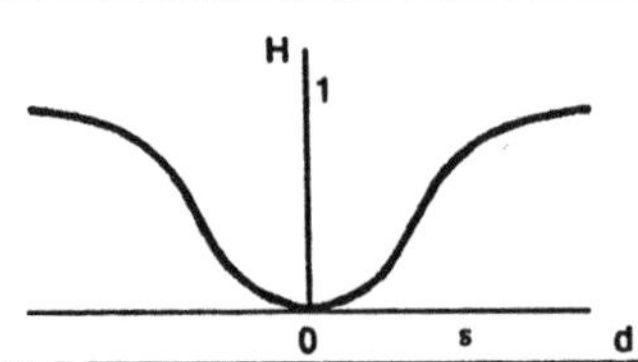

TYPE IV: LEVEL CRITERION: Two parameters have to be fixed: q (indifference area) and p (strict preference area), $q < p$. Between q and p an intermediate preference value is considered. The level criterion is particularly appropriate when f(.) is an application from A to a finite ordered set such as (very bad, bad, average, good, very good).

<u>TYPE V</u>: <u>V-SHAPE CRITERION WITH INDIFFERENCE AREA</u>: Again two parameters have to be fixed. Between q and p the intensity of preference increases linearly.

<u>TYPE VI</u>: <u>GAUSSIAN CRITERION</u>: The intensity of preference increases continuously with d without discontinuities nor in its shape nor in its derivatives. Only one parameter s has to be fixed. The value of s should be defined between q (threshold) and p (strict preference threshold).

It has been observed that the GAUSSIAN criterion has been the most selected by users for practical applications. After the GAUSSIAN criterion, the V SHAPE CRITERION WITH INDIFFERENCE AREA also has been often selected.

The six types of generalised criteria are not mathematically independent; For instance a level criterion for which p=q is a U SHAPE one, a type II criterion for which q=0 is a type III one,... However it is useful to keep the six shapes separately in order to provide six different visual graphs. By the way the notion of generalised criterion is easily understandable and easy to manage.

1.5. STEP 2: Outranking Graph

Let us consider again the multicriteria problem (1.1). Suppose for each criterion a generalised criterion has been defined. For each pair of alternatives and for each criterion j we dispose on:

$$\forall a,b \in A, \forall j=1,2,\dots,k, \ \{f_j(a), f_j(b), d_j=f_j(a)-f_j(b), P_j(a,b)\} \qquad (1.8)$$

Let us now define a preference index $\pi(a,b)$ of a over b over all the criteria, such that:

$$\pi(a,b) = \sum_{j=1}^{k} w_j P_j(a,b) \qquad \left(\sum_{j=1}^{k} w_j = 1\right) \qquad (1.9)$$

where w_j, j=1,2,...,k are weights associated to each criterion.

These weights are real numbers which do not depend on scales. A very clear interpretation of these weights will be given by the GAIA visual modelling technique.

It is often interesting to consider first equal weights, in that case $\pi(a,b)$ is simply the arithmetic mean of all the intensities of preference $P_j(a,b)$, $j=1,2,\ldots,k$.

$$\pi(a,b) = \frac{1}{k} \sum_{j=1}^{k} P_j(a,b) \tag{1.10}$$

The $\pi(a,b)$ values enjoy obviously the following properties:

$$\begin{cases} \pi(a,a) = 0 \\ 0 \le \pi(a,b) \le 1 \quad \forall a,b \in A \\ \pi(a,b) \approx 0 \text{ implies a weak global preference of a over b,} \\ \pi(a,b) \approx 1 \text{ implies a strong global preference of a over b,} \\ \pi(a,b) \text{ expresses how and with which intensity a dominates b,} \\ \text{while } \pi(b,a) \text{ how b dominates a, over all the criteria.} \end{cases} \tag{1.11}$$

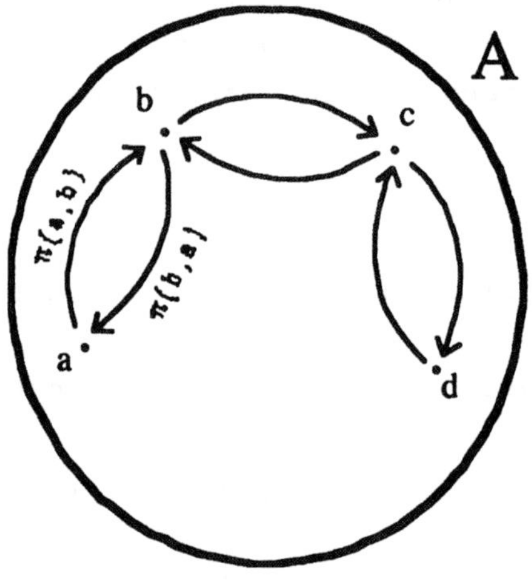

For each pair $a,b \in A$, the values $\pi(a,b)$ and $\pi(b,a)$ are calculated. This implies a complete fuzzy outranking graph on A. This graph is a considerable enrichment of the dominance graph because appropriate levels of intensities of preference were taken into account thanks to the associated generalised criteria.

1.6. STEP 3: Exploitation for Decision Aid

For each alternative $a \in A$, let us consider the two following outranking flows:

The **positive outranking flow:**

$$\phi^{+}(a) = \sum_{x \in A} \pi(a,x) \qquad (1.12)$$

The **negative outranking flow:**

$$\phi^{-}(a) = \sum_{x \in A} \pi(x,a) \qquad (1.13)$$

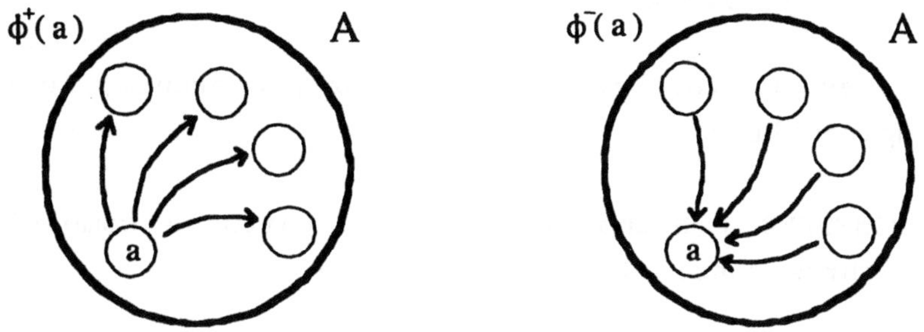

The positive outranking flow expresses how each alternative is outranking all the others. The higher $\phi^{+}(a)$, the better the alternative. $\phi^{+}(a)$ represents the power of a, it gives its **outranking** character.

The negative outranking flow expresses how each alternative is outranked by all the others. The smaller $\phi^{-}(a)$, the better the alternative. $\phi^{-}(a)$ represents the weakness of a, it gives its **outranked** character.

THE PROMETHEE I PARTIAL PREORDER

Two natural complete preorders can be deduced from the positive and the negative outranking flows. Let them be (S^{+},I^{+}) and (S^{-},I^{-})

$$\begin{cases} a\ S^{+}\ b & \text{iff} \quad \phi^{+}(a)>\phi^{+}(b) \\ a\ I^{+}\ b & \text{iff} \quad \phi^{+}(a)=\phi^{+}(b) \end{cases} \qquad (1.14)$$

$$\begin{cases} a\ S^{-}\ b & \text{iff} \quad \phi^{-}(a)<\phi^{-}(b) \\ a\ I^{-}\ b & \text{iff} \quad \phi^{-}(a)=\phi^{-}(b) \end{cases} \qquad (1.15)$$

The **PROMETHEE** I partial relation is the intersection of these two preorders

$$\begin{cases} a\ P^I\ b & \text{if} \begin{cases} a\ S^+\ b \text{ and } a\ S^-\ b \\ a\ S^+\ b \text{ and } a\ I^-\ b \\ a\ I^+\ b \text{ and } a\ S^-\ b \end{cases} \\ a\ I^I\ b & \text{if} \quad a\ I^+\ b \text{ and } a\ I^-\ b \\ a\ R\ b & \text{otherwise} \end{cases} \tag{1.16}$$

where P, I and R state respectively for preference, indifference and incomparability.

The **PROMETHEE** I Method consists in pairwise comparisons. In each case three conclusions are possible.

1- <u>First conclusion</u>: aP^Ib. <u>a outranks b</u>: In this case a higher power of a is associated to a lower weakness of a. The positive flow is confirmed by the negative flow. Both positive and negative flows are coherent. The information is sure.

2- <u>Second conclusion</u>: aI^Ib. <u>a indifferent to b</u>: Both positive and negative flows of both alternatives are respectively equal.

3- <u>Third conclusion</u>: aRb. <u>a incomparable to b</u>: In this case a higher power of one alternative is corresponding to a lower weakness of the other. This usually happens when a is good on a set of criteria on which b is weak and reciprocally b is good on another set of criteria on which a is weak. It seems quite natural in such a case to consider a and b as incomparable. The procedure will not decide mathematically which alternative is better; this is dangerous, it is up to the decision-maker to take his responsability and not to the method. In this case the positive flow is not confirmed by the negative one. The method abstain to express any preference.

THE PROMETHEE II COMPLETE PREORDER

When the decision-maker is requesting a complete ranking, the net outranking flow may be considered

$$\phi(a) = \phi^{+}(a) - \phi^{-}(a) \tag{1.17}$$

It is the balance of flow. The higher the net flow the better the alternative.

The **PROMETHEE** II complete ranking is then defined by (P^{II}, I^{II})

$$\begin{cases} a\ P^{II}\ b & \text{iff} \quad \phi(a)>\phi(b) \\ a\ I^{II}\ b & \text{iff} \quad \phi(a)=\phi(b) \end{cases} \tag{1.18}$$

All the actions are then comparable. Ex-aequo are possible. No incomparability remains, but on the other hand the information is more disputable. Part of the information get lost by considering the balance of flow.

2. THE GAIA VISUAL MODELLING TECHNIQUE [1]

According to Requisite 6, section 1.2., it is particularly important in the analysis of a multicriteria problem to provide information about the conflictual character of the criteria. Criteria can be in agreement and express similar preferences, they can be independent, they can be in conflict. This is not obvious from the basic data.

It is also important to obtain a clear view of the alternatives being particularly good (or not) on the different criteria, to obtain clusters of similar criteria and clusters of similar alternatives. A decision axis along which the decision maker should be invited to select the best alternatives, according to the weights allocated to each criteria, should also be of major interest.

----

1. GAIA: Geometrical Analysis for Interactive Aid

The GAIA visual modelling technique provides such results. It is based on the PROMETHEE approach for which a clear geometrical interpretation is given.

2.1. Analytical Decomposition of the Net Flow

Let us consider again the positive and the negative outranking flows of the PROMETHEE Methods as defined in (1.12) and (1.13). For each $a \in A$, we have:

$$\begin{cases} \phi^+(a) = \sum_{x \in A} \pi(a,x) = \sum_{x \in A} \sum_{j=1}^{k} P_j(a,x)\, w_j \\ \\ \phi^-(a) = \sum_{x \in A} \pi(x,a) = \sum_{x \in A} \sum_{j=1}^{k} P_j(x,a)\, w_j \end{cases} \tag{2.1}$$

so that the net flow $\phi(a)$ can be obtained as:

$$\phi(a) = \phi^+(a) - \phi^-(a) = \sum_{j=1}^{k} \sum_{x \in A} [P_j(a,x) - P_j(x,a)]\, w_j \tag{2.2}$$

or alternatively:

$$\phi(a) = \sum_{j=1}^{k} \phi_j(a)\, w_j \tag{2.3}$$

where

$$\phi_j(a) = \sum_{x \in A} [P_j(a,x) - P_j(x,a)] \tag{2.4}$$

For the sake of presentation, it is preferable to consider normed flows. As each alternative is facing (n-1) others, let

$$\gamma(a) = \frac{\phi(a)}{n-1} = \sum_{j=1}^{k} \gamma_j(a) w_j \tag{2.5}$$

where

$$\gamma_j(a) = \frac{1}{n-1} \sum_{x \in A} [P_j(a,x) - P_j(x,a)] \tag{2.6}$$

For each a, the quantities $\phi_j(a)$ only depend on j, we therefore call them the associated unicriterion flows. Each alternative is then characterised by k unicriterion flows. Therefore it can be represented as a point in a $\mathbb{R}^k$ space, the axes of which correspond to the different criteria. Let α be this point:

$$\alpha: \{\phi_1(a), \phi_2(a), \ldots \phi_j(a), \ldots \phi_k(a)\} \tag{2.7}$$

The following property obviously holds

$$\sum_{a \in A} \phi_j(a) = 0 \tag{2.8}$$

so that the unicriterion flows are centred at the origin of the $\mathbb{R}^k$ space.

Let us consider the matrix Φ of all the unicriterion flows. It includes all the information provided by the PROMETHEE Method. All the alternatives may be represented as a cloud of n points in the $\mathbb{R}^k$ space:

	$\phi_1(\cdot)$	$\phi_2(\cdot)$	...	$\phi_j(\cdot)$	...	$\phi_k(\cdot)$		α
a_1	$\phi_1(a_1)$	$\phi_2(a_1)$	...	$\phi_j(a_1)$	...	$\phi_k(a_1)$		α_1
a_2	$\phi_1(a_2)$	$\phi_2(a_2)$	...	$\phi_j(a_2)$	...	$\phi_k(a_2)$		α_2
..	...	...	...	...	...	...		...
a_i	$\phi_1(a_i)$	$\phi_2(a_i)$	...	$\phi_j(a_i)$	...	$\phi_k(a_i)$		α_i
..	...	...	...	...	...	...		...
a_n	$\phi_1(a_n)$	$\phi_2(a_n)$	...	$\phi_j(a_n)$	...	$\phi_k(a_n)$		α_n

The matrix Φ includes better information than the initial evaluation table because the intensities of preference given by the generalised criteria were taken into account.

In general the number k of criteria is larger than 2, so that it is impossible to obtain a direct visual view of the cloud. We therefore intend to project it orthogonally on a plane.

Let $\hat{\alpha}_i$ be the projection of α_i and γ_j the projection of the unit vector e_j of the j-axis of $\mathbb{R}^k$.

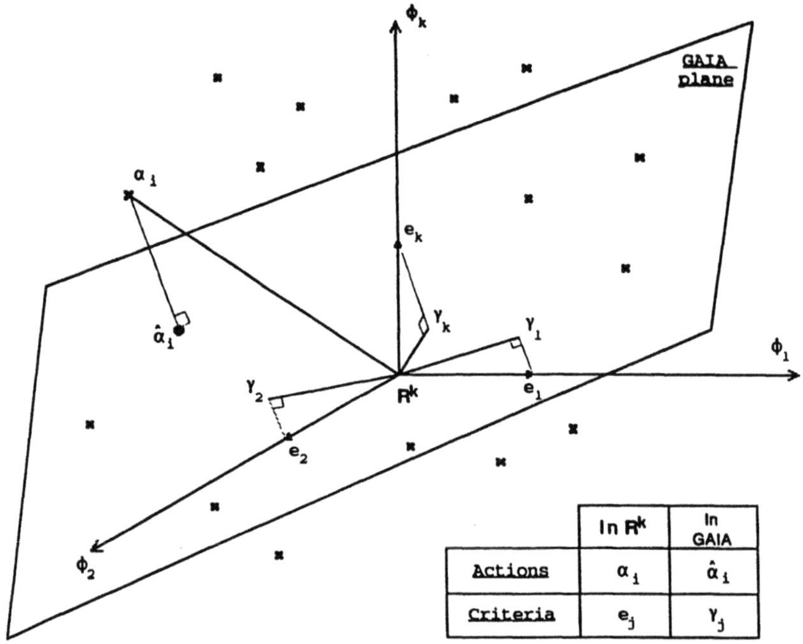

2.2. Principal Components Analysis

By using the Principal Components Analysis technique, it is possible to obtain the plane on which as few information as possible get lost by projection. For this purpose the expression u'Cu + v'Cv has to be maximised. It can be proved that:

$$\max_{u,v} \{u'Cu+v'Cv\} = \sum_{j=1}^{k} c_{jj}\|\gamma_j\|^2 + 2\sum_{j=1}^{k}\sum_{s\neq j} c_{js}(\gamma_j,\gamma_s) = n(\lambda_1+\lambda_2) \qquad (2.9)$$

where:

- C: is the covariance matrix of the γ_j, j=1,2,...,k.
- c_{jj}: is the variance of γ_j (criterion j)
- c_{js}: is the covariance between γ_j and γ_s (criteria j and s)
- $\|\gamma_j\|$: is the length of γ_j
- (γ_j, γ_s): is the scalar product between γ_j and γ_s

- λ_1 and λ_2 are respectively the largest and the second largest eigenvalues of C.
- u and v are the corresponding eigenvectors (u v).

The (u,v) plane or the GAIA plane is the one on which as less information as possible get lost by projection. A measure of the information having been preserved is given by

$$\delta = \frac{\lambda_1+\lambda_2}{\sum_{j=1}^{k} \lambda_j} \tag{2.10}$$

where λ_j, j = 1,2,...,k, are the k eigenvalues of C. (As C is symmetrical, all λ_j are real).

It is interesting to note that in all the Real World applications having been treated by GAIA, the value of δ has been larger than 60% and in most of the cases than 80%. This means that, even when the number of criteria is rather large (over 20), the GAIA plane provides reliable information.

2.3. Visualisation of the Criteria

Let us consider the projections γ_j, j=1,2,...,k, on the GAIA plane. These projections have different lengths and different orientations. Thanks to (2.9) it is possible to obtain a very clear interpretation of these lengths and orientations.

<u>1. Differentiation power of the criteria</u>

If a criterion j is strongly differentiating the alternatives, the variance of γ_j will be large so that c_{jj} will be large. As $c_{jj}\|\gamma_j\|^2$ is contributing to the maximum of (2.9), the GAIA plane will be chosen so that γ_j will be long.

The length of γ_j is therefore a measure of how criterion j differentiates the alternatives. The longer γ_j, the more criterion j differentiates the alternatives.

2. Similar criteria

Similar criteria have large positive covariances. Suppose c_{js} is large, as $c_{js}(\gamma_j, \gamma_s)$ is contributing to the maximum of (2.9), the GAIA plane will be positioned so that the scalar product will be large.

When γ_j and γ_s are oriented approximately in the same direction, the criteria j and s will be strongly correlated and therefore express similar preferences.

3. Independent criteria

If two criteria j and s are independent, their covariance c_{js} will be nearly zero. As, in this case, $c_{js}(\gamma_j, \gamma_s)$ cannot have a significant contribution to (2.9), the GAIA plane will be chosen so that the scalar product will be nearly zero. So, independent criteria will have γ vectors nearly orthogonal.

4. Conflicting criteria

Conflicting criteria have a negative covariance. If c_{js} is strongly negative, as $c_{js}(\gamma_j, \gamma_s)$ is contributing to the maximum of (2.9), the scalar product will also be strongly negative.

So, conflicting criteria will have γ vectors in opposite directions.

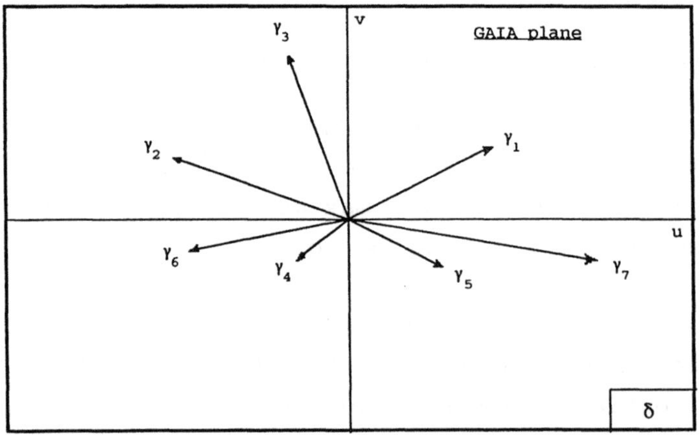

In the case of this example, criteria 1, 2, 3, 6 and 7 are strongly differentiating the actions, criteria 4 and 5 not. Criteria 2, 6 and 4 provide a cluster of criteria expressing similar preferences; Criteria 1 and 3 are independent. The cluster of criteria 2, 6 and 4 is strongly in conflict with the cluster including criteria, 1, 7 and 5.

The GAIA plane provides a powerful tool for the analysis of the differentiation power of the criteria and their conflicting character. However two restrictions have to be formulated:

(i) The GAIA plane only includes a proportion δ of the total information.

(ii) The conflicting character of the criteria is not given in abstracto on the criteria themselves but in concreto on the given data.

2.4. PROMETHEE Decision Axis

The allocation of weights to the different criteria is a crucial problem in all multicriteria techniques. The weights considered in the PROMETHEE Method are real numbers which do not depend on scales (see section 1.5.). A clear interpretation of these weights can be given by the GAIA visual modelling technique. Suppose the weights

$$w:(w_1,w_2,\dots w_j,\dots w_k) \qquad (2.11)$$

were allocated to the different criteria. The vector w can be considered in $\mathbb{R}^k$. Moreover we have:

$$(\alpha_i,w) = \sum_{j=1}^{k} w_j \gamma_j(a_i) = \gamma(a_i) \qquad (2.12)$$

This means that the projection of α_i (alternative a_i) on w gives the net flow of this alternative. As we have seen PROMETHEE II provides a complete ranking of the alternatives according to their net flows. The projections of the α_i on w therefore gives the complete PROMETHEE II ranking. The vector w may be considered as the decision axis. This axis lies in $\mathbb{R}^k$ but usually not in the GAIA plane. An image of it can be obtained by projecting the unit vector e of w on the GAIA plane. Let π be this projection and let us call it the PROMETHEE decision axis. If π is short, the PROMETHEE decision axis will have no strong decision power, in this case w is nearly orthogonal to the GAIA plane; the criteria are strongly conflicting and a good compromise should be selected in the neighbourhood of the origin.

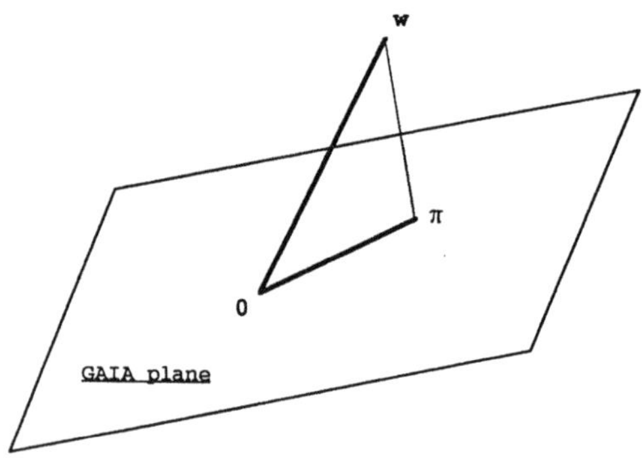

The weight vector w looks like a stick located above the GAIA plane, a stick that the decision maker can move according to his preferences in favour of particular criteria. When the stick w is moved, the PROMETHEE decision axis also moves in the GAIA plane where the consequences for decision making can be clearly observed.

When the PROMETHEE decision axis is long, the decision maker will be invited to select alternatives as far as possible in its direction. If the scalar product (π, γ_j) is large and positive (π and γ_j oriented in the same direction) criterion j will be in agreement with PROMETHEE II. If it is strongly negative (π and γ_j in opposite directions) criterion j will be in conflict with PROMETHEE II.

2.5. Visualisation of the Alternatives

Each alternative a_i has a projection $\hat{\alpha}_i$ in the GAIA plane so that a visualisation of the alternatives is obtained.

It is easy to prove that alternatives having a $\hat{\alpha}_i$ projection located in the direction of particular criteria are **good alternatives on these criteria**.

When the distance between two projections $\hat{\alpha}_r$ and $\hat{\alpha}_s$ is small, the alternatives a_r and a_s will have similar rows in the Φ matrix and therefore these actions will be rather similar for the decision maker.

Clusters of similar alternatives can easily be detected in the GAIA plane. In the example given below, the cluster of alternatives (a_1,a_2,a_3,a_4) is rather good on the criteria 4,5 and 6, but bad on the criteria 1,2 and 3. Opposite conclusions should be given for the cluster (a_5,a_6,a_7,a_8).

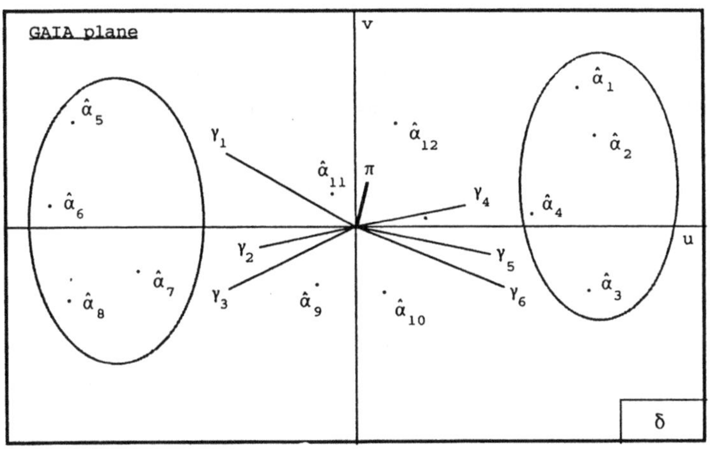

When the criteria are strongly conflictual, the PROMETHEE decision axis is rather short and **good compromises** are located close to the origin. In the case of the example a_{11} and a_{12} should be the best compromises.

Discordance between two alternatives can also easily be visualised. It is the case when one of them is good on some criteria and bad on others, while the opposite holds for the second one. A strong discordance is for example observed between a_2 and a_6.

The **recommended actions according to PROMETHEE II** are of course located in the direction of the PROMETHEE decision axis π.

A modification of the weights can modify seriously the conclusions. A **visual sensitivity analysis**, consisting in various modifications of the weights, is therefore strongly recommended before finalising the decision.

This sensitivity analysis is particularly easy to manage because for each set of weights only the PROMETHEE decisions axis π moves, the $\hat{\alpha}_i$ and γ_j projections of the alternatives and of the criteria remain unchanged in the GAIA plane. This calculation is done in real time by the GAIA microcomputer software.

3. APPLICATIONS

Several Real World multicriteria economical problems were treated by the PROMETHEE-GAIA methodology. For instance for the location of new electric power plants (Mladineo et al., 1989), for the location of concrete factories (Ribarovic and Mladineo, 1987), for the location of tourist facilities (Mladineo and Grabovac, 1988), for waste management (Briggs et al., 1990), for the allocation of resources to hospitals (D'Avignon and Mareschal, 1989), for automatic diagnosis in Health care (Dubois et al., 1989), for industrial evaluation in banking, for marketing problems, for manpower selection, ...

The methodology looks extremely successful because it is easy to understand, easy to manage and because it takes into account the most severe requisites that can be formulated for an appropriate multicriteria methodology.

As illustration let us consider again the didactic example of the location of an electric power plant. In a Real World location problem the evaluation criteria can be grouped in four classes:

- Economical criteria (costs, manpower, ...)
- Technical criteria (power, ...)
- Ecological criteria (environment, ...)
- Miscellaneous criteria (safety, ...)

Suppose it is the purpose to build an additional electric power plant in Europe. Six possible locations and six evaluation criteria are considered.

a_1: ITALY — $f_1(.)$: Manpower for running the plant
a_2: BELGIUM — $f_2(.)$: Power (in Megawatt)
a_3: GERMANY — $f_3(.)$: Constructions costs (in 10^6 U$)
a_4: U.K. — $f_4(.)$: Annual Maintenance costs (in 10^6 U$)
a_5: PORTUGAL — $f_5(.)$: Ecology. Number of villages to evacuate
a_6: FRANCE — $f_6(.)$: Safety level.

The basic data are given in the left hand side of the following table. For each criterion it is mentioned if it is to be maximised or to be minimised.

CRITERIA	min max	a_1	a_2	a_3	a_4	a_5	a_6	TYPE GEN.CRIT.	PARAMETERS
$f_1(.)$	min	80	65	83	**40**	52	94	II	q=10
$f_2(.)$	max	90	58	60	80	72	**96**	III	p=30
$f_3(.)$	min	600	**200**	400	1000	600	700	V	q=50,p=500
$f_4(.)$	min	54	97	72	75	**20**	36	IV	q=10,p=60
$f_5(.)$	min	8	1	4	7	3	5	I	-
$f_6(.)$	max	5	1	7	**10**	8	6	VI	s=5

The optimal values on each criterion are given in bold. It is impossible, even by looking carefully to these data, to have an idea of the best choice.

The PROMETHEE-GAIA Methodology is particularly appropriate to treat such location problems. The generalised criteria have first to be selected. The choice takes place interactively between the decision maker and the analyst. The parameters are fixed. Their values are given in the right hand side of the previous table. All the criteria have the same weight. The outranking graph can then be calculated. The following results are obtained for the $\pi(a_i,a_j)$ values and for the ϕ^+, ϕ^- and ϕ outranking flows:

	a_1	a_2	a_3	a_4	a_5	a_6	$\phi^+(.)$	$\phi^-(.)$	$\phi(.)$
a_1	0	0.296	0.250	0.268	0.100	0.185	1.099	1.827	-0.728
a_2	0.462	0	0.389	0.333	0.296	0.500	1.980	1.895	+0.085
a_3	0.236	0.180	0	0.333	0.056	0.429	1.234	1.681	-0.447
a_4	0.399	0.505	0.305	0	0.223	0.212	1.644	1.746	-0.102
a_5	0.444	0.515	0.487	0.380	0	0.448	2.274	0.808	+1.466
a_6	0.286	0.399	0.250	0.432	0.133	0	1.500	1.744	-0.274
$\phi^-(.)$	1.827	1.895	1.681	1.746	0.808	1.744			

The PROMETHEE I and II relations can then easily be drawn. In this case it appears clearly that the best choice is a_5.

In PROMETHEE I it appears that a_1 and a_2 are incomparable. a_1 is a powerful machine, rather expensive, requesting the evacuation of many villages while a_2 is a small plant, the cheapest one, requesting the evacuation of only one village. The alternatives are obviously discordant, it is the reason why they are incomparable.

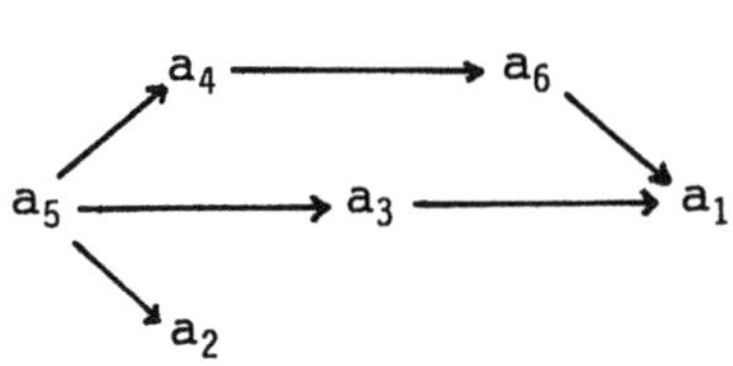

PROMETHEE I

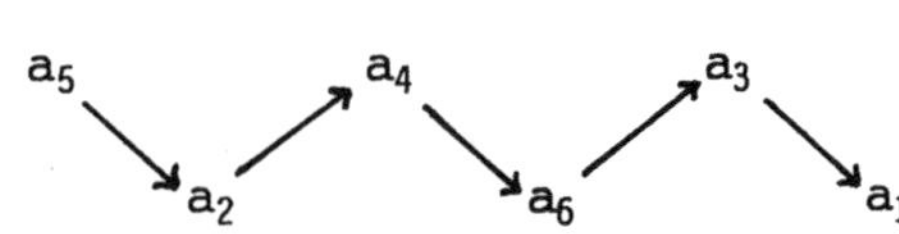

PROMETHEE II

Let us now calculate the GAIA plane. We first observe that δ=83%, only a few information get lost by projecting from $\mathbb{R}^6$.

The criteria Power, Manpower and Construction Costs strongly differentiate the alternatives.

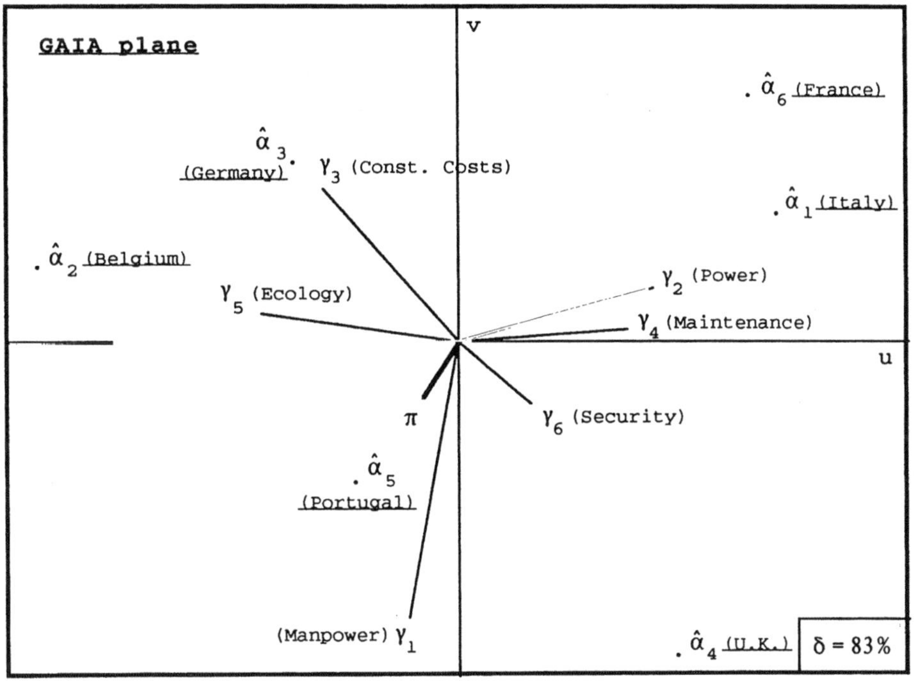

Globally the criteria look conflicting. The PROMETHEE decision axis π is not too strong. Sensitivity analysis on the weights could change the conclusions.

France and Italy are powerful plants. Belgium and Germany are not but they are cheap. Portugal is selected by PROMETHEE II, it is located in the direction of the PROMETHEE decision axis.

Some criteria are independent such as Ecology and Manpower, some pair of alternatives are discordant such as Italy and Belgium, Germany and U.K.,... And so on the GAIA visual interpretation.

4. THE BANKADVISER INDUSTRIAL EVALUATION SYSTEM

4.1. Introduction

BANKADVISER is a software for Industrial Evaluation based on the PROMETHEE Multicriteria Methodology.

The field of applications is very broad. In its first application it has been used to evaluate the industrial clients of a large bank. But it may also be applied to the evaluation of the industrial clients of an insurance company, to the evaluation of firms of a particular industrial sector (distribution, steel industry, motor car industry, chemistry, agriculture ...).

The items to evaluate will be called clients (factories, firms companies, institutions,...). The client bank (CLB) is the data bank including all the information available about the clients.

Any number of clients may be considered. Applications with about 50 clients were treated, but others were managing 100, 500, 1000 and even more clients.

The purpose of BANKADVISER is to provide assistance to evaluate each client by using two kinds of data:

(i) The own scores, the own economical performances of the client

(ii) The scores and economical performances of the other clients.

Most of the evaluation systems only take the first kind of data into account. In that case the client is evaluated only according to its own scores. In addition to this BANKADVISER provides information about the position of the client with regard to the others, to its competitors (positioning of the client).

Analysts are very fond of such information. A firm having good scores is not necessarily healthy; the whole sector can be particularly performant and the firm, despite its good scores, can be loosing market shares, its competitors being better.

The purpose will be reached by providing aggregated and disaggregated information. The aggregated information consists in a global appreciation of the client, while the disaggregated one will detect separately its weak and its strong features.

These results are obtained by the PROMETHEE multicriteria techniques. The analysis is decomposed in two phases: first the collection and the management of the data and then the evaluation itself.

4.2. Collection and Management of the Data

The basic data of each client consists in elements of its balance sheet and its income statement (profit and losses).

For instance fixed assets, equity, long term debt, net current assets, inventories, accounts receivable, other assets, accounts payable, other liabilities, requirements, liquid funds, short term borrowings, cash, total assets, ..., from the balance sheet; and sales, gross profit, value added, operations profit, financial profit, other profit, net profit, depreciation, cash-flow, generated cash-flow, ..., from the income statement.

The BANKADVISER microcomputer software can treat up to 20 data of the balance sheet and up to 20 of the income statement. Annual data, up to four years, are stored by the programme.

The collection of the data depends on the local liabilities. In some countries it is compulsory for each firm to have its balance sheet and its income statement published each year. In that case the information if often diffused by the Central or National Bank or by some specialised economical journal.

In the other case such information can often be obtained through other sources, the banks usually dispose on that kind of data.

The data can easily be implemented one by one, or through an interface with a central data bank.

BANKADVISER provides for each client several computer displays such as:

- Display of the balance sheet and the income statement
- Graphical displays of the balance sheet
- A decomposition of a management ratio into commercial, industrial and financial ratios
- The evolution over the last four years of some basic data (sales, net current assets, requirements, cash).

The display of this information is in itself very useful for financial analysts. A qualitative evaluation of the clients can be obtained. An overview of the balance sheet and the income statement is given. The relative importance of the different items is clearly presented, some interesting conclusions on the management of the client can be made.

4.3. Criteria for Industrial Evaluation

The second phase of BANKADVISER consists in the Industrial Evaluation itself. This evaluation depends on several criteria. The basic data of the balance sheet and the income statement are in general not evaluation criteria. For instance a firm having high sales is not necessarily healthy.

Criteria are obtained by means of ratios or more generally by means of mathematical formulas including the basic data. The decision-maker may define his criteria according to his own preferences. Generally four types of criteria are considered:

SOLVENCY RATIOS	LIQUIDITY RATIOS
Equity / Total Assets	Cash(n)-Cash(n-1) / Cash(n-1)
Cash Flow / Long Term Debt	365 x Account Receivable/Sales
...,...	...,...

PROFITABILITY RATIOS	MANAGEMENT RATIOS
Net Profit / Sales	Sales / Total Assets
Cash-Flow / Equity	...,...
...,...	

Several criteria of each type may be considered. Suppose k criteria $f_1, f_2, \ldots, f_j, \ldots, f_k$ have been defined by the decision-maker. Each client is then evaluated through k criteria. For the n clients of CLB we then have the following evaluation table.

	f_1	f_2	...	f_j	...	f_k
CLIENT 1	$f_1(c_1)$	$f_2(c_1)$	...	$f_j(c_1)$	...	$f_k(c_1)$
CLIENT 2	$f_1(c_2)$	$f_2(c_2)$	...	$f_j(c_2)$	...	$f_k(c_2)$
...			...		...	
CLIENT i	$f_1(c_i)$	$f_2(c_i)$	...	$f_j(c_i)$	...	$f_k(c_i)$
...			...		...	
CLIENT n	$f_1(c_n)$	$f_2(c_n)$	...	$f_j(c_n)$	...	$f_k(c_n)$

The evaluation of one particular client with regard to all the others of CLB is obviously a multicriteria problem. The PROMETHEE Methods are particularly appropriate to treat such problems. Generalised criteria have to be associated to each criterion. This can be done very easily as proposed in section 1. The Gaussian criterion and the V-shape one with indifference area are the most selected by decision-makers. The associated preference parameters p, q and s are defined easily because they have an economical significance. Also weights giving the importance of each criteria have to be defined as requested in (1.9).

4.4. Client Bank (CLB) - Reference Bank (RFB)

For several reasons, it is not useful to compare one particular client of CLB to all the others.

- CLB includes often too many clients (500, 1000, ...)
- CLB is a variable data bank due to entrances and departures
- The computation time should be prohibitive
- ...

It is therefore proposed to select a reference bank (RFB) on which each client should be evaluated. A reference bank is a limited set of representative clients for instance 10, 25, 50. It acts as a **standard**. This procedure offers many advantages:

- RFB is stable
- The RFB is the same standard for each client. After some use, its features are well known by the decision-maker.
- The evaluation of a client on the RFB requests a limited computation time. Real time evaluations become possible.
- Different RFB may be considered: representative clients, only good clients, target clients, ..., ...

4.5. The BANKADVISER Procedure

Suppose a CLB of n clients and an associated RFB of 50 clients have been defined. Suppose the basic data of the balance sheet and of the income statement have been introduced. Suppose the evaluation criteria, the associated generalised criteria, and the associated weights have been defined. Then the BANKADVISER procedure can provide a global and a disaggregated evaluation for each client:

1. The global evaluation

Suppose the decision-maker is interested in client i. The client is taken out of the CLB and introduced in the RFB.

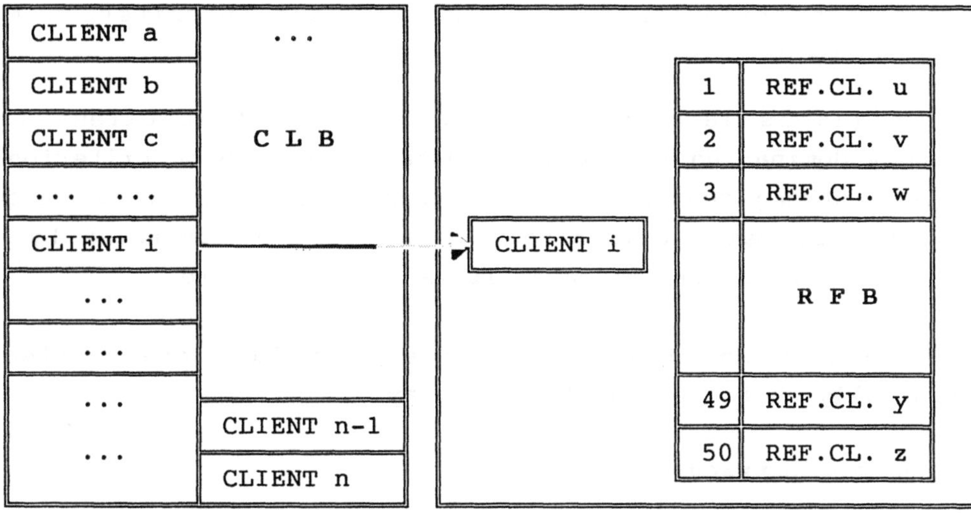

The RFB is now including 51 clients (RFB + client i). These clients are ranked by the PROMETHEE II Method. The computation with regard to the clients of the RFB are done in advance. Only the net flow $\gamma(i)$ (see section 2.1.) has to be calculated so that the computation can be done in real time.

1	REF.CL. u
2	REF.CL. v
3	REF.CL. w
12	REF.CL. i
50	REF.CL. y
51	REF.CL. z

The clients are ranked from the best to the worst one. The ranking of client i provides its **positioning** with regard to the standard. This ranking is obviously a global evaluation of the client. This global evaluation is displayed on a strip graduated from 1 to 50 in the BANKADVISER display.

2. The disaggregated evaluation

The global evaluation does not inform on the criteria on which the client is good (strong features) and those ones on which it is weak (weak features). A criterion by criterion analysis is therefore requested.

For each criterion f_j, $j=1,2,...,k$, the unicriterion flow $\gamma_j(i)$ can be calculated (see section 2.1.):

If $\gamma_j(i)$ is largely positive, it means that client i is strongly **dominating** the other clients of the RFB on criterion j. It is a strong feature. On the other hand if $\gamma_j(i)$ is largely negative, it means that client i is strongly **dominated** by the clients of RFB on criterion j. It is a weak feature.

The strong and the weak features are immediately detected graphically on the BANKADVISER display.

4.6. The BANKADVISER Display

The evaluation of the client is displayed on one screen. It includes four parts:

- The identification of the client
- The global evaluation. The ranking of the client is given on a strip or a Ruler with regard to the other clients of the RFB (Positioning)
- The disaggregated evaluation: A visual display of the profile of the client on each criterion
- The numerical data of the client on each criterion.

For each criterion an additional screen provides:

- The type of the criterion (solvency, liquidity, management profitability ratio,...)
- The mathematical formula of the criterion
- The type of the generalised criterion as foreseen by PROMETHEE
- The weight associated to the criterion
- Some additional statistics on the criterion values over all the clients of CLB (average, standard deviation, minimum and maximum value).

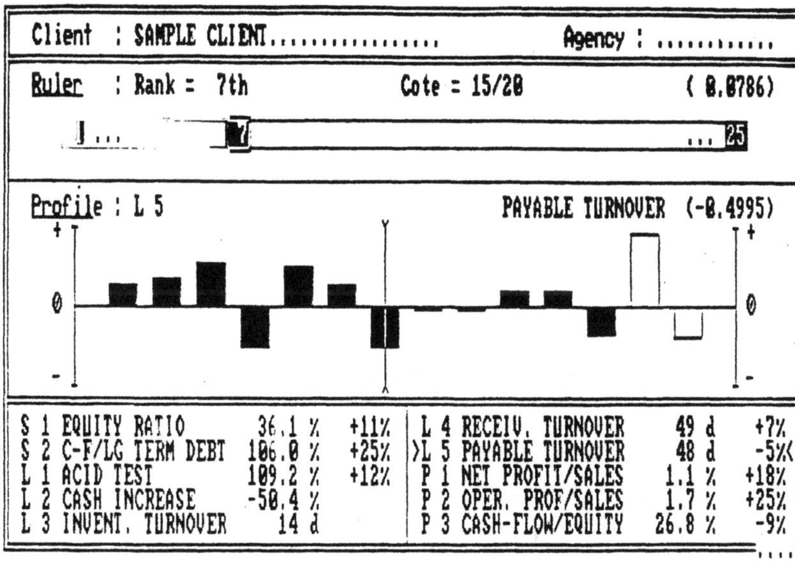

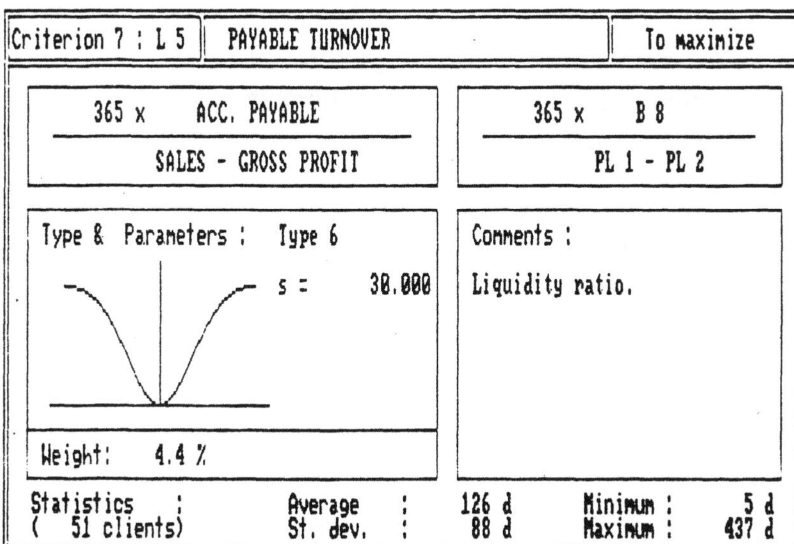

4.7. Conclusion

BANKADVISER is a powerful tool for financial analysts. It provides a clear information and an efficient evaluation on the industrial clients of a data bank. It is based on the powerful PROMETHEE Methodology.

BANKADVISER is an Expert System providing information, but it does not finalise the decision. The financial analyst remains actively involved in the decision procedure. It is his responsibility to finalise the decision. Decisions such as: to provide a loan or not, to fix the loan modalities (interest, terms, ...), to request some insurance or not, to cover some deficits or not, to provide guaranties on foreign activities, ... can be prepared effectively by using BANKADVISER.

Several Real World applications have been treated successfully.

5. ASSOCIATED MICROCOMPUTER SOFTWARE: PROMCALC, GAIA AND BANKADVISER

The following microcomputer software can be obtained from the authors:

- PROMCALC: PROMETHEE Methods (I and II)
- GAIA: Visual Modelling with PROMETHEE
- BANKADVISER: Industrial Evaluation.

PROMCALC and GAIA both have two functions. A didactic function to learn the methodology. A tool function to solve large problems up to 60 alternatives, 30 criteria. Both are particularly performant and user friendly. Their Main Menu includes successively:

1. Treat an example
2. Define a problem
3. View of the problem
4. Use the spreadsheet
5. Solve the problem
6. Modify the data
7. Save the problem on a file
8. Print all results of the problem
9. Quit PROMCALC (or GAIA)

BANKADVISER is also available. Each application requests some particular adaptations according to the nature of the data, the selected criteria, the items of the Reference Bank. Appropriate adapted versions can be obtained on request.

REFERENCES

Brans, J.P. (1982): "L'Ingéniérie de la Décision. Elaboration d'Instruments d'Aide à la Décision. Méthode PROMETHEE", *Université LAVAL, Colloque d'Aide à la Décision*, Quebec, Canada, 183-213.

Brans, J.P., Mareschal, B. and Vincke, Ph. (1984): "PROMETHEE. A new family of outranking methods in MCDM", *IFORS 84*, North Holland, 477-490.

Brans, J.P. and Vincke, Ph. (1985): "A preference ranking organisation method. The PROMETHEE method for MCDM", *Management Science*, Vol. 31, 6, 647-656.

Brans, J.P., Mareschal, B. and Vincke, Ph. (1986): "How to select and how to rank projects. The PROMETHEE method", *EJOR*, vol. 24, 228-238.

Briggs, Th., Kunsch, P.L. and Mareschal, B. (1990): "Nuclear Waste management. An application of the multicriteria PROMETHEE method", *EJOR*, vol. 44, 1-10.

D'Avignon, G.R. and Mareschal, B. (1989): "Specialisation of hospital services in Quebec. An application of the PROMETHEE and GAIA methods", *Mathematical and Computer Modelling*, vol. 12, 10/11, 1393-1400.

Dubois, Ph., Brans, J.P., Cantraine, F. and Mareschal, B. (1989): "MEDICIS: An expert system for computer-aided diagnosis using the PROMETHEE Method", *EJOR*, vol. 39, 284-292.

Mareschal, B. and Brans, J.P. (1988): "Geometrical representations for MCDM (GAIA)", *EJOR*, Vol. 34, 69-77.

Mareschal, B. (1986): "Stochastic PROMETHEE multicriteria decision-making under uncertainty", *EJOR*, 26, 58-64.

Mareschal, B. (1988): "Weight stability intervals in the PROMETHEE multicriteria decision-aid method", *EJOR*, 33, 54-64.

Mladineo, N., Margeta, J., Brans, J.P. and Mareschal, B. (1987): Multicriteria ranking of alternative locations for small scale hydroplants, *EJOR*, 31, 215-222.

Mladineo, N., Babic, Z., Milicic, M. and Plazibat, N. (1989): "Optimal location choice for power plants in Dalmatia by the PROMETHEE methods". Working paper. Submitted for publication.

Mladineo, N. and Grabovac, J. (1988): "The application of multicriteria analysis in the selection of the optimal renewable energy sources for tourist facilities", *Proceedings Zbornik Radova*, Yugoslavia, 110-121.

Ribavoric, Z. and Mladineo, N. (1987): "Application of multicriterional analysis to the ranking and evaluation of the investment programmes in the ready mixed concrete industry", *Engineering Costs and Production Economics*, 12, 367-374.

A PAIRWISE CRITERION COMPARISON APPROACH: THE MAPPAC AND PRAGMA METHODS

Benedetto Matarazzo

Istituto di Matematica
Facoltà di Economia e Commercio
Università di Catania, ITALY

1. INTRODUCTION

The aim of the two methods briefly outlined here is to aid the decision maker with **discrete problems** in the presence of more than one **cardinal criterion**. Methodologically the two have much in common with the original Pairwise Criterion Comparison Approach (PCCA) whose specific quality consists in the possibility to compare feasible actions with respect to all the possible unranked **pairs** of distinct **criteria** considered. The **partial** results thus obtained are then suitably **aggregated** and used in order to aid the decision maker in a variety of problems. From this point of view, therefore, the approach proposed may be considered as an attempt to make explicit the limited capacity of the human mind to make comparisons between numerous and often conflicting evaluations simultaneously; it offers, instead, a series of comparisons easy to execute one at a time.

The two methods considered may be differentiated in that MAPPAC (Multicriterion Analysis of Preferences by means of Pairwise Actions and Criterion comparisons) (Matarazzo, 1986), like the outranking methods, is based on the comparison of **one pair** of actions at a time, while in PRAGMA (Preference Ranking Global frequencies in Multicriterion Analysis) (Matarazzo, 1988a) all the feasible actions are compared simultaneously. Moreover, with reference to the information obtained, the former method supplies indices of **preference intensity** for each pair of actions, while the latter gives the **ranking frequencies** of each feasible action.

Suitable processing of the information obtained with both methods makes it possible to build complete or partial preorders.

The particular technique used by PCCA certainly increases the number of calculations made necessary; but the apparent complication, in our opinion, is justified by the further information obtained regarding the partial (that is, referred to each pair of criteria) dominance, preference and discordance. This information, which is not taken into consideration by the other methods, may be suitably interpreted and used by the decision maker. In particular it is possible to obtain partially compensatory or noncompensatory decisional models.

After defining our terms, we will describe the axiomatic system and the basic preference indices used in the MAPPAC method (section 2) and explain as a consequence how aggregated preference rankings can be built (section 3). We will then (section 4) illustrate the basic principles on which the PRAGMA method is based; how this method supplies information not obtainable with MAPPAC and how, with the use of this information, it is possible in particular to rank the actions (section 5). We shall conclude with a few brief final considerations (section 6).

2. THE PREFERENCE INDICES OF THE MAPPAC METHOD

Let $A = \{a_i \mid i \in I\}$ be the (finite) set of feasible actions, I the set of indices identifying each action, $|I| = n$, $G = \{g_j \mid j \in J\}$ a consistent family of cardinal criteria (see Jacquet-Lagrèze, 1985; Keeney and Raiffa, 1976; Roy, 1985) of evaluation of the feasible actions, and J the set of indices identifying each criterion, $|J| = m$; then let $g_j : A \to \mathbb{R}$, $j \in J$, according to a nondecreasing preference of the decision maker, and these preferences are expressed on interval scales (Vansnick, 1990) (that is, g_j is a cardinal utility function or a graduation (Roy, 1985)).

Introducing the real numbers

$$a_j^* \le \min_{i \in I} \{g_j(a_i)\} \text{ and } b_j^* \ge \max_{i \in I} \{g_j(a_i)\}$$

or two suitable parameters a_j^* and b_j^*, $j \in J$, which represent respectively, in the decision maker's view, the "neutral" and "best" levels on each attribute, it is possible to effect a positive affine transformation of the elements of the matrix of the evaluations $\underline{P} = [g_j(a_i)]$, $i = 1,2,\ldots,n$, $j = 1,2,\ldots,m$, obtaining the matrix $\underline{C} = [c_{ij}]$, $i = 1,2,\ldots,n$, $j = 1,2,\ldots,m$, where

$$c_{ij} = \begin{cases} \dfrac{g_j(a_{i4})-a_j^*}{b_j^*-a_j^*} & \text{if } a_j^* < b_j^* \\ & \qquad (0 \le c_{ij} \le 1 \text{ for all } i \in I,\ j \in J) \\ 0 & \text{if } a_j^* = b_j^* \end{cases}$$

If w_j, $w_j > 0$, $j \in J$, represents the relative importance of the jth criterion, given $\lambda_j = w_j \mid \Sigma_j w_j$, it is possible to express the normalized weighted matrix of the evaluations $\underline{C}_w = [\lambda_j C_{ij}]$, $i = 1, 2,\ldots,n$, $j = 1, 2,\ldots,m$. Therefore the action a_h **partially dominates** the action a_k (a_h, $a_k \in A$) with respect to the pair of criteria g_i, $g_j \in G$ ($a_h D_{ij} a_k$) if and only if one of the following relations is valid:

$$g_i(a_h) > g_i(a_k) \text{ and } g_j(a_h) > g_j(a_k) \iff c_{hi} > c_{ki} \text{ and } c_{hj} > c_{kj}$$
$$g_i(a_h) = g_i(a_k) \text{ and } g_j(a_h) > g_j(a_k) \iff c_{hi} = c_{ki} \text{ and } c_{hj} > c_{kj}$$
$$g_i(a_h) > g_i(a_k) \text{ and } g_j(a_h) = g_j(a_k) \iff c_{hi} > c_{ki} \text{ and } c_{hj} = c_{kj}$$

We remember that a_h **dominates** a_k ($a_h D a_k$) if and only if $g_j(a_h) \ge g_j(a_k)$, for all $j \in J$, at least one of these inequalities being strict. Of course, the dominance relation implies the preference relation P. We may also say that a_h **strictly dominates** a_k ($a_h D_s a_k$) if and only if $g_j(a_h) \ge g_j(a_k)$, for all $j \in J$, where at most only one equality is valid. Therefore, strict dominance implies partial dominance for every pair of feasible criteria, and vice versa:

$$a_h D_{ij} a_k, \text{ for all } i,j \in J \iff a_h D_s a_k \Rightarrow a_h D a_k \Rightarrow a_h P a_k.$$

With respect to a generic pair of feasible actions a_h and a_k and to an unranked pair of different criteria $g_i, g_j \in G$, we may define the following **basic preference indices** $\pi_{ij}(a_h, a_k)$ and $\pi_{ij}(a_k, a_h)$ (Matarazzo, 1984 and 1986):

(1)

$\pi_{ij}(a_h,a_k)$ =	$\pi_{ij}(a_k,a_h)$ =	
1	0	if $c_{hi} > c_{ki}$ and $c_{hj} > c_{kj}$
0	1	if $c_{hi} < c_{ki}$ and $c_{hj} < c_{kj}$
0.5	0.5	if $c_{hi} = c_{ki}$ and $c_{hj} = c_{kj}$
$\dfrac{\lambda_i(c_{hi}-c_{ki})}{\lambda_i(c_{hi}-c_{ki}) + \lambda_j(c_{kj}-c_{hj})}$	$\dfrac{\lambda_j(c_{kj}-c_{hj})}{\lambda_j(c_{kj}-c_{hj}) + \lambda_i(c_{hi}-c_{ki})}$	if $c_{hi} > c_{ki}$ and $c_{hj} \leq c_{kj}$; $c_{hi} = c_{ki}$ and $c_{hj} < c_{kj}$
$\dfrac{\lambda_j(c_{hj}-c_{kj})}{\lambda_j(c_{hj}-c_{kj}) + \lambda_i(c_{ki}-c_{hi})}$	$\dfrac{\lambda_i(c_{ki}-c_{hi})}{\lambda_i(c_{ki}-c_{hi}) + \lambda_j(c_{hj}-c_{kj})}$	if $c_{hi} \leq c_{ki}$ and $c_{hj} > c_{kj}$; $c_{hi} < c_{ki}$ and $c_{hj} = c_{kj}$

We observe that these indices constitute a complete valued preference structure (Roubens and Vincke, 1985; Vincke, 1990) and may be interpreted as *preference intensity* of a_h over a_k and of a_k over a_h, respectively.

The following relations are valid, for all $(a_h, a_k) \in A \times A$:

$$0 \leq \pi_{ij}(a_h, a_k) \leq 1 \quad \text{and} \quad \pi_{ij}(a_h, a_k) + \pi_{ij}(a_k, a_h) = 1; \tag{2}$$

in particular:

$$a_h \, D_{ij} \, a_k \iff \pi_{ij}(a_h, a_k) = 1 \iff \pi_{ij}(a_k, a_h) = 0.$$

As far as the axiomatic system of the basic preference indices introduced is concerned, we observe that these are of a **compensatory** type and functions of the values of the differences between the weighted evaluations of the two actions compared with respect to the pair of criteria considered, if these evaluations are discordant. On the other hand they are functions only of the mathematical sign of these differences in the case of concordant evaluations, therefore emphasizing the relations of partial dominance.

With reference to the index $\pi_{ij}(a_h,a_k)$ it must be noted that, in the case of discordant evaluation, the trade-off between the evaluations of a_h and a_k on the same criterion is equal to 1, while the marginal rate of substitution between the normalized weighted evaluations of the same action a_h with respect to the two criteria g_i and g_j is

$$\frac{\pi_{ij}(a_h,a_k)}{\pi_{ij}(a_k,a_h)} \quad \text{or} \quad \frac{\pi_{ij}(a_k,a_h)}{\pi_{ij}(a_h,a_k)}$$

The basic preference indices introduced may immediately be interpreted geometrically (see Matarazzo, 1986) by considering the **partial profiles** of the actions a_h and a_k with respect to the criteria g_i and g_j (figure 1):

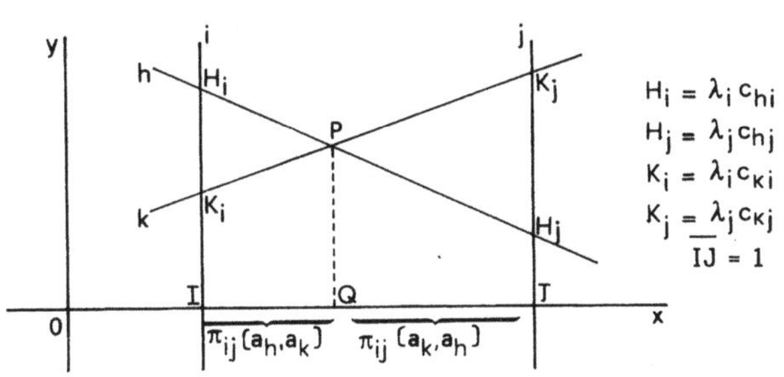

Fig.1 Geometrical interpretation of basic preference indices.

To avoid the possibility that evaluations which are slightly favourable to a_h may lead to the relation $a_h\ D_{ij}\ a_k$, it is possible to introduce different types of *indifference areas* as a function of suitable *marginal indifference thresholds*, one for each criterion considered (see Matarazzo, 1990). These thresholds are not assumed to be constant as in the PROMETHEE Method (Brans et al., 1984; Brans and Mareschal, 1990).

Designating by P_{ij} and I_{ij} respectively the partial preference and indifference relations, we obtain:

$a_h \, P_{ij} \, a_k \iff 0 \leq \pi_{ij}(a_k,a_h) < 0.5 < \pi_{ij}(a_h,a_k) \leq 1$

$a_h \, I_{ij} \, a_k \iff \pi_{ij}(a_h,a_k) = \pi_{ij}(a_k,a_h) = 0.5.$

It may be proved (see Matarazzo, 1984) that the pair of relations (P_{ij}, I_{ij}) thus defined creates a **complete preorder**.

Observe that $\pi_{ij}(a_h,a_k) = 0.5$ may occur both if a_h and a_k are **indifferent** with respect to the criteria g_i and g_j (that is, $g_i(a_h) = g_i(a_k)$ and $g_j(a_h) = g_j(a_k)$), and if a_h and a_k are **equivalent**, that is if:

$(c_{hi}-c_{ki}) \cdot (c_{kj}-c_{hj}) \neq 0$ and $(c_{hi}-c_{ki})/(c_{kj}-c_{hj}) = \lambda_j / \lambda_i.$

Observe also that if we consider a new set of feasible actions A', $A' \supset A$, the values of the elementary preference indices (1) introduced do not change if the parameters a_i, a_j, b_i and b_j used in the affine transformation of the evaluations remain unaltered (that is, if the original scenario of the decision maker remains unchanged).

Of course the partial indifference relation may be emphasized by introducing a real parameter δ, $\delta \in [\frac{1}{2}, 1]$, for which:

$a_h \, P^{\delta}_{ij} \, a_k \iff \pi_{ij}(a_h,a_k) > \delta$

and

$a_h \, I^{\delta}_{ij} \, a_k \iff 1 - \delta \leq \pi_{ij}(a_h,a_k) \leq \delta.$

It is also possible to obtain the weak partial preference relation Q^{τ}_{ij} by introducing another parameter ϵ, (see Matarazzo, 1990), $\frac{1}{2} \leq \delta < \epsilon \leq 1$, such that $a_h \, Q^{\tau}_{ij} \, a_k \iff \delta < \pi_{ij}(a_h,a_k) \leq \epsilon$.

With reference to two distinct criteria g_i, g_j it is possible to obtain n^2 indices of the type $\pi_{ij}(a_h,a_k)$, $(a_h, a_k) \in A \times A$, which may be collected in a square $n \times n$ matrix

$\mathbf{\Pi}_{ij} = [\pi_{ij}(a_h,a_k)]$, $(h = 1,2,\ldots,n;\ k = 1,2,\ldots,n)$.

In this matrix the elements of the principal diagonal will be equal to 0.5 for (1), and for the remaining n(n-1) preference indices it is sufficient to calculate half of each by means of (1) (for example the indices $\pi_{ij}(a_h,a_k)$, $h < k$) since the symmetric indices may be immediately obtained from (2).

Then, considering all the $\binom{m}{2}$ feasible unranked pairs of criteria g_i, $g_j \in G$, we obtain m(m-1)/2 matrices of the $\underline{\boldsymbol{\pi}}_{ij}$ type. The weighted sum of these will lead to the square n x n aggregated matrix of preferences $\underline{\boldsymbol{\pi}}$:

$$\underline{\boldsymbol{\pi}} = \sum_{\substack{i,j \\ (i<j)}} \underline{\boldsymbol{\pi}}_{ij} \frac{\lambda_i + \lambda_j}{m - 1}, \tag{3}$$

whose generic element $\pi(a_h,a_k)$ indicates the **aggregated preference intensity** between each pair of feasible actions. Naturally, it will again be $\pi(a_h,a_k) + \pi(a_k,a_h) = 1$, for all $(a_h,a_k) \in A \times A$ and with regard to the preference (P) and indifference (I) relations with reference to the family G of criteria considered will be, for all a_h, $a_k \in A$:

$$a_h \; P \; a_k \iff 0 \leq \pi(a_k,a_h) < 0.5 < \pi(a_h,a_k) \leq 1$$
$$a_h \; I \; a_k \iff \pi(a_h,a_k) = \pi(a_k,a_h) = 0.5.$$

Observe that the index $\pi(a_h,a_k)$ is equal to 1 if and only if $a_h \; D_s \; a_k$; it is equal to 0 if and only if $a_k \; D_s \; a_h$.

Of course, also with reference to the aggregated preference indices, it is possible to build particular complete valued preference structures. For example, we may consider the structure of semiorder, obtained, as mentioned above, by introducing a real parameter δ, $\delta \in [½, 1]$, which in this case emphasizes the aggregated indifference relation. Alternatively, by introducing two parameters δ, ϵ, $½ \leq \delta < \epsilon \leq 1$, it is possible to build a complete two-valued preference structure, which considers only two levels of aggregated preference intensity, represented by the two relations P (strict preference) and Q (weak preference) (see Matarazzo, 1988b; Roubens and Vincke, 1985; Vincke, 1990).

3. THE BUILDING OF PREFERENCE RANKINGS

The indices of preference intensity contained in the aggregated matrix $\underline{\pi}$ may, among other things, permit the building of specific partial or complete rankings of feasible actions. The various techniques possible (Matarazzo, 1984 and 1986) based on the concept of *qualification* of a feasible action, introduced by Roy, or on the sum of the aggregated preference indices referred to a feasible action, aim to obtain the partition of A into S equivalence classes $C_1, C_2, \ldots, C_S$, $S \le n$ (complete preorder), by means of a descending procedure (from the *best* action to the *worst*) or by an ascending procedure (from the *worst* to the *best*). In either case, the peculiar feature of these techniques is that at every step they select the action(s) assigned to a certain position in the ranking considered and then repeat the procedure for the subset of the remaining actions, eliminating with each iteration the actions selected in the preceding one. Here is a brief example of one of the techniques proposed. Calculate the **global preference index:**

$$\sigma_+^{(1)}(a_h) = \sum_{a_k \in A-\{a_h\}} \pi(a_h,a_k), \; h = 1,2,\ldots,n.$$

This will be:

$$0 \le \sigma_+^{(1)}(a_h) \le n - 1, \text{ for all } a_h \in A.$$

In particular we obtain:

$$\sigma_+^{(1)}(a_h) = n - 1 \quad \text{or} \quad \sigma_+^{(1)}(a_h) = 0,$$

if and only if a_h strictly dominates, or is strictly dominated by, respectively, all the remaining feasible actions. We then select the action(s) with the highest index $\sigma_+^{(1)}$ of all those obtained. This action, or these actions, will occupy the first place in the decreasing ranking, forming class C_1. Then, given $A^{(1)} = A - C_1$, we repeat the procedure with reference to the actions in this new subset, obtaining the indices:

$$\sigma_+^{(2)}(a_h) = \sum_{a_k \in A^{(1)}-\{a_h\}} \pi(a_h,a_k), \; a_h \in A^{(1)}.$$

This iteration will make it possible to form class C_2, and so on.

The *increasing solution* may be obtained by calculating the global indices

$$\sigma_-^{(1)}(a_h) = \sum_{a_k \in A-\{a_h\}} \pi(a_k,a_h), \; h = 1,2,\ldots,n,$$

and placing in the last class C_s the action(s) which present the highest value for this index. We then proceed with the calculation of the indices $\sigma_-^{(2)}$ related to the subset $A - C_s$ and so on.

From the intersection of the two rankings (complete preorders) thus obtained, it is possible to obtain the final ranking which, in general, is a **partial preorder** containing the *sure part* of the decision maker's preferences and accentuating the incomparabilities (decreasing and increasing conflictual solution) (Matarazzo, 1986; Roy, 1985).

If, on the other hand, we prefer to obtain a **complete preorder** from matrix $\underline{\pi}$, it is sufficient to associate with every feasible action a_h the index:

$$\sigma^{(t)}(a_h) = \sigma_+^{(t)}(a_h) - \sigma_-^{(t)}(a_h),$$

which expresses, with every iteration $t = 1,2,\ldots,$ the difference between how far a_h is globally preferred to the remaining actions and how far these are preferred to a_h.

Observe that from the global preference indices introduced, suitably normalized at each step, it is possible to obtain and interpret also graphically some reliability measures of the global preference relations among the feasible actions (see Matarazzo, 1988b).

Of course, in the building of all the complete preorders it is possible to introduce *global indifference thresholds*, to prevent small differences in the global indices considered at every iteration from assuming a discriminating role (Matarazzo, 1990).

It is also possible and useful to consider the existence of criteria in **discordance** with the relation a_h P a_k (Roy, 1985 and 1990). The criterion $g_j \in G$ will therefore be discordant if and only if $\pi(a_h, a_k) > 0.5$ and $g_j(a_h) < g_j(a_k)$. We may then measure the marginal intensity of this discordance, by introducing a specific index $d_j(a_h, a_k)$, $j \in J$, to compare with a marginal veto threshold v_j $(v_j > 0)$, $j \in J$, established by the decision maker. It is sufficient at this point to fulfil only one condition $d_j(a_h, a_k) \geq v_j$, $j \in J$, in order to create a situation of **veto preference** for a_h with respect to a_k. The unique feature of the fundamental principles of the MAPPAC method makes it also possible to introduce, for every pair of criteria (g_i, g_j), a discordance index $d_{ij}(a_h, a_k)$ and a corresponding threshold r_{ij} $(r_{ij} > 0)$, so that $d_{ij}(a_h, a_k) \geq r_{ij} \Rightarrow a_h \; R_{ij} \; a_k$.

The incomparability relation R_{ij}, therefore, introduced in order to reduce the compensatory effect of the basic preference indices, shows a definite refusal to make a decision in the preference modelling phase. There will therefore be **explicit incomparability** between the feasible actions a_h and a_k, if and only if at least one relation occurs of the $a_h \; R_{ij} \; a_k$, $i, j \in J$. Further details and the formulation of the indices mentioned here may be found in (Matarazzo, 1988b).

4. PREFERENCE RANKING FREQUENCIES IN THE PRAGMA METHOD

The PRAGMA method is also based on pairwise comparisons by means of distinct criteria. This method however does not aim to build preference indices for pairs of feasible actions, but rather to help the decision maker in his choice problems by supplying him with specific information: the partial and global fre-quencies with which each action is present at a given *level*.

The method constitutes a useful complement to MAPPAC, to which it is related in the methodology and instruments used; moreover it uses the MAPPAC basic preference indices for its calculations.

Let g_i, g_j be a pair of distinct criteria of the family G. Using the same symbols as above, the corresponding **partial profile** of an action a_h, $a_h \in A$, is given by the segment H_iH_j (figure 2), where the point H_i has the ordinate $\lambda_i c_{hi}$ and the point H_j has the ordinate $\lambda_j c_{hj}$.

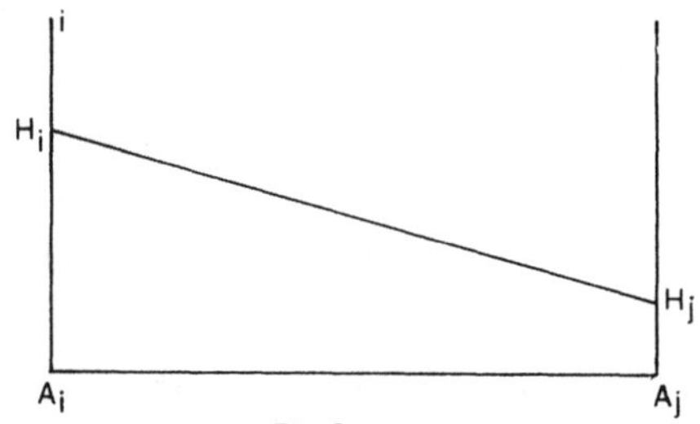

Partial profile of action a_h

Fig.2

Considering all unranked pairs of criteria, it is possible to obtain $\binom{m}{2}$ distinct partial profiles of a_h. We call **global profile** of a_h the set of these $\binom{m}{2}$ partial profiles.

We define as **partial broken line-k**, or **partial broken line of level k** of a_h, $k = 1,2,...,n$, the set of consecutive segments of its partial profiles, to which correspond, for each point, $k - 1$ partial profiles (distinct or coinciding) of greater ordinate. If, for example, it is $A = \{a_r, a_s, a_t\}$ and if the partial profiles are those represented in figure 3, we obtain:

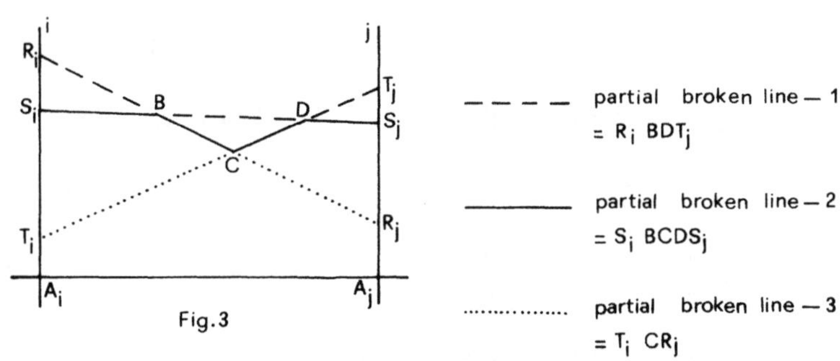

Fig.3

Partial profiles and partial broken lines of a_r, a_s, a_t.

We observe that, in general, the partial broken line-k, $k = 1, 2, \ldots, n$, coincides with the partial profile of a_h, $a_h \in A$, if and only if a_h is partially dominated by d actions $(0 \leq d \leq k-1)$ and dominates the remaining ones, and/or if $(0 \leq p \leq k-1,\ d + p = k-1)$ of distinct actions a_r, $a_s \in A$ exist such that the partial profiles of a_h, a_r and a_s intersect at the same point (Matarazzo, 1988a).

Further, we define as **global broken line-k** or **global broken line of level k** $(k = 1,2,\ldots,n)$ the set of $\binom{m}{2}$ partial broken lines-k obtained by considering all the unranked pairs of distinct criteria g_i, $g_j \in G$.

The global broken line-k coincides with the global profile of a_h if and only if all the partial broken lines of level k, obtained by considering each of the $\binom{m}{2}$ pairs of criteria, coincide with the corresponding partial profiles of a_h.

We define as the **partial frequency of level k** $(k = 1,2,\ldots,n)$ of a_h, with reference to the criteria g_i and g_j, the value of the orthogonal projection on the straight line A_iA_j (given $\overline{A_iA_j} = 1$) of the intersection of the partial profile of a_h with the corresponding partial broken line of level k. If we indicate this frequency as $f_{ij}^{(k)}(a_h)$, it will be $0 \leq f_{ij}^{(k)}(a_h) \leq 1$, for all $a_h \in A$, $k = 1, 2, \ldots, n$. Thus, for example, from the graph in figure 4:

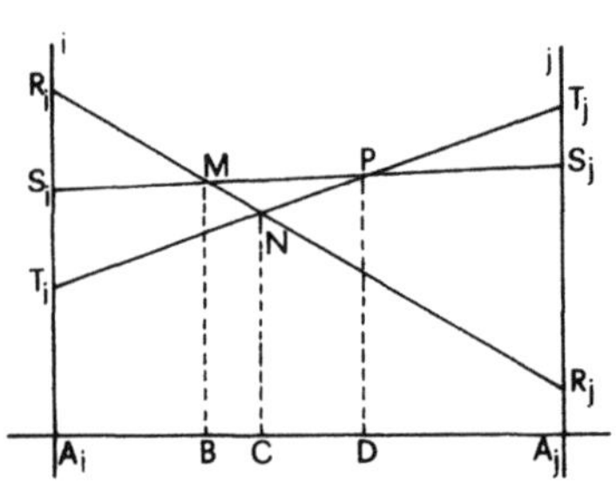

$\overline{A_iA_j} = 1$; $\overline{A_iB} = 0{,}3$; $\overline{BC} = 0{,}1$; $\overline{CD} = 0{,}2$; $\overline{DA_j} = 0{,}4$

$f_{ij}^{(1)}(a_r) = 0{,}3$; $f_{ij}^{(2)}(a_r) = 0{,}1$; $f_{ij}^{(3)}(a_r) = 0{,}6$;

$f_{ij}^{(1)}(a_s) = 0{,}3$; $f_{ij}^{(2)}(a_s) = 0{,}7$; $f_{ij}^{(3)}(a_s) = 0$;

$f_{ij}^{(1)}(a_t) = 0{,}4$; $f_{ij}^{(2)}(a_t) = 0{,}2$; $f_{ij}^{(3)}(a_t) = 0{,}4$.

Fig.4 Partial frequencies of a_r, a_s, a_t.

The partial frequencies may be represented in matrix form, obtaining the square n x n matrix $\underline{F}_{ij}$, which is the matrix of the partial ranking frequencies:

$$\underline{F}_{ij} = [f_{ij}^{(k)}(a_h)], \; h \in I; \; k = 1,2,\ldots,n; \; i,j \in J. \qquad \textbf{(4)}$$

The elements of the hth line of **(4)** indicate in order the fractions of the interval (A_i, A_j) for which the action a_h is in the kth position $(k = 1,2,\ldots,n)$, while the elements of the kth column of **(4)** indicate those fractions for which the kth position (in the partial preference ranking considered) is assigned to the actions $a_1, a_2, \ldots, a_n$, respectively. Obviously:

$$\sum_{k=1}^{n} f_{ij}^{(k)}(a_h) = 1, \; \forall a_h \in A \; \text{ and } \sum_{h=1}^{n} f_{ij}^{(k)}(a_h) = 1, \; k = 1,2,\ldots,n.$$

If $f_{ij}^{(k)}(a_h) \in \{0,1\}$, for all $a_h \in A$ and $k = 1,2,\ldots,n$, the partial profiles of all the actions will be non-coinciding, and there will be no inversions with respect to the preference relation in the complete preference preorders obtained distinctly by means of the criteria g_i and g_j. If v $(v = 2,3,\ldots,n)$ partial profiles are coinciding, the corresponding partial broken lines-k must be built taking distinctly into account the coinciding profiles v times (Matarazzo, 1988a).

Let us then define **global frequency of level k** $(k = 1,2,\ldots,n)$ of a_h as the weighted arithmetical average of all the $\binom{m}{2}$ partial frequencies of level k of a_h, obtained by considering all the (unranked) pairs of distinct criteria g_i and g_j. Therefore, designating this frequency by $f^{(k)}(a_h)$, we obtain:

$$f^{(k)}(a_h) = \sum_{\substack{i,j \\ (i<j)}} f_{ij}^{(k)}(a_h) \, \frac{\lambda_i + \lambda_j}{m-1}, \quad a_h \in A, \; k = 1,2,\ldots,n.$$

The linear combination of the matrices **(4)** with weights $(\lambda_i+\lambda_j)/(m-1)$ will therefore give the square n x n matrix $\underline{F} = [f^{(k)}(a_h)]$ $(h = 1,2,\ldots,n;\; k = 1,2,\ldots,n)$, called the global ranking frequency matrix. Its generic element $f^{(k)}(a_h)$ indicates the relative frequency with which $a_h \in A$ is present in the kth position ($k = 1,2,\ldots,n$) in the particular ranking obtained by considering all the criteria $g_j \in G$ and the global profiles of all the feasible actions. It will therefore be:

$$\sum_{k=1}^{n} f^{(k)}(a_h) = 1, \ \forall \ a_h \in A \text{ and } \sum_{h=1}^{n} f^{(k)}(a_h) = 1, \ k = 1, 2, \ldots, n.$$

It is possible to calculate the partial frequencies $f_{ij}^{(k)}(a_h)$ by means of an algorithm which instrumentalizes the indices $\pi_{ij}(a_h, a_k)$ **(1)** of the MAPPAC method (see Matarazzo, 1988a). It is therefore possible to consider marginal indifference thresholds and suitable indifference areas also when the PRAGMA method is used. In other words, the indices $\pi_{ij}(a_h, a_k)$ instrumentally used, may be calculated in advance by using all the techniques adopted with reference to the MAPPAC method (see Matarazzo, 1988b).

Apart from these calculations, it is useful in any case to remember among others some particular features belonging to the ranking frequencies obtained with the PRAGMA method:

1) The partial frequencies (and therefore also the global ones) of $a_h \in A$ are functions of the sign and the value of the weighted differences between the evaluations of a_h and those of the remaining feasible actions with respect to the criteria considered. The values of these weighted differences may be overlooked only in the case of partial dominance (for partial frequencies) or strict dominance (for global frequencies), active or passive, of the action a_h.

2) If a_h partially dominates n-k actions and it is partially dominated by the remaining k-1 actions, $k = 1,2,\ldots,n$, the result is $f_{ij}^{(k)}(a_h) = 1$, whatever the values of λ_i and λ_j.

3) If a_h strictly dominates n-k actions and is strictly dominated by the remaining k-1 actions, $k = 1,2,...,n$, the result is $f^{(k)}(a_h) = 1$, whatever the values of the weights w_j, $j \in J$.

4) If $f^{(k)}(a_h) = 1$, the action a_h occupies the kth position, $k = 1,2,...,n$, in every monocriterion ranking. In general, however, the converse is not true, because a_h may not be preceded (and followed) by the same subset of actions in every monocriterion ranking.

Therefore, the information obtained by means of an analysis of the global frequencies $f^{(k)}(a_h)$ is more complete and more accurate than that obtained from an examination of all the distinct monocriterion rankings of the feasible actions, or from a mixture of these.

5. THE USE OF RANKING FREQUENCIES TO SOLVE DECISIONAL PROBLEMS

We maintain that it is often sufficient to analyze the elements of matrix $\underline{F}$ to obtain those indications which can give concrete help to the decision maker in all multicriteria choice problems (Roy, 1985; Scharlig, 1985). To this purpose, in fact, the concise and accurate information regarding the positions each action may occupy in an aggregated and/or partial multicriteria ranking can prove extremely useful. A straightforward reading of the global frequencies of matrix $\underline{F}$ is often sufficient to indicate which action(s) will finally be chosen.

If we want to obtain complete or partial rankings of the feasible actions it is possible, for example, to proceed in this way. Calculate, for each action $a_h \in A$, the **accumulated frequencies** of order k, $k = 1,2,...,n$, summing the first k elements of the hth row of matrix $\underline{F}$, that is:

$$F^{(1)}(a_h) = f^{(1)}(a_h) \text{ and } F^{(k)}(a_h) = \sum_{i=1}^{k} f^{(i)}(a_h), \quad k = 2,3,...,n.$$

Then establish the **order** q (q = 1,2,...,n-1) of the frequencies which are considered relevant to the building of the ranking, that is indicate to what order q we intend to take into consideration the accumulated frequencies $F^{(k)}(a_h)$ for this purpose. The following index is then built:

$$S^{(q)}(a_h) = \sum_{k=1}^{q} \alpha_k F^{(k)}(a_h), \quad a_h \in A; \; 1 \geq \alpha_1 \geq \alpha_2 \geq \ldots \geq \alpha_q > 0. \qquad (5)$$

This gives the measure of the "strength" with which a_h occupies the first q positions in the aggregated ranking. This in practice will be $1 \leq q \leq n/2$, which regards the first positions in the ranking; the coefficients α_k indicate the relative importance of these (not increasing with k). In the first class C_1 of the decreasing ranking will be placed the action(s) to which the maximum value of $S^{(q)}(a_h)$ corresponds. In order to avoid ex aequo rankings, we proceed by selecting whichever actions have obtained an equal value of $S^{(q)}$ on the basis of the values of the indices $S^{(q+1)}$ and, in the case of further equality, on those of the indices $S^{(q+2)}$ and so on. In this case ex aequo actions would be accepted only if their corresponding indices $S^{(i)}$ proved equal, for i = q, q+1, ..., n. If, on the other hand, we desire to prevent small differences in the indices $S^{(q)}$ from having a discriminatory role in the building of the rankings, it is possible to consider global indifference thresholds (see Matarazzo, 1988b).

If we place $\alpha_k = 1$, for all k, in **(5)**, we do not emphasize the greater importance of the global ranking frequencies of the first positions. If we then accept q = 1, this takes into account only the global frequencies of the first position for the purpose of building the ranking.

After building class C_1, with reference to the subset of the remaining actions $A^{(1)} = A - C_1$, we again calculate the partial, global and accumulated frequencies and the index **(5)**, proceeding as above in order to build class C_2, and so on. We observe that at each iteration t the order q_t, on the basis of which the index

$S^{(q_t)}$ **(5)** is to be calculated, must be recalculated so that it is a non increasing whole number and, taking into account the number $|A^{(t)}|$ of actions of the evaluation set at each iteration t, the ratio $q_t/|A^{(t)}|$ is as near as possible to the ratio q/n of the first iteration (see Matarazo, 1988a). In general, the rankings obtained are a function of the value of the order q originally selected (see Matarazo, 1988b).

If at each useful iteration $F^{(K)}(a_r) \geq F^{(K)}(a_s)$ for all k=1,2,...,n and $F^{(k)}(a_r) > F^{(k)}(a_s)$ for some k, or if $\sum_{k=1}^{n} [F^{(k)}(a_r) - F^{(k)}(a_s)] \geq 0$ for all n and $F^{(k)}(a_r) \neq F^{(k)}(a_s)$ for some k, it is possible to speak of **first degree** or **second degree frequency dominance**, respectively, of a_r over a_s. In boh cases, if $\alpha_1 > \alpha_2 > \ldots > \alpha_q$, a_r will precede a_s in any of the rankings obtained, whatever value may be chosen for the other q.

Besides the partition of the actions of A into equivalence classes (complete preorder) obtained with the descending procedure (or procedure *from above*) described, it is also possible to build another complete preorder in the same way using the ascending procedure (or procedure *from below*), that is selecting the action(s) to be placed in the last, next to last, ... and finally in the first equivalence class.

In conclusion, it is possible to build a final ranking (partial preorder) of the feasible actions, as the intersection of the two decreasing and increasing rankings obtained by means of the two separate procedures described. Using the PRAGMA method for the building of rankings, it is possible not only to establish any implicit incomparability deriving from the inversion of preference in the preorders obtained by means of the two separate procedures, but also in this case to consider an **explicit incomparability** to be obtained if the relative tests give a positive result, during the preference modelling phase. Since, as we have said, the PRAGMA method makes instrumental use of the basic preference indices (1), it is possible to use once again the same discordance indices already introduced in the MAPPAC method (see Matarazzo, 1988b).

Besides these, moreover, it is also possible to consider other analogous **discordance indices** peculiar to the PRAGMA method, built that is using the partial and global ranking frequencies. Thus, for example, with respect to a pair of criteria both considered important, or to all the criteria simultaneously, a strengthening of the ranking frequencies of an action a_h, respectively partial or global, corresponding to the first and last positions in the ranking, can reveal strongly discordant evaluations of a_h by means of those criteria. Therefore this kind of situation, suitably analysed, could lead the decision maker to reconsider the nature of a_h, and, in the building phase of the rankings, this situation may then lead to a rapid choice of a_h both in the decreasing and in the increasing procedure, resulting in situations of conflictuality and implicit incomparability.

6. FINAL CONSIDERATIONS

This brief description of the two methods presented permits us to reflect a little on some of their more relevant aspects with reference to MCDA.

The partial information supplied by both methods can in our opinion prove very useful by helping the decision maker to face concrete problems. Often the possibility of PCCA to focus attention on one pair of criteria of particular relevance, and to make partial comparisons of various types, proves of fundamental importance in decisional problems and may be connected to the concepts of compensation, marginal substitution rate and trade-off typical of other classes of MCDA.

The comparison by means of pairs of criteria, peculiar to the two methods described, by accentuating partial concordance (dominance) and discordance, likens them to certain features of the outranking methods (Roy, 1990; Brans and Mareschal, 1990), additive and non-additive (Mareschal, 1988), of some soft discrete multicriteria methods (Hinloopen et al., 1983; Nijkamp and Voogd, 1985; Janssen et al., 1990) and of the multiattribute utility theory (Keeney and Raiffa, 1976; Vincke, 1985). In effect the basic preference indices $\pi_{ij}(a_h,a_k)$ (1) are a function of the value of the differences

$$\Delta_i(a_h,a_k) = g_i(a_h) - g_i(a_k) \text{ and}$$
$$\Delta_j(a_h,a_k) = g_j(a_h) - g_j(a_k) \text{ if and only if}$$
$$\Delta_i(a_h,a_k).\Delta_j(a_h,a_k) < 0;$$

in this case they are also of a compensatory type. If on the other hand $\Delta_i(a_h,a_k).\Delta_j(a_h,a_k) \geq 0$ results, they are only a function of the signs of these differences and therefore of a noncompensatory type (Jacquet-Lagrèze, 1985). If we make a partition of the set G of the criteria considered in subsets:

$$G^+(a_h,a_k) = \{g_i \in G \mid g_j(a_h) > g_j(a_k)\}$$
$$G^-(a_h,a_k) = \{g_j \in G \mid g_j(a_h) < g_j(a_k)\}$$
$$G^=(a_h,a_k) = \{g_j \in G \mid g_j(a_h) = g_j(a_k)\},$$

of the $\binom{m}{2}$ indices $\pi_{iJ}(a_h,a_k)$, $|G^+(a_h,a_k)|.\ |G^-(a_h,a_k)|$ will be of the first type and all the others of the second type (see Matarazzo, 1990-).

The MAPPAC method offers preference indices (partial and aggregated) which constitute a complete valued preference structure; the PRAGMA method offers ranking frequencies (partial and global). The two methods appear to us to be complementary in that the information they supply, which seems on the surface to be very different, in fact makes each the natural, logical and methodological complement to the other. In the phase of using this information for the building of a ranking of feasible actions, the techniques used in both methods have one feature in common: whichever procedure (descending or ascending) is adopted, each iteration considers only the feasible actions not yet selected necessary (especially in the PRAGMA method) but, in our opinion, offers an advantage not to be overlooked. In fact, by operating in this way it is possible to reduce the risk that a dominant or dominated action may have a discriminating role over the remaining actions in the building of a ranking (see Matarazzo, 1988b).

It is possible with each method to consider both an implicit incomparability, due to the inversion of preference in the decreasing and increasing rankings obtained, and an explicit incomparability, emerging in the preference modelling phase as a result of the presence of strongly discordant evaluations.

In both methods, finally, it is possible to consider suitable thresholds and to carry out the relevant tests in order to make them as flexible as possible with regard to the decision maker's operative needs and requirements. In our opinion, the two methods illustrated represent only an example - even if a significant one - of the considerable possibilities that PCCA can offer, on a theoretical and practical level, which deserve to be made the object of further research.

REFERENCES

Brans, J.-P., Mareschal, B. and Vincke. Ph. (1984): "PROMETHEE: a new family of outranking methods in multicriteria analysis", in J.-P. Brans (ed.), *Operational Research' 84*, North-Holland, 477-490.

Brans, J.-P. and Mareschal, B. (1990): "The PROMETHEE methods for MCDM, the PROMCALC, GAIA and BANKADVISOR software", in this volume.

Hinloopen, E., Nijkamp, P. and Rietveld, P. (1983): "Qualitative discrete multiple criteria choice models in regional planning", *Regional Science and Urban Economics*, 13, 77-102.

Jacquet-Lagrèze, E. (1985): "Basic concepts for multicriteria decision support", in G. Fandel, J. Spronk and B. Matarazzo (eds.), *Multiple Criteria Decision Methods and Applications*, Springer-Verlag, Berlin, 11-26.

Janssen, R., Nijkamp, P. and Rietveld, P. (1990): "Qualitative multicriteria methods in the Netherlands", in this volume.

Keeney, R.L. and Raiffa, H. (1976): *Decision with Multiple Objectives: Preferences and Value Trade-offs*, Wiley, New York.

Mareschal, B. (1988): "Weight stability intervals in multicriteria decision aid", *EJOR*, 33 (1), 54-64.

Matarazzo, B. (1984): "The MAPPAC Method", Università di Catania.

Matarazzo, B. (1986): "Multicriterion analysis of preferences by means of pairwise actions and criterion comparisons", *Applied Mathematics and Computation*, 18, 119-141.

Matarazzo, B. (1988a): "Preference ranking global frequencies in multicriterion analysis (PRAGMA)", *EJOR*, 36, 36-49.

Matarazzo, B. (1988b): "A more effective implementation of the MAPPAC and PRAGMA methods", *Foundations of Control Engineering*, 13, 155-173.

Matarazzo, B. (1990-): "MAPPAC as a compromise between outranking methods and MAUT", forthcoming in *EJOR*.

Nijkamp, P. and Voogd, H. (1985): "An informal introduction to multicriteria evaluation", in G. Fandel, J. Spronk and B. Matarazzo (eds.), *Multiple Criteria Decision Methods and Applications*, Springer-Verlag, Berlin, 61-84.

Roubens, M. and Vincke, Ph. (1985): *Preference Modelling*, Springer-Verlag, Berlin.

Roy, B. (1985): *Méthodologie Multicritère d'Aide à la Décision*, Economica, Paris.

Roy, B. (1990): "The outranking approach and the foundations of the ELECTRE methods", in this volume.

Schärlig, A. (1985): *Décider sur plusieurs critères*, Presses Polytechniques Romandes, Lausanne.

Vansnick, J.-C. (1990): "Measurement theory and decision aid", in this volume.

Vincke, Ph. (1985): "Multiattribute utility theory as a basic approach" in G. Fandel, J. Spronk and B. Matarazzo (eds.), *Multiple Criteria Decision Methods and Applications*, Springer-Verlag, Berlin, 27-40.

Vincke, Ph. (1990): "Basic concepts of preference modelling", in this volume.

CHAPTER III

VALUE AND UTILITY THEORY APPROACH

CONJOINT MEASUREMENT: THEORY AND METHODS

Rakesh K. Sarin

The Fuqua School of Business
Duke University
Durham, NC 27706
U.S.A.

1. INTRODUCTION

Which of the two hotels do you prefer?

* $105 per night, small room restaurant facilities, convenient location in downtown, impersonal service,

or

* $75 per night, large room, no restaurant facilities, one mile from downtown, warm and friendly service.

In some cases your preference between the two hotels may be determined by some simple rule; for example, a budget constraint. In general, however, you will consider tradeoffs among the levels of the multiple attributes in your decision. The process of assessing tradeoffs among attributes of interest may not be explicit; however, you do engage in some balancing of desirable and undesirable features.

How do people choose among multiattribute alternatives (products or services) is the objective in *conjoint measurement*? For managerial decision making, a procedure is needed to *measure* your preferences so that one may be able to assess which attributes are more important to you than the others and how you tradeoff a gain in one attribute to a loss in some other attribute.

In this paper we will discuss the theory of conjoint measurement and some methods to implement the theory in practical managerial settings. We begin by providing a brief measurement in our context and then proceed to discuss a special type of measurement-conjoint measurement in some detail. The discussion in this paper is largely based on Krantz et al. (1971).

2. WHAT IS MEASUREMENT?

A common meaning of «measurement» is to assign numbers to empirical objects according to some definite scheme. In *extensive measurement* a set of assumptions about qualitative ordering $\succsim$ and a concatenation operation o are made to obtain a scale ϕ satisfying

(i) $a \succsim b$ if and only if $\phi(a) \geq \phi(b)$

(ii) $\phi(a o b) = \phi(a) + \phi(b)$.

An example is «length» measurement where a and b could be two rods, $\succsim$ is «at least as long as» and aob denotes laying rods a and b end to end in a straight line so $\phi(aob)$ measures the combined length of rods a and b.

In behavioral and social sciences it is rare to find empirical concatenation operations o that meet the assumptions of extensive measurement. Some argued therefore that fundamental measurement - one that does not require prior measurement of other quantities - is impossible in social sciences. Of course in social sciences we have been measuring all sorts of attributes for a long time but to physicists these were not «kosher». This is what they had to say to some measurement processes employed in social sciences. «...To insist on calling these other processes measurement adds nothing to their actual significance but merely debases the coinage of verbal intercourse. Measurement is not a term with some mysterious inherent meaning, part of which may have been overlooked by physicists and may be in course of discovery by psychologists. It is merely a word conventionally employed to denote certain ideas. To use it to denote other ideas does not broaden its meaning but destroys it: we cease to know what is to be understood by the term when we encounter it; our pockets have been picked of a useful coin...»

Conjoint measurement that we cover in this paper is an example of a fundamental measurement and is applicable to a wide variety of social science situations. The basic technique of «counting of units» (see Krantz et al., 1971, 1.1.2) can be applied, though in an indirect way, to measure attributes like preference for brands of products, intelligence, aesthetic value of plays, loudness, brightness, fear, or power.

3. WHAT IS CONJOINT MEASUREMENT?

Suppose we are interested in measuring an attribute of an object (such as preference for a car). Further suppose the object is composed of *two or more* components, each of which affects the attribute in question. For example, the components price, size of engine, body type may influence our preference for a car.

Conjoint measurement deals with the construction of measurement scales for composite objects (such as cars) which preserve their observed order with respect to the relevant attribute (preference) and where the scale value of each object (car) is a function of the scale values of its components (price, size of engine, body type). The word conjoint is used to highlight that the objects and their components are measured simultaneously.

In *additive* conjoint measurement the scale value for the object is the simple sum of the scale values of the components. Thus, loosely speaking,

$$\phi(\text{car}) = \phi_1(\text{price}) + \phi_2(\text{size of engine}) + \phi_3(\text{body type}).$$

Note, and note well, that ϕ_1, ϕ_2, ϕ_3 are not constructed independently and then ϕ obtained by simply summing these three scales. Thus, you cannot interpret ϕ_1 by itself and nor can you construct it independent of other components. I hammer on this point because I have seen many mis-applications of conjoint measurement where ϕ_i's are constructed using independent processes such as rating scales and ϕ is obtained by simply summing these component scales.

4. PRELIMINARIES

Throughout this paper we will use the set up as follows:

There is a set of objects X and a decision maker (d.m.) is willing to make paired comparisons between any two objects x and y in X. These paired comparisons take the form «x is preferred or indifferent to y» denoted $x \succsim y$. If $x \succsim y$ and $y \succsim x$ then we simply write $x \sim y$ and say that «x is indifferent to y». Similarly, we call «x is strictly preferred to y» denoted $x \succ y$ if $y \not\succsim x$.

Our objective is to assign numbers to objects in set X (choice set) such that a higher number corresponds to a more preferred object. So we are looking for a function from the choice set to the real numbers. The name of the game is to somehow make it «easy» to assign numbers to the objects. But, to do so we will have to make some assumptions (called axioms) about the preferences of the d.m. among objects in X. These are qualitative or behavioral assumptions about the relation $\succsim$.

The theory therefore is specific to a d.m. who accepts these assumptions about her own behavior. Occasionally, we will argue, maybe in vain, that all «rational» d.m.'s should accept some of the assumptions. I frankly do not quite know the definition of «Rational» so we will not use this word too often. You should ask yourself whether upon sufficient introspection do you accept a particular assumption or not.

A representation theorem will be stated that will show us how numbers are assigned to objects (define a function ϕ that maps X on to $\mathbb{R}$). Some axioms are called *necessary*, that is, these must be satisfied for a representation to be possible. Some are called *sufficient*, that is, if these hold then a representation is possible.

Suppose we employ a set of axioms to prove a representation theorem that defines a function ϕ that is used to order objects in X by the relation $\succsim$. An example in conjoint measurement is the additive case, where

$$\phi(x) = \phi_1(x_1) + \phi_2(x_2) + \phi_3(x_3).$$

Now, think of our car example and we can order all the cars in set X using above expression.

Our ranking of cars using ϕ is based on our preferences and we do not frankly care if some other ϕ' is employed so long as it provides us the same ranking as we obtained by using ϕ. In the above example every $\phi' = \alpha\phi + \beta$, $\alpha > 0$ (α and β are two constants) will provide the same ranking of cars in X as ϕ does.

Such results are called *uniqueness* results and we use the phrase ϕ is unique up to ... (in our example above a positive linear or affine transformation). We also use the term ϕ is an *interval scale*. Other examples are *ordinal scale* where ϕ and $f(\phi)$ (f is strictly increasing real-valued function of a real variable ϕ) provide the same ranking, *ratio scale* where ϕ and $\alpha\phi$, $\alpha > 0$, provide the same ranking, and *log-interval* scale where ϕ and $\alpha\phi^{\beta}$, $\alpha > 0$, $\beta > 0$ provide the same ranking. For each of our key results we will state these uniqueness results as well.

In summary our set up has the following elements:

(i) a set of objects X called a choice set;

(ii) a preference relation $\succsim$ and axioms about d.m.'s preferences among objects in X;

(iii) a representation theorem that specifies a function for assigning numbers to objects in X;

(iv) a uniqueness result which specifies the family of functions which are equally valid representation since they all provide the same ranking of objects in X.

5. NOTATION AND GOAL IN CONJOINT MEASUREMENT

A composite object x belongs to a set X ($x \in X$), where

$$X = X_1 \times X_2 \times \ldots \times X_n,$$

and $n \geq 2$. The interpretation for X_i is the set of scores, outcomes, or payoffs on component or criterion i. Thus, an object $x = (x_1, x_2, \ldots, x_n)$ is represented by its scores on n criteria. A car costing $15,000, body type hatch back, and 4-cylinder engine will be represented as ($15,000, hatch-back, 4-cylinder). If there are 6 possible body types and 5 possible engine types, then X_2 will have 6 elements and X_3 will have 5 elements.

We are looking for a function ϕ so that a car $x = (x_1, x_2, x_3)$ is preferred or indifferent to a car $y = (y_1, y_2, y_3)$. We need assumptions about $\succsim$ so that there exists $\phi: X \to \mathbb{R}$ such that

$$x \succsim y \text{ iff } \phi(x) \geq \phi(y), \text{ for all } x, y \in X.$$

The above representation holds if X is finite and *(i)* $x \succsim y$ or $y \succsim x$ for all $x, y \in X$ *(ii)* if $x \succsim y$, $y \succsim z$, then $x \succsim z$, for all $x, y, z \in X$.

A preference relation that satisfies (i) and (ii) above is called a *weak order*. With modest axioms (i) and (ii) we can get ϕ but it is not very useful in most applications. To construct ϕ in our car example, assign any number to a car x. Now pick another car y if it is preferred to car x assign it a higher number and if x is preferred to it assign it a lower number. Now pick the third car z and use the same procedure; except, if it is in between x and y (preferred to x and less preferred to y) then assign it a number between $\phi(x)$ and $\phi(y)$. The process continues till all cars in X are assigned numbers. If two cars are equal in preference then these are assigned the same number. This is pure ordinal measurement and as evident (I hope) not so helpful. After all you cannot get more powerful measurement with such weak inputs (axioms) as (i) and (ii) above. So we want more. At the very least we want to further restrict ϕ so that

$$\phi(x) = F(\phi_1(x_1), \phi_2(x_2), \ldots, \phi_n(x_n)),$$

where F is increasing in each of its arguments so that we can interpret ϕ_i's as measuring the «goodness» of the ith component. This is useful since we can talk about $x_i \succsim_i y_i$ if $\phi_i(x_i) \geq \phi_i(y_i)$ for the ith component by fixing other components at common values. However, to simplify measurement still further we need to impose additional assumptions so that

$$\phi(x) = F(\phi_1(x_1), \phi_2(x_2), \ldots, \phi_n(x_n)) = \sum_{i=1}^{n} \phi_i(x_i).$$

This then is the goal of *additive conjoint measurement* and now the «goodness» of x is the sum of the «goodness» of each of its components.

We will first discuss necessary and sufficient conditions for additive conjoint representation. Then will be examine how ϕ_i's can be constructed in practical applications.

6. NECESSARY CONDITIONS FOR ADDITIVE CONJOINT MEASUREMENT

Since we are getting a numerical representation the first condition is simply that $\succsim$ is a weak order

<u>**Axiom 1**</u> $\succsim$ is a weak order.

This axiom requires that the d.m. is able to compare any two objects $x, y \in X$ (indicate $x \succsim y$ or $y \succsim x$) and these preferences are transitive (if $x \succsim y$, $y \succsim z$, then $x \succsim z$ for all $x, y, z \in X$).

We now state an independence condition that we call *mutual preferential independence* which states that preference order over any two components does not depend on the fixed levels of the remaining n-2 components. For n-3, this condition will require that if

$$(x_1', x_2', x_3) \succsim (x_1'', x_2'', x_3),$$

then for all $y_3 \in X_3$,

$$(x_1', x_2', y_3) \succsim (x_1'', x_2'', y_3).$$

In the above we fixed the level of component 3, similar condition should hold if we fix the level on component 1 or component 2. Another way to interpret this condition is that indifference curves on $X_i \times X_j$ do not change with the fixed levels of the remaining components.

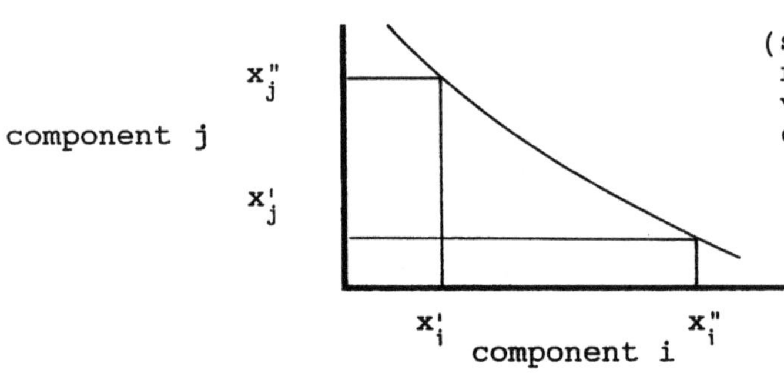

(same indifference curve regardless of where the values on other (n-2) componentes are fixed)

A violation of this condition will occur if for example, (x_i', x_j') is preferred to (x_i'', x_j'') when the remaining (n-2) components are fixed at their best level. However, the preference order reverses i.e., (x_i'', x_j'') is preferred to (x_i', x_j') when the remaining (n-2) components are fixed at their worst level.

Actually we need to verify independence for only (n-1) pairs $\{x_1, x_2\}, \{x_2, x_3\}, \ldots, \{x_{n-1}, x_n\}$. If it holds for these (n-1) pairs then it will hold for any pair $\{x_i, x_j\}$. Further, it will hold for any subset M of (1, 2, ..., n) that is preference order over $\mathbf{X}_{n \in M} X_n$ is independent of fixed level on $\mathbf{X}_{n \notin M} X_n$.

<u>Axiom 2 Mutual Preferential Independence</u>

For any $x_i', x_i'' \in X_i$ and $x_{i+1}', x_{i+1}'' \in X_{i+1}$, suppose it is true that

$$(x_1, x_2, \ldots, x_i', x_{i+1}', \ldots, x_n) \succsim (x_1, x_2, \ldots, x_i'', x_{i+1}'', \ldots, x_n)$$

then

$$(y_1, y_2, \ldots, x_i', x_{i+1}', \ldots, y_n) \succsim (y_1, y_2, \ldots, x_i'', x_{i+1}'', \ldots, y_n),$$

for all $y_j \in X_j$, $j \neq i, i+1$. Further, the condition holds for $i = 1$ to $n-1$.

An implication of this condition is that we can talk about preference order over any components (more generally, any subset of components) without worrying about the fixed levels on the remaining components.

Note that a weaker condition that requires each component to be independent of the remaining (n-1) components will not imply axiom 2. This weaker independence is stated as follows:

If for some i, $x_i', x_i'' \in X_i$, $x_j \in X_j$, $j \neq i$, it is true that

$$(x_1, \ldots, x_{i-1}, x_i', x_{i+1}, \ldots, x_n) \succsim (x_1, \ldots, x_{i-1}, x_i'', x_{i+1}, \ldots, x_n)$$

then

$$(y_1, \ldots, y_{i-1}, x_i', y_{i+1}, \ldots, y_n) \succsim (y_1, \ldots, y_{i-1}, x_i'', y_{i+1}, \ldots, y_n)$$

for all $y_j \in X_j$, $j \neq i$. It induces an ordering $\succsim_i$ on each component.

This weak independence merely requires that ordering on a given component X_i with all other components fixed does not depend on the level where these components are fixed at.

Note that if $\phi(x) = F(\phi(x_1), \phi_2(x_2), \ldots, \phi_n(x_n))$, where F is strictly increasing, then weak independence must be satisfied. When we specialize F further so that $\phi(x) = \phi_1(x_1) + \phi_2(x_2) + \ldots + \phi_n(x_n)$ then Axiom 2 becomes necessary.

The final necessary condition is the Archimedean axiom that rules out lexicography difficulties.

Pick a component say X_1. Now suppose that x_1^m and x_1^{m-1} are equidistant in the sense that for each $m = 2, 3, \ldots$, $x_1^m \succ x_1^{m-1}$ and $(x_1^m, x_2, x_3, \ldots, x_n) \sim (x_1^{m-1}, y_2, y_3, \ldots, y_n)$, for two fixed sets of points $x_j, y_j \in X_j$, for $j > 1$.

Thus, if additive conjoint measurement holds then for every m

$$\phi_1(x_1^m) - \phi_1(x_1^{m-1}) = \left[\sum_{j=2}^{n} \phi_j(y_j) - \sum_{j=2}^{n} \phi_j(x_j)\right] > 0$$

Thus $\phi_1(x_1^m) \to \infty$ a $m \to \infty$. Hence, there could not exist $\bar{x}_1 \succsim_1 x_1^m$ for all m. We call $\{x_1^1, x_1^2, \ldots, x_1^m, \ldots\}$ as a standard sequence. This standard sequence is increasing as for every m, $x_1^m \succ_1 x_1^{m-1}$.

If standard sequence were to be decreasing $x_1^{m-1} \succ x_1^m$ for every m, then $\phi_1(x_1^m) \to -\infty$ as $m \to \infty$ and hence, there could not exist $\underline{x}_1$ such that $x_1^m \succsim \underline{x}_1$ for all m. Of course standard sequence could be on any component.

Axiom 3 Archimedean If $\{x_k^m\}$ is a standard sequence, then there do not exist both $\bar{x}_k$ and $\underline{x}_k$ from X_k such that $\bar{x}_k \succsim_k x^m \succsim_k \underline{x}_k$ for all m.

The archimedean axiom must hold if $\succsim$ is an independent preference relation and if we are to obtain an additive conjoint representation of $\succsim$.

7. SUFFICIENT CONDITIONS FOR ADDITIVE CONJOINT MEASUREMENT

The next axiom is based on a solvability assumption which requires that given any three elements x_1', x_2', x_1'', we can always find x_2'' so that $(x_1', x_2') \succsim (x_1'', x_2'')$. This is called unrestricted solvability. We need a weaker condition called restricted solvability which says that the solution x_2'' exist if and only if there exist $\bar{x}_2$ and $\underline{x}_2 \in X_2$ such that $(x_1'', \bar{x}_2) \succsim (x_1', x_2') \succsim (x_1'', \underline{x}_2)$. This is demonstrated in the figure below:

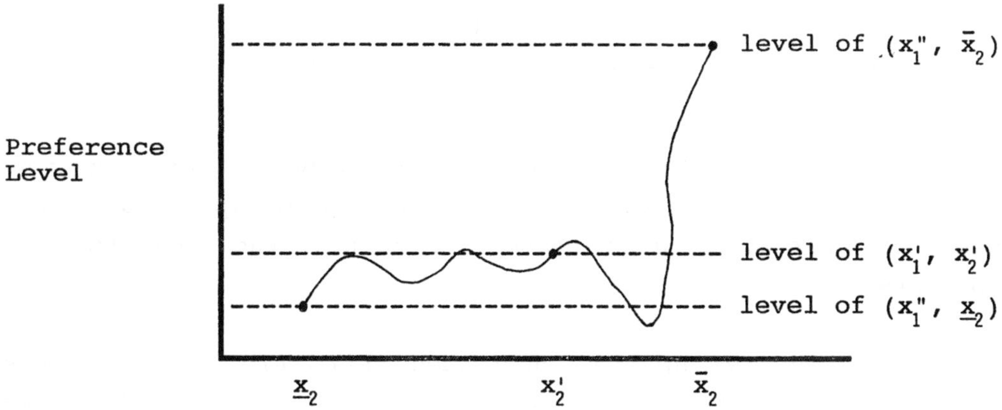

__Axiom 4 Restricted Solvability__ For all i=1, ..., n, whenever $(x_1,\ldots,x_{i-1},\bar{x}_i,x_{i+1},\ldots,x_n) \succsim (y_1,\ldots,y_n) \succsim (x_1,\ldots,x_{i-1},x_i,x_{i+1},\ldots,x_n)$, then there exists $x_i^0 \in X_i$ such that

$$(y_1, y_2, \ldots, y_n) \sim (x_1, x_2, \ldots, x_{i-1}, x_i^0, x_{i+1}, \ldots, x_n).$$

The second sufficient condition requires that each component actively affects the preference. A component X_i is *essential* if and only if there exist x_i' and x_i'' such that

$$(x_i, \ldots, x_i', \ldots, x_n) \not\sim (x_i, \ldots, x_i'', \ldots, x_n).$$

Thus essentialness requires that $x_i' \succ_i x_i''$ for some x'_i and x''_i. Clearly a non essential component can be excluded in practical application since all values on this component are judged to be indifferent.

8. REPRESENTATION THEOREM FOR ADDITIVE CONJOINT MEASUREMENT

The main result is stated in Theorem 1.

__Theorem 1__ If there are three or more essential components, a sufficient set of conditions for there to be an additive conjoint representation is that axioms 1-4 are satisfied.

$$x \succsim y \text{ if and only if } \sum_{i=1}^{n} \phi_i(x_i) \geq \sum_{i=1}^{n} \phi_i(y_i), \text{ for all } x, y \in X.$$

Further, $\phi_i' = \alpha\phi_i + \beta_i$, $\alpha > 0$ will also provide additive conjoint representation. Thus, ϕ_i's are unique up to positive linear transformation with change in unit α the same for all components.

The above Theorem applies only when there are three or more essential components. If only two components are essential then we must add a condition called *Thomsen condition* to the list of axioms.

9. PROBLEMS WITH FINITE DATA

The solvability assumption ensures that our object set or choice set X is sufficiently rich so that certain constructions can be carried out. We need the capability that given any three elements x_1', x_2', y_1', we can solve for the fourth so that

$$(x_1', x_2') \sim (y_1', y_2').$$

Standard sequences give is the «measuring rod» and solvability allows us to make these measuring rods of sufficient fineness.

In social science applications two problems occur if we impose solvability to obtain additive conjoint structure.

1. When one of the components is discrete then additive conjoint model may be satisfied but solvability may be violated.

2. Even if solvability is satisfied, the factorial design for collecting data may render the process of solving the equations practically impossible.

So we need a way to obtain additive representation without solvability and we use a 3 x 3 x 2 factorial design to obtain data:

x_3	x_1	y_1	z_1
x_2	12	14	18
y_2	4	10	16
z_2	2	6	8

y_3	x_1	y_1	z_1
x_2	11	13	17
y_2	3	9	15
z_2	1	5	7

The above data show the rankings of 18 possible elements (say cars). We need to answer two questions:

1. Does the data collected satisfy necessary conditions for additivity?
2. What is the additive representation that fits the data?

Now if a necessary condition such as independence fails then additive representation cannot hold. But what if the necessary conditions hold and solvability is reasonable for the underlying data generating process?

Unfortunately, we may still not be able to obtain additive representation. This means that we cannot find ϕ_1, ϕ_2, ϕ_3 that satisfy a set of linear inequalities implied by the collected data. So with a finite amount of data the whole question of obtaining additive representation boils down to our ability to solve systems of equations and inequalities. What then are the necessary and sufficient conditions that guarantee the solvability of such equations? The answer to this question is tedious and basically involves all kinds of cancellation conditions such as Thomsen condition.

Fortunately, there are several computer programs described in (Krantz, 1971) that help you find an additive representation for a given set of data. Further, some programs help you find the most reasonable additive representation even when, because of response errors or otherwise, additivity conditions are only approximately met.

10. CONSTRUCTIVE PROCEDURE FOR ADDITIVE CONJOINT MEASUREMENT

Now we will discuss a procedure that is applicable when you can sit down with a d.m. (with a cup of coffee) and interrogate her about her preferences. Let us suppose there are two components (the procedure applies for n components as well) X_1 and X_2 and you have verified all the conditions so that an additive conjoint representation is possible. We need to assess or elicit ϕ_1 and ϕ_2 so that

$$\phi(x_1, x_2) = \phi_1(x_1) + \phi_2(x_2)$$

We start with an origin $(\underline{x}_1, \underline{x}_2)$ and set its value to be zero (for simplicity). Thus $\phi(\underline{x}_1, \underline{x}_2) = 0$, $\phi_1(\underline{x}_1) = 0$, $\phi_2(\underline{x}_2) = 0$. Now we select a unit $(x_1, \underline{x}_2)$ and set its value 1 (again for simplicity). Thus $\phi(x_1, \underline{x}_2) = 1$, $\phi_1(x_1) = 1$.

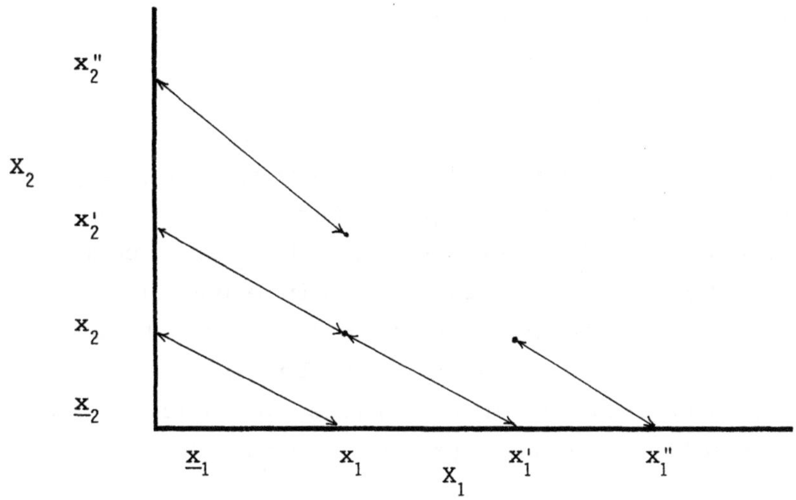

Now we find x_2 so that $(x_1, \underline{x}_2) \sim (\underline{x}_1, x_2)$, which gives

$$\phi_1(x_1) + \phi_2(\underline{x}_2) = \phi_1(\underline{x}_1) + \phi_2(x_2)$$

We now know $\phi(x_1, x_2) = \phi_1(x_1) + \phi_2(x_2) = 1 + 1 = 2$. So we next find x_2' so that

$(x_1, x_2) \sim (\underline{x}_1, x_2')$, which give $\phi_2(x_2') = 2$.

Similarly, we find x_2'' so that

$(x_1, x_2') = (\underline{x}_1, x_2'')$, which gives $\phi_2(x_2'') = 3$.

In this manner we can go on determining values of ϕ_2 at several points and fair a curve through these points or fit an analytical functional form.

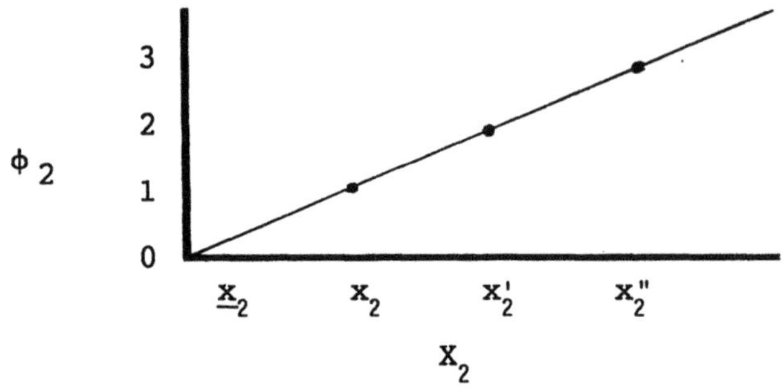

In this process we have used the interval $[\underline{x}_1, x_1]$ as the measuring rod. By choosing an appropriate x_1 we can make the *standard sequence* on X_2 sufficiently fine. The common sense suggests that x_1 be so chosen that we obtain enough points for ϕ_2 so that we feel comfortable in fitting a curve through these points. On the other hand x_1 should not be so close to $\underline{x}_1$ that a d.m. finds it difficult to discriminate the options presented to her (e.g., specify x_2 so that $(x_1, \underline{x}_2) \sim (\underline{x}_1, x_2)$).

In a similar manner we can reverse the process and use the interval $[\underline{x}_2, x_2]$ as a measuring rod to specify ϕ_1. We seek x_1', x_1'', etc. so that

$(x_1, x_2) \sim (x_1', \underline{x}_2)$
$(x_1', x_2) \sim (x_1'', \underline{x}_2)$, and so on.

Thus, $\phi_1(x_1') = 2$, $\phi_1(x_1'') = 3$, ...

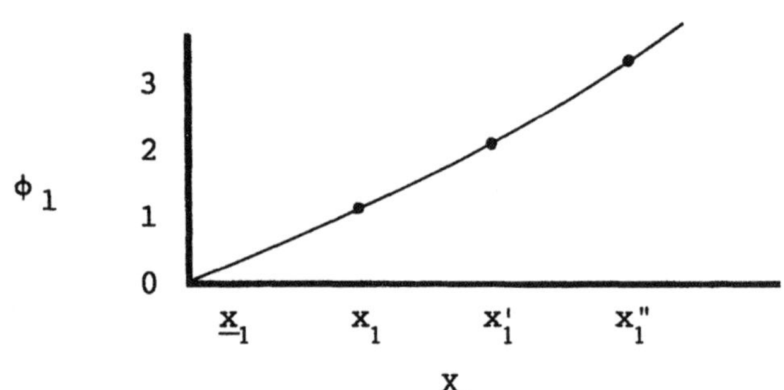

The constructive procedures do suffer from some problems:

a) These are subject dependent and require interaction between analysts (or computer) and the subject;
b) random error may be magnified at each successive stage;
c) there may be an order bias or time bias (fatigue factor) that may introduce systematic errors.

For this reason some statistical procedures have been developed to obtain conjoint representation.

11. STATISTICAL PROCEDURES FOR CONJOINT MEASUREMENT

In the statistical estimation procedure a set of alternative (product profiles) is selected by the analyst based on a appropriate experimental design consideration. For example, if there are three attributes and each attribute i takes on three levels x_i, y_i, z_i then a full factional design will involve 3 x 3 x 3 = 27 product profiles. An example of a product profile is (x_1, x_2, z_3). In most applications one would employ a fractional factorial design since the number of possible product profiles in a full factorial design are numerous. A useful experimental design in such cases is an *orthogonal array*, in which the combinations of levels on different attributes are selected so that the independent contributions of all attributes are balanced.

The subjects are asked to provide their preference over the selected product profiles (27 profiles in the above example). Often ranking or rating scales are used to measure preference. In ranking a subject simply rank orders the product profiles from the most preferred to the least preferred profile. In a rating scale the subject provides the strength of preference for alternative product profiles. The most preferred product is assigned a rating of say 100 and the least preferred product a rating of 0. If a product is half way between the most and the least preferred in the subject's preference, then it is assigned a rating of 50.

For any arbitrary product profile, in our example, the total utility is computed using the model:

$$\alpha + \beta_{11} x_1 + \beta_{12} y_1 + \beta_{21} x_2 + \beta_{22} y_2 + \beta_{31} x_3 + \beta_{32} y_3,$$

where variables x_i and y_i are dummy variables that take on values 1 or 0 depending on whether the level is present or absent in the product profile. For example the total utility of a product profile (y_1, z_2, x_3) is computed by substituting $x_1 = 0$, $y_1 = 1$, $x_2 = 0$, $y_2 = 0$, $x_3 = 1$, $y_3 = 0$ in the model above. It equals: $\alpha + \beta_{12} + \beta_{31}$.

The task in statistical estimation procedure is to estimate the parameters α and β_{ij}'s. This is done by using MONANOVA or multiple regression. These techniques treat the ranking or ratings provided by the subject as a dependent variable and provide parameter values so that estimated ranking or ratings correspond to the empirical ranking or ratings as closely as possible. Since the number of responses obtained from the subject are larger than the number of parameters to be estimated, the measurement error is permitted.

The key advantage of a statistical approach over a constructive approach is ease in implementation and smoothing the errors in responses. These techniques have found a wide application in a variety of setting.

We should note that while practical difference between using ranking or a rating scale may be small; theoretically, the rating scale and multiple regression technique assumes a *difference measurement*. For several other choices in experimental design and statistical estimation techniques for conjoint model the reader should consult books and survey articles in the field of marketing science.

REFERENCE

Krantz, D.H., Luce, R.D., Suppes, P. and Tversky, A. (1971), *Foundations of Measurement. Vol. 1. Additive and Polynomial Representations*, Academic Press, New York.

MULTI CRITERIA DECISION MAKING

AND

THE ANALYTIC HIERARCHY PROCESS

Ernest H. Forman

School of Government and Business Administration
George Washington University
Washington D.C. 20052
U.S.A.

1. INTRODUCTION

The *Analytic Hierarchy Process* (Saaty, 1977 and 1980) is a multiple criterion evaluation methodology that is both descriptive and prescriptive. The Analytic Hierarchy Process (AHP) is, in many ways, similar to Multi Attribute Utility Theory. However, unlike MAUT, AHP does not prescribe that judgments be perfectly consistent, nor does it prescribe when or when not to allow for rank reversals. AHP allows the decision makers to decide how much inconsistency is reasonable, if any, and whether nor not rank reversal (a reflection of relative rather absolute worth) should be permitted.

AHP is comprised of a few powerful and widely accepted concepts:

- Structuring complexity in a hierarchy

- Making pairwise, relative comparisons

- Using redundancy of judgments to improve accuracy and deal with «fuzziness».

Psychologists have proven the human brain is limited to about seven items in both its short term memory capacity and its discrimination ability (channel capacity)[1]. Although we do not like to admit it, our brains are quite limited in certain respects. Humans have had to <u>learn</u> how deal with complexity[2]. The «hierarchical» arrangement has been found to be the best way for human beings to cope with complexity[3,4].

----

1. Miller, G.A. "The Magical Number Seven, Plus or Minus Two: Some Limits on Our Capacity for Information Processing", *Psychological Review*, Vol. 63, No. 2, 81-97, March 1956.

2. We discover this time and time again. For example, if you try to recall a sequence of nine or eleven digits as someone reads them, you will probably find yourself grouping (psychologists call this chunking) the digits into groups in an effort to overcome the limitations of your short term memory. If you prepare a presentation with ten or fifteen bullet items, you will probably find yourself organizing them into categories and subcategories so that you do not lose your audience.

3. In his book on Hierarchical Structures (American Elsevier, New York, 1969) L.L. Whyte expressed this thought as follows: «The immense scope of hierarchical classification is clear. It is the most powerful method of classification used by the human brain-mind in ordering experience, observations, entities and information. ...The use of hierarchical ordering must be as old as human thought, conscious and unconscious...»

4. Novel laureate Herbert Simon, father of the field of Artificial Intelligence, observed is his book *The New Science of Management Decision* that: «Large organizations are almost universally hierarchical in structure. That is to say, they are divided into units which are subdivided into smaller units, which are, in turn, subdivided and so on. ...Hierarchical subdivision is not a characteristic that is peculiar to human organizations. It is common to virtually all complex systems of which we have knowledge. ...The near universality of hierarchy in the composition of complex systems suggest that there is something fundamental in this structural principle that goes beyond the peculiarities of human organization. ...An organization will tend to assume hierarchical form whenever the task environment is complex relative to the problem-solving and communicating powers of the organization members and their tools. Hierarchy is the adaptive form for finite intelligence to assume in the face of complexity».

Relative judgments are usually easier to make and more meaningful than absolute judgments. An absolute judgment is, in a sense, made relative to what has been previously stored in one's brain, possibly a long time ago[1].

AHP uses redundancy to reduce errors and provide a measure of the consistency of judgments. Redundancy permits accurate priorities to be derived from verbal judgments even though the words themselves are not very accurate[2]. This ability opens up a new world of possibilities - we can use words to compare qualitative factors and derive ratio scale priorities that can be combined with quantitative factors!

Weights or priorities are not arbitrarily «assigned» with the AHP pairwise comparison process, but are derived from a set of judgments, either verbal or numerical using the scale shown in Figure 1.

Elements are compared to their peers at each level of the hierarchy. A comparison between a pair of peers is made about the relative importance, preference, or likelihood of the elements with respect to their parent element. A model should be constructed such that peers are approximately within an order of magnitude of one another. Figure 2 shows a matrix of comparisons about the relative importance of four criteria in selecting a site for a retail establishment.

The priorities above are derived by calculating the eigenvector corresponding to the principal eigenvalue of the matrix of judgments.

----

1. Or possibly, as when one is operating in «new waters» not at all.

2. AHP also has numerical mode which, for numerical aspects of a problem would be even more «accurate». But it is not always appropriate to use numbers in such a direct fashion because priorities derived directly from accurately measured factors do not take into account the decision makers utility!

EC Pairwise Comparison Scale

NUMERICAL SCALE	VERBAL SCALE	EXPLANATION
1.0	Equal importance of both elements	Two elements contribute equally
3.0	Moderate importance of one element over another	Experience and judgment favor one element over another
5.0	Strong importance of one element over another	An element is strongly favored
7.0	Very strong importance of one element over another	An element is very strongly dominant
9.0	Extreme importance of one element over another	An element is favored by at least an order of magnitude of difference
2.0,4.0 6.0,8.0	Intermediate values between two adjacent judgments	Used for compromise between two judgments
Increments of 0.1	Intermediate values in increments of .1 (Example: 6.3)	Used for even finer gradations of judgments

Figure 1. AHP Scale

JUDGMENTS AND PRIORITIES WITH RESPECT TO
GOAL TO SELECT THE BEST RETAIL SITE

	COST	VISIBLE	CUST.FIT	COMPET'N
COST		3.0	3.0	5.0
VISIBLE			1.0	5.0
CUST.FIT				1.0
COMPET'N				

Matrix entry indicates that ROW element is
1 EQUALLY 3 MODERATELY 5 STRONGLY 7 VERY STRONGLY 9 EXTREMELY
more IMPORTANT than COLUMN element
unless enclosed in parenthesis.

COST :COST PER SQUARE FOOT OF RETAIL SPACE
VISIBLE :VISIBILITY OF STORE FRONT
CUST.FIT :CUSTOMER FIT--SITE'S CUSTOMER TRAFFIC VS. TARGET MARKET SPEC'S
COMPET'N :COMPETITION--# OF COMPETITIVE STORES IN SAME TRADING AREA

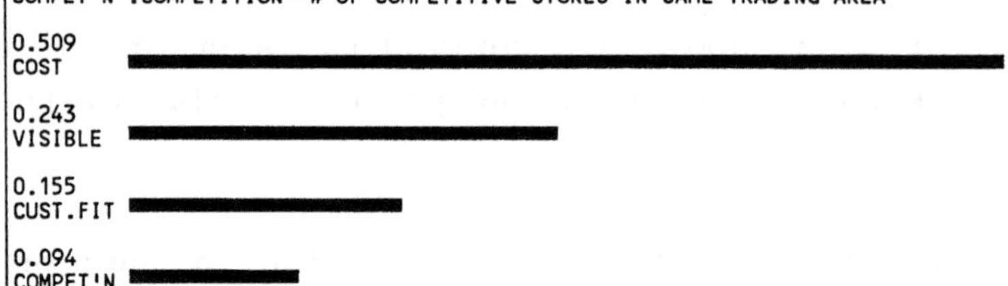

INCONSISTENCY RATIO = 0.098

Judgments and Resulting Priorities

Figure 2 - Matrix of Comparisons and Resulting Priorities

2. A LITTLE OF MATH: WHY AHP USES EIGENVALUES AND EIGENVECTORS

Suppose we knew the relative weights of a set of «activities». We can express them in a pairwise comparison matrix as follows:

$$\underline{A} = \begin{bmatrix} w_1/w_1 & w_1/w_2 & w_1/w_3 & \cdots & w_1/w_n \\ w_2/w_1 & w_2/w_2 & w_2/w_3 & \cdots & w_2/w_n \\ w_3/w_1 & w_3/w_2 & w_3/w_3 & \cdots & w_3/w_n \\ . & . & . & & . \\ . & . & . & & . \\ . & . & . & & . \\ w_n/w_1 & w_n/w_2 & w_n/w_3 & \cdots & w_n/w_n \end{bmatrix}$$

If we wanted to «recover» or find the vector of weights, $[w_1, w_2, w_3, \ldots w_n]$ given these ratios, we can take the matrix product of the matrix $\underline{A}$ with the vector $\underline{w}$ to obtain:

$$\begin{bmatrix} w_1/w_1 & w_1/w_2 & w_1/w_3 & \cdots & w_1/w_n \\ w_2/w_1 & w_2/w_2 & w_2/w_3 & \cdots & w_1/w_n \\ w_3/w_1 & w_3/w_2 & w_3/w_3 & \cdots & w_1/w_n \\ . & . & . & & . \\ . & . & . & & . \\ . & . & . & & . \\ w_n/w_1 & w_n/w_2 & w_n/w_3 & \cdots & w_n/w_n \end{bmatrix} \begin{bmatrix} w_1 \\ w_2 \\ w_3 \\ . \\ . \\ . \\ w_n \end{bmatrix} \begin{matrix} = \\ = \\ = \\ \\ \\ \\ = \end{matrix} \begin{bmatrix} nw_1 \\ nw_2 \\ nw_3 \\ . \\ . \\ . \\ nw_n \end{bmatrix}$$

$$\underline{A} \qquad \underline{w} = n\ \underline{w}$$

If we knew $\underline{A}$, but not $\underline{w}$, we could solve the above for $\underline{w}$. The problem of solving for a non-zero solution to this set of equations is very common in engineering and physics and is known as an eigenvalue problem:

$$\underline{A}\ \underline{w} = \lambda\ \underline{w}.$$

The solution to this set of equations is, in general found by solving and nth order equation for λ. Thus, in general, there can be up to n unique values for λ, with an associated $\underline{w}$ vector for each of the n values.

In this case however, the matrix $\underline{A}$ has a special form since each row is a constant multiple of the first row. For such a matrix, the rank of the matrix is one, and all the eigenvalues of $\underline{A}$ are zero, except one. Since the sum of the eigenvalues of a positive matrix is equal to the trace of the matrix, or the sum of the diagonal elements, the non zero eigenvalue has a value of n, the size of the matrix. This eigenvalues is referred to as λ_{max}.

Notice that each column of $\underline{A}$ is a constant multiple of $\underline{w}$. Thus, $\underline{w}$ can be found by normalizing any column of $\underline{A}$. The matrix $\underline{A}$ is said to be strongly consistent in that

$$a_{ik}a_{kj} = a_{ij} \text{ for all } i,j.$$

Now let us consider the case where we do not know $\underline{w}$, and where we have only estimates of the a_{ij}s in the matrix $\underline{A}$ and the strong consistency property most likely does not hold. (This allows for small errors and inconsistencies in judgments). It has been shown that for any matrix, small perturbations in the entries imply similar perturbations in the eigenvalues, thus the eigenvalue problem for the inconsistent case is:

$$\underline{A}\ \underline{w} = \lambda_{max}\ \underline{w},$$

where λ_{max} will be close to n (actually greater than or equal to n) and the other λ's will be close to zero. The estimates of the weights for the activities can be found by normalizing the eigen-vector corresponding to the largest eigenvalue in the above matrix equation [1].

The closer λ_{max} is to n, the more consistent the judgments. Thus, the difference, $\lambda_{max} - n$, can be used as a measure of inconsistency (this difference will be zero for perfect consistency). Instead of using this difference directly, Saaty defined a *consistency index* as:

$$C.I. = (\lambda_{max} - n)/(n-1)$$

since it represents the average of the remaining eigenvalues [2].

In order to derive a meaningful interpretation of either the difference or the consistency index, Saaty simulated a random pairwise comparisons for different size matrices, calculating the consistency indices, and arriving at an average consistency index for random judgments for each size matrix. He then defined the consistency ratio as the ratio of the consistency index for a particular set of judgments, to the average consistency index for random comparisons for a matrix of the same size.

Since a set of perfectly consistent judgments produces a consistency index of 0, the consistency ratio will also be zero. A consistency ratio of 1 indicates consistency akin to that which would be achieved if judgments were not made intelligently, but rather at random. This ratio is called the <u>in</u>consistency ratio in

---

1. Because of the reciprocal property of the comparison matrix, the eigenvector problem can be solved by raising the matrix to the nth power, and taking the limit as n approaches infinity. The matrix will always converge. Saaty has shown that this corresponds to the concept of dominance walks. The dominance of each alternative along all walks of length k, as k goes to infinity, is given by the solution to the eigenvalue problem.

2. Note: Other methods to estimate activity weights, such as least squares and log least squares have been suggested. While these methods produce results that are similar to the eigenvector approach, no other method maintains the reciprocal property of the pairwise comparison matrix (know as weak consistency), nor produces a comparable measure of inconsistency.

Expert Choice since the larger the value, the more inconsistent the judgments.

If we are perfectly consistent we can not say that our judgments are good, just as we can not say that there is nothing wrong with us physically if our body temperature is 98.6 degrees. On the other hand, if our inconsistency is say 40 or 50% (an inconsistency ratio of 100% is equivalent to random judgments), we can say there is something wrong, just as we can say that there is something wrong if our body temperature is 105 degrees.

An inconsistency ratio of about 10% or less is usually considered «acceptable», but the particular circumstance may warrant the acceptance of a higher value[1]. Let us look at some of the reasons why inconsistency occurs as well as the useful information that the inconsistency ratio conveys, and how to reduce it. The most common cause of inconsistency is a clerical error. When entering one or more judgments into a computer, the wrong value, or perhaps the inverse of what was intended is entered. Clerical errors can be very detrimental and often go undetected in many computer analyses[2]. When using Expert Choice, one can easily find and correct such errors.

A second cause of inconsistency is lack of information. If one has little or no information about the factors being compared, then judgments will appear to be random, and a high inconsistency ratio will result[3]. Sometimes we fool ourselves into thinking

----

1. For example, a body temperature of 102 degrees may be taken as normal if we know that the person has just completed a marathon run on a hot, sunny day.

2. For example, just one clerical error in a multiple regression of one million data points can cause the resulting regression parameter estimates to be considerably different.

3. That is unless one attempts to hide the lack of information by making judgments that appear to be consistent. One is reminded of Ralph Waldo Emerson's saying «Foolish consistency is the hobgoblin of small minds».

that we know more than we really do. It is useful to find out that a lack of information exists, although sometimes we might be willing to proceed without immediately spending time and money gathering additional information in order to ascertain if the additional information is likely to have a significant impact on the decision.

Another cause of inconsistency is lack of concentration during the judgment process. This can happen if the people making judgments become fatigued[1] or are nor really interested in the decision.

Still another cause of a high inconsistency ratio is an actual lack of consistency in whatever is being modeled. The real world is rarely perfectly consistent and is sometimes fairly inconsistent. Professional sports is a good example. It is not too uncommon for Team A to defeat Team B, after which Team B defeats Team C, after which Team C defeats Team A! Inconsistencies such as this may be explained as being due to random fluctuations, or to underlying causes (such as match-ups of personnel), or to a combination. Regardless of the reasons, real world inconsistencies do exist and thus will appear in our judgments.

A final cause of inconsistency is «inadequate» model structure. Ideally, one would structure a complex decision in a hierarchical fashion such that factors at any level are comparable, within an order of magnitude or so, of other factors at that level. Practical considerations might preclude such a structuring and it is still possible to get meaningful results. Suppose for example, we compared several items that differed by as much as two orders of magnitude. One might erroneously conclude that the AHP scale is incapable of capturing the differences since the scale ranges[2] from 1 to 9. However, because the resulting prior-

----

1. At which point it is time to stop and resume at a later time.

2. Actually 9.9 using the Expert Choice numerical mode.

ities are bases on second, third, and higher order dominances, AHP can produce priorities far beyond an order of magnitude[1]. A higher than usual inconsistency ratio will result because of the extreme judgments necessary. If one recognizes this as the cause, (rather than a clerical error for example), one can accept the inconsistency ratio even though it is greater than 10%.

It is important that the a low inconsistency not become the goal of the decision making process. A low inconsistency is necessary but not sufficient for a good decision. It is possible to be perfectly consistent but consistently wrong. It is more important to be accurate than consistent.

3. AN ILLUSTRATION OF THE USE OF AHP/EXPERT CHOICE: A SITE LOCATION PROBLEM

Assume that we want to determine the best retail site within a geographic area for a small ice store catering to young children and families. We have narrowed down the site alternatives to three locations: the first one is located in a suburban shopping center. The second site is located on Main Street (the main business district area of the city), and the third is a suburban mall location called the Mall. Details regarding each of these sites are presented below:

Suburban Shopping Center

A vacant store location that was formerly a pizza shop is available for $28/sq.ft. per month in a neighbourhood «strip» shopping center at a busy highway intersection. The area is populated with 45,000 (mostly middle income, young family) residents of a community who live in townhouses and single family dwellings in the area. The strip center is constantly busy with retail customers of the major supermarket chain, a drug store, a

---

1. For example, if A is nine times B, and B is nine times C, then the second order dominance of A over C is 81 times.

hardware store, a hair stylist/barber shop, and several other small businesses sharing the location. No ice cream shops are located in the community.

Main Street

For $50.00/sq.ft. per month we can locate our store in the ground level of a large high rise office and retail complex. The shop would be in a moderately out of the way corner of the building. The majority of the people frequenting the building and the surrounding area are young professionals who are in the area Monday through Friday only. There is one ice cream store within a ten block radius of this location.

The Mall

This location would cost $75.00/sq.ft. per month. We would be in the main food area of a major suburban mall with 75 retail shops and three «magnate» stores, Sears and two large department stores. The mall is frequented by teens, young mothers, and families usually on weekend days and weekday nights. There are three ice cream stores at various locations within the mall.

The information on the three candidate sites can be used to build a basic model. There can be many variations to this model depending on how you choose to structure the problem. There is no one specific right or wrong way to model any decision problem. A basic approach to modeling the site location problem is outlined next.

3.1. Decompose the Problem

The first step in using AHP and the Expert Choice software is to develop a hierarchy by breaking the problem down into its components. The three major levels of the hierarchy shown in Figures 3 and 4 are the goal, criteria, and alternatives.

- *A goal* - a statement of the overall objective.
  In our example, to Select the Best Retail Site

- *Criteria* - the factors one must consider in reaching the ultimate decision.
  In our example: Cost, Visibility, Customer Fit, Competition.

- *Alternatives* - the feasible alternatives that are available to reach the ultimate goal.
  In our example: Suburban Shopping Center, The Mall, and Main Street.

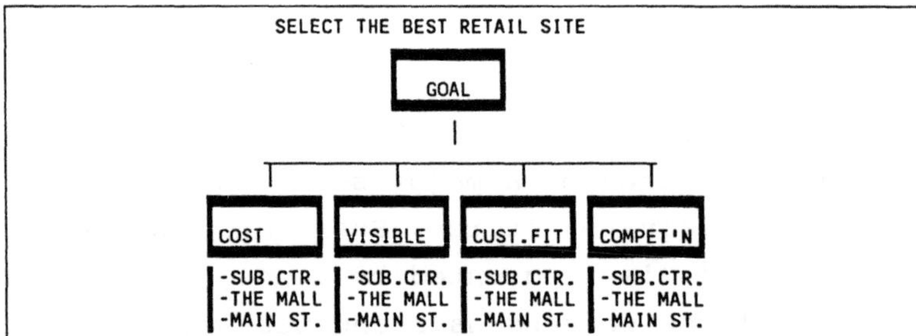

GOAL --- SELECT THE BEST RETAIL SITE
COMPET'N --- COMPETITION--# OF COMPETITIVE STORES IN SAME TRADING AREA
COST --- COST PER SQUARE FOOT OF RETAIL SPACE
CUST.FIT --- CUSTOMER FIT--SITE'S CUSTOMER TRAFFIC VS. TARGET MARKET SPEC'S
MAIN ST. --- MAIN STREET--CENTER CITY, OFFICE & RETAIL COMPLEX SITE
SUB.CTR. --- SUBURBAN STRIP SHOPPING CENTER
THE MALL --- SUBURBAN SHOPPING MALL SITE
VISIBLE --- VISIBILITY OF STORE FRONT

Figure 3 - Basic EC Model with Goal, Criteria and Alternatives

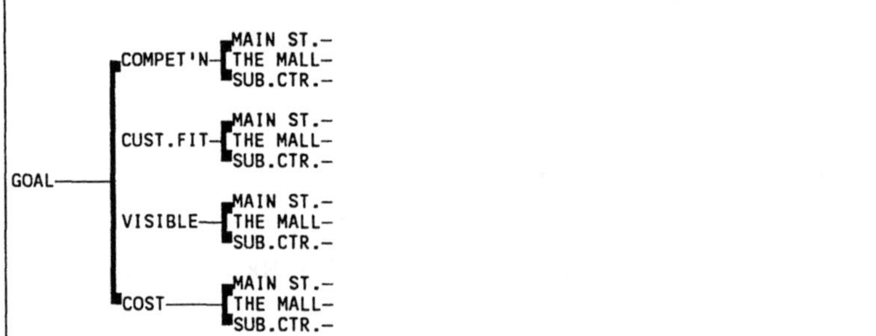

Figure 4 - Basic EC Model: An Alternative View

Expert Choice can easily support more complex hierarchies ontaining:

- *Subcriteria* - This allows more specificity in the model. By adding subcriteria you can refine you criteria and further detail your objectives. Figure 5 shows the subcriteria <u>Maintenance</u> and <u>Rent</u> under the Cost criterion.

GOAL: SELECT BEST RETAIL SITE

Glossary of Terms Used in Model

```
  GOAL    --- SELECT BEST RETAIL SITE
COMPET'N  --- COMPET'N
COST      --- COST
CUST.FIT  --- CUST.FIT
MAIN ST.  --- MAIN ST.
MAINTNCE  --- MAINTENANCE COSTS--JANITORIAL FEES & COSTS IMPOSED BY LANDLORD
RENT      --- COST PER SQUARE FOOT OF RETAIL SPACE
SUB.CTR.  --- SUB.CTR.
THE MALL  --- THE MALL
VISIBLE   --- VISIBLE
```

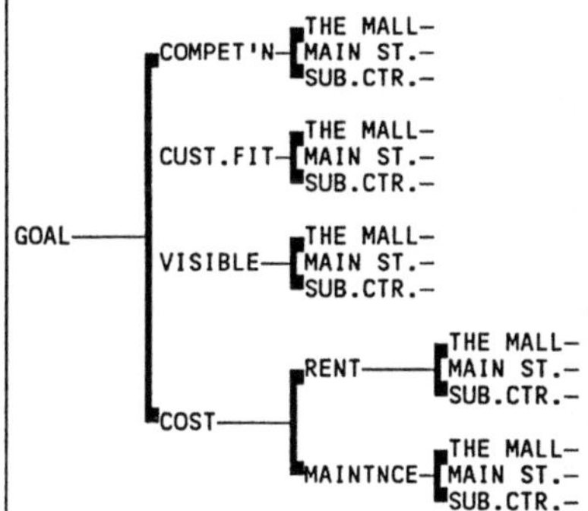

Figure 5 - EC Model with Subcriteria

- *Scenarios or Uncertainties* - The importance of different criteria and alternatives may depend on the specific future conditions which are often difficult to predict. Scenarios can be modeled with Expert Choice allowing you to consider decision alternatives under a variety of circumstances.

Scenarios representing the three possible states of the economy, Gloomy Economy, Boom Economy, and Status Quo are shown in Figure 6.

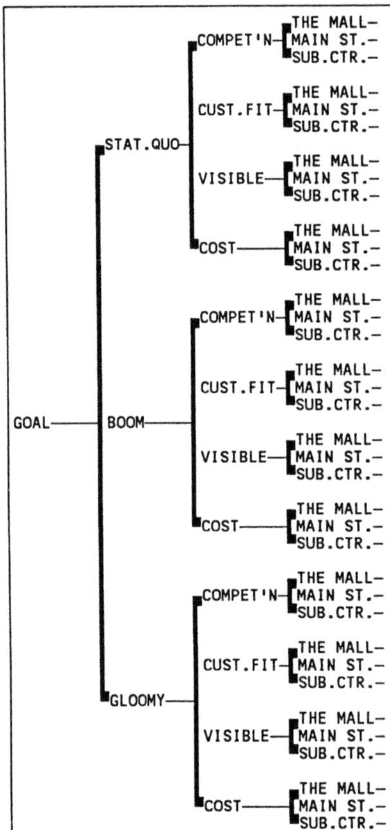

Figure 6 - EC Model with Environmental Scenarios

Players Decision are often made through group consensus, yet it is often difficult for all members of a group to meet, or for each member's opinions to be heard during a meeting. By including a level for players in an EC model, each member's views can be incorporated into the decision making process. Figure 7 illustrates players that include a <u>Vice-President</u>, <u>Marketing Director</u>, and <u>Consultant</u>.

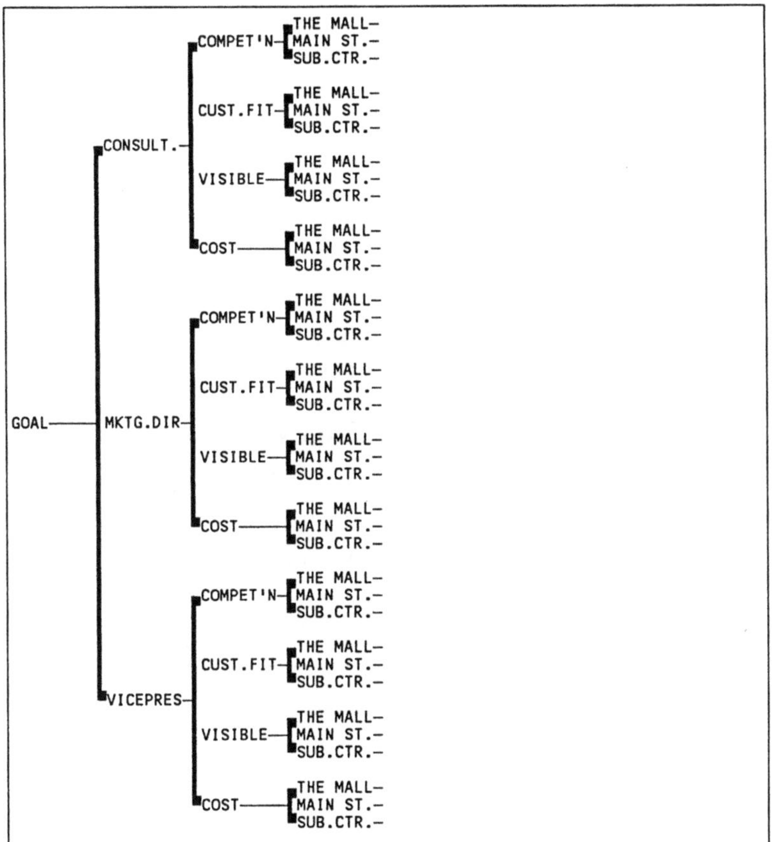

Figure 7 - EC Model with Company Players

- *A Large Number of Alternatives* - Some decisions inherently involve a large number of alternatives which need to be considered. When this is true, the Expert Choice ratings utility easily accommodates a large number of alternatives, such as dozens of potential sites to compare in a large metropolitan area. Figure 8 and Table 1 illustrate the ratings approach.

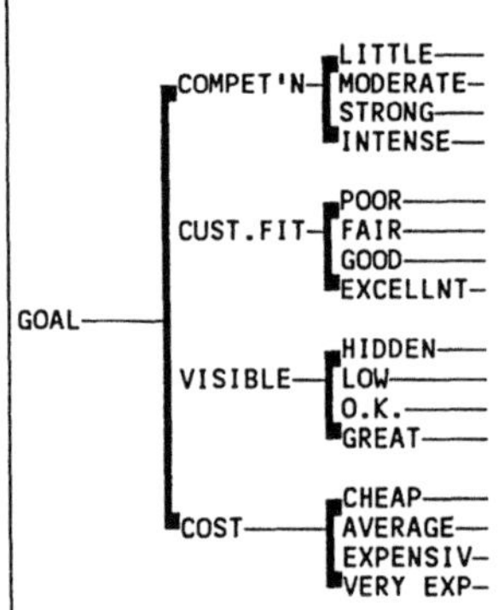

Figure 8 - EC Model with Large Number of Alternatives

Table 1 Ratings for a Large Number of Alternatives

	COST .5088	VISIBLE .2427	CUST.FIT .1550	COMPET'N .0935	TOTAL
1 SUBURBAN CNTR#1	CHEAP	O.K.	GOOD	STRONG	0.431
2 SUBURBAN CNTR#2	CHEAP	LOW	GOOD	MODERATE	0.364
3 OLD TOWN AREA	VERY EXP	GREAT	EXCELLNT	INTENSE	0.270
4 THE MALL	VERY EXP	GREAT	EXCELLNT	STRONG	0.266
5 MAIN ST/HI RISE	EXPENSIV	O.K.	GOOD	STRONG	0.232
6 NEAR APT.CLUSTER	AVERAGE	LOW	GOOD	MODERATE	0.229
7 OFF INTERSTATE	EXPENSIV	O.K.	EXCELLNT	MODERATE	0.221
8 SUBURBAN CNTR#3	AVERAGE	HIDDEN	FAIR	MODERATE	0.194

The ratings approach consists of defining «intensities» of achievement or preference for each of the criteria. These intensities are used in place of alternatives in the first stage of the evaluation. For example, instead of comparing the relative preference of two specific alternatives with respect to **VISIBILITY,** we would compare the relative preference of a non-specific alternative that possesses **GREAT** visibility to some other alternative that has **LOW** visibility. This results in measures of preference for the intensities. A ratings «spreadsheet» is then used to evaluate each alternative as to its intensity on each criterion [1].

3.2. Establish Priorities

After arranging the problem in a hierarchical fashion, the next step is to evaluate each element of the problem. Each node is evaluated against each of its peers in relation to its parent mode; these evaluations are called pairwise comparisons. Referring back to the our basic site selection model in Figure 3:

SELECTING THE BEST RETAIL SITE is the parent mode of **COST, VISIBILITY, CUSTOMER FIT, AND COMPETITION.**

COST is a parent to **MAIN STREET, THE MALL,** and **SUBURBAN CENTER.**

COST, VISIBILITY, CUSTOMER FIT, and **COMPETITION** are peers.

MAIN STREET, THE MALL, and **SUBURBAN CENTER,** are peers.

----

1. With the ratings approach, pairwise comparisons are made for the criteria, as well as for the intensities under each criterion. The results are ratio scale priorities for the importance of each criterion, as well as ratio scale priorities for the intensities below each criterion. Then, using the ratings «spreadsheet», each alternative is evaluated as to its intensity for each criterion. The ratio scale priorities are summed to give an overall ratio scale measure of the preference for the alternative.

Although a decision hierarchy is defined «top down», it should usually be evaluated «bottom up» because the thinking involved in the lower level comparisons of alternative preferences helps us make better judgments about the relative importance of the criteria [1].

Thus, before prioritizing the criteria, we examine the relative preferences for the alternatives with respect to each criterion. In our example, we would determine our preferences for **MAIN STREET**, **THE MALL**, or **SUBURBAN CENTER** with respect to COST. Since cost is an objective criterion, we can use our financial data. Monthly rent on **MAIN STREET** is **$50.00** per square foot, while rent in **THE MALL** is **$75.00** per square foot, and the **SUBURBAN CENTER** only **$28.00** per square foot. Since we want to minimize cost, **SUBURBAN CENTER** is about 2.6 times more preferable that **THE MALL**, and about 1.8 time more preferable than **MAIN STREET**, and **MAIN STREET** is about 1.5 times more preferable than **THE MALL** [2]. The judgments and resulting priorities are shown in Figure 9 [3].

----

1. Suppose we considered two alternative bridge designs, and had two criteria in mind: safety and aesthetics. If we evaluated top down, most people would say that safety was more important than aesthetics. However, if the evaluation is performed «bottom up», we might find that when comparing the bridge designs with respect to safety that design A is preferred to design B, although **both** designs far exceed all published guidelines. When comparing the designs with respect to aesthetics, we might feel that design B is elegant and much preferred to the «ugly» design of bridge A. Given these considerations, we would then probably conclude that, for this decision, aesthetics are more important than safety.

2. This assumes a linear utility function which is not always the case. A non-linear utility function can easily be accommodated by specifying ratios in accordance with the perceived utility of dollars, or by using the verbal comparison mode.

3. In actual practice, we might make judgments considering the relative preference as determined by cost and our utility, and not just cost alone. The preference could be expressed numerically or verbally. For example, we might consider an alternative with a monthly cost of $50 per square foot to be strongly more preferable than an alternative costing $75 per square foot.

JUDGMENTS AND PRIORITIES WITH RESPECT TO
COST < GOAL

	SUB.CTR.	THE MALL	MAIN ST.
SUB.CTR.		2.6	1.8
THE MALL			(1.5)
MAIN ST.			

Matrix entry indicates that ROW element is___
1 EQUALLY 3 MODERATELY 5 STRONGLY 7 VERY STRONGLY 9 EXTREMELY
more PREFERABLE than COLUMN element
unless enclosed in parenthesis.

SUB.CTR. :SUBURBAN STRIP SHOPPING CENTER
THE MALL :SUBURBAN SHOPPING MALL SITE
MAIN ST. :MAIN STREET--CENTER CITY, OFFICE & RETAIL COMPLEX SITE

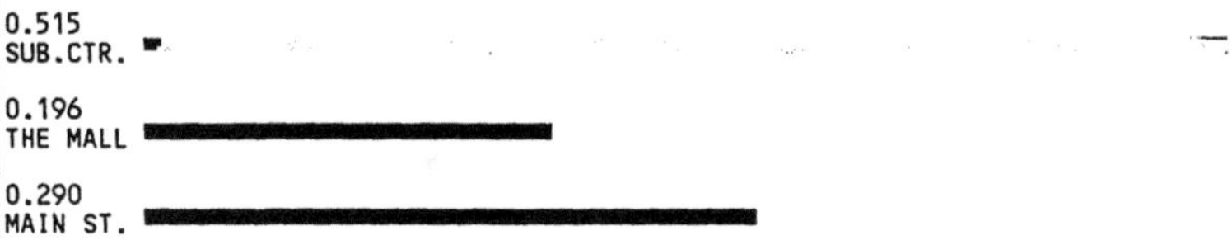

INCONSISTENCY RATIO = 0.000

Figure 9 - Preference for Alternatives with Respect to Cost

After judgments about the preferences for the alternatives have been made with respect to the **COST** criterion, we compare the alternative sites with respect to each of the remaining criteria in the same manner. Then we would compare the relative importance of the criteria.

3.3. Synthesis

Once judgments have been entered for each part of the model, the information is synthesized to achieve an overall preference. The synthesis produces a report which ranks the alternatives in relation to the overall goal. This report includes a detailed ranking showing how each alternative was evaluated with respect to each criterion. Figure 10 shows the details followed by a ranking of the alternatives.

```
                     SELECT THE BEST RETAIL SITE
          TALLY FOR SYNTHESIS OF LEAF NODES WITH RESPECT TO   GOAL

LEVEL 1          LEVEL 2          LEVEL 3          LEVEL 4          LEVEL 5
-------          -------          -------          -------          -------

COST     =0.509
.                SUB.CTR. =0.262
.                MAIN ST. =0.147
.                THE MALL =0.099
VISIBLE  =0.243
.                THE MALL =0.158
.                SUB.CTR. =0.068
.                MAIN ST. =0.017
CUST.FIT =0.155
.                SUB.CTR. =0.070
.                THE MALL =0.070
.                MAIN ST. =0.014
COMPET'N =0.094
.                SUB.CTR. =0.053
.                MAIN ST. =0.031
.                THE MALL =0.009

                 SYNTHESIS OF LEAF NODES WITH RESPECT TO   GOAL

                         OVERALL INCONSISTENCY INDEX =  0.07
```

```
MAIN ST. --- MAIN STREET--CENTER CITY, OFFICE & RETAIL COMPLEX SITE
SUB.CTR. --- SUBURBAN STRIP SHOPPING CENTER
THE MALL --- SUBURBAN SHOPPING MALL SITE
```

Figure 10 - Synthesis for Site Location Problem

For our example, the Synthesis shows **SUBURBAN CENTER** to be the **BEST RETAIL SITE**. We can examine the details of this decision to see that this site alternative was chosen because it offered better customer fit at a lower cost, thus favourably satisfying two important criteria. Although **THE MALL** location provided better visibility, it was very expensive and had heavier competition. Expert Choice has helped us determine that the better visibility was not worth the added cost.

3.4. Sensitivity

A sensitivity analysis can be performed to see how sensitive the alternatives are to changes in the importance of the criteria. Figure 11 shows a sensitivity analysis of the results with respect to the importance of the **COST** criterion. The graph displayed shows that the current priority for cost is a little greater than .50 (see vertical dashed line). The height of the intersection of this dashed line with the alternative lines shows the alternatives priorities. Thus, **SUB. CTR.** is the preferred alternative. If cost were to become less important (i.e. if you were to decrease its priority), your overall preference for **SUB. CTR.** decreases while that of **THE MALL** increases. If the priority of cost were to decrease below about .5, then **THE MALL** would be the preferred alternative. However, since it would take a significant change in the priority of cost in order to change the ranking of the alternatives, we can say that the results are not very sensitive to small changes in the priority of cost.

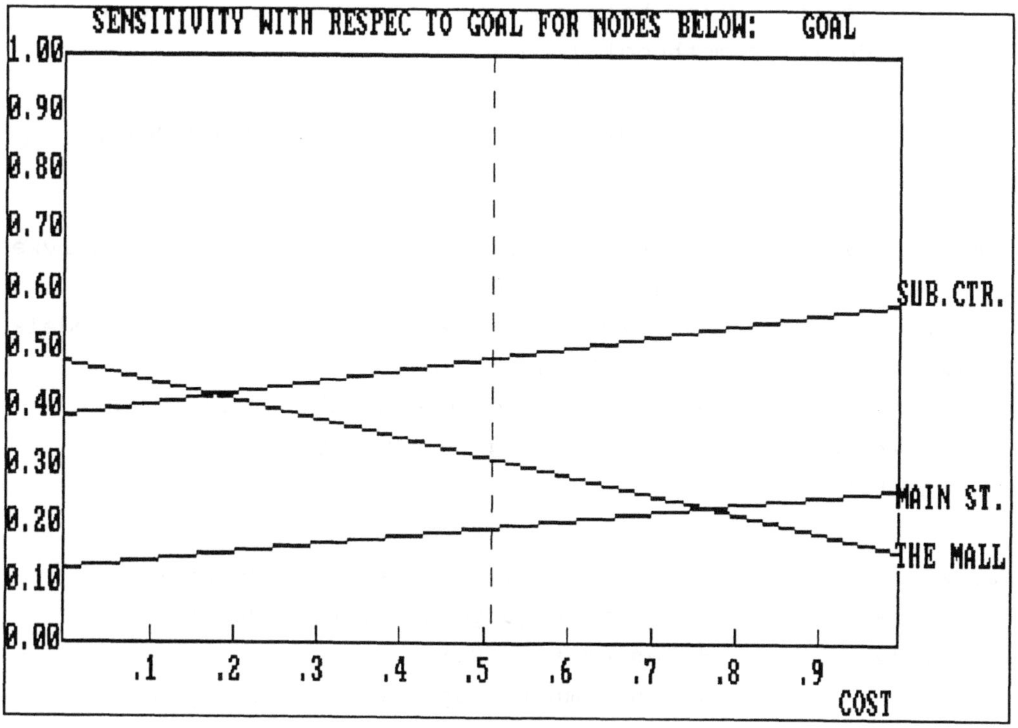

FIGURE 11 - Sensitivity Analysis for the Cost Criterion

4. OTHER ASPECTS

Other aspects of AHP and Expert Choice may interest the reader, but are beyond the scope of this paper. These include the use of subsidiary models to evaluate the importance of players or the likelihoods of scenarios, geometric averages for group judgments, structural adjustment of priorities based on the number of descendants, product adjustment for priorities based on multiplicative factors and other sensitivity analyses modes.

5. SUMMARY

Several multiple criteria decision methodologies have been presented at this School. The choice of which method to use is itself a multi criteria decision. The following summary of AHP characteristics will help the reader when choosing a multiple criteria decision methodology.

Structure	+ Effective way to deal with complexity
Complexity	+ Flexible
As a	
Hierarchy of:	- Need for learning how to structure/analyze
	- Requires some effort
Criteria	- There may be reasonable alternative models - which to use?
Sub-criteria	
Players	- Different to some people - may resist thinking in new ways
Scenarios	
Alternatives	
Pairwise	+ Easier to make relative judgments
(Relative)	+ Relative judgments more accurate
Judgments	+ Easier to justify
	+ Can compare «apples» and «oranges»
	- Different to some people - may resist thinking in new ways

Redundancy of Judgments and used of Words or Numbers	+ Increased accuracy + Allows for «fuzzy» judgments words less demanding - Takes more time - Use of words can be dangerous if there is not much redundancy (e.g. when there are only three factors being compared) - Different to some people - may have to overcome skepticism
Measure of Inconsistency (Does not prohibit inconsistency)	+ Helps identify: . clerical errors . lack of information . «inadequate» model structure + Real world is seldom perfectly consistent - Decision makers can become preoccupied with inconsistency and make a low inconsistency their goal - Goes against some people's notion of expecting/demanding perfect consistency (e.g. an axiom of MAUT)
Produces Ratio Scale Priorities	+ Results (synthesis) is «meaningful» + Results can be used in other decision models, e.g. resource allocation with Linear Programming
Rank Reversal (can allow or prevent)	+ Rank reversal is a real world phenomenon, resulting from preferences based on scarcity - Goes against some peoples pre-conceived notions

Addresses only the Choice phase of decision making

- AHP is not a complete decision support system. Other phases must be part of the decision making process:
  - . information gathering
  - . alternative generation
  - . constraints (musts)
  - . combinations of alternatives

+ Choice phase is both difficult and important
+ AHP can be used with many other decision aids

BIBLIOGRAPHY

Forman, E.H. (1985), "Decision Support for Executive Decision Makers", *Information Strategy; The Executives Journal*, Vol. 1, No. 4, Auerbach Publishers, Pennsauken NJ.

Forman, E.H., Saaty T.L., Selly, M.A., Waldron, R. (1983), *Expert Choice*, Decision Support Software, McLean, VA.

Harker, P.T. (1987), "Alternative Modes of Questioning in the Analytic Hierarchy Process", *Mathematical Modeling*, Vol. 9, No. 3-5, 353-360.

Saaty, T.L. (1977), "A Scaling Method for Priorities in Hierarchical Structures", *J. Math. Psychology*, Vol. 15, 234-281.

Saaty, T.L. (1980), *The Analytic Hierarchy Process*, McGraw-Hill, New York.

Simon, H.A. (1960), *The New Science of Management Decision*, Harper and Brothers, New York, 40-43.

Whyte, L.L. (1969), *Hierarchical Structures*, American Elsevier, New York, 1969.

Zahedi, F. (1986), "The Analytic Hierarchy Process - a Survey of the Method and Its Applications", *Interfaces*, Vol. 16, 96-108.

USE OF A SIMPLE MULTI-ATTRIBUTE VALUE FUNCTION INCORPORATING VISUAL INTERACTIVE SENSITIVITY ANALYSIS FOR MULTIPLE CRITERIA DECISION MAKING

Valerie Belton
University of Strathclyde
and
Stephen Vickers
Heriot-Watt University

1. INTRODUCTION

V·I·S·A is a computer program for multiple criteria decision aid, based on a simple weighted multi-attribute value function, incorporating a hierarchical structure of criteria and visual interactive sensitivity analysis. The use of a model of this kind as an aid to multi-attribute decision making is nothing new. Long before MCDM became an established field of study in the 1970's use of models of this kind been reported in the literature, for example, Churchman and Ackoff, 1954. Many popular approaches to decision aiding adopt the same framework, for example, Edwards' SMART (Edwards, 1982), the Analytic Hierarchy Process developed by Saaty (Saaty, 1980), Social Cost Benefit Analysis (Lichfield et al., 1975). The approach is best suited to the problem of choosing a preferred alternative from a set of well defined alternatives, or to indicate a preference ordering over such a set of alternatives. There are many other approaches suited to this type of problem, for example, the Electre methods and Promethee. However, few approaches have been developed to handle a situation in which the decision maker wants to take account of very many criteria; in such a situation those methods incorporating a hierarchical structure of criteria are most appropriate, namely the multi-attribute value function and the Analytic Hierarchy Process.

In this paper we describe briefly the steps involved in the development and use of such a model. After describing the basic model we discuss briefly the motivation for the development of **V·I·S·A**, explain how to use the visual interactive component, and discuss the potential strength of such a tool. We conclude with a discussion of the strengths and weaknesses of this particular approach to decision aiding. The use of a such a model in practice to assist a large organisation in the choice of a contractor for the development and installation of a computerised international cash management system is described in a separate paper (Belton, 1985). It would be helpful to read the case study together with this paper to understand how the use of a model such as this is integrated into the decision making process.

2. THE DEVELOPMENT AND USE OF A SIMPLE MULTI-ATTRIBUTE VALUE FUNCTION

This approach assumes that the decision maker, or group, with the assistance of a decision analyst, is able to identify a number of discrete alternatives for evaluation and to structure the criteria by which the alternatives are to be evaluated in a suitable hierarchical form. This aspect of decision aid, namely problem structuring, is one which has received little attention in the MCDM literature, but it is clearly of paramount importance. You should refer to Watson and Buede (1987) for a discussion of the issues involved. Occasionally, the process of problem structuring is sufficient for the problem owners to come to understand it in sufficient detail to reach a decision without the need to make use of a formal model. Below we identify the steps involved in the development and use of a model of this kind. These are necessarily presented sequentially, but in practice the process is an iterative one, as illustrated in Figure A.

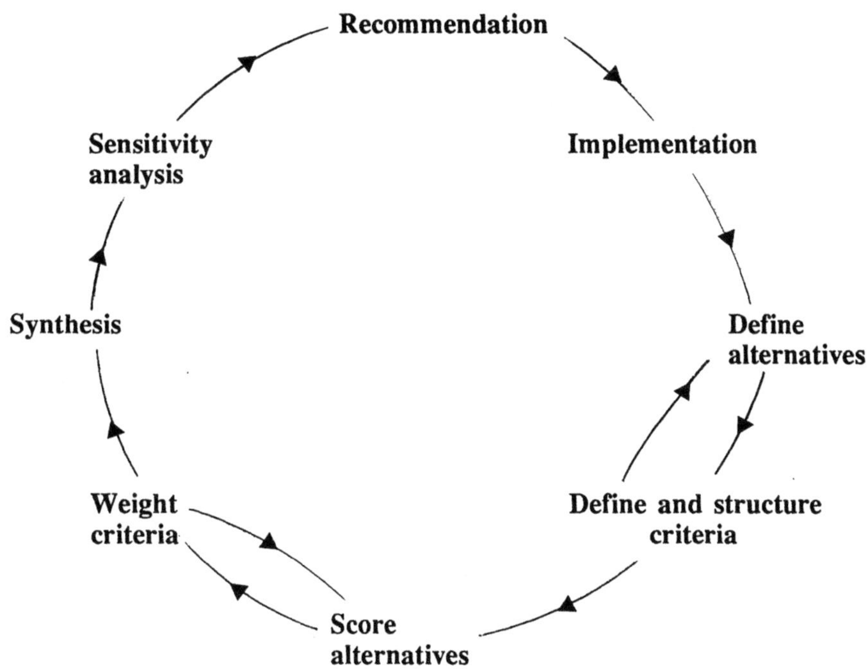

Step 1: Define the alternatives

In some cases it will be clear what the alternatives are. In others it will be necessary to define the alternatives. In others it may be necessary to reduce a long list of alternatives to a manageable shortlist. This can be done in a number of ways; for example, by screening the alternatives to eliminate those which do not meet a pre-specified level on certain criteria, by selecting a representative set of alternatives, or by determining a few critical criteria for evaluation and selecting those alternatives which perform best according to those criteria. Although there is no theoretical limit to the number of alternatives that can be evaluated using this approach it should be borne in mind, especially if the hierarchy of criteria is an extensive one, that collecting information about a large number of alternatives can be an extremely time consuming task.

Step 2: Define the relevant criteria

The definition of alternatives and criteria will probably be an iterative process, as illustrated in Figure A, new alternatives suggesting new criteria and vice-versa. Formal techniques may be used, for example, brainstorming or repertory grids.

If the problem is a complex one then it will probably be necessary to structure the criteria in a hierarchy. The usual hierarchy is of the tree form, that is, higher level criteria are progressively decomposed to greater levels of detail. A criterion which has sub-criteria will be referred to as a *parent* criterion and a set of sub-criteria sharing the same parent will be referred to as a *family*. As indicated above there is little in the way of formal procedure to assist in structuring a hierarchy, it is a skill acquired with practice. There is, of course, no «right» hierarchy for any particular problem and it may be possible to developed alternative structures.

In defining the criteria and refining them into a hierarchical structure a number of factors need to be kept in mind:

Independence: For an additive model of the kind being described here to be an appropriate model of preferences, those preferences should satisfy certain independencies assumptions. Basically, the level of trade-off the decision maker is willing to accept between any two criteria should be independent of the value of any other criterion. If this is not the case then such a model is inappropriate. We will not go into the theoretical detail of these assumptions here, you are referred to Keeney and Raiffa (1976) or to French (1986) for that detail. If an initial set of criteria do not satisfy that assumption it is sometimes possible to redefine criteria, by composition or by further decomposition, in a way which does meet the requirement.

Composition of «families» of criteria: As a rough guideline, a family should contain at most 10 criteria, and criteria belonging to the same family should be of the same order of importance (i.e. the criteria weights should not differ by much more than a factor of 10).

Step 3: Evaluate the alternatives with respect to the criteria

This part of the process is generally referred to as *scoring*. There are many possible ways of doing this, one of which is described in detail. At this stage we want to assess the value of each alternative with respect to the criteria. The value scale is not necessarily a linear function, nor a monotonic function, of the scale on which the attribute is naturally measured. Consider, for example, two criteria which may be relevant in a house-purchase decision: distance from shops and size of garden. Distance from shops may be as illustrated in Figure B(i), that is, intermediate distances are preferable to being too close or too far away. Preference for size of garden may be an S-shaped function of the actual size, i.e. the value of additional space decreases with the size of the garden, as illustrated in Figure B(ii).

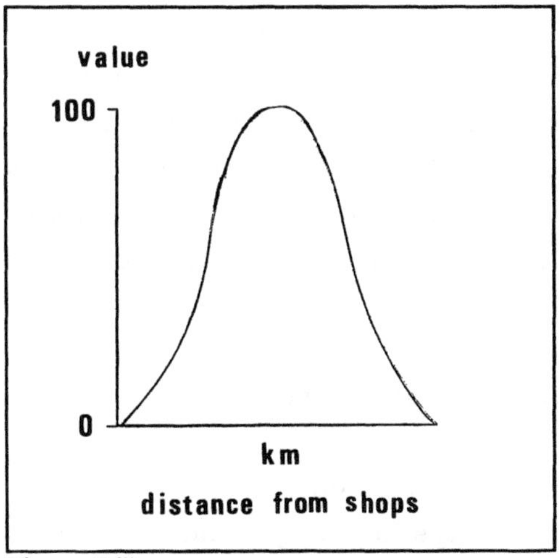

Figure B(i)

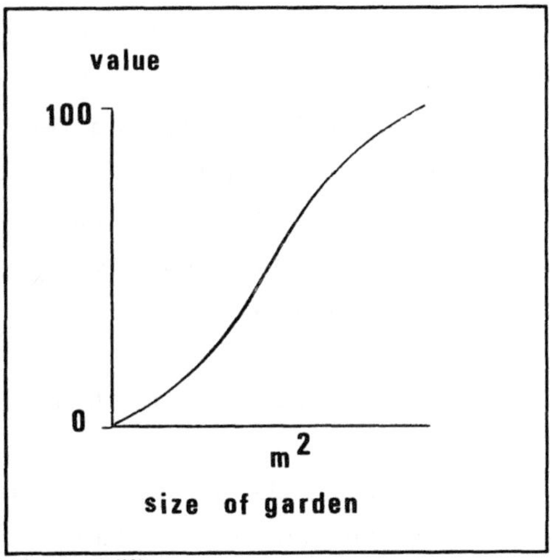

Figure B(ii)

In some cases there may not be a natural measurement scale. In such cases it may be possible to define a proxy measure scale, failing that a subjective scale must be used.

A suggested procedure for scoring is as follows:

(a) Assign a value of 0 to the lowest rated alternative on each criterion.
(b) Assign a value of 100 to the highest rated alternative on each criterion.
(c) Assign intermediate values to other alternatives to reflect their value relative to 0 and 100.

[Note that these values are measured on an *interval scale*.]

This procedure defines a *local scale* for the evaluation of the alternatives with respect to each criterion, that is, it is completely defined by the alternatives under consideration. Such a scale has the advantage that it is relatively easily defined without any need to refer to global or absolute standards. It has the disadvantage that if one wishes to extend the analysis to include an additional alternative it may be the case that the new alternative performs better or worse than any of the existing ones on some criterion, thus necessitating rescaling of all criteria values to maintain a 0 to 100 scale and a consequent reevaluation of criteria weight. An alternative approach is to define a *global scale* on which the 0 and 100 points are defined by some globally worst and best possibility for each criterion. The definition of these points clearly calls for additional work but results in a more flexible model. It is important to be aware that the scales used for scoring will have influence in the next step of the analysis.

Step 4: Assess the relative importance of criteria

This part of the process is generally referred to as *weighting*. As for scoring there are many possible ways of weighting criteria, one of which is described below. The weights are <u>trade-off values</u>.

They say how much of one criterion you would be prepared to give up in return for an improvement on another criterion. The actual values of the weights are related to the measurement scales used for the scores. Thus the weight of a criterion captures both the psychological concept of importance and the discriminatory power of the scale on which the criterion is measured. It is possible for a criterion which is very important to the decision maker to have a low weight because it does not discriminate between the alternatives under consideration, i.e, the scores of 0 and 100 represent something very similar. This is particularly so when using local scales. In order to distinguish between the psychological concept of importance and the concept captured by criteria weights, we will refer to the former as importance and the latter as *significance* . A suggested procedure for weighting is as follows:

(a) Rank the criteria in order of significance.
(b) Assign a weight of 10 to the most highly ranked (the reference criterion).
(c) Weight the other criteria relative to the reference criterion.
(d) Normalise the weights to sum to unity.
[Note that these values are measured on a ratio scale.]

One builds a repertoire of methods to help decision makers assess criteria weights in practice. One approach, akin to the very formal methods described in (Keeney and Raiffa, 1976), which is useful when there are not too many criteria in a family, is to ask the decision maker first to imagine a scenario in which the only available alternative has the characteristics of the lowest rated alternative on every criterion, i.e. zero score on all criteria, then to imagine that she is offered the opportunity of increasing just one criterion to its maximum value. The one the chooses is the one which should receive greatest weight. The process can be repeated, increasing one criterion to 100 at each step, until all the criteria have been ranked.

Step 5: Determine the overall evaluation of each alternative

Using the model described below determine an overall value for each alternative.

Model: $V_i = \Sigma_j w_j v_{ij}$

where V_i is the overall value of alternative i

v_{ij} is the evaluation of alternative i on criterion j (score)

w_j is the relative importance of criterion j (weight).

At this stage it is helpful to look at alternative profiles. The figure below shows a set of alternative profiles displayed using **V·I·S·A**. Each criterion in the selected family is represented by a vertical line and the performance of each alternative is illustrated by the point at which the line depicting its performance crosses the criterion line (0 at the bottom, 100 at the top). Thus we can trace the performance of each alternative identifying those which are principally dominating and those which are principally dominated. We can pick out alternatives, which are good all-rounders and those which rely on significant strengths to compensate for weaknesses elsewhere. In Figure C(i) below we see that the red car and the pink car have significant strengths and weaknesses, while the blue car and the gold car perform intermediately on all criteria.

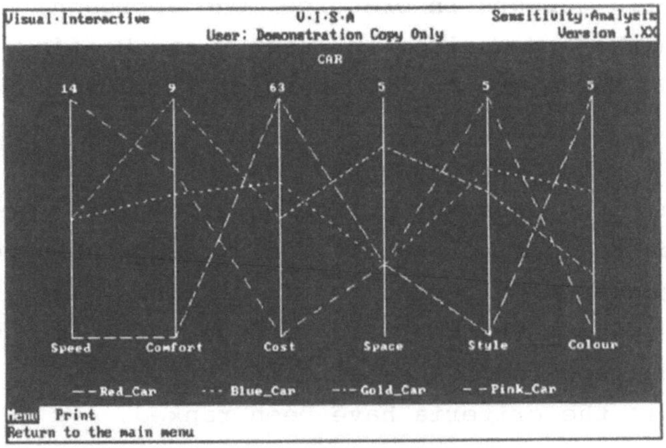

Figure C(i) - Alternative Profile Graph

It may be helpful to give an indication of the significance of the criteria in presenting the alternative profiles. This is achieved in Figure C(ii); here the height of the vertical line corresponding to a criterion reflects the significance of the criterion. This allows us to identify alternatives which may not be dominating overall but which are dominating on the most significant criteria.

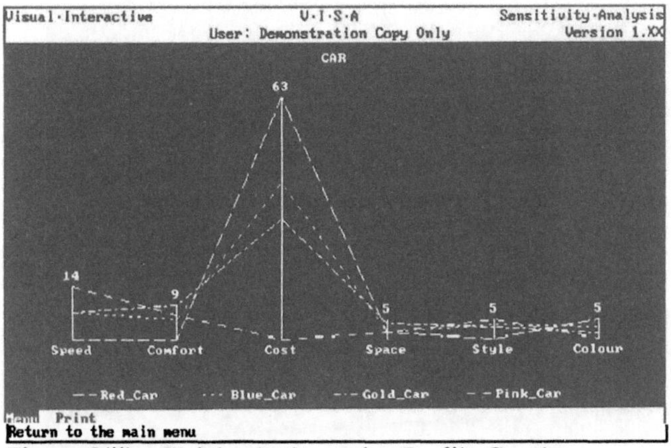

Figure C(ii) - Weighted Alternative Profile Graph

Step 6: Sensitivity analysis

It is very important to carry out a thorough sensitivity analysis, particularly on the criteria weights. There are many ways in which sensitivity analysis can be presented in attractive graphical format, for example, the graphical display of the effect of changing a single criterion weight shown in Figures D(i) and D(ii). The horizontal axis represents the weight assigned to the selected criterion, the vertical axis the overall score. A line for each alternative shows how its overall score changes as the criterion weight changes. The vertical line away from the axis shows the current criterion weight. We see from Figure D(i) that the weight assigned to colour has no effect on the overall ranking of the alternatives. Figure D(ii) illustrates that the outcome is more sensitive to the weight assigned to cost. In the following section we discuss the interactive sensitivity analysis.

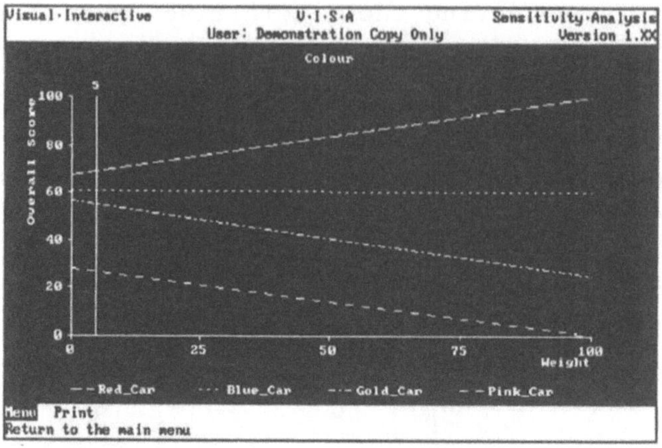

Figure D(i) - Sensitivity Analysis Graph for Colour

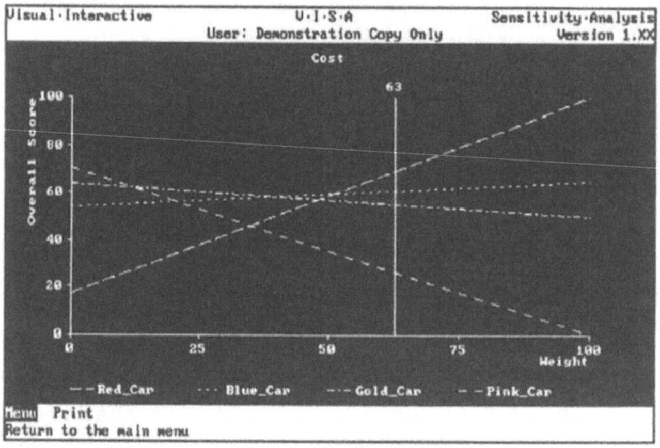

Figure D(ii) - Sensitivity Analysis Graph for Cost

Step 7: Recommendation and presentation of report

The analysis described above must be considered together with other information relevant to the decision making process in order to reach a final recommendation. It is often the case that this must then be presented to a higher level authority. In doing so it is important to remember to capture the important points of the analysis and to present them effectively, without overwhelming by too much detail.

Step 8: Implementation

As in so many OR textbooks this step is last on the list. However, it must be stressed that this should not be the first time it is considered. Throughout the whole analysis the decision makers should be aware of the factors affecting the implementation of the alternatives under consideration, indeed, such factors may be considered as criteria in the analysis.

3. V·I·S·A
VISUAL INTERACTIVE SENSITIVITY ANALYSIS FOR MCDM

The motivation to develop **V·I·S·A** came from research into a means of extending sensitivity analysis within the simple MAVF model beyond the one dimensional analyses of the type described above (Belton, 1986). Extensive experience in the use of simple multiple criteria models to advise decision makers indicated a need for more sophisticated sensitivity analyses and an effective way of presenting the information gathered from such analyses to the decision maker. Even in problems involving few criteria the amount of information yielded by a complete multi-dimensional sensitivity analysis can be overwhelming and the analyst is still faced with the problem of extracting what is useful to the decision maker. With existing software it was not possible to do this analytically and to do so in an ad-hoc manner was a time consuming process which could lead to loss of interest by the decision maker and conceivably to a feeling that the analysis was no longer under her control. In grappling with how to deal with this problem the idea for **V·I·S·A** was born. The program makes use of a simple presentation together with interactive graphics to allow the user to investigate the effect on the overall scores of any changes she may wish to make to the allocation of criteria weights within a particular family. The concept is very simple. The user sees a visual display as illustrated in Figure E.

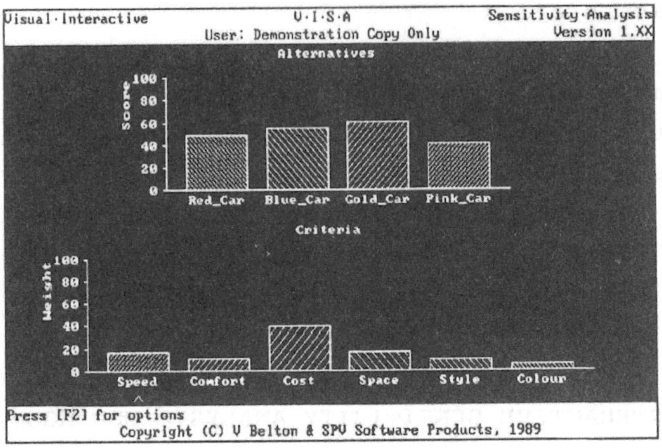

Figure E - Interactive Sensitivity Analysis Graph

The top bar chart represents the overall scores of the alternatives being evaluated, given the current status of the model. The bottom bar chart represents the distribution of criteria weights within the selected family. By using the cursor control keys the user can select a particular criterion and change the weight allocated to it. All other weights change to preserve the normalisation, their relative values being retained. The effect of the change on the overall scores is seen immediately as the bars grow or shrink. The user can display the numeric values on either or both of the bar charts, obtain a printout of the visual display, obtain a printout of the current status of the model in tabular form, opt to retain the current set of values or return to the initial set of values.

4. STRENGTHS AND WEAKNESSES OF THE SIMPLE MULTI-ATTRIBUTE VALUE FUNCTION APPROACH

The simple weighted value function model has often been criticised as a decision aid. In this section we explore some of those criticisms and discuss the strengths of the approach.

As the description of the model suggests it is <u>simple</u>, its critics may say that it is simple-minded and naive, however, its supporters would claim its simplicity and ease of understanding as a strength. There is no doubt that it is founded on some strong assumptions and is demanding of the decision maker in terms of the amount of information required. The decision maker is required to provide a complete evaluation of all alternatives under consideration in the light of all the relevant criteria. Sometimes is may not be possible to do this with a satisfactory degree of accuracy. The role of sensitivity analysis is to answer questions of precisely this nature.

It is very important to be aware that the overall scores resulting from the analysis should not be unquestioningly accepted as «the answer». If the result of the analysis is that two or more alternatives perform similarly well overall, then the user should not blindly accept the resulting ranking but should explore the results in greater depth. Other MCDM approaches, such as Electre and Promethee, deliberately avoid the provision of a complete ranking of alternatives, providing only a partial order in which some alternatives may be incomparable. When using a simple model such as this the onus is on the analyst or the decision maker herself to be aware of the factors which may lead to incomparability in other models, in particular when two alternatives have similar overall scores. The richness of the graphical presentation of alternative profiles and sensitivity analysis, together with the interactive analysis, provide a good environment for the exploration of such factors. Specifically, the user should be aware if an alternative performs well because it is a reasonable «all-rounder», or if it is highly dependent on a few strengths compensating for weaknesses elsewhere.

A number of criticism have been directed at the use of linear models, a discussion of which can be found in (Zeleny, 1982). These focus on the potential pitfalls of using a linear approximation to a non-linear preference function. If the decision maker's preference function is truly non-linear over the range of possibilities described by the alternatives under consideration and cannot be rescaled to make a linear approximation a reasonable one, then it should be accepted that the simple weighted value function model is inappropriate. However, one situation described by Zeleny which should be borne in mind, is that the linear model will never select a convex dominated alternative as the preferred one. This is illustrated in Figure F below:

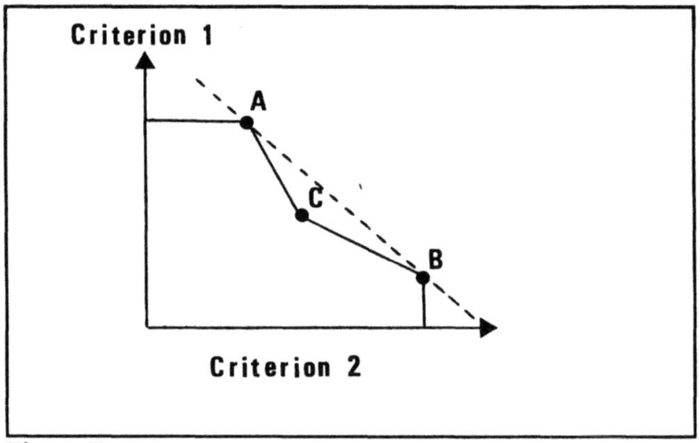

Figure F

Alternatives A and B both perform well on one criterion and poorly on the other. Alternative C performs intermediately on both criteria, but whatever the criteria are weighted it will never obtain the highest overall score.

The advantages of using any multiple criteria approach to provide a structure to the process of decision making, a forum for discussion and a better basis for the justification and explanation of decisions are well documented in the literature, for example, see the case study (Belton, 1985). However, over and above these the weighted value function approach has the advantage that it is transparent and easily understood by any decision maker. Of all multiple criteria methods it has probably been the most widely used and the most widely criticised, as a result of which its strengths and its limitations are well documented, thus the user can be well prepared for problems which may arise.

Overall, if used as an aid, and not as a route to finding a non-existent «right answer», it is an extremely flexible tool to explore and to learn about the problem under study. It does not incorporate some of the more sophisticated procedures of other approaches as a matter of course, but one can learn from other methods and incorporate features from them within the framework of this model. It is possible that in the near future we will see a fully integrated software system that allows the user to select and incorporate features which today characterise different methodologies, but until then for all practical purposes it is necessary to work with the framework and software of one approach.

5. IS VISUAL INTERACTIVE MODELLING AN EFFECTIVE TOOL FOR MCDM?

As MCDM matures as a field of study some attention is moving focus from the development of new methodologies to the development of effective software for implementation, with considerable emphasis on the use of attractive graphics and interactivity. The Mini-Euro Conference on Visual Interactive Modelling held in November 1988 included three papers on multiple criteria decision making. There is no doubt that such tools are «fun» and capture the attention of the user, at least initially. However, as practitioners as well as researchers we ought to question the effectiveness of such developments. First we need to define effectiveness. We believe that there are two important, and very different, components of increased effectiveness: improved decision making and an increased likelihood that the tool will be used by practising managers. These are clearly two very separate objectives although they may be achieved by the same vehicle. We do not believe that they conflict in any way. That is, the increased likelihood of implementation through the use of the visual interactive interface is not at the expense of a reduction in the quality of decisions, and vice versa. We are fortunate in MCDM not to have to concern ourselves with some of the issues facing visual interactive simulation in this respect. We do not have to worry that the user may be seduced by a visually attractive presentation into accepting a decision lacking in statistical validity. We can even brush aside concerns about sub-optimality; in a multiple criteria context the «optimal» choice is defined by the decision maker's preferences and thus is the decision maker's preferred choice. However, this does mean that we have no external criteria against which to validate a decision or assess its quality. It is our feeling, currently based on no more than intuition, that the use of a visual interactive method will enhance the decision makers's learning about the problem and will to some extent overcome the feeling that the first «answer» is the «right» one by encouraging exploration of the problem. An experimental programme to investigate the validity or otherwise of this claim is currently being prepared.

REFERENCES

Belton, V. (1985), "The use of a simple multiple-criteria model to assist in the selection from a shortlist", *Journal of the Operational Research Society*, Volume 36(4), 265-274.

Belton, V. (1986), A Comparative Study of Methods for Multiple Criteria Decision Aiding, PhD Thesis, Cambridge.

Churchman, C.W. and Ackoff, R.L. (1954), "Towards an approximate measure of value", *Operations Research*, 1954, 172-181.

Edwards, W. and Newman, J.L. (1982), *Multiattribute Evaluation*, Sage.

French, S. (1986), *Decision Theory*, Wiley.

Keeney, R.L. and Raiffa, H. (1976), *Decisions with Multiple Objectives: Preferences and Value Tradeoffs*, Wiley.

Lichfield, N., Kettle, P. and Whitbread, M. (1975), *Evaluation in the Planning Process*, Pergamon.

Saaty, T.L. (1980), *The Analytic Hierarchy Process*, McGraw Hill.

Watson, S.R. and Buede, D. (1987), *Decision Synthesis*, Cambridge University Press.

Zeleny, M. (1982), *Multiple Criteria Decision Making*, McGraw Hill.

INTERACTIVE ASSESSMENT OF PREFERENCES USING HOLISTIC JUDGMENTS THE PREFCALC SYSTEM

Eric Jacquet-Lagrèze
Euro-Decision and Lamsade
Euro-Decision, 64 rue Royale
78000 VERSAILLES FRANCE

1. INTRODUCTION AND BACKGROUND

In 1979 we published in French a paper (Jacquet-Lagrèze, 1979) where we presented an interactive scheme for assessing preference models.

The interactive process includes iterative phases:

- classical aggregation phases where we ask the user (decision-maker and/or analyst) to estimate directly the parameters of the model (*i.e.* weights, trade-offs, indifference values,...)
- disaggregation phases where we asked for holistic judgments (*i.e.* a global preference order on a subset of the alternatives) enabling an indirect estimation of the parameters of the model.

In 1982 we published with J. Siskos (Jacquet-Lagrèze and Siskos, 1982) a paper where we presented the **UTA** method to estimate the parameters of an additive value function using as input data the evaluation matrix and a preference order on the alternatives. An ordinal regression model based on a Linear Programming formulation enables to estimate the parameters of an additive utility function with linear piecewise marginal utility functions. A post-optimal analysis enables to produce some sensitivity analysis and yields a confidence on the estimated parameters.

In 1984 we published with M. Shakun (Jacquet-Lagrèze and Shakun, 1984) a paper where we presented the main features of an interactive approach which yields the **PREFCALC** software package. About 100 copies of the system which is proposed in four languages (French, English, German and Italian) have been sold since 1983. The French version of the user guide has been published in a PC user journal (Jacquet-Lagrèze, 1984).

Writing this new paper, we aim towards two goals:

i) present in English the main features of the existing **PREFCALC** method and computer program,

ii) give some light on what should be in our view the next version of **PREFCALC**, taking into account the new programming facilities existing on the micro-computers and some of the feed-back we got from the users of the present version of the method.

2. AGGREGATION AND/OR DISAGGREGATION?

Most of the existing methods proposed in the literature are aggregation methods of the criteria in the following sense (see for instance Roy, 1985 or Vincke, 1989):

a) The analyst considers a set of feasible alternatives (we assume here a finite set for which a list can be produced), a set of criteria (sometimes attributes for which a monotone relation to the preference is not assumed). This first step yields an evaluation matrix (table or data file). Let us call x_{ij} the value of alternative j on criterion (attribute) i and x_i any value on the scale chosen for the criterion (nominal, ordinal or cardinal).

Let us recall here that in many MCDA situations, the formal modelling of the problem stops here, leaving the Decision Maker (DM) making his own opinion without using any mathematical model to aggregate the criteria.

b) When further aid is required, the analyst chooses a mathematical model structure to represent the global preference. This structure can be:

- a simple linear additive model: $u(x) = \sum_i w_i x_i$
- a non linear additive model: $u(x) = \sum_i w_i u_i(x_i)$
- an outranking aggregation model which yields a relation S enabling partial comparisons of the alternatives: j S j' if j is at least as good as j'.
- ...

c) The analyst aggregates the criteria using the mathematical model. He asks the Decision Maker to give information on the parameters (*i.e.* the weights and/or other parameters required by the model structure).

Using this information and the evaluation matrix $\mathbf{X} = (x_{ij})$ he computes (nowadays using a computer program) the overall preference (global utilities, a partial order,...).

The main question is HOW to aggregate the criteria. The answer being choose a mathematical model and ask for the parameters (see figure 1).

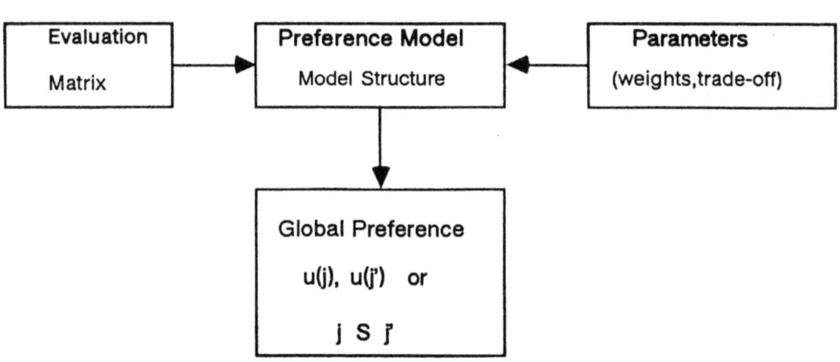

Figure 1 - Aggregation of criteria

What is a DISAGGREGATION method and why should one use it?

In quite many situations, it is very difficult for the DM to answer the questions asked by the analyst in order to get weights. The process involved to assess an overall preference using a decomposed analytical procedure is not natural to him. Easily enough the DM can say that such criterion is more important that such other, but most aggregation methods require a more precise information.

On the other hand, especially when some alternatives are known to the DM, he feels that he can more easily compare these alternatives in a holistic manner. "I do not precisely know why, but I do believe that j is better than j'. Am I wrong? Do I implicitly use a criterion I do not want to take into account? Do I overweight operating costs compared to investments costs?, etc..."

Trying to answer these questions is fundamental in such situations. And if we do not give operational tools to let the DM perform such an analysis, he might reject any model based on weights which are less meaningful to him.

The main question here is WHY does the DM prefers j to j'? The answer being choose a mathematical model and ask for holistic judgments or preferences of the DM and estimate indirectly the parameters of the model.

This is what we called DISAGGREGATION (see figure 2).

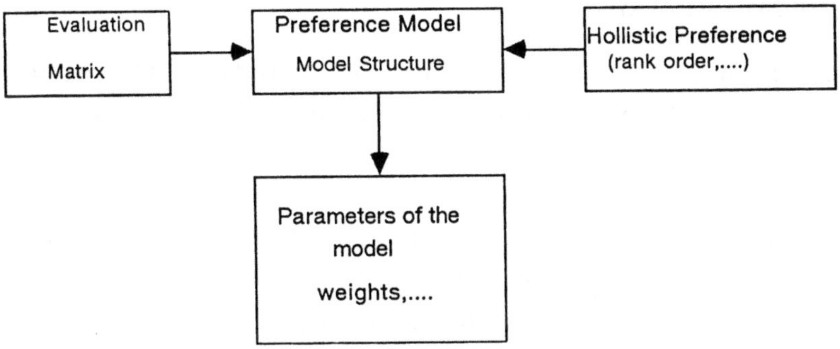

Figure 2 - Disaggregation of a holistic preference

Now let's be interactive!..

We do not claim that disaggregation should replace aggregation. The key point is that in many situations the two types of information (*i.e.* relative importance of criteria on one hand and holistic comparisons of some known alternatives on the other hand) are at the same level in the DM's mind. One is not the consequence of the other, but they are related to each other in a dialectic manner.

To propose an operational method to implement this key assumption we suggested to use aggregation and disaggregation iteratively. This is what we called the AGGREGATION-DISAGGREGATION approach which was first implemented in the **PREFCALC** system.

In the next section (3) we describe **PREFCALC** and in the following one (4) we suggest important improvements.

3. THE PREFCALC SYSTEM

3.1. Assumptions and Parameters

- <u>Additivity</u>

The preference model is an additive linear utility function, each marginal utility function $u_i(x_i)$ being piecewise linear.

- Monotonicity

The x_i are criteria (and not attributes) therefore $u_i(x_i)$ is a monotone function (increasing for a "benefit" and decreasing for "cost").

Let us introduce x_{i*} the limit of the non desirable level on criterion i (if $x_{ij} < x_{i*}$, then j is not feasible in the sense it is not acceptable for the DM). Note here that x_{i*} can evolve during the interactive process. Above x_{i*}, compensation between criteria is possible, under x_{i*}, compensation is not possible.

Let us introduce x_i^* the limit of the most desired level on criterion i. x_i^* can be a theoretical value like value 100 on a scale [0, 100] or the maximum value obtained by a feasible alternative. In the latter case, the set of values x_i^* define the ideal point in the criteria space.

- Cardinality

We assume in **PREFCALC** that the notion of intensity of preference is meaningful (it is allowed to compare intervals of preferences). Therefore u(x) is defined up to a linear transformation and we can choose arbitrarily

$$u(x_*) = \sum_i u_i(x_{i*}) = 0$$

$$u(x^*) = \sum_i u_i(x_i^*) = 1$$

For sake of convenience, we can use the equivalent form where the weights are explicitly introduced:

$$u(x) = \sum_i w_i v_i(x_i)$$

$$v_i(x_i^*) = 1 \quad \text{for all } i$$

$$\sum_i w_i = 1$$

- <u>piecewise linear marginal utility functions</u>

$\beta_i + 1$ points are chosen to divide in equal β_i intervals the interval $[x_{i*}, x_i^*]$

$$x_{ik} = x_{i*} + (x_i^* - x_{i*})\ (k-1)\ /\ \beta_i \qquad \text{for } k = 1,\ldots,\beta_i + 1$$

The utility u_{ik} of these points define the set of unknown parameters of the model.

In order to assess a model, one has first to choose the following parameters:

- the extreme values x_{i*} and x_i^*
- the number of linear pieces β_i.

3.2. Aggregation: Direct Assessment of the Parameters

When choosing this option in Prefcalc, the user is asked first to give numerical values for the weights which are automatically normalized to 1.

Then in order to avoid the user to build directly a marginal utility function an arbitrary quadratic function for the break points k (x_{ik}, u_{ik}) is proposed to him:

$$u_{ik} = [(x_i - x_{i*})\ /\ (x_i^* - x_{i*})]^{1/2}$$

The user can then modify the values u_{ik} to draw a marginal utility function representing in a better way his preferences.

<u>Remark 1</u>
This model remains simple (additivity). Users not familiar with MCDM concepts understand possibilities offered by the model to express judgments such as "after such value x_{ik}, the importance of the criterion decreases (*i.e.* the slope of the curve decreases)". They appreciate the extreme possibility to draw a flat curve beyond a particular value x_{ik}.

Remark 2

The initial quadratic function is often kept because it is concave and represents the situation for which a criterion is important when the value is low on this criterion and once we are good on this criterion it is much less important.

Remark 3

For an ordinal qualitative scale, it is recommended to choose only one linear piece because in a certain way the initial values are already defined as a preference scale. On the other hand it is most important to choose more than one linear piece for numerical scales (**PREFCALC** is limited to 5 linear pieces).

3.3. Disaggregation: Indirect Assessment of the Parameters Using Linear Programming

We ask the user of **PREFCALC** to choose a subset (limited to 30) of the set of alternatives (limited to 200). This subset is called the reference set and should include if possible alternatives known to the DM (good ones and bad ones).

We ask the DM to rank order the r alternatives of the reference set in a holistic manner:

$1 \succ 2 \succ \ldots \succ j \succ j+1 \succ \ldots \succ r.$

In **PREFCALC** we consider as in **UTA** that the unknown utility function should respect this rank-order, so we introduce error variables z_j that we wish to get null.

If M is a high number giving priority to give a 0 value to the z_j, the linear program solved in **PREFCALC** is:

$$\text{Max } \delta - M \sum_j z_j$$

PL1 $\quad u(j) - u(j+1) + z_j > \delta \quad \text{for all } j = 1, 2, \ldots, r-1$

$$\sum_i u_i^* = 1$$

Remind that $u(j)$ is a linear combination of the unknown variables u_{ik}, and therefore PL 1 is a linear program with variables δ, z_j, u_{ik}.

We do not have to introduce explicitly the constraints of monotonicity:

$u_{ik} - u_{ik+1} \geq 0$

since the variables introduced in **PREFCALC** are the differences:

$y_{ik} = u_{ik} - u_{ik+1}$.

The non negativity of these new variables is automatically assumed in the linear programming formulation.

Note that if we have r = 10, n = 10 criteria with 3 linear pieces for each criterion ($\beta_i = 3$ for i = 1 to 10) PL 1 has:

1 + 10 + 3 x 10 = 41 variables

9 + 1 = 10 constraints.

The result of such a small LP is given in a few seconds on a PC. Solving this LP yields two types of solutions:

a) The preference order is not consistent with the additive utility function. This situation occurs when one at least of the z_j is strictly positive. Then $\delta = 0$ and if we get for instance a value $z_j = 0.05$, this means that although $j \succ j+1$, the utility function yields $u(j) - u(j+1) + 0.05 \geq 0$ (the error is equal to 0.05). In order to try to get consistency **PREFCALC** suggests one of the following options:

- increase the number of linear pieces,
- modify the preference order
- modify the criteria and/or the evaluation matrix $\mathbf{X} = (x_{ij})$.

Note also that inconsistency can occur because the assumption of additivity (preferential independence) is not satisfied.

b) The rank-order is consistent with at least one additive utility function. This situation occurs when all the z_j are null, then its possible to get $\delta > 0$ (see figure 3). The set of admissible utility functions is a polyhedron in the space of the unknown parameters. We try search for a point O which maximizes δ. It is a MaxMin criterion: Maximize the minimum difference of utility between two alternatives.

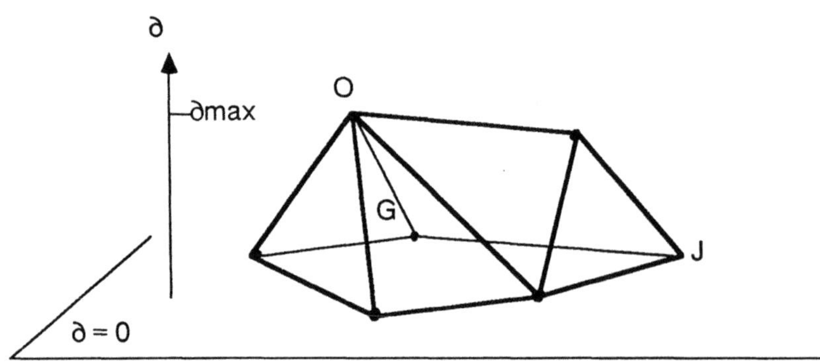

Figure 3 - Choosing an average model (point G)

In order to give to the user an idea of how large the polyhedron is and therefore how imprecise the estimation of the parameters is, we compute all the adjacent vertices of the optimal solution (O). This computation is very fast (p pivots if p is the number of parameters) and much easier to achieve compared to the computation required in order to give a better image of the polyhedron, which would be more satisfactory from a theoretical point of view but more difficult to implement.

Using the p points (utility functions) we compute the average one (G) which is proposed to the user as a model. Under his demand, the user can also display for each marginal utility an upper limit and a lower limit in addition of the average utility function, giving him so an idea of the fuzziness of the estimation.

Remark 1

The error function is not very satisfactory in case *a)* (no solution). A better error function implies to introduce two error variables per alternative, in order to minimize the sum of the deviations. Such a modification has been introduced in the extension **MINORA** of **UTA** proposed by Siskos and Yannacopoulos (1983).

Remark 2

A point such as J is never computed by **PREFCALC**. This could be misleading in some situations. The fuzziness could actually be worse than the one computed by the program.

Remark 3

Only a few criteria (up to 10 in **PREFCALC**) can be taken into account otherwise the disaggregation would yield too imprecise estimation or the user would have to rank-order too many alternatives.

Concluding Remark

Probably the weakest point in **PREFCALC** is that we have no memory of an aggregation phase when we go into a disaggregation phase. Any information such as the shape of a marginal utility function or the relative importance of the criteria is not taken into account when we start the estimation procedure in the disaggregation phase.

4. TOWARDS A NEW VERSION OF PREFCALC

4.1. Why a New Version?

During the last two years we have been thinking it would be nice to have a new version of **PREFCALC** for the two very different reasons:

a) **PREFCALC** was written in 1983, its design is old and does not take advantage of the new possibilities in programming on microcomputers (full colors for the graphics, windows, maybe mouse,...)

b) more important, in order to implement in a more satisfactory way the aggregation-disaggregation approach, we need to keep memory of a previous phase when we start a new one.

4.2. Assumptions for the New Model Structure

- Additivity. It is simple to understand and yields LP formulation in a disaggregation phase.

- Piecewise utility curves are nice, but probably one should need other functions such as polynomial ones.

- For a piecewise utility function, any value x_{ik} should be allowed as a breaking point. None monotone functions should be permitted in order to relax the assumption of using criteria (attributes should be allowed).

4.3. General Design

The input data (evaluation matrix) $X = (x_{ij})$ should be accessible at any time with a spreadsheet presentation. The names of the columns (*i.e.* criteria) and the names of the rows (*i.e.* alternatives) should remain visible as titles.

The 4 first rows could be used to display and eventually modify the following values:

- x_i^*: most desired value
- x_{i*}: least desired value
- w_i': weight of the criterion given by the user
- w_i : weight of the criterion computed after a disaggregation phase.

The 3 first columns could be used to display and eventually modify the following values:

- $u(j)$: overall utility of alternative j computed by the model
- u'_j: objective value for alternative j given by the user.

4.4. Aggregation Phase: Direct Assessment

A default option would be for each criterion a simple linear marginal utility function:

$$X_i = u_i(x_i) = (x_i - x_{i*})/(x_i^* - x_{i*}).$$

An option to assess differently this function should be available. Support with icons to suggest types of curves could be included. Final drawing with a mouse would be a nice feature. For instance, in the piecewise linear case, the user could plot directly to the screen with the mouse the breaking points (x_{ij}, u_{ik}).

The user should not give directly numerical values to the weights (w'_i) but should adjust them graphically using the arrows or the mouse. The relative importance would be displayed in a piechart form or as a barchart graphic.

4.5. Disaggregation Phase Using a Goal-Programming Formulation

Although it would be possible to estimate indirectly the shape of the marginal utility functions (as in **UTA** and **PREFCALC**) we would suggest to restrict ourselves to the case of the weights.

As for the weights, any editing of objective values u'_j should be done graphically using the mouse for instance (see figure 4).

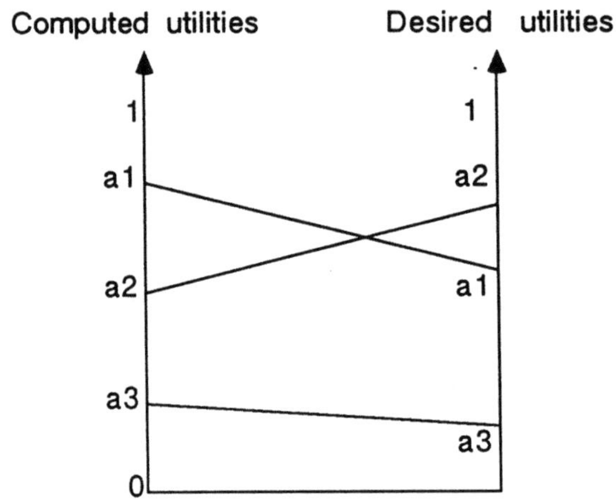

Figure 4 - Computed and desired utilities

We want to take into account in a disaggregation phase the values of the weights w'_i given by the user in the previous phase. If we start with a disaggregation phase, then we chose $w'_i = 1/n$ for all i.

We estimate only the weights w_i. Therefore let us call $X_i = u_i(x_i)$ and $X_{ij} = u_i(x_{ij})$ the transformed values.

The model becomes:

$$u(x) = \Sigma\, w_i X_i = \Sigma\, w_i u_i(x_i).$$

The utility of an alternative j is:

$$u(j) = \Sigma\, w_i x_{ij} = \Sigma\, w_i u_i(x_{ij}).$$

We then minimize a compromise distance between:

- the two set of weights: $d(\mathbf{w},\mathbf{w'}) = \Sigma\ |w_i - w'_i|$
- the two set of utilities: $d(\mathbf{u},\mathbf{u'}) = \Sigma\ |u_j - u'_j|$.

If we given a higher priority to the criterion Minimize the distance d(u,u') which corresponds to a disaggregation objective, then we can estimate the parameters in solving the following linear program PL 2 where M is a penalty:

$$\text{Min } M \sum_{j \in J} z_j + \sum_{i=1}^{n} t_i$$

PL 2

$$-z_j \leq \sum_i w_i X_{ij} - u_i' \leq z_j \qquad \text{for all } j \in J$$

$$-t_i \leq w_i - w_i' \leq t_i \qquad i = 1 \text{ to } n$$

The size of this PL 2 is very small yielding results in a very short time on a microcomputer.

If r is the number of alternatives in J, the number of variables is r + 2n and the number of constraints 2r + 2n.

Let us note here the complete symmetry of this formulation in the case of the aggregation phase. If we give in this case a high priority to the distance d(**w**,**w'**), minimizing the sum d(**u**,**u'**) + M.d(**w**,**w'**), for M big enough a trivial solution of PL2 yields:

$$t_i = 0 \text{ and } w_i = w_i' \qquad \text{for } i = 1 \text{ to } n$$

$$z_j = \sum_i w_i X_{ij} - u_i' \qquad \text{for all } j \in J.$$

This solution corresponds exactly to the situation of an aggregation phase for which we do not take into account any desired value u_j' for the result, but simply compute directly $u(j) = \sum w_i X_{ij}$, using the desired weights w_i'.

REFERENCES

Jacquet-Lagrèze, E. (1979), "De la logique d'agrégation de critères une logique d'agrégation - désagrégation de préférences et de jugements", *Cahiers de l'ISMEA*, Série Sciences de Gestion, 13, 839-859.

Jacquet-Lagrèze, E. and Siskos J. (1982), "Assessing a set of additive utility functions for multicriteria decision making, *European Journal of Operational Research*, 10, 2, 151-164.

Jacquet-Lagrèze, E. and Shakun, M.F. (1984), "Decision support systems for semistructured buying decisions", *European Journal of Operational Research*, 16, 1, 48-56.

Jacquet-Lagrèze, E. (1984), "PREFCALC: évaluation et décision multicritère", *Revue de l'Utilisateur de l'IBM PC*, 3, 38-55.

Roy, B. (1985), *Mèthodologie Multicritère d'Aide à la Dècision*, Economica, Paris.

Siskos, J. and Yannacopoulos, D. (1983), "Amélioration de la méthode **UTA** par introduction d'une double fonction d'erreur", *Cahier du Lamsade* No 49, Université de Paris Dauphine.

Vincke P. (1989), *L'Aide Multicritère à la Dècision*, SMA Editions de l' Université de Bruxelles.

AN ADDITIVE VALUE FUNCTION TECHNIQUE WITH A FUZZY OUTRANKING RELATION FOR DEALING WITH POOR INTERCRITERIA PREFERENCE INFORMATION

Carlos A. Bana e Costa

IST / CESUR-INIC, Technical University of Lisbon
Av. Rovisco Pais, 1000 Lisbon - PORTUGAL

1. POOR INTER-CRITERIA PREFERENCE INFORMATION: WHEN AND WHY?

In any decision aid process a dialogue must occur between the analyst and the decision actors[1] during the phase of modelling their inter-criteria preferences[2]. Generally, it consists in a questioning process along which the analyst tries, step by step, to get increasingly precise answers from the actors, in his efforts to obtain preference information as complete as possible, allowing him to specify the role to be devoted to each criterion (j, $j=1,\ldots,n$) in discriminating between potential actions. In MAUT and AHP[3] operational approaches the analyst measures that role by assessing *tradeoff ratios or substitution rates*; with an identical purpose, in the context of an Outranking Approach[4] *coefficients of relative importance* (and probably *veto thresholds* also) are used.

----

1. By «decision actor» we refer to those actors whose preference systems must be taken into account (such as the decision-makers themselves or other agents directly involved in the decision making process). From now on we will simply refer to them as «actors».

2. Rules for the development of this dialogue have to be established, accepted and well understood by all the participants. In particular, the procedure supporting the dialogue must be consistent with the type of aggregation model previously accepted.

3. MAUT - Multiattribute Utility Theory (see Keeney and Raiffa, 1976).
AHP - Analytic Hierarchy Process (see Saaty, 1980 and Forman, 1990).

4. We refer here to the Outranking Approaches based in the concepts of concordance and discordance, which strictly speaking are the «true» outranking concepts (see Roy, 1990).

Those familiar with the practice of decision aid shall recognise that, in any of the above referred operational approaches, the questioning procedures appropriate to specify the weights w_j of the criteria are intrinsically complex and hard to tackle [1]. This fact by itself can create uncertainty and imprecision and/or may cause the interruption of the process close to initial stages.

Indeed, it is not rare the analyst to be confronted with situations where it is difficult or even unrealistic to completely specify the weights, due to a large variety of practical circumstances. «For example, the decision maker may not be able to spend the time to carry out the procedures needed for a complete assessment. (...) In other situations, there is no single "decision maker" [2], and the various involved parties *(actors)* disagree about the values of w_j» (Kirkwood and Sarin, 1985). Sometimes only ordinal information is available, the output of the dialogue being only partial preference relations between subsets or coalitions of criteria, or at most a rank-order of the weights w_j. In other situations, it is only appropriate to determine lower and upper interval bounds for each weight (or for each tradeoff or substitution rate), for reasons of imprecision, uncertainty, and/or inaccurate determination (Roy, 1988), and/or because of the different preference systems of the actors. We return to the subject of this section in paragraph 2.4.1.

----

1. «Weights» w_j $(j=1,...,n)$ are used to refer either to «scaling factors» or to «coefficients of importance», assuming that it is clear in the reader's mind that their meaning and characteristics depend upon on the type of operational approach being used (see Vansnick, 1986). As Roy (1990) states, the weights have an intrinsic character in an ELECTRE method, but they are non-intrinsic in the AHP and MAUT methods because they depend on the nature of the scale chosen for evaluation (which further complicates the weighting process). To ignore this fact can give rise to meaningless practices, like it is to combine (apparently simpler) weighting techniques with aggregation models with which they are not theoretically compatible. As an example, between an (unfortunately) large collection of cases, one can refer to applications of Saaty's Method to specify the weights and in sequence to make use of those substitution rates as coefficients of importance within an ELECTRE Method.

2. The absence of a single and/or well identified decision maker is not so rare as one could imagine, mainly in applications in the public sector.

In this paper an additive value function technique combined with a fuzzy outranking relation is developed for aiding decision making in situations involving «poor» (incomplete, partly available) inter-criteria preference information.

We call it the *«Outweigh Approach»* and we confine the model to the case of certainty.

2. BASIC ASSUMPTIONS OF THE ADDITIVE MODEL

In order to understand the methodological context of application of any MCDA method it is essential to well define its basic theoretical assumptions. Those underlying the additive model used in our approach are described in this section following the sequence:

1 - Assumptions related with the phase of analysis of the decision aid process;
2 - assumptions related with the phase of intra-attribute preference modelling;
3 - assumptions related with the preference aggregation phase;
4 - assumptions related with the phase of inter-criteria preference modelling.

2.1. Assumptions Related with the Phase of Analysis

i) There is a well-defined discrete set of alternatives A:

$$A = \{a_1, \ldots, a_i, \ldots, a_m\}.$$

ii) The «decision problem formulation» can be of the *choice or classification types* 1 . *I.e.,* the final goal of the decision process is either to select «the best alternative» or to rank the alternatives by order of global preferability.

----

1. See Roy (1985) for a formal definition.

iii) $X = \{X_1, \ldots, X_j, \ldots, X_n\}$ is a set of quantitative and/or qualitative attributes, an attribute X_j being a set of at least two elements x_j expressing different levels of some underlying dimension. An alternative can be typified by a tuple $(x_1, \ldots, x_j, \ldots, x_n)$ representing its evaluations on the attributes $X_1, \ldots, X_j, \ldots, X_n$ (assumption of certainty case).

iv) X exhaustively represents an a priori defined set of points of view covering all the aspects, consequences or components of a decision that each actor individually and/or all the actors in consensus accept to be important and meaningful for discriminating between each pair of alternatives in A. Furthermore:

v) X satisfies the preference independency condition, *i.e.*, it must be possible to rank-order the elements x_j of any attribute X_j, according to the preference systems of the decision-maker, independently of the levels on the other (n-1) attributes. We designate by "$\succ_j$" the (strict weak order) preference relation associated with X_j $(j=1, \ldots, n)$.

vi) Each attribute X_j is bounded with respect to the relation $\succ_j$, being x_j° the lower bound (worst level) and x_j^* the upper bound (best level).

2.2. Assumptions Related with the Phase of Intra-attribute Preference Modelling

i) As it is well known (Roberts, 1979), given $\succ_j$ it is possible to assign to each evaluation level x_j of X_j a real number $v_j(x_j)$ representing $\succ_j$ [1]. We assume this measure of x_j is unique, *i.e.*, $v_j(x_j)$ is a single real number and the application $v_j: X_j \rightarrow \mathbb{R}$ is

----

1. I.e., being x_j and y_j any two evaluation levels on the attribute X_j, if the decision-maker prefers x_j to y_j than the real numbers $v_j(x_j)$ and $v_j(y_j)$ must be so that $v_j(x_j)$ be greater than the value $v_j(y_j)$.

an *interval scale of measurement* (*i.e.*, unique up to positive linear transformations)[1].

Thus v_j represents the intra-attribute preference information given on X_j. v_j is called the (criterion) *value function* associated with the attribute X_j and $v_j(x_j)$ the *value* (score) corresponding to the evaluation level x_j.

Let $(x_1,\ldots,x_j,\ldots,x_n)$ and $(y_1,\ldots,y_j,\ldots,y_n)$ represent the evaluation profiles of two alternatives a and b of A. On each attribute $X_j \in X$ one, and only one, of the two following preference situations can hold:

Strict preference ($\succ_j$): $a \succ_j b$ iff $v_j(x_j) > v_j(y_j)$ or
$b \succ_j a$ iff $v_j(y_j) > v_j(x_j)$

or

Indifference ($\sim_j$): $a \sim_j b$ iff $v_j(x_j) = v_j(y_j)$.

ii) It is assumed that the measures within each v_j value scale vary between 0 and 1, the measure $v_j(x_j^*) = 1$ being the value of the best evaluation level x_j^* of X_j and $v_j(x_j^o) = 0$ the value of the worst level x_j^o.

iii) The Outweigh Approach has been conceived to be useful in practice after the phase of intra-attribute preference modelling, *i.e.*, after the construction of the criteria functions - the preference scales v_j $(j=1,\ldots,n)$ - which are assumed to be known (complete intra-attribute preference information)[2].

----

1. For a basic introduction to Measurement Theory concepts see (Vansnick, 1990).

2. This paper does not detail the development of the phase of modelling intra-attribute preferences. Nevertheless, we want to remark that as a rule the construction of the value functions v_j is not a easy task. Basic references on this subject are Fishburn (1967), Keeney and Raiffa (1976), Keeney (1977) and Dyer and Sarin (1979).

2.3. Assumptions Related with the Preference Aggregation Phase

We confine our model to an *additively decomposable measurement context under certainty*, with the following assumptions:

i) If complete inter-criteria preference information would exist, it should be possible to define a real-valued function v: $X \to \mathbb{R}$

$$v(x_1,\ldots,x_j,\ldots,x_n) = f\,[v_1(x_1),\ldots,v_j(x_j),\ldots,v_n(x_n)],$$

such that, for all pairs of alternatives a and b of A:

$$a \succsim b \text{ iff } v(a) \geq v(b)$$

where " $\succsim$ " is a weak order binary relation and a $\succsim$ b means that a is at least globally as good as b.

v is a «single synthetical criterion», and v(a) is a level of the v scale (that we will designate by global value of the alternative a) modelling (expressing) the overall preferability of a.

ii) Of course, the weak preference relation $\succsim$ is consistent with (respects) the dominance relation (Δ), *i.e.*, if a dominates b (aΔb) a is surely globally preferred to b ($\Delta \subset \succsim$).[1]

iii) The function v is assumed to have an additively decomposed form (weighted-additive evaluation function):

$$v(a) = \sum_{j=1}^{n} w_j v_j(a).$$

In this model, the weighting constants w_j (j=1,...,n) are tradeoff values reflecting in terms of global preference the increase on a criterion value necessary to compensate (in the decision maker preference system) a decrease on another criterion.

----

1. a dominates b whenever a is at least as good as b (a $>_j$ b or a $\sim_j$ b) for all attributes $X_j \in X$ (j = 1,...,n), and a is strictly better than b (a $>_k$ b) for at least one attribute $X_k \in X$. An alternative is said to be optimal if it dominates all the other alternatives.

They cannot be interpreted as direct indicators of attribute importance, being intrinsically related with the discrimination power of the criterion scales v_j [1].

iv) As it is well known, in the context of a *descriptive approach* several basic conditions must hold for the validation of the additive model [2], one of the most important being the mutual preference independency axiom (Debreu, 1960). These conditions are often assumed to hold in applications, although in many situations this may be a rough simplification. Like those cases in which independence assumptions are violated by the simple fact that the tradeoff between two criteria depends on the value on another criterion. [3]

Nevertheless, we do not adopt the descriptive approach but rather a *constructive* perspective. Thus, we do not a priori assume the existence of a well-defined global preference function in the decision maker's mind but only that we intend to construct that function supported by a *simple* [4] model as it is the weighted sum [5].

----

1. "Weight is simply a transformation of a scale, and that scale is meaningful only if its end points are clearly defined" (Edwards et al., 1988). This is why it is more appropriate to call them scaling constants. See illustration in Keeney and Raiffa (1976, section 5.9).

2. Two basic references are Fishburn (1970) and Krantz et al. (1971).

3. See Chankong and Haimes (1983) for a theoretical review of the additive model conditions and Roy and Bouyssou (1987) for a critical discussion. See also the discussion in Bell et al. (1977, chapter 19) and French (1988).

4. "Simplicity helps in many ways. It helps a nontechnician (or even a technician) to understand what is going on. It makes analysis briefer and easier to perform. It makes judgments easier to make. Above all it helps the decision maker to use and communicate about the tools that have been developed and their results, and this promotes adoption and use" (Edwards et al., 1988).

5. As Edwards (1977) remarks, "probably the most widely used, and certainly the simplest, aggregation rule (...) consists of simply taking a weighted linear average".

v) The levels of the v scale are assumed to vary between 0 and 1, being $v(a^*) = 1$ the global value of the «ideal alternative» $a^* = (x_1^*, \ldots, x_j^*, \ldots, x_n^*)$ and $v(a^\circ) = 0$ the global value of the «anti-ideal alternative» $a^\circ = (x_1^\circ, \ldots, x_j^\circ, \ldots, x_n^\circ)$. This implies the following conditions defining the domain of variation of the scaling constants:

vi) $\sum_{j=1}^{n} w_j = 1$ and $w_j > 0$ for all j.

vii) Given the prior assumptions, if complete inter-criteria preference information would exist, one and only one of the two classic situations of global preference should hold:

Strict preference ($\succ$): $a \succ b$ iff $\sum_{j=1}^{n} w_j.[v_j(x_j)-v_j(y_j)] > 0$ or

$b \succ a$ iff $\sum_{j=1}^{n} w_j.[v_j(x_j)-v_j(y_j)] < 0$

or

Indifference ($\sim$): $a \sim b$ iff $\sum_{j=1}^{n} w_j.[v_j(x_j)-v_j(y_j)] = 0$.

2.4. Assumptions Related with the Phase of Inter-criteria Preference Modelling

2.4.1. Types of poor weighting information

we assume that the phase of inter-criteria preference modelling has taken place following a well-defined and consistent questioning procedure during which the weights were *not precisely determined*.

We are than interested in cases of *decision making under poor weighting information* which can take a large variety of forms, grouped below in three main categories:

i) Some (not exhaustive) conditions relating the w_j variables [1].

ii) A rank-order of the w_j variables [2], for instance:

$$w_1 > \ldots > w_j > \ldots > w_n.$$ [3]

iii) Lower and upper preference bounds for value tradeoffs, for substitution rates or for weights [4].

iii.1) In the context of an operational approach with a «single synthetical criterion» $v(x_1,\ldots,x_j,\ldots,x_n)$, such as MAUT, the classic questioning procedures for determining the scaling constants w_j $(j=1,\ldots,n)$ involve the assessment of multiattribute tradeoffs [5]. But a decision-maker may not be able to respond specifically enough to hypothetical tradeoff questions [6] and/or the answers of multiple actors are not consensual. In these cases, it can be only possible or realistic to define lower and upper interval bounds for the value tradeoffs, which give directly rise to (n-1) independent inequalities of the type:

$$\alpha_j < w_j/w_r < \beta_j \quad (j = 1,\ldots,n \text{ and } j \neq r)$$

----

1. Although developed within undoubtedly different operational MCDA approaches, both the questioning procedures proposed by Keeney and Raiffa (1976, section 3.7) and by Roy et al. (1986) for the assessment of weights (scaling constants or coefficients of importance, respectively) are based upon on comparisons between subsets or coalitions of criteria. If for some reason the dialogue is broken at initial stages, their output will be nothing better than a few inequalities relating weights, for example, "one weight in bigger (or smaller) and the other weights are not rank-ordered".

2. Here included not only the case of a complete rank-order of the weights, but also some variations of practical interest, like "the constants are rank-ordered with two constants equal" or "subsets of constants are rank-ordered" (these two cases are both considered by Kirkwood and Sarin (1985)).

3. Two MAUT applications both dealing with a complete rank-order of the scaling constants are described in (Kirkwood and Sarin, 1985) and (Bana e Costa and Neves, 1989).

4. Intervals (bands) of no preference, i. e., no value between the bounds is more representative in terms of inter-criteria preferences than any other, such that narrow bands correspond to richer information (less uncertainty, imprecision and/or conflict).

5. See Keeney and Raiffa (1976), Keeney and Nair (1977) and Keeney (1980) for description and practical applications of these procedures.

6. "Often a decision maker is unable or unwilling to provide precise multiattribute tradeoffs" (Hazen, 1986).

where α_j and β_j ($0 \leq \alpha_j \leq \beta_j \leq 1$) are respectively lower and upper interval bounds and r is a reference criterion, typically (but not necessarily) the one which has been determined to have the largest weight w_r.[1]

Similarly, in some average sum cases the tradeoff or substitution rates (s_{ij}) between a reference criterion r and each of the other j (j≠r) criteria are directly examined in order to set for each rate (s_{jr}) an upper and a lower boundary between which all the actors agree that the rate may vary:

$$s^{\circ}_{jr} \leq s_{jr} \leq s^{*}_{jr} \quad \text{(for all } j \neq r\text{)}.$$ [2]

iii.2) In the context of an outranking approach identical situations can hold, so that only an interval I_j will be available for the coefficient of relative importance of each criterion:

$$w^{\circ}_{j} \leq w_j \leq w^{*}_{j} \quad \text{(for all } j \neq r\text{)}.$$

This type of incomplete inter-criteria preference information can also occur in MAUT applications when the scaling constants are directly specified. In many applications, direct assessment can be forced by the actors' inability to compare fictitious alternatives, and analysts often tend to follow more simple weighting procedures (like direct rating techniques) to avoid the intrinsic complexity of tradeoff questions[3].

----

1. Hazen (1986) illustrates this type of poor information with an hypothetical example constructed from the real-world decision analysis case of selecting nuclear power plant sites described by Keeney and Nair (1977) and Keeney (1980).

2. This is the case of the choice problem of a suburban highway design in the Paris forest environment described by Bertier and de Montgolfier (1974). A simplified version of this problem is used as an example by Bana e Costa (1989 b) to illustrate the applicability of the Outweigh Approach for supporting decision-making in situations characterised by the fact that an interval for each tradeoff rate is independently defined by each of the actors involved.

3. It is important to remark that (practical) simplicity and (theoretic) adequacy are often incompatible aims, and, unfortunately, many applications of simple weighting techniques ignore this fact.

Moreover, special peculiarities of decision situations can by themselves rule out procedures requiring the assessment of multi-attribute tradeoffs, such as when the scaling constants must be defined before knowing the evaluations of a well defined set of alternatives [1].

All the types of reasons already referred along this paper and, particularly, the use of less robust questioning procedures than the classic ones create imprecision, and/or uncertainty, and/or inaccurate determination. These effects justify the practical interest of defining bands of variation for the weights, the amplitude of which can be smaller or larger, depending on the conjunction of a smaller or greater number of effects and on their degrees of intensity. As Roy (1975) pointed out, «a too precise specification of the weights creates false certainties».

2.4.2. The space of analysis and the dominance aggregation rule

Depending on the type of prior partial weighting information assessed, the analysis takes place in the $\mathbb{R}^{n-1}$ «space of the tradeoff rates» (s_{jr}, $j \neq r$) or in the «space of the scaling constants». In this paper only the latter will be considered [2].

The set of conditions $\Sigma_j\ w_j = 1$ and $w_j > 0$, $X_j \in X$, defines a polyhedron that we will call ***«general feasible set»*** (Ω). Figures 1-*i)* to 1-*iii.2)* show graphical overviews of three attribute examples of the several types of poor weighting information above typified. The respective conditions constrain the form of Ω to a new convex polyhedron - the ***«feasible set of weights»*** $\mathbf{W}$ ($W \subset \Omega$).

----

1. A very illustrative example of this type of situation can be found in Edwards et al. (1988). They describe the design of a system for evaluation and prioritisation of research projects, to be systematically used from year to year, thus being easily understandable that the weights needed to be available before the value measures were.

2. See Bana e Costa (1989 b) for an application developed in the space of the tradeoff rates.

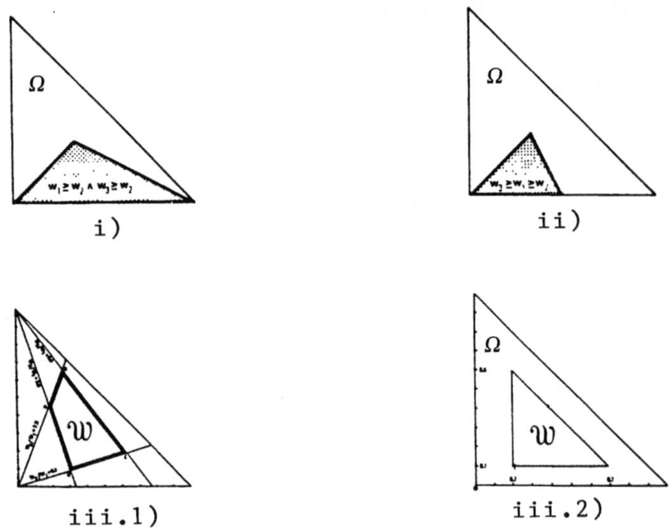

Figures 1-*i)* to 1-*iii.2)*

Let us consider, for the moment, a situation of total absence of inter-criteria preference information. Thus, **W** = **Ω** and it will be only possible to build an aggregation rule based on the concept of dominance (Δ):

- If a dominates b (a Δ b), a is (surely) globally better than b;
- if b dominates a (b Δ a), b is (surely) globally better than a;
- otherwise, a and b are incomparable (a ? b).

«Incomparability» (?) reflects the fact that there are not enough strong arguments to make a choice (strict preference or indifference) between a and b, because in some points of Ω a can be preferred to b but in other points b will preferred to a.

Suppose from now on that inter-criteria preference information is available, but only partially. That is, we have not a single point P but all a feasible set of weights **W**. The crucial question we will have to answer is: **How to construct a consistent pairwise comparison rule under those circumstances?**

In the next sections we will explain our answer to this question. As any other multicriteria method, our approach is an attempt to enrich the results of the dominance analysis, by searching for additional arguments allowing to decrease the great number of incomparability cases [1].

3. SOME PRECEDING RELATED RESEARCH

Several researchers have been devoting attention to the problem of poor weighting information. We will restrict this section to a short indication of the works more directly related with our model [2].

The additive method REGIME ANALYSIS, described in this volume by Janssen et al. (1990) (see also Hinloopen et al., 1983), only requires a rank-order of the weights (and also a complete ranking of the alternatives by each criterion).

The «Method of Progressive Reduction of Incomparability» (Bertier and de Montgolfier, 1974; Gibert and Gagey, 1976) was developed to face compensatory situations with multiple actors involved where consensual lower and upper interval bounds are defined for the each substitution rate (s_{jr}) between each criterion (j) and a previously chosen reference criterion (r). This approach was extended by Jacquet-Lagrèze (1975) and Bana e Costa (1989 b). If no weighting information is available, the TRIDENT Technique of Tavares (1984) (see also Bana e Costa, 1988) analyses all the possible complete rankings of the alternatives produced by a weighted linear average (for three criteria, only).

----

1. Nevertheless, we do not assume it is possible to totally overcome incomparability. As the outranking approaches, the outweigh approach admits as realistic that «indecision» situations can not be totally eliminated in many (most of) decision aid processes.

2. See (Bana e Costa, 1989 a) for a more detailed list of references, both in the framework of the so-called Outranking Family and in bayesian probabilistic or expected-utility contexts with incomplete knowledge of probabilities in MAUT.

Researchers such as Hannan (1981) or Kirkwood and Sarin (1985) were inspired by the pioneer work of Fishburn (1964 and 1965), namely on Fishburn's Method of Equating Coefficients developed to analyse uncertainty problems with incomplete knowledge of probabilities. Assuming additive or multiplicative multiattribute value (or utility) functions as partially known, those and other works (such as Sarin, 1977 a and b; Chelsea et al., 1983; White et al., 1983; White et al., 1984; Malakooti, 1985; Hazen, 1985 and 1986) can be viewed as attempts to establish results about «dominance» [1] and/or optimality, and/or to dèveloped pairwise procedures to establish the degree to which a set of alternatives can be rank-ordered given different forms of poor inter-criteria preference information.

For the cardinal additive model, Kirkwood and Sarin (1985) define a theorem (and its corollary) which establish the conditions for an alternative to be surely preferred to another one («dominance» [1]) for several sets of conditions on the weighting constants, such as the «constants are rank-ordered» (also analysed by Hannan (1981)) and some variations of practical interest, like «subsets of constants are rank-ordered». Sarin (1977 a, b) uses a linear programming approach to analyse the case where lower and upper interval bounds are available for the scaling constants. Malakooti (1985) follows a similar approach to study the same case and also the case of ordinal information on the scaling constants. Hazen (1986) extends its research to the case of ordinal additive utility, and he also analyses the case of multiplicative cardinal utility. He derives «potential optimality and dominance» results for the cardinal additive and multiplicative cases under the two following types of prior incomplete information: the scaling constants are rank-ordered, and lower and upper interval bounds are known for the tradeoffs assessed between alternatives that differ only on a reference attribute X_r and each attribute X_j ($j=1,\ldots,n$ and $j\neq r$).

----

1. More precisely "restricted dominance" as defined in section 4.1.

4. RESTRICTED DOMINANCE ANALYSIS

4.1 The Concept of Restricted Dominance

According to the assumptions in section 2.3, at each point P $(w_1^P, \dots w_j^P, \dots w_n^P)$ in **W** one, and only one, of the following situations of global preference holds:

$$a \succ^P b \text{ iff } v^P(a) > v^P(b)$$

$$a \sim^P b \text{ iff } v^P(a) = v^P(b)$$

$$b \succ^P a \text{ iff } v^P(a) < v^P(b)$$

being $v^P(a) = \sum_{j=1}^{n} w_j^P v_j(a)$ the global value of a at P.

Let $\Gamma(a,b)$ be the subset of points of Ω to which corresponds a situation of indifference between two alternatives a and b:

$$\Gamma(a,b) = \{P \in \Omega : \sum_{j=1}^{n} w_j^P . [v_j(x_j) - v_j(y_j)] = 0\}.$$

It can happen that $\Gamma(a,b)$ does not intersect the set of feasible weights **W** ($W \cap \Gamma(a,b) = \phi$). In this case, one of the two alternatives will be always globally preferred to the other in **W**, and we will say that a situation of *«restricted dominance»* (Δ^r) occurs, even if neither a dominates b nor b dominates a (*i.e.*, $\Omega \cap \Gamma(a,b) \neq \phi$) - see figure 2-*i*) [1]:

$$a \, \Delta^r \, b \text{ iff } \sum_{j=1}^{n} w_j^P . [v_j(x_j) - v_j(y_j)] > 0 \quad \text{for all } P \in \mathbf{W}.$$ [2]

---

1. The "restricted dominance relation" is also transitive.

2. Following Hazen (1986), we can distinguish between «additive (restricted) dominance» for the case of an additive multiattribute aggregation function and «multiplicative dominance» if the multiplicative model is used. Let us also point out that the «restricted dominance relation» is equivalent to the outranking relation built in the framework of the Method of Progressive Reduction of Incomparability. So, taking the concept of outranking relation in a non strict sense, we could use «additive outranking» to refer to the same concept of «additive dominance».

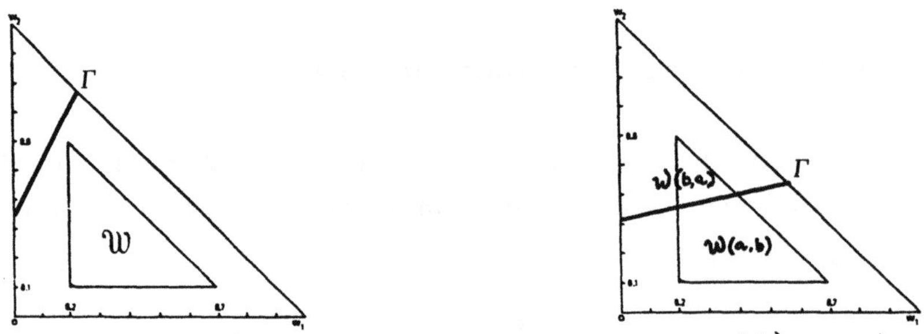

i) ii)

Figures 2.*i)* and 2.*ii)*

Similarly, restricted dominance does not hold (neither $a \Delta^r b$ nor $b \Delta^r a$) when $\Gamma(a,b)$ intersects somewhere the feasible set **W** ($\Gamma(a,b) \cap W \neq \phi$), thus defining on **W** two sub-sets W(a,b) and W(b,a) (see figure 2-*ii*)), such that:

$$W(a,b) = \{P \in W: \sum_{j=1}^{n} w_j^P . [v_j(x_j) - v_j(y_j)] > 0\}$$

and

$$W(b,a) = \{P \in W: \sum_{j=1}^{n} w_j^P . [v_j(x_j) - v_j(y_j)] < 0\}.$$

If $W(b,a) = \phi$, a restrictedly dominates b ($a \Delta^r b$). Similarly, if $W(a,b) = \phi$, b restrictedly dominates a ($b \Delta^r a$).

Normally, restricted dominance is a richer relation than dominance because it tends to decrease the number of cases of incomparability, although usually not rich enough to completely rank-order the alternatives in A.

4.2. Progressive Restricted Dominance Analysis

A progressive enrichment of the conclusions of a restricted dominance analysis can be obtained by means of an interactive process for successively contracting the size of the feasible set **W**, with the aim of decreasing the number of incomparability cases, as figure 3 exemplifies.

The process consists on the construction of a sequence of z ($z \geq 1$) «restricted dominance relations» Δ^r ($r=1,\ldots,z$) each of them modelling the actors' global preferences within a reduced feasible set W^r ($r = 1,\ldots,z$)

$$W^1 \supset \ldots \supset W^r \supset \ldots \supset W^z,$$

being z as great as it would be allowed by the conditions of each particular application.[1]

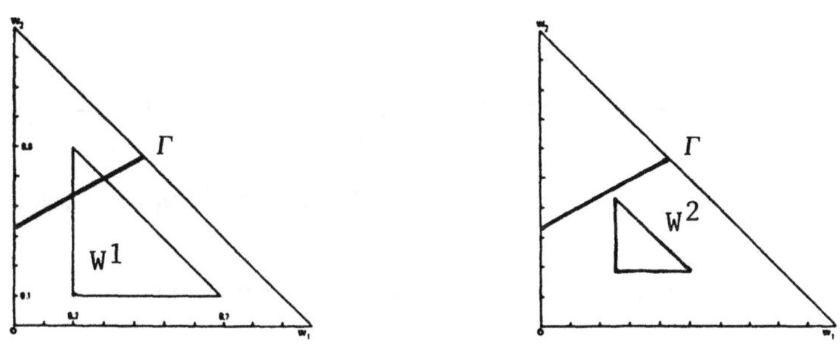

Figure 3

4.3. Partial Ranking Based on the Restricted Dominance Concept

Based on the concept of restricted dominance we can formulate the following pairwise rule for comparing and ranking two alternatives $a = (x_1,\ldots,x_j,\ldots,x_n)$ and $b = (y_1,\ldots,y_j,\ldots,y_n)$:

---

1. This way of proceeding will (very probably) enrich the dominance analysis, provided that it can be possible to go back to previous phases. As a matter of fact, many times the evolution of the decision aid process will not permit to contract the initial feasible set W^1, as for instances, when it is not possible to establish new consensus between the actors.

Under conditions of poor information available about the inter-criteria preferences of the actors, represented by a feasible set of weights **W**, a is surely preferred to b if $W(b,a) = \phi$, that is, if the global value of a is always greater than the global value of b:

$$v^P(a) - v^P(b) > 0, \quad \text{for all } P \in W,$$

which is identical to:

$$a \Delta^r b \text{ if and only if } \min_{\mathbf{W}} v^P(a) - v^P(b) > 0.$$

Thus, for the additive value function model, the restricted dominance analysis between the two alternatives a and b can be made by a linear programming approach:

1) Solve the linear program: $\underset{\mathbf{W}}{\text{Min}} \sum_{j=1}^{n} w_j . [v_j(x_j) - v_j(y_j)] = \mathbf{Z}^*,$

subject to: $(w_1, \ldots, w_j, \ldots, w_n) \in W$

2) if $\mathbf{Z}^* > 0$, then a is surely preferred to b.

The linear program 1) is well studied in the literature for types of poor information where the respective sets of feasible weights are always sets of linear constraints on Ω, as those defined in section 2.4. We refer to the works of Sarin (1977 a and b), Malakooti (1985) and Hazen (1986). For example, in cases where interval bounds are available for the scaling constants the program 1) is a linear knapsack problem solved by inspection by Sarin (1977 a and b).

Approaches other than linear programming are also well-suited. For example, if a complete rank-order of the scaling constants is available, Hannan (1981) and Kirkwood and Sarin (1985) prove a very useful theorem for analysing additive restricted dominance between pairs of alternatives:

THEOREM - If $w_1 > \ldots > w_j > \ldots > w_n$, then a is guaranteed to be preferred to b if and only if

$$\sum_{j=1}^{m} [v_j(x_j)-v_j(y_j)] \geq 0, \quad m = 1,\ldots,n$$

with at least one of the inequalities being strict, i. e.,

$$\sum_{j=1}^{m} [v_j(x_j)-v_j(y_j)] > 0, \text{ for at least one } m = 1,\ldots,n.$$

Hazen (1986, page 300) proves an analogous theorem for the case where interval bounds are available for tradeoff values [1].

Depending on the type of poor inter-criteria preference information available, one can choose the more adequate of the above methods for analysing restricted dominance between all the n(n-1) pairs of alternatives in **A**.

Finally, the set of alternatives **A** can be (partially) rank-ordered, for which we propose to use the so-called Rank-ordering Algorithm developed by Kirkwood and Sarin (1985).

4.4. Illustrative Example

Progressive dominance analysis and the Rank-ordering Algorithm will be illustrated with an example involving five alternatives (a_1 to a_5) and three attributes (X_1 to X_3) (we refer to the housing policy problem treated in Bana e Costa, 1988 and 1989 a).

In this problem an initial set of feasible weights ($\mathbf{W}^1$) was established (figure *1-iii.2)*):

$$\mathbf{W}^1 = \{(w_1,w_2,w_3): 0.2\leq w_1\leq 0.7 \text{ and } 0.1\leq w_2\leq 0.6 \text{ and } 0.2\leq w_3\leq 0.7\}.$$

----

1. That is, if $W = \{(w_1,\ldots,w_j,\ldots,w_n) \in \Omega: \alpha_j < w_j/w_r < \beta_j \ (j=1,\ldots,n \text{ and } j\neq r)\}$ - case iii.1) in section 2.4.

Under these conditions alternative a_2 dominated alternative a_5 and this is the only conclusion provided by (classic) dominance analysis. Moreover, a restricted dominance analysis did not add any further result. This is not surprising because the intervals for the scaling constants are very large.

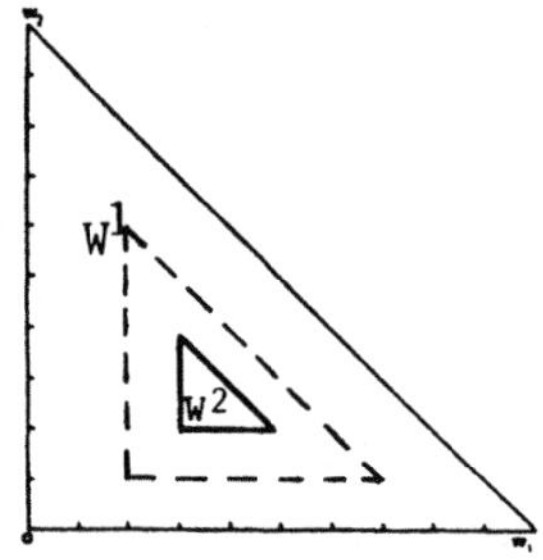

Figure 4

Suppose now, for exemplification, that the actors agreed with contracting $\mathbf{W}^1$, and a new consensus was established upon a second set of weights $\mathbf{W}^2$ (see figure 4) smaller than the initial one ($W^2 \subset W^1$):

$$\mathbf{W}^2 = \{(w_1,w_2,w_3):\ 0.3 \le w_1 \le 0.5 \text{ and } 0.2 \le w_2 \le 0.4 \text{ and } 0.3 \le w_3 \le 0.5\}.$$

Under these new conditions, the output of the restricted dominance analysis is this time much more rich:

a_2 is preferred to a_1, a_4 and a_5,
a_3 is preferred to a_1, a_4 and a_5,
a_5 is preferred to a_1.

These results can be summarised in the pairwise ranking matrix of table 1, where a *1* means that the respective row alternative is preferred to the respective column alternative. At the bottom of the matrix we added a row with the number of alternatives which restrictedly dominates each column alternative.

TABLE 1 - RESTRICTED DOMINANCE MATRIX

	a_1	a_2	a_3	a_4	a_5
a_1	0	0	0	0	0
a_2	1	0	0	1	1
a_3	1	0	0	1	1
a_4	0	0	0	0	0
a_5	1	0	0	0	0
Column Totals	3	0	0	2	2

TABLE 2 - RANK-ORDERING TABLE

Required No. of Alternatives	Column Totals	Alternatives with Column Totals	No. of Altern. with Column Totals	Cumulative No. of Alter. with Column Totals
1	4	-	0	0
2	3	a_1	1	1
3	2	a_4, a_5	2	3
4	1	-	0	3
5	0	a_2, a_3	2	5

The Restricted Dominance Matrix of table 1 will be the basis for illustrating the Rank-Ordering Algorithm. The following description is nothing more than a direct adaptation of the words of Kirkwood and Sarin (1985, section 2):

RANK-ORDERING ALGORITHM

1. «Prepare a rank-ordering table as shown in table 2.
   a. Under «Required No. of Alternatives» enter the numbers 1, 2,..., n.
   b. Under «Column Totals» enter the number n-1, n-2,..., 2, 1, 0 in successive rows.
   c. Under «Alternatives with Column Totals» list all the alternatives with a «Column Total» on the pairwise ranking table (table 1) that is equal to the number in the second column of the rank-ordering table.
   d. Under «No. of Alternatives with Column Totals» enter the number of alternatives listed in the third column. (Note that the sum of all entries in the fourth column is n).
   e. Under «Cumulative No. of Alternatives with Column Total», column down to and including that row.

2. In the rank-ordering table, draw a line across the table immediately below each row where the entries in the first and fifth columns are equal. The sets of alternatives in pairs of lines are successively more preferable going down the table. Within each set, it is not possible to rank-order the alternatives given only the preference information contained in the pairwise ranking table.»

The last column of table 2 gives the final partial ranking of the alternatives based only on the restricted dominance analysis: the sub-set of alternatives $\{a_2, a_3\}$ is preferred to the sub-set of alternatives $\{a_1, a_4, a_5\}$. (Note from table 1 that a_5 is preferred to a_1).

5. THE OUTWEIGH COMPARISON PROCEDURE: DEGREE OF CREDIBILITY AND FUZZY OUTRANKING RELATION

5.1. Objective and Theoretical Framework

Whatever the last restricted dominance analysis done, its output, although probably richer than the one from classic dominance, will be in general still too poor in terms of final comparisons.

But it is possible to go further in the decision aid process. With the Outweigh Approach we propose a pairwise comparison procedure to exploit in greater depth the available information within **W** . It consists mainly on the construction and exploitation of a *fuzzy outranking relation* in A, representing the degree of credibility that can be assigned to the statement «a is at least as good as b», a and b being any two alternatives of A.

The concept of fuzzy outranking relation embodies the contents of this section. We will define it making use of Siskos et al. (1984, section 2.1):

«We call the fuzzy outranking relation (S^d) in A x A a membership function d: A x A $\rightarrow$ [0,1] in which the different values d(a,b) denote the strength of the relationship between any two actions a and b in **A** (...). Thus, as Roy (1977) indicates, d(a,b) is the degree of credibility of the outranking of the action b by the action a. (...) The fuzzy outranking relation is reflexive (d(a,a) = 1, for all a $\in$ **A**) and can induce other better known preferential structures (...)».

5.2. Construction of the Fuzzy Outweigh Relation

We call *«fuzzy outweigh relation»* the fuzzy outranking relation that we will construct in this section to exploit the information given in a feasible set of weights **W**.

The concept of fuzzy outweigh relation can be easily introduced with the support of a very simple reasoning: let us hypothesize, for a moment, that **W** is a discrete set of points in Ω, with a number of points equal to **# W**. Determine the number of points in **W** for which a is preferred to b, that is:

$$\# W(a,b) = \# \{P \in \mathbf{W}: v^P(a) - v^P(b) > 0\}.$$ [1]

If # W(a,b) = # W, a (restrictedly) dominates b ($a \Delta^r b$), and so, the degree of credibility d(a,b) associated with the statement «a is at least as good as b» is maximum. On the other hand, if # W(a,b) = 0, d(a,b) is minimum. But, if neither $a \Delta^r b$ nor $b \Delta^r a$, a natural measure of the degree of credibility is given by the fraction of points of **W** where a is at least as good as b, that is:

$$d(a,b) = \frac{\#W(a,b)}{\#W}$$

As **W** is by definition continuous in the space $\mathbb{R}^{n-1}$, by analogy the measure of the credibility degree d(a,b) between any two alternatives a and b of A will be given by:

$$d(a,b) = \frac{\int_{W(a,b)} \delta w_1, \ldots, \delta w_j, \ldots, \delta w_{n-1}}{\int_{W} \delta w_1, \ldots, \delta w_j, \ldots, \delta w_{n-1}}$$

or, more simply:

$$d(a,b) = \frac{V[W(a,b)]}{V[W]},$$

where, for the general n criteria problem, V[W] is the volume of the convex polyhedron **W** and V[W(a,b)] is the volume of the convex polyhedron W(a,b) [2]. These volumes can be computed using the algorithm presented in (Lasserre, 1983).

----

1. More rigorously, we have to include also the points for which a and b are indifferent.

2. Within the context of the REGIME Analysis Hinloopen et al. (1983) also proposed to associate the value of the credibility degree d(a,b) with the relative size of w(a,b).

The following are the basic characteristics and properties of the «fuzzy outweigh relation» above constructed:

- d: $A \times A \rightarrow [0,1]$, that is, $0 \le d(a,b) \le 1$, for all (a,b);
- the fuzzy outweigh relation is reflexive: $d(a,a) = 1$, $\forall a \in A$;
- $d(a,b) + d(b,a) = 1$, for all pairs (a,b);
- $d(a,b) = 1$ iff a restrictedly dominates b ($a \neq b$);
- $d(a,b) = 0$ iff b restrictedly dominates a ($a \neq b$);
- d is a (max min) transitive relation, that is:
  $d(a,b) \ge \max_{c \in A} \min [d(a,c), d(c,b)]$, for all $a,b \in \mathbf{A}$.
- $\Delta \subset \Delta^r \subset S^d$.

Let us come back to the application. Remember from section 4.4 that, for the contracted set of feasible weights W^2 (see figure 4), the restricted dominance analysis gave the following partial (global) ranking in A: $\{a_2,a_3\}$ preferred to $\{a_1,a_4,a_5\}$ (and a_5 preferred to a_1).

Applying now the concept of fuzzy outweigh relation to that problem, the credibility degrees matrix of table 3 results. Of course, all the elements of this matrix corresponding to restricted dominance situations are equal to 1, but now the pair-wise comparison information is much more rich than the one offered by the restricted dominance matrix of table 1.

TABLE 3 - CREDIBILITY MATRIX

	a_1	a_2	a_3	a_4	a_5
a_1	1	0	0	.133	0
a_2	1	1	.991	1	1
a_3	1	.009	1	1	1
a_4	.867	0	0	1	.678
a_5	1	0	0	.322	1

5.3. Exploiting the Fuzzy Outweigh Relation

Based on the fuzzy outweigh relation S^d, and to exploit, in **W**, the pairwise comparison information given by the membership function d, one can construct a sequence of crisp (non-fuzzy) outweigh (outranking) relations, $S^1 \subset S^2 \subset \ldots \subset S^s \subset \ldots$,[1] of the following type:

$$a\ S^s\ b \text{ iff } d(a,b) \geq s, \text{ with } s \in]0.5,\ 1],$$

being the threshold (cut value) **s** as greater as it is weaker the strength of the arguments required to validate the assertion «a is at least as good as b».[2]

If «a outweighs b» (a S^s b) we will say that the actors are willing to accept that «a is at least as good as b», this act involving some «risk», a measure of which can be the value of (1-d(a,b)). Under this reasoning, if d(a,b) and d(b,a) are both simultaneously lower than **s**, there are not sufficiently strong arguments to pairwise rank the alternatives a and b, and so, a and b are incomparable.

In the presence of the values of the credibility degrees (see table 3) the analyst can start by remarking to the decision maker that, in spite of a_2 being not surely globally better than b (because a does not restrictedly dominates b), nevertheless $d(a_2,a_3) = 0.991$ is very close to 1, that is, a_3 is only preferred to a_2 in a very small (and perhaps worthless) part of the feasible set of weights (W^2). This fact can give (weaker but) sufficient arguments to validate, in the decision maker preference system, the assertion «a_2 is at least globally as good as a_3»,

----

1. Being the fuzzy outweigh relation max-min transitive, the derived crisp relations are obviously transitive.

2. The nature of the conditions which must be satisfied to validate the assertion «a outweighs b» (in the sense of «a is at least as good as b») is obviously influenced by «the strength of the arguments required: the strongest we can imagine is a dominates b but the outranking concept is an interesting one only because weak arguments can be sufficient. This is way S is usually a notably richer relation than dominance.» (Roy, 1990).

provided that he is willing to accept a risk of 1% (thus enriching the partial ranking in A). If that is the case, this modelling procedure can be in sequence applied to pairwise compare the pair of alternatives with the second greater value of credibility, that is, a_4 and a_1. Now, to validate «a_4 is at least as good as a_1» the decision maker will have to accept a risk around 15%. If he considers this value too high, the procedure stops here. If not, the next pair will be analysed, and so on until a threshold s has been fixed.

Suppose s = 0.85 is the minimum «degree of strength» accepted by the actors in our example. The crisp outweigh relation ($S^{0.85}$) modelling the final preferences in A is represented by the pairwise comparison matrix of table 4.

TABLE 4 - PAIRWISE COMPARISONS (s = 0.85)

	a_1	a_2	a_3	a_4	a_5
a_1	-	0	0	0	0
a_2	1	-	1	1	1
a_3	1	0	-	1	1
a_4	1	0	0	-	0
a_5	1	0	0	0	-
Column Totals	4	0	1	2	2

Finally, again applying the Rank-ordering Algorithm of Kirkwood and Sarin the partial global ordering of figure 5 results:

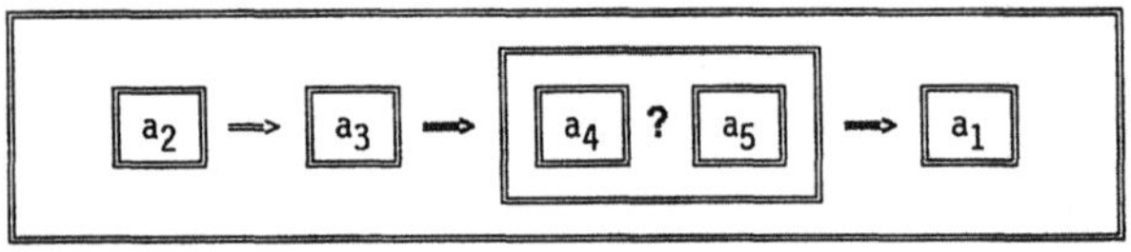

Figure 5

6. FINAL COMMENTS

Situations involving poor (incomplete or partial) inter-criteria preference information are very common in the practice of decision aid. Nevertheless, there exists a significant lack of operational MCDA approaches explicitly devoted to support decision making under those circumstances.

As a matter of fact, traditional multicriteria methods only operate with stable and single weights. To overcome this drawback in the application of the traditional methods, it is usual, in the practice, to choose a first vector of weights, and, afterwards, to make a *sensitivity analysis* using other vectors of weights, in order to conclude about the robustness and stability of the results obtained with the initial weights [1]. But, one has to recognise that this procedure do not directly and specifically deals with imprecise weights, being only, a way of bypassing the problem.

The Outweigh Analysis is a contribution towards the development of procedures able of directly facing situations of prior partially available weighting information. We have assumed an additively decomposable multiattributed value function has the basic aggregation model, under which our approach takes place in two main phases.

First, we introduce the restricted dominance concept for analysing which alternatives are guaranteed to be preferred to other alternatives, under several types of incomplete inter-criteria information. This problem is well studied in the MAUT literature. Second, to exploit in greater depth the information available, and to investigate if it is possible to enrich the conclusions of the restricted dominance analysis, usually too poor to support, by itself, a final decision, we have proposed

----

1. See the VISA approach, presented in this volume by Belton and Vickers (1990).

the concept of fuzzy outweigh relation, derived from the concept of fuzzy outranking relation but significantly different from it, in terms of methodological context, to justify a new designation, thus avoiding miss-understanding. Basically, it search for additional arguments to decrease incomparability. The «outweigh concerns» was introduced in (Bana e Costa, 1988) and further developed in (Bana e Costa, 1989 a and b).

This research was partially supported by the Cultural Program between the Flemish Community of Belgium and Portugal. A short version of this paper was presented at the *International Workshop on Multiple Criteria Decision Support*, Helsinki, Finland, August, 1989.

7. REFERENCES

Bana e Costa, C.A. (1988), "A methodology for sensitivity analysis in three-criteria problems: a case study in municipal management", *EJOR*, Vol.33-2 (159-173).

Bana e Costa, C.A. (1989 a), "The outweigh approach for multicriteria decision ad under partial intercriteria preference information", *Centrum Voor Statistiek en Operationeel Onderzoek, STOOTW/246*, Vrije Universiteit Brussel.

Bana e Costa, C.A. (1989 b), "Une méthode pour l'aide à la Décision en situations multicritères et multi-acteurs", *Documet du Lamsade n.59*, Université de Paris-Dauphine, Paris.

Bana e Costa, C.A., Neves, C.D. (1989), "Describing and formalizing the evaluation process of Portuguese Navy officers", in A.G. Lockett and G. Islei (eds.), *Improving Decision Making in Organizations*, Springer-Verlag, Berlin.

Bell, D.E., Keeney, R.L. and Raiffa, H. (eds.) (1977), *Conflicting Objectives in Decisions*, Wiley, New York.

Belton, V. and Vickers, S. (1990), "Use of a simple multi-attribute value function incorporating visual interactive sensitivity analysis for multiple criteria decision making", in this volume.

Bertier, P. and de Montgolfier, J. (1974), "On Multicriteria Analysis: an application to a forest management problem", *Revue Metra*, Vol. 13 (33-45).

Chankong, V. and Haimes, Y.Y. (1983), *Multiobjective Decision Making: Theory and Methodology*, North-Holland, New York.

Chelsea, C., White, III and Hany, K. (1983), "Multi-stage decisionmaking with imprecise utilities", in P. Hansen (ed.), *Essays and Surveys on Multiple Criteria Decision Making*, Springer-Verlag, Heidelberg.

Debreu, G. (1960), "Topological methods in cardinal utility theory", in K.J. Arrow, S. Karlin and P. Suppes (eds.), *Mathematical Methods in the Social Sciences*, (16-26).

Dyer, J.S. and Sarin, R.K. (1979), "Measurable multiattribute value functions", *Oper.Res.*, 27 (810-822).

Edwards, W. (1977), "Use of multiattribute utility measurement for social decision making", in D.E. Bell, R.L. Keeney and H. Raiffa (eds.) (247-276).

Edwards, W., von Winterfeldt, D. and Moody, D.L. (1988), "Simplicity in decision analysis: an example and a discussion", in D.E. Bell, H. Raiffa, and A. Tversky (eds.), *Decision Making: Descriptive, Normative, and Prescriptive Interactions*, Cambridge University Press, Cambridge (443-464).

Fishburn, P.C. (1964), *Decision and Value Theory*, Wiley, New York.

Fishburn, P.C. (1965), "Analysis of decisions with incomplete knowledge of probabilities", *Oper.Res.*, 13 (217-237).

Fishburn, P.C. (1967), "Methods for estimating additive utilities", *Management Science*, Vol.13 (435-453).

Fishburn, P.C. (1970), *Utility theory for Decision Making*, Wiley.

Forman, E. (1990), "Multi criteria decision making and the analytic hierarchy process", in this volume.

French, S. (1988), *Decision Theory*, Wiley, 1986.

Gilbert, R. and Gagey, D. (1976), "Décisions avec critères multiples: une méthode de reduction progressive de l'incomparabilité (principes et applications)", *Choix No. 10*, Ministére de l'Agriculture de France, Paris.

Hannan, E.L. (1981), "Obtaining nondominated priority vectors for multiple objective decisionmaking problems with different combinations of cardinal and ordinal information", *IEEE Trans. Syst., Man, Cybern.*, Vol. SMC-11, No.8 (538-543).

Hazen, G.B. (1985), "Partial preference information and first order differential optimality: an illustration", in Y.Y. Haimes and V. Chankong (eds.), *Decision Making with Multiple Objectives*, Springer-Verlag, Heidelberg.

Hazen, G.B. (1986), "Partial information, dominance, and potential optimality in multiattribute utility theory", *Oper. Res.*, Vol.34, No.2 (296-310).

Hinloopen, E., Nijkamp, P. and Rietveld, P. (1983), "Qualitative discrete multiple criteria choice models in regional planning", *Regional Science and Urban Economics*, 13 (77-103).

Jacquet-Lagrèze, E. (1975), "How can we use the notion of semiorders to build outranking relations in multi-criteria decision making", in D.Wendt and C. Vlek (eds.), *Probability and Human Decision Making*, Reidel, Dordrecht (87-112).

Janssen, R., Nijkamp, P. and Rietveld, P. (1990), "Qualitative multicriteria methods in the Netherlands", in this volume.

Keeney, R.L. (1977), "The art of assessing multiattribute utility functions", *Organiz. Behav. Human Perform.*, 19 (267-310).

Keeney, R.L. (1980), *Siting Energy Facilities*, Academic Press, New York.

Keeney, R.L. and Nair, K. (1977), "Selecting nuclear power plant sites in the Pacific northwest using decision analysis", in D.E. Bell, R.L. Keeney and H. Raiffa (eds.) (298-322).

Keeney, R.L. and Raiffa, H. (1976), *Decisions with Multiple Objectives: Preferences and Value Tradeoffs*, Wiley, N.Y.

Kirkwood, C.W. and Sarin, R.K. (1985), "Ranking with partial in formation: a method and an application", *Oper. Res.*, Vol.33, No.1 (38-48).

Krantz, R.L., Luce, R.D., Suppes, P. and Tversky, A. (1971), *Foundations of Measurement, Volume1*, Academic Press, N.Y.

Lasserre, J.B. (1983), "An analytical expression and an algorithm for the volume of a convex polyhedron in R^{n-1}, *JOTA*, Vol.39, No.3 (363-377).

Malakooti, B. (1985), "A nonlinear multi-attribute utility theory", in Y.Y. Haimes and V. Chankong (eds.), *Decision Making with Multiple Objectives*, Springer-Verlag, Berlin.

Roberts, F.S. (1979), *Measurement Theory with Applications to Decisionmaking, Utility, and the Social Sciences*, Addison-Wesley, Reading, Massachusetts.

Roy, B. (1975), "Vers une méthodologie générale de l'aide à la décision", *Revue METRA*, Vol.14, No.3 (459-497).

Roy, B. (1977), "Partial preference analysis and decision aid: the fuzzy outranking relation concept", in D.E.Bell, R.L. Keeney and H. Raiffa (eds.) (40-75).

Roy, B. (1978), "ELECTRE III: un algorithme de classement fondé sur une représentation floue des préferences en présence de critéres multiples", *Cahiers du CERO*, 20 (3-24).

Roy, B. (1985), *Méthodologie Multicritère d'Aide à la Décision*, Economica, Paris.

Roy, B. (1988), "Main sources of inaccurate determination, uncertainty and imprecision in decision models, in B.R. Munier and N.F. Shakun (eds.), *Compromise, Negotiation and Group Decision*, Reidel, Dordrecht, (43-62).

Roy, B. (1990), "The outranking approach and the foundations of ELECTRE methods", in this volume.

Roy, B. and Bouyssou, D. (1987), "Procedures d'agrégation multicritère conduisant à un critère unique de synthèse", *Document du Lamsade No.42*, Université de Paris-Dauphine, Paris.

Roy, B. and Hugonnard, J.C. (1982), "Ranking of suburban line extension projects on the Paris metro system by a multicriteria method", *Transportation Research*, 16A - 4 (301-312).

Roy, B., Présent, M. and Silhol, D. (1986), "A programming method for determining which Paris metro stations should be renovated", *EJOR*, Vol.24-2 (318-335).

Saaty, T.L. (1980), *The Analytic Hierarchy Process*, McGraw-Hill, New York.

Sarin, R.K. (1977a), "Interactive evaluation and bound procedure for selecting multi-attributed alternatives", *TIMS Studies in Management Sciences*, Vol.6 (211-224).

Sarin, R.K. (1977 b), "Screening of multiattribute alternatives", *OMEGA*, Vol.5, No.4 (481-489).

Siskos, J., Lochard, J. and Lombard, J. (1984), "A multicriteria decision-making methodology under fuzziness: application to the evaluation of radiological protection in nuclear power plants", in H.-J. Zimmermann, L.A. Zadeh and B.R. Gaines (eds.).

Tavares, L.V. (1984), "The TRIDENT approach to rank alternative tenders for large engineering projects", *Foundations of Control Engineering*, Vol.9-4 (181-191).

Vansnick, J.-C. (1986), "On the problem of weights in multiple criteria decision making (the non compensatory approach)", *EJOR*, Vol.24-2 (288-294).

Vansnick, J.-C. (1990), "Measurement theory and decision aid", in this volume.

White, C.C., Dozono, S. and Scherer, W.T. (1983), "An interactive procedure for aiding multiattribute alternative selection", *OMEGA*, Vol.11.

White, C.C., Sage, A.P. and Dozono, S. (1984), "A model of multiattribute decisionmaking and tradeoff weight determination under uncertainty", *IEEE Trans. Syst., Man, Cybern.*, Vol. SMC-14 (223-229).

Zimmermann, H.-J., Zadeh, L.A. and Gaines, B.R. (eds.) (1984), *Fuzzy Sets and Decision Analysis*, North-Holland, Amsterdam.

QUALITATIVE MULTICRITERIA METHODS IN THE NETHERLANDS

Ron Janssen, Peter Nijkamp, and Piet Rietveld

Free University, 1007 MC Amsterdam
THE NETHERLANDS

1. INTRODUCTION

The evaluation of any type of economic development requires a broad (i.e. multi-dimensional) analytical framework in order to capture all relevant socio-economic aspects. For instance, the assessment of monument conservation strategies, of environmental management projects or medical health care programmes includes a wide variety of external effects (which are often also of an intangible nature) which cannot be adequately covered by means of conventional pricing systems; the «measuring rod of money» can only be used as the analytical tool par excellence, if the assumption of a perfectly operating market system is plausible. Only in such cases the notion of a willingness to pay and of a consumer surplus (necessary to estimate the social benefits of a proposed action) is meaningful.

In other cases, artifical methods may have to be used, such as the shadow price approach, the contingent valuation approach or the hedonic pricing method. But in all these cases the multi-dimensional nature of proposed actions can hardly be taken into account. Seen from this perspective, a multi-attribute utility approach seems to be more appropriate, as this approach takes for granted that the (individual or social) value of a comodity depends on a set of multiple attributes of a commmodity, which cannot be reduced to one numerical expression. Thus the value of a commodity has to be expressed in a **multidimensional profile**.

In this context, the development of multicriteria analysis in the Netherlands has to be interpreted. There is a wide recognition that in many areas of policy analysis a multidimensional representation is an inherent feature. Besides, there has been a growing awareness of the incommensurable (or often intangible) nature of various aspects of such policy choices, so that as a logical follow-up the evaluation towards **qualitative multicriteria analysis** has arisen. This approach is essentially based on a compound evaluation of the advantages and disadvantages of a choice alternative, in which various aspects cannot be expressed in financial terms nor in a cardinal metric system.

In the Netherlands, a wide variety of qualitative multicriteria analysis has emerged, with application to many fields, such as physical planning, environmental management, resource policy, transportation planning, etc. (see for an overview among others Nijkamp et al. 1989). In the framework of the present paper a compact overview and some empirical evidence of some of these qualitative multicriteria methods will be given.

2. DEALING WITH ORDINAL PRIORITY INFORMATION

One of the ingredients of multicriteria analysis is priority information. Such information can be expressed in various forms such as: lexicographic orders, minimum requirements, aspiration levels (goals, targets) as well as weights.

In this section we will focus on expressing information on weights. Weights are usually employed in the context of linearized utility functions, but there is no necessity to do so. For example, concordance analysis is a multicriteria method where weights are used outside the context of the utility concept (cf. Roy, 1968 and Van Delft and Nijkamp, 1977).

There are essentially two ways of assessing weights: **indirect** methods and **direct** methods.

With indirect methods, weights are derived in an analytical ay from preferences among alternatives as expressed by a deci-ion-making unit. In the case of revealed preference, the weights re derived from previous choices. An alternative approach is to se stated preferences by confronting a decision making unit with ome alternatives and asking for preference statements about airs of alternatives (cf. Jacquet-Lagrèze and Siskos, 1982). nother possibility is to assess weights in an interactive way, s has been suggested for example by Frisch (1976). For a review nd critique of these approaches refer to (Nijkamp et al., 1989).

In the case of direct methods, decision making units express he relative importance of weights in a direct way. This can be one for example by using a trade-off interpretation or a rating ethod (distribute a given number of points among the set of riteria). These approaches are often experienced as too demand-ng by decision-makers and therefore, other methods have been roposed for less demanding statements on priorities. Such ethods include the use of ranked criteria, verbal statements e.g. by means of fuzzy sets) or Saaty's analytical hierarchy pproach where weights are assessed by means of paired compari-ons using again verbal statements about their relative impor-ance. A concise review can be found in (Nijkamp et al., 1989).

In the present paper, attention will be focussed on **ranking** riteria in order of importance. Three approaches have been eveloped to deal with ranked criteria:

- Extreme value method,
- Random sampling method,
- Expected value method.

These methods share the property that the weights are assumed o be non-negative and add up to 1. Thus, if criteria have been anked in decreasing order of importance, the set of weights S hich is consistent with the information on the ranking reads as ollows:

$$S = \{(\lambda_1, \ldots, \lambda_J) \mid 0 \le \lambda_1 \le \lambda_2 \le \ldots \le \lambda_J; \Sigma_j \lambda_j = 1\} \quad (1)$$

where, λ_j denotes the weight of criterion j. The set S is a convex polyhedron with J vertices. Thus, ordinal information on weights gives rise to a large set of possible quantitative values of weights. The problem is how to make the set S tractable in the context of multicriteria analysis.

The **Extreme value method** focuses on the vertices of the set S. For example, if there are 3 criteria the weight combinations taken into consideration are: (1, 0, 0), (1/2, 1/2, 0), and (1/3, 1/3, 1/3), as also illustrated in Figure 1. Examples of this approach can be found in (Paelinck, 1977), (Kmietovich and Pearman, 1981) and (Voogd, 1983).

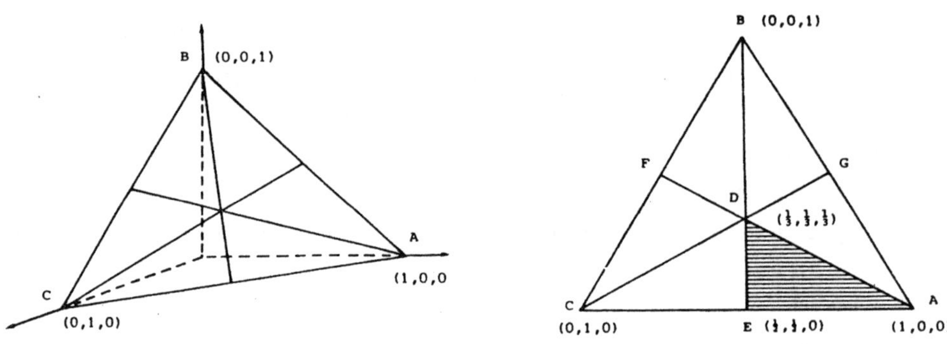

Figure 1. Set of feasible weights in the case of 3 criteria

It is not difficult to see that if, in the case of a linear utility function, alternative A is preferred to B according to the vertices of S, it is preferred to B for all elements of S. Thus, in this case, the extreme value method is quite easy to apply to investigate preference relationships between alternatives. However, in the case of non-linear utility functions this approach is no longer applicable.

In addition, if for some vertices of S alternative A is preferred to B and for other vertices the reverse holds true, the results of this approach are not so easy to handle. There is still another objection against the extreme value approach. The ranking of criteria on which it is based is usually nothing more than a rough approximation of the set of criterion weights the decision maker has in mind. As illustrated by Figure 2, the vertices of S may be a rather bad sample of the set S^* the decision maker really has in mind. Therefore, there is reason to focus on the interior points of the set S, as is done in the following approaches.

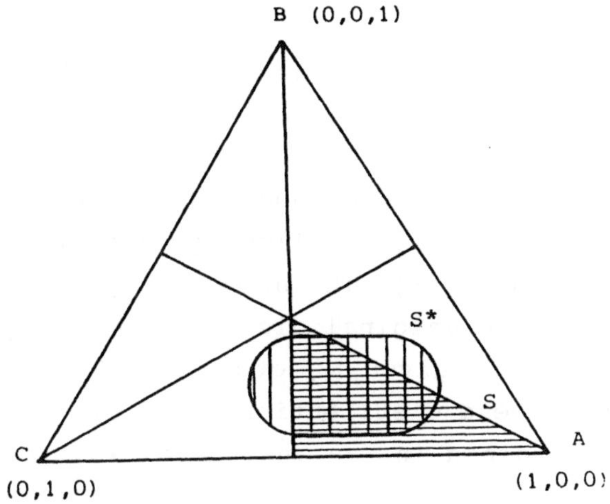

Figure 2. Ranking-based weights (S) as an approximation of the set of weights S^* implicitly assumed by decision maker

The **Random sampling method** is based on a statistical distribution of weights combinations in S. The distribution which is most easily to defend in the absence of further priority information is the uniform distribution: all elements in S are equally probable. This gives rise to the following distribution:

$$f(\lambda_1, \ldots, \lambda_{J-1}) = c \text{ if: } \begin{array}{l} 0 \le \lambda_1 \le 1/J \\ \lambda_1 \le \lambda_2 \le 1/(J-1) - \lambda_1/(J-1) \\ \ldots \\ \lambda_{J-2} \le \lambda_{J-1} \le 1/2 - \lambda_1/2 - \ldots - \lambda_{J-2}/2 \end{array} \quad (2)$$

$$= 0 \text{ elsewhere}$$

where c can be shown to be equal to (J-1)!J! (Rietveld, 1989). Once the values of $\lambda_1, \ldots, \lambda_{J-1}$ are know, the value of λ_J can be found as: $1 - \lambda_1 - \ldots - \lambda_{J-1}$.

It may be tempting to draw random combinations of weights on the basis of the following approach. Draw J numbers a_j (j=1,..., J) from a uniform distribution on the interval (0,1) and define λ_1 as the smallest volume of a_j divided by Σ a_j. This does not lead to the uniform distribution defined in (2), however.

An operational approach to generate a random sample of weight combinations consistent with (2) is the following (see Rietveld, 1988 and Rietveld and Janssen, 1989). The approach consists of two steps. In the first step the marginal distribution of λ_1, the conditional distribution of λ_2 given λ_1, etc. are derived. In Appendix 2 it is shown how these distributions $f(\lambda_1)$, $f(\lambda_2|\lambda_1)$, $f(\lambda_3|\lambda_1,\lambda_2)$, etc. can be obtained in an analytical way on the basis of (2).

In the second step a random generator is used to draw subsequently a value of λ_1, based on $f(\lambda_1)$, a value of λ_2 based on $f(\lambda_2|\lambda_1)$, etc. In Appendix 2 it is indicated that no standard random generators exist to do this job because of the special forms the conditional distribution functions assume. It can be shown, however, that a standard random generator can be used after appropriate transformations of the weights.

The random sampling method can be used in combination with any multi-criteria method. Its results read in stochastic terms:

the probability that alternative A is preferred above alternative B is x%. Thus, the random sampling method gives rise to detailed information on pairwise comparisons of alternatives. If one is in need of a more concise result, one may use the third approach to ranked criteria: the expected value method.

The **Expected value method** aims at determining the mean values of the J weights given distribution (2). As shown in Rietveld (1984, 1989), it is possible to derive the expected values (E) in an analytical way. The following results are found:

$$
\begin{aligned}
E(\lambda_1) &= 1/J^2 \\
E(\lambda_2) &= 1/J^2 + 1/[J(J-1)] \\
&\dots \\
E(\lambda_{J-1}) &= 1/J^2 + 1/[J(J-1)] + \dots + 1/[J(J-2)] \\
E(\lambda_J) &= 1/J^2 + 1/[J(J-1)] + \dots + 1/[J(J-2)] + 1/[J(J-1)].
\end{aligned}
\qquad (3)
$$

In table 1, the outcomes of (3) are presented for some selected values of J. The table clearly reveals that the expected value approach gives rise to a cardinalization which is different from the usual «naive» approach to ordinal numbers. The naive approach - interpreting rank numbers as if they were cardinal - would, for example, in the case of J=3 amount to cardinal weights equal to 1/6, 2/6, 3/6, a result different from that in Table 1.

Table 1. Expected values of criterion weights

J	Expected values of criterion weights					
(number of criteria)	$E(\lambda_1)$	$E(\lambda_2)$	$E(\lambda_3)$	$E(\lambda_4)$	$E(\lambda_5)$	$E(\lambda_6)$
2	.25	.75	-	-	-	-
3	.11	.28	.61	-	-	-
4	.06	.15	.27	.52	-	-
5	.04	.09	.16	.26	.46	-
6	.03	.06	.10	.16	.26	.41

The results in (3) have a very simple interpretation: the expected value of the weights is identical to the centroid of the polyhedral set S. For example, when J = 3, the extreme points are (0, 0, 1), (0, 1/2, 1/2) and (1/3, 1/3, 1/3). The centroid of S is found by computing the unweighted mean of these points. Thus we arrive at (2/18, 5/18, 11/18), which is identical to the outcome of (3).

The expected value method can also be used in interactive procedures. For example, a decision maker formulates a ranking of criteria, after which the Expected value method is used to generate the mean values of the weights. Then the decision maker has the option to judge whether these values give an adequate representation of his implicit priorities. If not, the decision maker may propose changes in numerical values. This procedure has been implemented in the DEFINITE program mentioned above (cf. Herwijnen and Janssen, 1988). Applications of the Expected value method can be found in (Janssen and Rietveld, 1985) and (Rietveld and Janssen, 1989).

The methods described above deal with the standard case of ranked data. For modifications to be applied in the case of ties or incomplete rankings, we refer to (Rietveld, 1989).

3. QUALITATIVE MULTICRITERIA METHODS

As mentioned before, information on relevant judgement aspects of a compound choice problem is often qualitative in nature. This problem of «measuring the unmeasurable» (Nijkamp et al., 1985) is an intriguing issue in evaluation research. There is a wide variety of qualitative multicriteria methods, such as Concordance analysis, Multi-dimensional scaling method, Prioritization method, Permutation method and Regime method (see for surveys among others Nijkamp, 1981; Nijkamp et al., 1989; Rietveld, 1980; and Voogd, 1983).

In the limited context of the present paper, a few methods (i.e., Expected value method, Evamix method and Permutation method) will be discussed briefly, followed by an in depth discussion of a new method, i.e. Regime analysis.

In the **Expected value method**, ordinal criterion scores are replaced by quantitative scores using a transformation procedure similar to the procedure described in section 2 (see also Rietveld, 1984). Weighted summation can now be used to rank the alternatives.

The **Evamix method** is a generalization of Concordance analysis in the case of mixed (qualitative/quantitative) information on the performance of alternatives according to the judgement criteria. Thus a pairwise comparison is made for all pairs of alternatives to determine so called concordance and discordance indices. The difference with standard concordance analysis is that separate indices are constructed for the qualitative and the quantitative criteria. The final ranking of alternatives is the result of a combination of the concordance and discordance indices for both qualitative and quantitative criteria (cf. Voogd, 1983).

The **Permutation method** (see Paelinck, 1976) addresses in particular the question: which rank order of alternatives is - after a series of permutations - in harmony with the ordinal information contained in the project effect scores and the weights? For instance, in case of n alternatives, there are n! permutations possible. This large number of permutations is then judged by using Kendall's rank correlation coefficient in order to find that alternative which has a maximum correlation between any of the n! rank orders and the information contained in the project effect matrix. This approach is mathematically rather cumbersome and presupposes a specific interpretation of weights, viz. via the kendall coefficient.

There is unfortunately often a discrepancy between simple but analytically wrong methods, and sophisticated but analytically sound methods. In recent years, a new method has emerged which tries to meet reasonable criteria such as methodological soundness, mathematical and statistical accessibility, and easy computer use. This method is called the **Regime method** and will be concisely presented in this paper (see also Hinloopen et al., 1983; and Hinloopen and Nijkamp, 1988). Suppose a problem with I choice options or alternatives i (i = 1, ..., I) is characterized by J judgement criteria j (j = 1, ..., J). The basic information we have is composed of qualitative data about the ordinal value of all J judgement criteria for all I choice options. In particular, we assume a partial ranking of all I choice options for each criterion j, so that the following effect matrix can be constructed,

$$E = \begin{bmatrix} e_{11} & \cdots & e_{1J} \\ \vdots & & \vdots \\ e_{I1} & \cdots & e_{IJ} \end{bmatrix} \qquad (4)$$

The entry e_{ij} (i = 1, ..., I; j = 1, ..., J) thus represents the rank order of alternative i according to judgement criterion j. Without loss of generality, we may assume a rank order characterized by the condition «the higher, the better»; in other words, if $e_{ij} > e_{i'j}$, then choice option i is preferable to option i' for judgement criterion j.

As there is not usually a single dominating alternative, additional information is needed on the relative importance of (some of) the judgement criteria. In the case of weighting methods this information is given by means of preference weights attached to the successive criteria. If information is ordinal, the weights are represented by means of rank orders λ_1 (j = 1, ..., J) in a weight vector λ:

$$\lambda = (\lambda_1, \ldots, \lambda_J)^T \qquad (5)$$

Clearly, it is again assumed that $\lambda_j > \lambda_{j'}$ implies that criterion j is regarded as a more important criterion than j'.

Next, the Regime methods uses a pairwise comparison of all choice options, so that the mutual comparison of two choice options is not influenced by the presence and effects of other alternatives. Of course, the eventual rank order of any two alternatives is co-determined by remaining alternatives (compare the independence of irrelevant alternatives problem).

In order to explain the mechanism of the regime method, the concept of a regime is first explained in an ordinal sense. Consider two alternative choice options, i and i'. If for criterion j a certain choice option i is better than option i' (that is, $s_{ij'j} = e_{ij}-e_{i'j} > 0$), then it should be noted that in the case of ordinal information, the order of magnitude of $s_{ij'j}$ is not relevant, but only its sign. Consequently, if $r_{ij'j} = \text{sign}\, s_{ii'j} = +$, then alternative i is better than alternative i' for criterion j. Otherwise, $r_{ii'j} = -$, or (in the case of ties) $r_{ii'j} = 0$. By making such a pairwise comparison for any two alternatives i and i' for all criteria j (j = 1, ..., J) we may construct a J x L regime vector $r_{ii'}$, defined as:

$$r_{ii'} = (r_{ii'1}, \ldots, r_{ii'J})^T, \tag{6}$$

Thus, the regime vector contains only + and - signs (or, in the case of ties, 0 signs as well), and reflects a certain degree of (pairwise) dominance of choice option i with respect to option i' for the unweighted effects for all J judgement criteria. Clearly, there are I(I-1) pairwise comparisons altogether, and hence also I(I-1) regime vectors. These regime vectors can be included in an J x I(I-1) regime matrix R:

$$R = \begin{bmatrix} \underbrace{r_{12}, r_{13}, \ldots, r_{1I}}_{I-1} & \cdot\ \cdot & \underbrace{r_{I1}, \ldots, r_{I(I-1)}}_{I-1} \end{bmatrix} \tag{7}$$

It is evident that if a certain regime vector $r_{ii'}$ contained only + signs, alternative i would dominate alternative i' in all aspects. Usually, however, a regime vector contains both + and - signs, so that additional information in the form of the weights vector (5) is required.

In order to treat ordinal information on weights, the assumption is now made that the ordinal weights λ_j (j=1, ..., J) are a rank order representation of an (unknown) underlying cardinal stochastic weight vector λ^*, viz. $\lambda^* = (\lambda_1^*, \ldots, \lambda_J^*)^T$, with max $(\lambda_j^*) = 1$, $\lambda_j^* \geq 0$.

The ordinal ranking of the weights is then supposed to be consistent with the quantitative information incorporated in an unknown cardinal vector, λ^*; in other words, $\lambda_j > \lambda_{j'} \rightarrow \lambda_j^* > \lambda_{j'}^*$. Next, it is assumed that the weighted dominance of choice option i with regard to option i' can be represented by means of the following stochastic expression based on a weighted summation of cardinal entities (implying essentially an additive linear utility structure):

$$v_{ii'} = \sum_{j=1}^{J} r_{ii'j}\lambda_j^* \qquad (8)$$

If $v_{ii'}$ is positive, choice option i is clearly preferred to option i'. However, in this case no information is available on the cardinal value of λ_j^* but only on the ordinal value of λ_j (which is assumed to be consistent with λ_j^*).

Therefore, a certain probability, $p_{ii'}$ is introduced, for the dominance option i with respect to option i', i.e.

$$p_{ii'} = \text{prob}\ (v_{ii'} > 0) \qquad (9)$$

and define as an aggregate probability measure,

$$p_i = \frac{1}{I - 1} \sum_{i' \neq i} p_{ii'} \tag{10}$$

Then it can easily be seen that p_i is the probability that alternative i is on average higher valued that any other alternative, based on a pairwise confrontation with an arbitrary alternative. Consequently, the eventual rank of choice options is determined by the rank order (or the order of magnitude) of the p_i.

However, the crucial problem here is to assess $p_{ii'}$ and p_i. This implies that an assumption needs to be made about the probability distribution function both of the λ_j^* and of the $r_{ii'j}$. In view of the ordinal nature of the λ_j, it is plausible to assume for the whole relevant area a uniform density function for the λ_j^* (see also section 2). The motive is that if the ordinal weights vector, λ, is interpreted as originating from a stochastic weight vector, λ^*, there is, without any prior information, no reason to assume that a certain numerical value of λ^* has a higher probability than any other value. In other words, the weights vector, λ^*, can adopt with equal probability each value that is in agreement with the ordinal information implied by λ. This argument is essentially based on the «principle of insufficient reason», which also constitutes the foundation stone for the so-called Laplace criterion in the case of decision making under uncertainty (Taha, 1976). However, if prior information in a specific case suggests there is reason to assume a different probability distribution function (a normal distribution, for example), there is no reason to exclude this new information. Of course, this may influence the values of $p_{ii'}$ and hence the ranking of alternatives. The precise way in which rank order results can be derived from a probability distribution when there is qualitative information will not be discussed further here, as this topic has been extensively described elsewhere (Hinloopen and Nijkamp, 1988). But it may suffice to mention that, in principle, the use of stochastic analysis in combination with computer simulations, which is consistent with an originally ordinal data set, may help which is consistent with an originally ordinal data set, may help to overcome the methodological problem emanating from impermissible numerical operations on qualitative data. The Regime method is also able to handel ties in the effect matrix and in the weight vector.

A situation with mixed information emerges if either the impact matrix or the weight vector (or both) contain both cardinal and ordinal information. In that case the ordinal information has to be transferred first into appropriate cardinal units, so as to make the two different data systems mutually compatible. Both the ordinal and the cardinal versions of the regime method aim at finding a dominant choice option based on a multivariate data set. Although the numerical operations are of course different, the underlying methodology is analogous, as in both cases an attempt is made at identifying from a discrete set of choice possibilities an as yet unknown choice option which is as close as possible to an ideal point or ideal ranking. The technicalities of this method can be found in (Hinloopen et al., 1983) and (Hinloopen and Nijkamp, 1988), and will not be discussed here any further.

4. AN APPLICATION: THE SITING OF NUCLEAR PLANTS IN THE NETHERLANDS

4.1. Introduction

At present in the Netherlands the share of nuclear power in total power production is very small. In 1985 the Dutch government has expressed the intention to build two new nuclear power plants with a capacity of 1000 MWe each. Given the decision to build these two nuclear plants an important decision is where to locate these two plants. After some initial scoping, first thirteen and later nine potential locations for the plants were selected (Tweede Kamer 18830, 43-44). These locations are shown in figure 3.

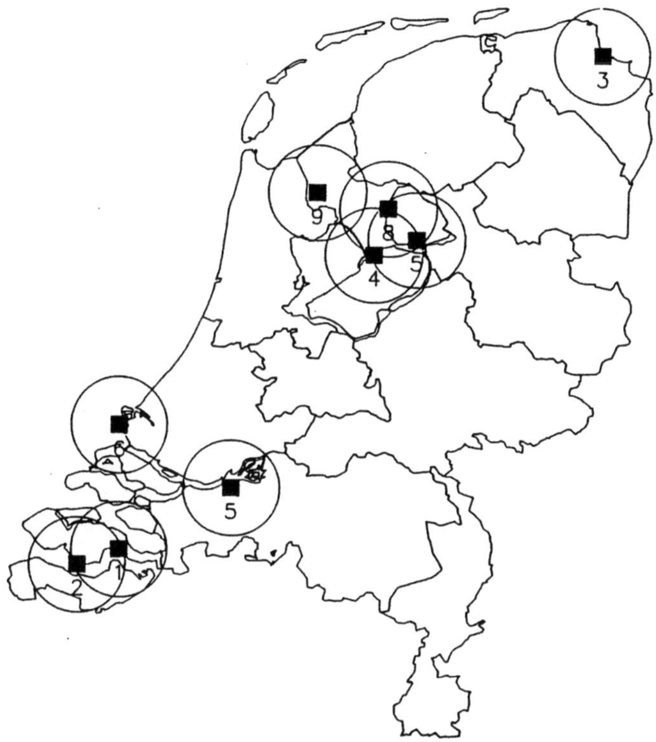

Figure 3. Potential locations for nuclear plants

In this section we will rank these remaining nine locations using 15 appraisal criteria. We will use our support system for decisions on a finite set of alternatives (DEFINITE) to produce this ranking. The evaluation procedure comprises the following steps:

- Problem definition.
- Problem presentation.
- Problem evaluation.
- Sensitivity analysis.

In this application we will only briefly discuss problem definition and presentation. We will focus on problem evaluation using the methods described in section 2 and 3.

4.2. Problem Definition

The impact matrix of this evaluation problem is shown in table 2 where nine potential locations are scored according to 15 criteria. Only the score for population around a site is measured on a cardinal scale. All other scores are measured on an ordinal scale: a score 1 is assigned to the best alternative, 2 to the second best, etc. (see Appendix 1 for a definition of the criteria).

Table 2. Impact matrix.
(Source: Tweede Kamer 18830, 43-44; advice to the government)

	Bath	Borssele	Eems	Flevo	Ketel	Maasvlak	Moerdijk	NO polder	Wiering
Population	51	49	16	27	30	43	100	19	21
Evacuation	1	2	1	1	1	2	1	1	1
Agricult at risk	2	2	2	2	2	1	3	2	2
Industry at risk	1	4	3	2	1	5	3	1	1
Fr water at risk	1	1	1	2	2	1	2	2	2
Cool-water quant	2	1	1	1	1	1	3	1	1
Cool-water qual	2	1	2	3	3	1	2	3	3
Air pollution	2	2	2	2	2	1	2	2	2
Thermal poll.	3	2	2	2	3	1	2	3	3
Indirect landuse	2	3	3	2	1	2	4	1	1
Landscape	3	1	1	1	3	1	2	3	3
Nat environment	3	1	3	1	2	1	1	1	1
National grid	2	2	3	1	1	2	2	2	3
Infrastructure	2	1	1	2	2	1	1	2	2
Coal-location	3	6	4	3	2	7	5	1	1

4.3. Problem Presentation

A graphical presentation of the impact matrix is shown in figure 4. This figure was derived by standardising all criterion scores between 0 and 1 (see section 3). The highest column in each row represents a score of 1 corresponding to the best alternative for that row. As a next step the criteria are ordered from most important (top) to least important (bottom). These priorities were determined by experts of the government advisory board on physical planning (Tweede Kamer 18830, 43-44).

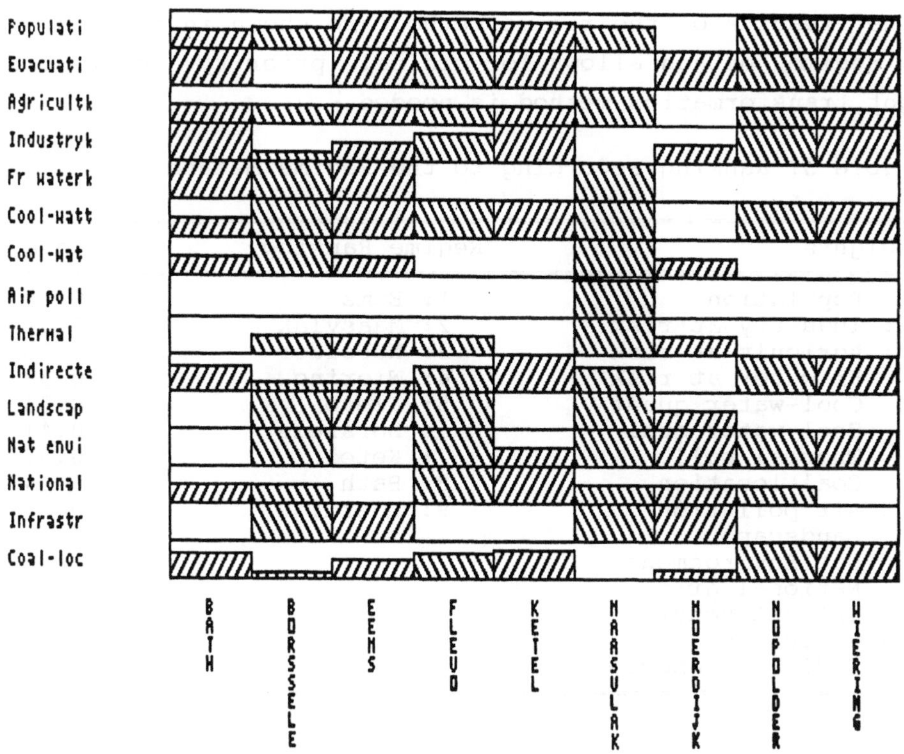

Figure 4. A graphical presentation of the impact matrix

Using a combination of the expected value method and the weighted summation method (see sections 2 and 3) the information on priorities and scores can be used to rank the alternatives from best (left) to worst (right). It is clear from this figure that the impact matrix contains many ties and that differences between alternative locations are fairly small.

4.4. Problem Evaluation

The methods described in section 2 and 3 are available to transform the mixed information on impacts and ordinal information on priorities into a ranking of the alternatives. (Since software for the Permutation method is not available results for this method are not included).

The results of the regime method are shown in table 3. Since the regime method allows for ordinal priority information no weight transformation method is needed.[1]

Table 3. Ranking according to the Regime method

Weights	Regime Ranking	Score
1: Population	1: Eems	0.86
2: Industry at risk	2: Maasvlak	0.81
3: Agricult at risk	3: NOPolder	0.75
Fr water at risk	4: Wiering	0.58
5: Cool-water quant	5: Flevo	0.55
Cool-water qual	6: Borssele	0.41
Thermal poll.	7: Ketel	0.37
Coal-location	8: Bath	0.14
9: Air pollution	9: Moerdijk	0.02
Landscape		
Nat Environment		
National grid		
Infrastructure		
14:Evacuation		
Indirect landuse		

The results of the Expected value and the Evamix method are shown in table 4. Since these methods require quantitative information on weights, the expected value method was used to transform the ordinal priority information into quantitative weights.

All methods result in an almost complete ranking of the alternatives. Figure 5 shows the ranking of five alternatives according to the three methods. The three methods give rise to rather similar rankings. There is one notable difference, however: in the Expected value method, Maasvlakte performs much worse than in the other two methods.

----

1. The procedure included in the regime method is similar to the random weight method as described in section 2.

Table 4. Ranking according to the Expected value and the Evamix methods

Weights		Expected Value Ranking	Score	Evamix Ranking	Score
1: Population	0.221	1:Eems	0.86	1:Maasvlak	0.11
2: Industry at risk	0.155	2:NOpolder	0.83	2:Eems	0.05
3: Agricult at risk	0.110	3:Flevo	0.82	3:NOpolder	0.03
Fr water at risk	0.110	4:Wiering	0.81	4:Bath	0.02
5: Cool-water quant	0.064	5:Ketel	0.79	5:Wiering	0.01
Cool-water qual	0.064	6:Bath	0.76	Borssele	0.01
Thermal poll.	0.064	7:Maasvlak	0.75	7:Ketel	0.00
Coal-location	0.064	Borssele	0.75	Flevo	0.00
9: Air pollution	0.027	9:Moerdijk	0.52	9:Moerdijk	-0.24
Landscape	0.027				
Nat Environment	0.027				
National grid	0.027				
Infrastructure	0.027				
14:Evacuation	0.007				
Indirect landuse	0.007				

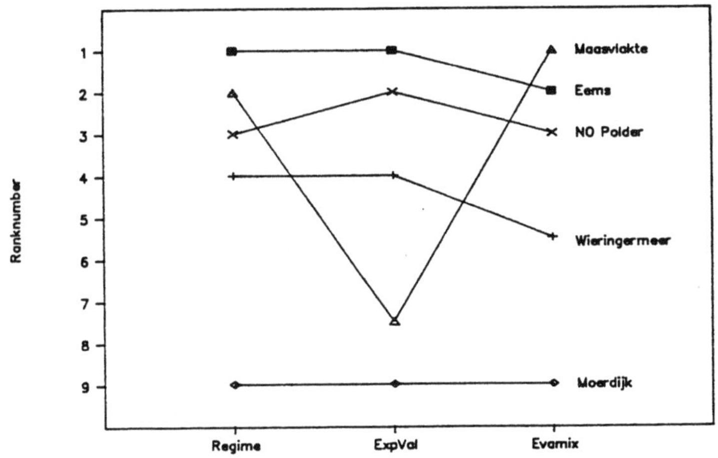

Figure 5. Ranking of five alternatives according to three multicriteria methods

To produce the rankings in table 4 the Expected value method is used to change ordinal priority information into quantitative weights. Other methods to perform this transformation are the Random and Extreme weight methods (see section 2). Table 5 shows the results of the Evamix method using all three of the transformation methods.

Table 5. Ranking of the alternatives using three different procedures to transform ordinal weights in combination with the Evamix Method.

Weights	Expected value	Random	Extreme Value
1: Population	1: Maasvlak	1:NOpolder	1:Bath
2: Industry at risk	2: Eems	2:Wiering	Borssele
3: Agricult at risk	3: NOPolder	3:Eems	Eems
Fr water at risk	4: Bath	4:Ketel	Flevo
5: Cool-water quant	5: Wiering	5:Flevo	Ketel
Cool-water qual	Borssele	6:Bath	Maasvlak
Thermal poll.	7: Ketel	7:Maasvlak	NO Polder
Coal-location	Flevo	8:Borssele	Wiering
9: Air pollution	9: Moerdijk	9:Moerdijk	9:Moerdijk
Landscape			
Nat Environment			
National grid			
Infrastructure			
14:Evacuation			
Indirect landuse			

The extreme weight method produces, as described in section 2, a ranking that complies with all possible quantitative values of the ordinal weights. This results, as can be seen from table 5, in an incomplete ranking of the alternatives: eight alternatives share the first position. It appears that, except for Moerdijk, all alternatives may arrive at the top of the ranking for one of the extreme weight combinations. This underlines that the extreme value method is often of little practical use in situations with a large degree of conflict among criteria.

4.5. Sensitivity analysis

More interesting than the overall stability of the derived ranking is very often the sensitivity of the ranking to specific scores. Since the government wishes to select two locations we will try to analyse how firmly the alternatives Eems and Maasvlakte hold the first two positions.

In table 6 robustness intervals are calculated for the pairs Eems - NO Polder and Maasvlakte - NO Polder.

Table 6. Robustness intervals; Regime method

Ranking	Score Industry at risk Eems
Eems > NO Polder	1 → 9
NO Polder > Eems	empty

Ranking	Score Cool water quality NO Polder
Maasvlak > NO Polder	2 → 9
NO Polder > Maasvlak	1 → 2

Table 6 shows that Eems ranks higher then NO Polder for any score assigned to the criterion Industry at risk. Rank 3 is assigned to Coal water quality for alternative NO Polder. Table 6 shows that with a rank of 1 for this criterion the ranking of Maasclak and NO Polder is reversed. Rank 2 for this criterion results in a tie of both alternatives.

A similar procedure can be applied to establish robustness intervals for ordinal weights.

5. CONCLUDING REMARKS

This paper focussed on qualitative multicriteria methods developed in the Netherlands. Various methods are described that can deal with qualitative information both on priorities and on effects. Special emphasis was given to a fairly recently developed method: the Regime method.

The methods described were applied to site nuclear plants in the Netherlands. They produced fairly similar rankings. The Extreme value method to deal with qualitive information on priorities proved to be of little practical use in this application.

The methods described are available in the decision support system DEFINITE. This increases acess of these methods to various types of users. This is important. Although the exact calculation procedures will not be clear to all users the principles of the methods are fairly straightforward and easy to explain.

REFERENCES

Frisch, R. (1976), "Cooperation between politicians and econometricians on the formalization of political preferences", *Economic Planning Studies*, Reidel, Dordrecht, 41-86.

Herwijnen, M. van and Janssen, R. (1989), "DEFINITE, a support System for decisions on a finite set of alternatives", in A.G. Lockett and G. Islei (eds.), *Improving Decision Making in Organisations*, Lecture Notes in Economics and Mathematical Systems, vol. 335, Springer-Verlag, Heidelber, 534-543.

Hinloopen, E., Nijkamp, P. and Rietveld, P. (1983), "Qualitative discrete multiple criteria choice models in regional planning", *Regional Science and Urban Economics*, vol. 13, 77-102.

Hinloopen, E. and Nijkamp, P. (1988), "Qualitative multiple choice analysis: the dominant Regime method", *Quality and Quantity*, forthcoming.

Hogg, R.V. and Craig, A.T. (1970), *Introduction to Mathematical Statistics*, MacMillan, London, 1970.

Jacquet-Lagrèze, E. and Siskos, J. (1982), "Assessing a set of additive utility functions for multicriteria decision making, the UTA method", *European Journal of Operational Research*, vol. 10, 151-164.

Janssen, R. and Hafkamp, W. (1986), "A decision support system for conflict analysis", *The Annals of Regional Science*, vol. 20, 3, 67-85.

Janssen, R. and Rietveld, P. (1985), "Multicriteria evaluation of land-reallotment plans: a case study", *Environment and Planning* A, vol. 17, 1653-1668.

Kmietowics, Z.W. and Pearman, D.A. (1981), *Decision Theory and Incomplete Knowledge*, Gower, Alderschot, U.K.

Nijkamp, P. (1971), *Environmental Policy Analysis*, Wiley, New York.

Nijkamp, P., Leitner, H. and Wrigley, N. (eds.) (1985), *Measuring the Unmeasurable*, Martinus Nijhoff, Dordrecht.

Nijkamp, P., Rietveld, P. and Voogd, H. (1989), *Multicriteria Evaluation in Physical Planning*, North Holland, Amsterdam.

Paelinck, J. (1976), "Qualitative multiple criteria analysis, environmental protection and multiregional development", *Papers of the Regional Science Association*, vol. 36, 59-74.

Paelinck, J.H.P. (1977), "Qualitative multiple criteria analysis: an application to airport location", *Environment and Planning*, vol. 9, 883-895.

Rietveld, P. (1980), *Multiple Objective Decision Methods in Regional Planning*, North Holland, Amsterdam.

Rietveld, P. (1984), "The use of qualitative information in macro economic policy analysis", in M. Despontin et al. (eds.), *Macro Economic Planning with Conflicting Goals*, Springer-Verlag, Berlin.

Rietveld, P. (1988), *Beslissings Ondersteunend Systeem voor Discrete Alternatieven (BOSDA). Gevoeligheidsanalyses bij Multicriteria Beslissingsmethoden*, Institut voor Milieu-vraagstukken, Vrije Universiteit, Amsterdam.

Rietveld, P. (1989), "Using ordinal information in decision making under uncertainty", *Systems Analysis, Modeling, Simulation*, vol. 6, forthcoming.

Rietveld, P. and Janssen, R. (1989), "Sensitivity Analysis in Discrete Multiple Criteria Decision Problems; on the Siting of Nuclear Power Plants", Research Memorandum, Department of Economics, Free University, Amsterdam, 1989.

Roy, B. (1968), "Classement et choix en presence de points de vue multiples (la methode ELECTRE)", *RIRO*, vol. 2, 57-75.

Taha, H.A. (1976), *Operations Research*, MacMillan, New York.

Van Delft, A. and Nijkamp, P. (1977), *Multicriteria Analysis and Regional Decision Making*, Martinus Nijhoff, The Hague.

Voogd, H. (1983), *Multicriteria Evaluation for Urban and Regional Planning*, Pion, London.

APPENDIX 1 - Definition of evaluation criteria

Population. To calculate this score a weighted sum of population around a location was calculated. The weight assigned decreases with distance. The result is standardized by division through the maximum score. A minus sign is added to indicate that the criterion is a cost criterion.

Evacuation. The score reflects the availability of sufficient transport infrastructure.

Agriculture at risk. This score reflects the vicinity to the location of agricultural land.

Industry at risk. This score reflects the size and importance of industry near the location.

Fresh water at risk. This score reflects the quantity of fresh water that may be affected by a nuclear plant at each location.

Cool-water quant. This score represents the quantity of available water for cooling the nuclear plant.

Cool-water qual. This score represents the capacity of the coolant to flush out pollution originating from a nuclear plant at each location.

Air pollution. It is assumed that the nuclear plant is an alternative to a coventional coal power plant. This is assumed to have the most beneficial effect at the most polluted location.

Thermal pollution. The amount of pollution is lower if users of the health generated are available. The score reflects the availability of such users.

Indirect landuse. This score reflects limitations on the potential landuses around a nuclear plant.

Landscape. This score reflects the visual effects the landscape around the landscape and the extent to which a nuclear plant fits in with existing activities.

Natural environment. This score reflects expected damage to the natural environment.

National grid. This score reflects the availability near of at the location of high voltage lines and connector stations.

Infrastructure. This score reflects the availability of transport and other infrastructure around the location.

Coal-location. It is assumed that the nuclear plant is an alternative to a conventional coal power plant. The score reflects the cost of the lost opportunity to build a coal plant at the location if a nuclear plant is constructed.

APPENDIX 2

The starting point is the joint density function of weights:

$$f(\lambda_1, \ldots, \lambda_{J-1}) = c \text{ for } \begin{array}{l} 0 \le \lambda_1 \le 1/J \\ \lambda_1 \le \lambda_2 \le (1-\lambda_1)/(J-1) \\ \lambda_2 \le \lambda_3 \le (1-\lambda_1-\lambda_2)/(J-2) \\ \ldots \\ \lambda_{J-2} \le \lambda_{J-1} \le (1-\lambda_1-\ldots-\lambda_{J-2})/2 \end{array}$$

$$= 0 \text{ elsewhere}$$

where $c = (J-1)!J!$

Then the marginal density function of λ_1 can be derived as:

$$f(\lambda_1) = (J-1)J(1-J\lambda_1)^{J-2} \text{ for } 0 \le \lambda_1 \le 1/J$$
$$= 0 \qquad \text{elsewhere.}$$

Further, the conditional density functions can be shown to read as follows for $j = 2,\ldots,J-1$:

$$f(\lambda_j | \lambda_1, \ldots, \lambda_{j-1})$$

$$= (J-j)(J-j+1) \frac{[1-\lambda_1-\ldots-\lambda_{j-1}-(J-j+1)\lambda_j]^{J-j-1}}{[1-\lambda_1-\ldots-(J-j+2)\lambda_{j-1}]^{J-j}}$$

where $\lambda_{j-1} \le \lambda_j \le (1-\lambda_1-\ldots-\lambda_{J-1})/(J-j+1)$.

Then, a random weight vector can be generated by drawing a value for λ_1 on the basis of $f(\lambda_1)$, followed by drawing a value for λ_2 on the basis of $f(\lambda_2|\lambda_1)$, etc. Finally, λ_J can be computed as $1-\lambda_1-\ldots-\lambda_{J-1}$.

The conditional distributions mentioned above are nor included in standard statistical packages. Therefore, random weight vectors cannot de directly created by means of random generators.

A solution for this problem is given by the theorem which says that if $F(x)$ is the distribution function of x, then $u = F(x)$ is uniformly distributed on the interval $0 \leq u \leq 1$ (Hogg and Craig, p. 349). For the latter uniform distribution, standard random generators are available. Then, if u_1 is uniformly distributed on the interval [0, 1], $\lambda_1 = F^{-1}(u_1)$ can be shown to be distributed according to the density function $f(\lambda_1)$ corresponding with the distribution function $F(\lambda_1)$. Thus random values for λ_1 can be found by using the following transformation:

$$\lambda_1 = (1/J).[1-(1-u_1)^{1/(J-1)}].$$

For $\lambda_2,\ldots,\lambda_{J-1}$ the following transformation has to be used:

$$\lambda_j=[(1-\lambda_1-\ldots-\lambda_{j-1})-(1-\lambda_1-\ldots-\lambda_{j-2}-(J-j+2)\lambda_{j-1})(1-u_j)^{1/(J-j)}]/(J-j+1).$$

Finally, λ_J can be computed as $1-\lambda_1-\lambda_{J-1}$.

CHAPTER IV

INTERACTIVE MULTIPLE OBJECTIVE PROGRAMMING

INTERACTIVE MULTIPLE OBJECTIVE PROGRAMMING: CONCEPTS, CURRENT STATUS, AND FUTURE DIRECTIONS

Ralph E. Steuer
Department of Management Science & Information Technology
Brooks Hall, University of Georgia
Athens, Georgia 30602, USA

Lorraine R. Gardiner
Department of Management
Auburn University
Auburn, Alabama 36849, USA

This paper discusses 32 topics that summarize interactive multiple objective programming, the area represented by STEM by Benayoun, de Montgolfier, Tergny and Larichev (1971), the Geoffrion-Dyer-Feinberg Procedure (1972), the Zionts-Wallenius procedure (1976 and 1983), Wierzbicki's Reference Point Method (1977, 1982 and 1986), etc. The concensus is that a spectrum of interactive procedures is necessary because the procedure to use on a given application is usually problem and user dependent.

In traditional (single objective) mathematical programming, we must settle on a single objective such as minimizing cost or maximizing profit. However, if we think long enough about any problem, we will frequently be able to identify multiple objectives. The following illustrate the range of problems that may be more adequately modeled with multiple objectives:

Oil Refinery Scheduling

min {cost}
min {imported crude}
min {high sulfur crude}
min {deviations from demand state}

Production Planning

max {total net revenue}
max {minimum net revenue in any period}
min {backorders}
min {overtime}
min {finished goods inventory}

Forest Management

max {sustained yield of timber production}
max {visitor days of dispersed recreation}
max {visitor days of hunting}
max {wildlife habitat}
max {animal-unit-months of grazing}
min {overdeviations from budget}

Water Resources

max {recreation benefits at reservoir 1}
max {power generation at reservoir 1}
max {flood control benefits below reservoir 1}
max {recreation benefits at reservoir 2}
max {power generation at reservoir 2}
max {irrigation benefits below reservoir 2}

Examples in the literature describing the application of interactive multiple objective programming include articles by Lee and Shim (1986), Nakayama and Furukawa (1985), Steuer and Wood (1986), Silverman, Steuer and Whisman (1988), Gardner, Huefner and Lotfi (1989), and Stam, Joachimsthaler and Gardiner (1989).

The multiple objective program (with all objectives in maximization form) to be solved interactively is

$$\begin{aligned} &\max \{f_1(x) = z_1\} \\ &\quad\vdots \\ &\max \{f_k(x) = z_k\} \\ &\text{s.t.} \quad x \in S \end{aligned}$$

where k is the number of objectives and S is the feasible region in decision space. Let $Z \subset R^k$ be the feasible region in criterion space where $z \in Z$ if and only if there exists an $x \in S$ such that $z = (f_1(x),\ldots,f_k(x))$. Let $K = \{1,\ldots,k\}$. Criterion vector $\bar{z} \in Z$ is nondominated if and only if there does not exist another $z \in Z$ such that $z_i \geq \bar{z}_i$ for all $i \in K$ and $z_i > \bar{z}_i$ for at least one $i \in K$. The set of all nondominated criterion vectors is designated N and is called the nondominated set. A point $\bar{x} \in S$ is efficient if and only if its criterion vector $\bar{z} = (f_1(\bar{x}),\ldots,f_k(\bar{x}))$ is nondominated. The set of all efficient points is designated E and is called the efficient set. Let $U: R^k \to R$ be the utility function of the decision maker (DM). A point $x^0 \in S$ that maximizes U over S is called an optimal point and its criterion vector $z^0 = (f_1(x^0),\ldots,f_k(x^0))$ is an optimal criterion vector. Our interest in the efficient set E and the nondominated set N stems from the fact that $x^0 \in E$ and $z^0 \in N$. If we had our druthers, the best way to solve a multiple objective program would be to assess the DM's U and then solve the (single criterion) program

$$\begin{aligned} \max\ & \{U(z_1,\ldots,z_k)\} \\ \text{s.t.}\ & f_i(x) = z_i \qquad i \in K \\ & x \in S. \end{aligned}$$

Any solution that solves this program is an optimal solution of the multiple objective program. Multiple objective programs, however, are not solved this way because of the difficulty in obtaining an accurate enough representation of U for use in the above program. As a result, we use approaches, principally interactive ones, that only require implicit, rather than explicit, knowledge about the DM's utility function. Such methods produce final solutions. A final solution is either optimal, or close enough to being optimal, to satisfactorily terminate the decision process.

Topics 1 to 16 characterize the modest mathematical background necessary for the study of interactive multiple objective programming. Topics 17 to 24 describe the current status of the field. Topics 25 to 32 indicate future directions. The 32 topics are as below:

<u>Concepts</u>

1. Decision Space Versus Criterion Space
2. Graphical Detection of Nondominated Criterion Vectors
3. Payoff Tables
4. Utopian Criterion Vectors
5. Aspiration Criterion Vectors
6. Normalization of the Objectives
7. Quasiconcave Utility Functions
8. Weighted-Sums Program
9. Unsupported Nondominated Criterion Vectors
10. Vector-Maximum Codes
11. Groups of Maximally Dispersed λ-Vectors
12. Lexicographic and Augmented Weighted Tchebycheff Sampling Programs
13. Contours of the Lexicographic and Augmented Weighted Tchebycheff Sampling Programs
14. Projecting an Unbounded Line Segment onto N
15. Determining an L-Contour Vertex λ-Vector
16. Contracting Λ -Space

<u>Current Status</u>

17. Interactive Procedures
18. Descriptions of the Non-Reference Point Procedures
19. Descriptions of the Reference Point Procedures
20. Implementation Similarities
21. Consolidation and the Unified Sampling Program
22. Customizing the Unified Sampling Program
23. Cold Starting Versus Warm Starting
24. Ideal Workhorse Software

<u>Future Directions</u>

25. Switching
26. Preserving Convergence Achievements
27. Trajectory Applications
28. Polyscreen Workstations
29. Cognitive Equilibrium
30. Network Optimization Applications
31. Russian Interactive Research
32. Mainframes Versus Microcomputers

1. DECISION SPACE VERSUS CRITERION SPACE

Whereas the study of traditional (single objective) mathematical programming is conducted in decision space, interactive multiple objective programming is mostly studied in criterion space. To illustrate the difference between a multiple objective program's representation in <u>decision space</u> and its representation in <u>criterion space</u>, consider

$$\begin{aligned} &\max \{x_1 - 1/2x_2 = z_1\} \\ &\max \{\qquad\quad x_2 = z_2\} \\ &\text{s.t.} \qquad\quad x \in S \end{aligned}$$

where S in decision space is given in Figure 1, and Z in criterion space is given in Figure 2. For instance z^4, which is the image of $x^4 = (3,4)$, is obtained by plugging (3,4) into the objective functions to generate $z^4 = (1,4)$. In this way, Z is the <u>image</u> of S under the f_i. In Figure 1 we see that the efficient set E is the set of boundary points x^3 to x^5 to x^6, and that the image of E is N which consists of the set of boundary points z^3 to z^5 to z^6 in Figure 2. Note that Z is not necessarily confined to the nonnegative orthant.

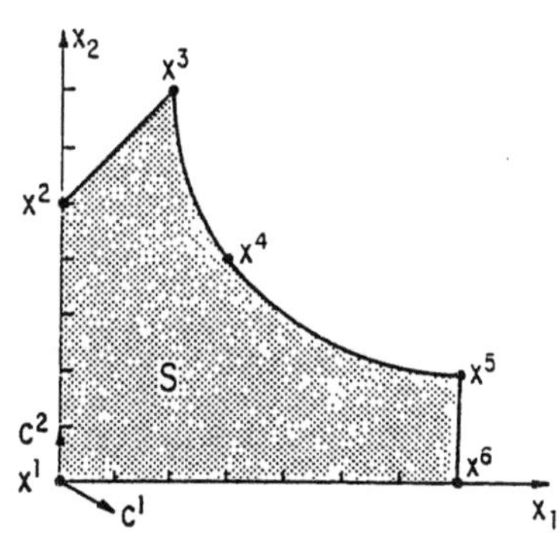

Figure 1. Representation in Decision Space.

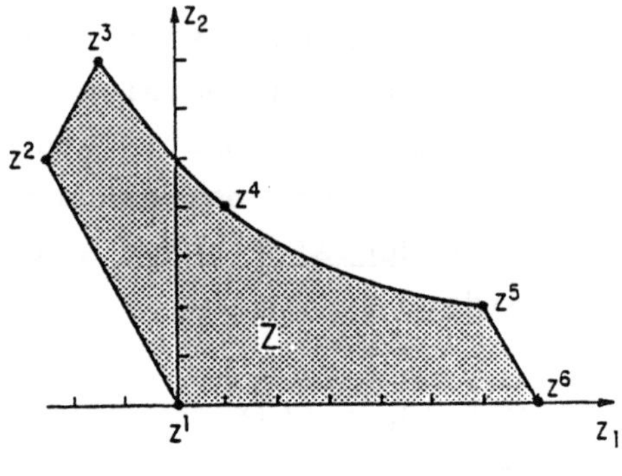

Figure 2. Representation in Criterion Space.

2. GRAPHICAL DETECTION OF NONDOMINATED CRITERION VECTORS

Let $\bar{z} \in Z$. Let $D_{\bar{z}}$ be the set formed by translating the nonnegative orthant of R^k to $\bar{z}$. Then $\bar{z}$ is a <u>nondominated</u> criterion vector if and only if $Z \cap D_{\bar{z}} = \{\bar{z}\}$. Since the intersection of Z and $D_{\bar{z}}$ always contains $\bar{z}$, this means that criterion vector $\bar{z}$ is <u>dominated</u> if and only if the intersection contains points in Z other than $\bar{z}$. Thus, in Fig. 2, z^4 is seen to be nondominated, and z^2 is seen to be dominated (because z^3, for example, is in $Z \cap D_{z^2}$).

3. PAYOFF TABLES

A <u>payoff table</u> is of the form

	z_1	z_2		z_k
z^1	z_1^*	z_{12}		z_{1k}
z^2	z_{21}	z_2^*		z_{2k}
			$\cdot$ $\cdot$ $\cdot$	
z^k	z_{k1}	z_{k2}		z_k^*

where the rows are the criterion vectors resulting from individually maximizing the objectives. If one is worried about alternative optima, a way to assure that all row criterion vectors are nondominated is to lexicographically maximize the objectives.

Lexicographic maximization is described as follows. Suppose we have alternative optima after maximizing one of the objectives (conducting the first maximization stage). Then, holding the optimal achievement of this objective fixed, we maximize another objective (conduct the second maximization stage). If there are still alternative optima, we additionally fix the maximal achievement of the second objective, and then maximize a third objective (conduct a third maximization stage). We continue in this fashion until either a unique solution is obtained or all objectives are exhausted. In this way, unless it is not possible to complete a maximization stage because of unboundedness, a nondominated criterion vector is assuredly obtained.

The z_i^* entries along the main diagonal of the payoff table are the maximal criterion values for the different objectives over the nondominated set. The minimum value in the i-th column of the payoff table is an estimate of the minimum criterion value of the i-th objective over N. Often these column minimums are used in place of the minimum criterion values over N because the minimum criterion values over N are difficult to obtain (see Isermann and Steuer (1988)).

4. UTOPIAN CRITERION VECTORS

We will define a utopian criterion vector $z^{**} \in R^k$ to be an (infeasible) criterion vector that strictly dominates every nondominated criterion vector. The components of z^{**} are formed as follows:

$$z_i^{**} = z_i^* + \epsilon_i$$

where the z_i^* are the maximal criterion values obtained from the main diagonal of the payoff table, and the ϵ_i are moderately small positive values. An ϵ_i value that raises z_i^{**} to the smallest integer greater than z_i^* is normally sufficient.

5. ASPIRATION CRITERION VECTORS

An aspiration criterion vector $q \in R^k$ is a criterion vector requested from a DM that is to reflect his or her expectations from the model. Several interactive procedures require aspiration criterion vectors. Aspiration criterion vectors are usually specified in the light of a payoff table such that $q \leq z^*$. Also, an aspiration criterion vector is often specified so that on the next iteration the solution procedure will project q onto N (i.e., compute the nondominated criterion vector closest to q).

6. NORMALIZATION OF THE OBJECTIVES

Suppose we have an objective such as profit that is measured in millions, and another such as safety that is measured on a scale of 0 to 10. In order to prevent the objectives with the largest criterion values from biasing the search process in their behalf, it is never a bad idea to normalize the objective functions. A good way to normalize the objectives is to equalize the R_i's, where R_i is the range of z_i over N (i.e., the difference between z_i^* and the minimum criterion value of the i-th objective over N). A way to roughly accomplish this purpose is to multiply each objective by its π_i range equalization factor

$$\pi_i = (1/\bar{R}_i) \; [\Sigma_j \; 1/\bar{R}_j]^{-1}$$

where $\bar{R}_i$ is the difference between z_i^* and the minimum value in the i-th column of the payoff table. Ideally, we would want to use R_i's instead of $\bar{R}_i$'s, but the R_i values are typically not available.

7. QUASICONCAVE UTILITY FUNCTIONS

In many papers on interactive multiple objective programming, the authors assume the DM's utility function U to be quasiconcave over the domain M of the problem, where $M \subset R^k$ is some superset of the feasible region in criterion space Z.

Definition: Let $U: M \to R$ be convex and let

$$\text{lev}_\alpha U = \{z \mid U(z) \geq \alpha, \ z \in M\}$$

denote the α-level set of U. Then U is quasiconcave over M if and only if all α-level sets are convex, $\alpha \in R$.

Although a weakened concavity notion, quasiconcavity is a nice assumption because it still assures that all local optima are global optima when the feasible region in criterion space Z is convex.

8. WEIGHTED-SUMS PROGRAM

When all objectives and constraints are linear, we have a multiple objective linear program (MOLP)

$$\max \{c^1 x = z_1\}$$
$$\vdots$$
$$\max \{c^k x = z_k\}$$
$$\text{s.t.} \quad x \in S = \{x \in R^n \mid Ax = b, \ x \geq 0, \ b \in R^m\}.$$

Let $\Lambda = \{\lambda \in R^k \mid \lambda_i \in (0,1), \Sigma_i \lambda_i = 1\}$. Λ is called weighting vector space, and it contains all strictly positive weighting vectors whose components add to one. Assume an MOLP. Then $\bar{z} \in Z$ is nondominated if and only if there exists a $\lambda \in \Lambda$ such that $\bar{z}$ is a maximal solution of the weighted-sums program

$$\max \{\lambda^T z \mid c^i x = z_i, \ x \in S\}$$

This statement is not valid if the multiple objective program is other than an MOLP. In the non-MOLP case, the weighted-sums program would be written

$$\max \{\lambda^T z \mid f_i(x) = z_i,\ x \in S\}$$

9. UNSUPPORTED NONDOMINATED CRITERION VECTORS

A nondominated criterion vector $\bar{z} \in Z$ is unsupported if and only if it does not maximize the weighted-sum program

$$\max \{\lambda^T z \mid f_i(x) = z_i,\ x \in S\}$$

for some $\lambda \in \Lambda = \{\lambda \in R^k \mid \lambda_i \in (0,1),\ \Sigma_i \lambda_i = 1\}$. Unsupported nondominated criterion vectors are in "nonconvex depressions" of the nondominated set. In Figure 2, the set of unsupported nondominated criterion vectors is the set of criterion vectors from z^3 through z^4 to z^5, exclusive of z^3 and z^5. The set of supported nondominated criterion vectors is the set that consists of z^3 plus the line segment z^5 to z^6, inclusive. Unsupported nondominated criterion vectors can only occur in non-convex feasible regions. Thus, unsupported nondominated criterion vectors cannot occur in an MOLP.

10. VECTOR-MAXIMUM CODES

Sometimes an MOLP is written as a vector-maximum problem

$$\text{v-max}\{Cx = z \mid x \in S\}$$

where C is the k x n criterion matrix whose rows are the c^i. A solution to the above vector-maximum problem is an efficient point.

In an MOLP, the efficient set E is either (i) all of S or (ii) a portion of the surface of S. Case (ii) is the usual situation.

Algorithms for computing (characterizing) E are called vector-maximum algorithms (see Evans and Steuer (1973), Yu (1974), Zeleny (1974), Isermann (1977), Gal (1977), and Ecker, Hegner, and Kouada (1980)).

There are two types of vector-maximum codes: (a) algorithms that compute all efficient extreme points and all unbounded efficient edges and (b) algorithms that compute all efficient extreme points, all unbounded efficient edges, and all maximally efficient facets. A maximally efficient facet is an efficient facet of S that is not contained in another efficient facet of higher dimension.

Introduced in 1974, ADBASE (Steuer (1989)) is the most widely distributed vector-maximum code. It is of type (a). For international transportability, ADBASE has been written in low level Fortran and runs on all kinds of mainframe and PC equipment. EFFACET (Isermann and Naujoks (1984)) is another Fortran vector-maximum code. It is of type (b), and it has been designed to run on mainframes and PCs as well.

11. GROUPS OF MAXIMALLY DISPERSED λ-VECTORS

Suppose we are using an interactive procedure that requires, on iteration h, a group of maximally dispersed λ-vectors from the following interval defined subset of weighting vector space

$$\Lambda^{(h)} = \{\lambda \in R^k \mid \lambda_i \in (\ell_i^{(h)}, \mu_i^{(h)}), \Sigma_i \lambda_i = 1\}$$
$$\text{where } (\ell_i^{(h)}, \mu_i^{(h)}) \subset (0,1) \text{ for all } i \in K$$

By maximally dispersed we mean λ-vectors that are as evenly distributed over $\Lambda^{(h)}$ as far as it is practicable to compute. Such groups of dispersed λ-vectors can be obtained by using the LAMBDA and FILTER codes from Steuer (1989). Suppose we need 10 maximally dispersed λ-vectors from $\Lambda^{(h)}$. Then the LAMBDA code would be used to produce an overabundance of, say 200, randomly generated λ-vectors from $\Lambda^{(h)}$. Then, from the 200, FILTER would be called upon to select 10 λ-vectors that are as far apart from one another as it is practicable to compute.

12. LEXICOGRAPHIC AND AUGMENTED WEIGHTED TCHEBYCHEFF SAMPLING PROGRAMS

For sampling the nondominated set, we have the <u>lexicographic weighted Tchebycheff sampling program</u>

$$\begin{aligned} &\text{lex min } \{\alpha - \Sigma_i z_i\} \\ &\text{s.t.} \quad \alpha \geq \lambda_i(z_i^{**} - z_i) && i \in K \\ &\quad f_i(x) = z_i && i \in K \\ &\quad x \in S \\ &\alpha \in R \text{ unrestricted, } z \in R^k \text{ unrestricted} \end{aligned}$$

where z^{**} is a utopian criterion vector and $\lambda \in \Lambda = \{\lambda \in R^k \mid \lambda_i \in (0,1), \Sigma_i \lambda_i = 1\}$. The lexicographic sampling program is important because:

(1) If $\bar{z} \in Z$ is a minimizing criterion vector solution of the lexicographic sampling program, $\bar{z}$ is nondominated.

(2) If $\bar{z} \in Z$ is nondominated, there exists a $\lambda \in \Lambda$ such that $\bar{z}$ <u>uniquely</u> solves the lexicographic sampling program.

Result (1) states that the lexicographic sampling program will only generate nondominated criterion vectors. Result (2) means that no nondominated criterion vector can <u>hide</u> from the lexicographic sampling program.

Often used in place of the lexicographic sampling program is the <u>augmented weighted Tchebycheff sampling program</u>

$$\begin{aligned} &\min \{\alpha - \rho\, \Sigma_i\, z_i\} \\ &\text{s.t. } \alpha \geq \lambda_i\, (z_1^{**} - z_i) && i \in K \\ &\quad f_i(x) = z_i && i \in K \\ &\quad x \in S \\ &\alpha \in R \text{ unrestricted, } z \in R^k \text{ unrestricted} \end{aligned}$$

where ρ is a sufficiently small positive scalar. The advantage of the augmented sampling program is that it avoids lexicographic optimizations, but its disadvantage is having to deal with the sufficiently small ρ in a computational environment.

13. CONTOURS OF THE LEXICOGRAPHIC AND AUGMENTES WEIGHTED SAMPLING PROGRAMS

The level sets (in criterion space) of the objective function of the first optimization stage of the lexicographic sampling program are translated nonnegative orthants whose vertices lie along the line that goes through z^{**} in the direction

$$(1/\lambda_1,\ 1/\lambda_2,\ \ldots,\ 1/\lambda_k)$$

In this way, the contours (sets of points of constant value) of the objective of the first optimization stage are "L-shaped" sets whose "vertices" lie along the same line through z^{**}. Thus it can be visualized that the first optimization stage determines the L-shaped contour furthest along the line that goes through z^{**} in the direction given above. If there is only one feasible criterion vector on this L-shaped contour, this is the nondominated criterion vector returned by the lexicographic sampling program. If there is more than one feasible criterion vector on this L-shaped contour, some may be dominated. In this case, the second optimization stage is employed to determine a feasible criterion vector on the L-shaped contour that is closest to z^{**} according to the L_1-metric. In this way, the lexicographic sampling program terminates with a criterion vector that is assured to be nondominated.

The contours of the objective function of the augmented weighted Tchebycheff program are as shown in Steuer (1986, Figures 14.4 and 15.9).

14. PROJECTING AN UNBOUNDED LINE SEGMENT ONTO N

Let us project an unbounded line segment emanating from $\bar{z}$ in the direction $d \in R^k$ onto N. Rearranging the augmented weighted Tchebycheff sampling program and substituting $\bar{z} + \theta d$ for z^{**}, where θ goes from $0 \to +\infty$ we have the augmented parametric sampling program

$$\begin{array}{lll} & \min \{\alpha - \Sigma_i z_i\} & \\ \text{s.t.} & z_i + \alpha \geq \lambda_i (\bar{z}i + \theta d - z_i) & i \in K \\ & f_i(x) = z_i & i \in K \\ & x \in S & \\ \multicolumn{3}{c}{\alpha \in R \text{ unrestricted, } z \in R^k \text{ unrestricted}} \end{array}$$

whose solution as θ goes from $0 \to +\infty$ is the projection of the unbounded line segment onto N.

15. DETERMINING AN L-CONTOUR VERTEX λ-VECTOR

Let z^{**} be a utopian criterion vector and $q < z^{**}$ be an aspiration criterion vector. Then the L-contour vertex λ-vector pertaining to z^{**} and q is the λ-vector that causes the vertices of the L-shaped contours of the first optimization stage of the lexicographic sampling program to lie along the line that goes through q and z^{**}. The components of the L-contour vertex λ-vector are given by

$$\lambda_i = (1/|z_i^{**} - q_i|) \; [\Sigma_j \; 1/|z_j^{**} - q_j|]^{-1}$$

16. CONTRACTING Λ -SPACE

In some of the interactive procedures, convergence to a final solution is controlled by contracting Λ -space. One way to contract Λ -space is as follows. Let $\lambda^{(h)} \in \Lambda$ be the L-contour vertex λ-vector of the current solution after iteration h. Suppose we want to position a subset of weighting vector space about $\lambda^{(h)}$ of the form

$$\Lambda^{(h+1)} = \{\lambda \in \Lambda \mid \lambda_i \in (\ell_i^{(h+1)}, \mu_i^{(h+1)}), \Sigma_i \lambda_i = 1\}$$

for use on iteration h + 1. According to a routine by Liou (1984), we can control the size of $\Lambda^{(h+1)}$. For instance, if we want $\Lambda^{(h+1)}$ to be, say, 36% of the size of Λ, Liou's routine will compute the half-width r so that we can construct the intervals

$$(\ell_i^{(h+1)}, \mu_i^{(h+1)}) = \begin{cases} (0, r^h) & \text{if } \lambda_i^{(h)} - r^h/2 \le 0 \\ (1 - r^h, 1) & \text{if } \lambda_i^{(h)} + r^h/2 \ge 1 \\ (\lambda_i^{(h)} - r^h/2, \lambda_i^{(h)} + r^h/2) & \text{otherwise} \end{cases}$$

where r^h is r raised to the h-th power such that $\Lambda^{(h+1)}$ is 36% of the size of Λ.

17. INTERACTIVE PROCEDURES

Originally, it was thought that an MOLP could be solved by computing all efficient extreme points and then asking the DM to examine their criterion vectors to select the best one. However, after vector-maximum codes became available, it was found that MOLPs had far more efficient extreme points than anyone had imagined. For instance, a 5 objective, 50 constraint, 100 structural variable MOLP can easily have over 2000 efficient extreme points. Although it was known that the efficient set E is usually a portion of the surface of S, no one expected it to be this large. As a result, people's attention turned to interactive procedures for exploring the efficient set for the best solution. The most prominent interactive procedures to have been developed are categorized as follows:

Non-Reference Point Procedures

1. e-Constraint Method (traditional)
2. Geoffrion-Dyer-Feinberg Procedure (1972)
3. Zionts-Wallenius Procedure (1976 and 1983)
4. Interactive Weighted-Sums Method (see Steuer (1986, Section 9.5))
5. Interactive Surrogate Worth Tradeoff Method (Chankong and Haimes (1978 and 1983))

Reference Point Procedures

6. STEM (Benayoun, de Montgolfier, Tergny and Larichev (1971))
7. Wierzbicki's Reference Point Method (1977, 1982 and 1986)
8. Interactive Goal Programming (see Franz and Lee (1980))
9. Tchebycheff Method (Steuer and Choo (1983))
10. Satisficing Tradeoff Method (Nakayama and Sawaragi (1984))
11. VIG: Visual Interactive Approach (Korhonen and Laakso (1986) and Korhonen and Wallenius (1988))

In the reference point procedures, aspiration and utopian criterion vectors are used to sample or pinpoint criterion vectors from the nondominated set N.

18. DESCRIPTION OF NON-REFERENCE POINT PROCEDURES

In the e-Constraint Method one of the objectives is selected for maximization subject to e_i lower bounds on each of the other objectives to form the j-th objective e-constraint program

$$\begin{aligned} &\max \{f_j(x) = z_j\} \\ &\text{s.t. } f_i(x) \geq e_i \qquad i \neq j \\ &\qquad x \in S \end{aligned}$$

Then, from the solution generated, it is hoped that another problem configuration will suggest itself that leads to a better solution, and so forth. The method is ad hoc because it is never precisely clear how to configure the problem (which objective to select and what lower bounds to use on the others).

On each iteration of the Geoffrion-Dyer-Feinberg Procedure, the weighted-sums program, with weights specified by the DM, is solved. The current point and the solution of the weighted-sums program then define a line of search. The most preferred solution along the line of search provides a new current solution. With the DM updating the weights, another weighted-sums program is formed. From the solution of this program and the new current solution, another line of search is defined. Then the most preferred solution along this line of search provides the current solution for the next iteration, and so forth.

Let $\Lambda^{(1)}$ be all of weighting vector space Λ. The Zionts-Wallenius Procedure begins by selecting a λ-vector from the middle of $\Lambda^{(1)}$. Then, using this λ-vector, the weighted-sums program is solved. By asking the DM questions about points nearby the solution generated, constraints are placed on $\Lambda^{(1)}$ to form a smaller set $\Lambda^{(2)}$. Then a λ-vector is selected from the middle of $\Lambda^{(2)}$. By asking questions about points nearby the solution of the new weighted-sums program, new constraints are placed on $\Lambda^{(2)}$ to form a yet smaller subset $\Lambda^{(3)}$, and so forth.

In the Interactive Weighted-Sums Method, a group of dispersed λ-vectors is drawn from $\Lambda^{(1)} = \Lambda$. Then the weighted-sums program is solved for each of the dispersed λ-vectors. The DM examines the solutions produced and selects his or her most preferred. About the λ-vector that produced the most preferred solution, a subset $\Lambda^{(2)}$ of weighting vector space is formed. Then a group of dispersed λ-vectors is drawn from $\Lambda^{(2)}$ and for each of the dispersed λ-vectors the weighted-sums program is again solved. About the λ-vector associated with the most preferred of the solutions, a smaller subset $\Lambda^{(3)}$ of weighting vector space is formed, as so forth.

The Interactive Surrogate Worth Tradeoff Method uses the j-th objective e-constraint program to make single probes of the nondominated set. The e_i lower bounds, at each iteration, are computed from the DM's responses to surrogate worth value questions concerning marginal tradeoff information among the objectives at the current solution. Then, with new lower bounds, the procedure repeats.

19. DESCRIPTIONS OF REFERENCE POINT PROCEDURES

STEM begins by constructing a payoff table and using z^* as a utopian criterion vector. Using its own rules to compute a λ-vector, the first stage only of the lexicographic sampling program is solved to make a single probe of the nondominated set. Determining which components of the resulting criterion vector are to be relaxed so that others can be improved, constraints are constructed to reduce the feasible region and a new λ-vector is computed. Using the new λ-vector, the first stage of the lexicographic sampling problem is again solved to produce a new solution. Determining which new components are to be relaxed to afford improvement in the others, the feasible region is again reduced and another λ-vector is computed, and so forth.

Wierzbicki's Reference Point Method begins by constructing a payoff table, establishing a z^{**} utopian criterion vector, and having the DM specify an aspiration criterion vector $q^{(1)}$. Then the augmented weighted Tchebycheff sampling program is solved using the L-contour vertex λ-vector defined by $q^{(1)}$ and z^{**}. After examining the solution produced, the DM is asked to specify an updated aspiration criterion vector $q^{(2)}$. Using $q^{(2)}$ and z^{**} to determine a new L-contour vertex λ-vector, the augmented weighted Tchebycheff program is again solved. After examining the solution produced, the DM is asked to specify another updated aspiration criterion vector $q^{(3)}$, and so forth.

In Interactive Goal Programming we set priority levels, penalty weights, and target values. This is another ad hoc method because it is never precisely clear how to configure the problem. Then, from the solution generated, it is hoped that some revised problem configuration will suggest itself that will lead to a better solution, and so forth.

We begin the Tchebycheff method by establishing a z^{**} utopian criterion vector and obtaining a group of dispersed λ-vectors from $\Lambda^{(1)} = \Lambda$. Then the lexicographic sampling program is solved for each of the dispersed λ-vectors. Examining the resulting solutions, the DM selects his or her most preferred. Using the most preferred solution and z^{**}, an L-contour vertex λ-vector is computed. About this λ-vector, a subset $\Lambda^{(2)}$ of weighting vector space is formed. Then a group of dispersed λ-vectors is drawn from $\Lambda^{(2)}$, and for each of the dispersed λ-vectors, the lexicographic sampling problem is again solved. About the L-contour vertex λ-vector pertaining to the most preferred of the solutions, a smaller subset $\Lambda^{(3)}$ of weighting vector space is formed, and so forth.

In the Satisficing Tradeoff Method, it is only necessary for the DM to consider a table which contains the components of the current solution, the current aspiration criterion vector, and z^{**}. From one iteration to the next, components of the aspiration criterion vector are altered in order to search for improved solutions.

With reference to some starting solution, VIG begins by asking the DM to specify an aspiration criterion vector $q^{(1)}$. Then, using the augmented parametric sampling program, the unbounded line segment emanating from the starting solution through $q^{(1)}$ is projected onto the nondominated set. Using computer graphics, the trajectories of criterion values describing the projection of the line segment onto N are displayed. After viewing the trajectories, the DM specifies his or her most preferred point along the projected line segment to yield a new current solution. At this point the DM specifies a new aspiration criterion vector $q^{(2)}$. Then, again using the augmented parametric sampling program, the unbounded line segment emanating from the new current solution through $q^{(2)}$ is projected onto N. After viewing the trajectories pertaining to the new projected line segment, the DM selects the current solution for the next iteration. The DM now specifies a third aspiration criterion vector $q^{(3)}$, and so forth.

20. IMPLEMENTATION SIMILARITIES

While the literature has stressed differences among the various interactive procedures, most of the procedures have remarkable implementation similarities in that they more or less fit the following general algorithmic outline:

Step 1: Set the controlling parameters for the 1-st iteration.

Step 2: Solve the weighted-sums, lexicographic, augmented parametric, or augmented weighted Tchebycheff sampling programs one or more times to probe the nondominated set.

Step 3: Examine the criterion vector results.

Step 4: If the DM wishes to continue, go to Step 5. Otherwise, exit with the current solution.

Step 5: Set the controlling parameters for the next iteration and go to Step 2.

Controlling parameters are parameters in the sampling programs that are varied in order to sample different points from the nondominated set. The only procedure that does not easily fit the general algorithmic outline is the Zionts-Wallenius procedure. Note that all Steps, except Step 2, are free of heavy number-crunching demands.

21. CONSOLIDATION AND THE UNIFIED SAMPLING PROGRAM

Consolidation refers to the combining of different interactive procedures into a common computer package. This is possible because (a) all but one of the interactive procedures of this paper fit the general algorithmic outline and (b) all of the Step 2 sampling programs are special cases of the unified sampling program

$$\text{lex min } \{s_1(z,x,\alpha),\ s_2(z,x,\alpha),\ldots,s_L(z,x,\alpha)\}$$

$$\begin{array}{lrcll}
\text{s.t.} & \alpha & \geq & \eta_i(z_i^{**} + \theta d_i - z_i) & i \in G \\
 & z_i & \geq & e_i & i \in H \\
 & z_i + d_i^- & \geq & t_i & i \in I \\
 & z_i - d_i^+ & \leq & u_i & i \in J \\
 & f_i(x) & = & z_i & i \in K \\
 & x & \in & S & \\
 & d^-, d^+ & \geq & 0 &
\end{array}$$

$\alpha \in R$ unrestricted, $z \in R^k$ unrestricted

where:

$s_m(z,x,\alpha)$	<u>scalar-valued function</u> of m-th lexicographic priority
L	number of lexicographic levels
η	minimax λ-vector vector
z^{**}	utopian criterion vector (or current solution in VIG)
d	direction
t	<u>satisficing</u> target vector
u	<u>saturation</u> target vector

$K = \{1,2,\ldots,k\}$, $G \subset K$, $H \subset K$, $I \subset K$, and $J \subset K$

The scalar-valued function of m-th priority is given by

$$s_m(z,x,\alpha) = \alpha_{(m)}\alpha - \rho_{(m)} \Sigma_i \mu_{(m)i} z_i + \tau_{(m)}[\Sigma_{i \in I} w^-_{(m)i} d^-_i + \Sigma_{i \in J} w^+_{(m)i} d^+_i]$$

where $\sigma_{(m)}$, $\rho_{(m)}$, and $\tau_{(m)}$ are <u>procedure-dependent constants</u> which serve primarily as "switches" to include portions of the scalar-valued functions; $\mu_{(m)}$ is a vector of weighting coefficients; and the $w^-_{(m)}$ and $w^+_{(m)}$ are vectors of penalty weights for deviations from target values. The controlling parameters which may be manipulated in the unified sampling program are η, q, d, θ, e, t, u, and $\mu_{(m)}$, $w^-_{(m)}$, $w^+_{(m)}$, for $m = 1,\ldots,L$.

22. CUSTOMIZING THE UNIFIED SAMPLING PROGRAM

To illustrate how the unified sampling program can be <u>customized</u> for different procedures, consider the e-Constraint Method. By letting

(a) $L = 1$
(b) $\sigma_{(1)} = 0$, $\rho_{(1)} = 1$ and $\tau_{(1)} = 0$
(c) $G = I = J = \phi$
(d) $H = K - \{j\}$

the unified sampling program reduces to the j-th objective e-constraint program.

$$
\begin{aligned}
&\min \{-z_j\} \\
&\text{s.t. } z_i \geq e_i && i \in (K - \{j\}) \\
&\quad f_i(x) = z_i && i \in K \\
&\quad x \in S \\
&z \in R^k \text{ unrestricted}
\end{aligned}
$$

With regard to Wierzbicki's Reference Point Method, let

(a) $L = 1$
(b) $\sigma_{(1)} = 1$, $\rho_{(1)} = 1$, and $\tau_{(1)} = 0$
(c) $G = K$, $H = I = J = \phi$
(d) $\eta = \lambda$ where $\lambda \in \Lambda$
(e) $\theta = 0$ and $q = z^{**}$ where $z_i^{**} = \max \{f_i(x) \mid x \in S\} + \epsilon_i$

Thus we have the augmented weighted Tchebycheff sampling program

$$
\begin{aligned}
&\min \{\alpha - \rho \textstyle\sum_i z_i\} \\
&\text{s.t. } \alpha \geq \lambda_i(z_i^{**} - z_i) && i \in K \\
&\quad f_i(x) = z_i && i \in K \\
&\quad x \in S \\
&\alpha \in R \text{ unrestricted, } z \in R^k \text{ unrestricted}
\end{aligned}
$$

The unified sampling program is customized similarly for the sampling programs of the other procedures.

23. COLD STARTING VERSUS WARM STARTING

Suppose we have a series of similar optimization problems. If the first is solved from "scratch", that problem is solved from a cold start. If a subsequent problem is solved from the optimal solution of a previously solved problem, that problem is solved from a warm start. The following suggests the savings that can be realized when solving a series of similar problems. Cold starting the first problem, suppose it takes 500 pivots. Warm starting the second problem, it may then take 5 pivots to restore feasibility along with another 50 pivots to achieve optimality. In this way, it may be possible to solve 5 to 10 subsequent problems in the time it takes to solve the first from scratch. The warm starting of subsequent problems is particularly advantageous in the Interactive Weighted-Sums and the Tchebycheff procedures because the sampling programs in each of these procedures have to be solved a series of times, the only difference one problem to the next being the λ-vector.

24. IDEAL WORKHORSE SOFTWARE

A common computer package for interactive multiple objective programming would essentially be a supervisory program for conducting Steps 1, 2, 3 and 5 of the general algorithmic outline. For Step 2 the supervisory program would call workhorse software as a subroutine to obtain solutions from the unified sampling program. MPSX (IBM (1979)), although the international standard for linear and integer programming, is perhaps an example of the type of workhorse software not to use in interactive multiple objective programming. While it possesses warm starting, its disadvantages are its high fee, that external control is only allowed through PL/1, and that the user is denied access to the source code. Thus the product is of limited usefulness in interactive multiple objective programming research.

On the other hand, MINOS (Murtaugh and Saunders (1980)), for linear and nonlinear programming, and GRG2 (Lasdon and Waren

(1986)), for nonlinear programming, are examples of the types of products more useful in interactive multiple objective programming. They can be obtained at cost for academic use, are written in Fortran, possess warm starting, and are distributed with the source code. Thus, they can be modified and embedded in interactive procedures, and then redistributed for further academic use without license restrictions.

Unfortunately, many users have experienced difficulty with the nonlinear part of MINOS, and neither MINOS nor GRG2 possess RHS parametrics. With no codes even as good as MINOS or GRG2 available in integer programming, ideal workhorse software for interactive multiple objective programming does not yet exist.

25. SWITCHING

One of the notable results of the empirically-based research of Brockhoff (1985) and Buchanan and Daellenbach (1987) is that a user knowledgeable of interactive multiple objective programming may wish to use different procedures on different iterations. We will refer to this as switching. For instance, a user may wish to start with the Tchebycheff Procedure to sample the entire nondominated set, then switch to VIG to explore a portion of the nondominated set, and finally switch to Wierzbicki's Method to pinpoint a final solution. To accomodate switching, Steps 1 and 5 of the general algorithmic outline are modified as follows:

Step 1: Set the controlling parameters of the selected procedure for the 1-st iteration.

Step 5: Set the controlling parameters of the selected procedure for the next iteration and go to Step 2.

26. PRESERVING CONVERGENCE ACHIEVEMENTS

Switching out of any of the procedures into the e-Contraint Method, STEM, the Geoffrion-Dyer-Feinberg Procedure, Interactive

Goal Programming, Wierzbicki's Reference Point Method, or VIG is straightforward. We simply use the current point after iteration h as the starting solution for iteration h + 1 (just as if we were starting the new procedure from scratch). When switching into the Interactive Weighted-Sums or the Tchebycheff Methods, we must preserve convergence achievements. Let $z^{(h)}$ designate the current solution after iteration h. Then, about $\lambda^{(h)}$, the L-contour vertex λ-vector pertaining to $z^{(h)}$ and z^{**}, we construct $\Lambda^{(h+1)}$ of a given size according to Section 16. Switching from any procedure other than the e-Constraint Method into the Surrogate Worth Tradeoff Method is not included in the current vision of the common computer package because of the work involved.

27. TRAJECTORY APPLICATIONS

Perhaps the most challenging applications in the mathematical programming are multiple criteria <u>trajectory</u> optimization applications. These are multiple criteria problems with <u>multiple observation points</u> (<u>multi-time periods</u> or <u>multi-surveillance points</u>). Suppose we have such a problem in which we wish to monitor k criteria over T observation points. In such problems we often have a goal level of achievement for each criterion at each observation point. Then, for each objective, the "path" of goal criterion values forms a <u>goal trajectory</u> over the T observation points. By the same token, for each solution, there is a "path" of criterion values for each objective over the T observation points. In these problems, the purpose is to find the solution whose k <u>criterion value trajectories</u> most closely match the k goal trajectories. To illustrate, we have the following:

<u>Manpower Planning</u> (with multi-time periods)

min {overdeviations from salary budget}
min {underdeviations from manpower requirements}
min {deviations from promotion targets}
min {underdeviations from workforce experience goals}

River Basin Management (with multi-surveillance points)

min {overdeviations from BOD standards}
min {overdeviations from nitrate standards}
min {pollution removal costs}
min {underdeviations from municipal water demands}
min {groundwater pumping}

With the techniques of interactive multiple objective programming, we should see a growth in the number of such applications in the 1990s.

28. POLYSCREEN WORKSTATIONS

Although computer workstations with more than two screens are not available as standard equipment, they certainly are feasible. The issue is "less display space and more windowing" versus "more display space and less windowing". Polyscreen workstations with up to six screens would enable us to see more and remember less (from windows that might otherwise be hidden). Having more on display at the same time would be especially helpful in trajectory applications because of the amount of information generated. For example, at each iteration, two screens might be used to display criterion value range information over the nondominated set, model parameter values, and bounds values currently in effect on each variable. The remaining four screens might then be used to cross-reference different candidate solutions. Polyscreen workstations are used in running nuclear power plants and in military applications, so why not in business decision making?

29. COGNITIVE EQUILIBRIUM

The conventional decision-making paradigm is that the DM possesses a utility function that he or she wishes to maximize over the feasible region. Difficulties have been experienced with this paradigm in interactive multiple objective programming. We know that the DM's aspirations change in many multiple criteria problems as more is learned about the problem. Does this mean that the DM's utility function can make major changes over a short period of time and is thus unstable? Or does this mean that a DM doesn't truely understand his or her utility function without significant interaction with the problem implying that utility functions are much more difficult to assess a priori that previously thought.

A new decision-making paradigm proposed by Zeleny (1989) called cognitive equilibrium seems more appropriate to interactive multiple objective programming. In this paradigm, decision making is the process of recursively redefining the problem and redefining one's aspirations until a form of stability or equilibrium is obtained at which time the problem is solved. In other words, decision making is the process of searching for harmony among chaos. It appears that this concept of decision making goes a lot further to explain what the field of interactive multiple objective programming has been encountering then conventional utility theory.

30. NETWORK OPTIMIZATION APPLICATIONS

Consider the linear program (LP)

$$\max \{c^T x \mid x \in S\}$$
$$\text{s.t.} \quad S = \{x \in R^n \mid Ax = b,\ x \geq 0,\ b \in R^m\}$$

If, apart from upper and lower bounds on the variables, A has (i) no more than two nonzero elements per column and (ii) the nonzero elements are either 1's or -1's, the LP is a network. Because of the special structure of A, networks can be solved using high-speed solution procedures.

Let us now consider a multiple objective network. Unfortunately, with the interactive procedures discussed in this paper, the unified sampling program of Section 20 is not a network because of (non-network) side-constraints. However, with the development of high-speed codes that can handle limited numbers of side-constraints (Kennington (1980) and Aronson (1989)), multiple objective networks now offer another growth area for applications in the 1990s.

31. RUSSIAN INTERACTIVE RESEARCH

There is a significant body of research on interactive multiple objective programming in the Russian literature about which little is known in the West. Fortunately, an English-written survey of this research appears in Lieberman (1990). It is expected that with changing world relations, the separateness of research in the East and West will move to full integration in the next decade.

32. MAINFRAMES VERSUS MICROCOMPUTERS

For instruction and research as well as in many small applications, a microcomputer installation of a common computer package would be ideal. But for large-scale applications, an integrated mainframe/microcomputer installation might be the only way to proceed. The common computer package would be on the mainframe and the microcomputer, with the assistance of graphical techniques, would then be used at the interface to conduct computer/user communications.

REFERENCES

Aronson, J.E. (1989), "A survey of dynamic network flows", *Annals of Operations Research*, to appear.

Benayoun, R., de Montgolfier, J., Tergny, J. and Larichev, O. (1971), "Linear programming with multiple objective functions: step method (STEM)", *Mathematical Programming*, Vol. 1, No. 3, 366-375.

Brockhoff, K. (1985), "Experimental test of MCDM algorithms in a modular approach", *European Journal of Operational Research*, Vol. 22, No. 2, 159-166.

Chankong, V. and Haimes, Y.Y. (1978), "The interactive surrogate worth trade-off (ISWT) method for multiobjective decision-making", *Lecture Notes in Economics and Mathematical Systems*, Vol. 155, Springer-Verlag, 42-67.

Chankong, V. and Haimes, Y.Y. (1983), *Multiobjective Decision Making: Theory and Methodology*, North-Holland, New York.

Ecker, J.G., Hegner, N.S. and Kouada, I.A. (1980), "Generating all maximal efficient faces for multiple objective linear programs", *Journal of Optimization Theory and Applications*, Vol. 30, No. 3, 353-381.

Franz, L.S. and Lee, S.M. (1980), "A goal programming based interactive decision support system", *Lecture Notes in Economics and Mathematical Systems*, Vol. 190, Springer-Verlag, 110-115.

Gal, T. (1977), "A general method for determining the set of all efficient solutions to a linear vectormaximum problem", *European Journal of Operational Research*, Vol. 1, No. 5, 307-322.

Gardiner, L.R. (1989), "Unified Interactive Multiple Objective Programming", Ph.D. Dissertation, Department of Management Science & Information Technology, University of Georgia, Athens, Georgia, USA.

Gardner, J.C., Huefner, R.J. and Lotfi, V. (1989), "A multi-period audit staff planning model using multiple objectives: development and evaluation", *Decision Sciences*, to appear.

Geoffrion, A.M., Dyer, J.S. and Feinberg, A. (1972), "An interactive approach for multicriterion optimization, with an application to the operation of an academic department", *Management Science*, Vol. 19, No. 4, 357-368.

IBM Document No. GH19-1091-1 (1979), "IBM Mathematical Programming System Extended/370: Primer", IBM Corporation, Data Processing Division, White Plains, New York, USA.

Ignizio, J.P. (1982), *Linear Programming in Single- & Multiple-Objective Systems*, Englewood Cliffs, New Jersey: Prentice-Hall.

Isermann, H. and Steuer, R.E. (1988), "Computational experience concerning payoff tables and minimum criterion values over the efficient set", *European Journal of Operational Research*, Vol. 33, No. 1, 91-97.

Isermann, H. and Naujoks, G. (1984), "Operating Manual for the EFFACET Multiple Objective Linear Programming Package", Fakultät für Wirtschaftswissenschaften, Universität Bielefeld West Germany.

Isermann, H. (1977), "The enumeration of the set of all efficient solutions for a linear multiple objective program", *Operational Research Quarterly*, Vol. 38, No. 3, 711-725.

Kennington, J.L. (1980), *Algorithms for Network Programming*, New York: John Wiley & Sons, 291 pp.

Korhonen, P.J. and Laakso, J. (1986), "A visual interactive method for solving the multiple criteria problem", *European Journal of Operational Research*, Vol. 24, No. 2, 277-287.

Korhonen. P.J. and Wallenius, J. (1988), "A Pareto Race", *Naval Research Logistics*, Vol. 35, No. 6, 615-623.

Lasdon, L.S. and Waren, A. D. (1986), "GRG2 User's Guide", University of Texas, Austin, Texas, USA.

Lee, S.M. and Shim, J.P. (1986), "Interactive goal programming on the microcomputer to establish priorities for small business", *Journal of the Operational Research Society*, Vol. 37, No.6, 571-577.

Lewandowski, A., Kreglewski, T., Rogowski, T. and Wierzbicki, A.P., (1987), "Decision support systems of DIDAS family (dynamic interactive decision analysis and support)", *Archiwum Automatyki i Telemechaniki*, Vol. 32, No. 4, 221-246.

Lieberman, E.R. (1990), *Multi-Objective Programming in the USSR*, book manuscript, School of Management, State University of New York at Buffalo, Buffalo, New York, USA.

Liou, F. H. (1984), "A routine for Generating Grid Point Defined Weighting Vectors", Masters Thesis, Department of Management Science & Information Technology, University of Georgia, Athens, Georgia, USA.

Murtaugh, B.A. and Saunders, M. A. (1987), "MINOS 5.1 User's Guide", Report SOL 83-20R, Department of Operations Research, Stanford University, Stanford, California, USA.

Nakayama, H. and Furukawa, K. (1985), "Satisficing trade-off method with an application to multiobjective structural design", *Large Scale Systems*, Vol. 8, No. 1, 47-57.

Nakayama, H. and Sawaragi, Y. (1984), "Satisficing trade-off method for multiobjective programming." *Lecture Notes in Economics and Mathematical Systems*, Vol. 229, Springer-Verlag, 113-122.

Ramesh, R., Karwan, M.H. and Zionts, S. (1989), "Interactive multicriteria linear programming: an extension of the method of Zionts and Wallenius", *Naval Research Logistics*, Vol. 36, No. 3, 321-335.

Silverman, J., Steuer, R.E. and Whisman, A.W. (1988), "A multi-period, multiple criteria optimization system for manpower planning", *European Journal of Operational Research*, Vol. 34, No. 2, 160-170.

Stam, A., Joachimsthaler, E.A. and Gardiner, L.R. (1989), "A Multiobjective Model for Sales Force Sizing and Deployment", Department of Management Science & Information Technology, University of Georgia, Athens, Georgia, USA.

Steuer, R.E. (1989), "Operating Manual for the ADBASE Multiple Objective Linear Programming Package", Department of Management Science & Information Technology, University of Georgia, Athens, Georgia, USA.

Steuer, R.E. (1986), *Multiple Criteria Optimization: Theory, Computation, and Application*, (originally published by John Wiley & Sons, New York, now being republished by Krieger Publishing, Melbourne, Florida), 546 pp.

Steuer, R.E. and Choo, E.U. (1983), "An interactive weighted tchebycheff procedure for multiple objective programming", *Mathematical Programming*, Vol. 26, No. 1, 326-344.

Steuer, R.E. and Wood, E.F. (1986), "On the 0-1 implementation of the tchebycheff solution approach: A water quality illustration", *Large Scale Systems*, Vol. 10, No. 3, 243-256.

Wierzbicki, A.P. (1977), "Basic properties of scalarizing functionals for multiobjective optimization", *Mathematische Operationsforschung und Statistik - Series Optimization*, Vol. 8, No. 1, 55-60.

Wierzbicki, A.P. (1982), "A mathematical basis for satisficing decision making", *Mathematical Modelling*, Vol. 3, 391-405.

Wierzbicki, A.P. (1986), "On the completeness and constructiveness of parametric characterizations to vector optimization problems", *OR Spektrum*, Vol. 8, No. 2, 73-87.

Winkels, H.-M. and Meika, M. (1984), "An integration of efficiency projections into the geoffrion approach for multiobjective linear programming", *European Journal of Operational Research*, Vol. 16, No. 1, 113-127.

Yu, P.L. (1974), "Cone convexity, cone extreme points, and nondominated solutions in decision problems with multi-objectives", *Journal of Optimization Theory and Applications*, Vol. 14, No. 3, 319-377.

Zeleny, M. (1989), "Stable patterns from decision-producing networks: new interfaces of DSS and MCDM", *MCDM WorldScan*, Vol. 3, Nos. 2 & 3, 6-7.

Zeleny, M. (1974), "Linear multiobjective programming", *Lecture Notes in Economics and Mathematical Systems*, No. 95, Berlin: Springer-Verlag.

Zionts, S. and Wallenius, J. (1976), "An interactive programming method for solving the multiple criteria problem", *Management Science*, Vol. 22, No. 6, 652-663.

Zionts, S. and Wallenius, J. (1983), "An interactive multiple objective linear programming method for a class of underlying nonlinear utility functions", *Management Science*, Vol. 29, No. 5, 519-529.

A COMPARISON OF MICROCOMPUTER IMPLEMENTED INTERACTIVE MOLP METHODS BASED ON A CASE STUDY

João N. Clímaco and Carlos H. Antunes

Department of Electrical Engineering
University of Coimbra
3000 Coimbra - PORTUGAL

SUMMARY

In this study we discuss some results and draw some conclusions from a comparative analysis of the application of three interactive multiobjective linear programming (MOLP) methods to an outline study of the expansion policy for a power generation system. The selection of a power generation expansion plan has been modelled as a tricriteria linear programming problem, where the three objective functions quantify the total system cost, the risk of supply failure and the environmental impact.

The Zionts-Wallenius and STEM interactive MOLP methods are revisited and compared with the TRIMAP method for an electric power system expansion planning case study. The TRIMAP method combines the weighting space decomposition, the introduction of constraints on the objective functions space and on the weighting space. Furthermore, the introduction of constraints on the objective functions values can be translated into the weighting space.

In our computer implementation of these methods special attention is paid to the user interface (a key issue to the successful application of the interactive algorithms). All the programs are implemented on Macintosh II (sharing the same data files), trying to make the most of the computer's user-friendly environment through the use of windows for displaying the graphics, pull-down menus for choosing the available actions and dialogue boxes for exchanging information with the user in order to facilitate the decision maker (DM) tasks.

The methods are briefly described in section 1. The case study is summarised in section 2. In section 3 some results of the application of the methods are presented. The conclusions based on a comparative analysis of the results are drawn in section 4.

1. THE MOLP METHODS

1.1. The TRIMAP Method

The TRIMAP method is based on the progressive and selective "learning" of the set of Pareto optimal solutions. The aim is not to "converge" on any "best compromise" solution but to help the decision maker (DM) to eliminate the subsets of the Pareto optimal solutions which are of no interest to him/her. The method combines three main procedures: weighting space decomposition, introduction of constraints on the objective functions space and on the weighting space. Furthermore, the introduction of constraints on the objective functions values can be translated into the weighting space.

In each step of the man/machine dialogue (when a new phase of computation is being prepared) the DM just has to give indications about the regions where the search for the Pareto optimal solutions should be carried out. At this point additional constraints on the values of some of the objective functions may also be introduced by the DM. There are no irrevocable decisions along the process being always admissible to go "backwards". The method is not too demanding with respect to the information required from the DM in each interaction, the main concern being to have a simple and flexible procedure.

The TRIMAP method is intended for tricriteria problems. Although being a limitation, this fact allows for the use of graphical means which are particularly suited for the dialogue with the DM. This undoubtedly increases his/her capacities of information processing and simplifies the dialogue phase.

The interactive process begins with the automatic computation of a number of Pareto optimal extreme points, following a procedure similar to one developed in (Reeves and Franz, 1985). Next, a new computation phase is prepared through a dialogue phase with the DM. These two phases are repeated until the DM considers to have "sufficient" knowledge about the set of Pareto optimal solutions.

The computation process of the Pareto optimal feasible polyhedron extreme points used in TRIMAP consists in solving problems of the type:

$$\max_{\underline{x} \in F} \{\lambda_1 f_1(\underline{x}) + \lambda_2 f_2(\underline{x}) + \lambda_3 f_3(\underline{x})\} \tag{1}$$

where $\Sigma \lambda_i = 1$ and $\lambda_i > 0$, $i = 1,2,3$.

Note that $\underline{x} \in F$ is an efficient solution if and only if it is an optimal solution to (1). Generally, the concept of nondominance or Pareto optimality refers to the objective function space whereas the concept of efficiency refers to the decision variable space.

Furthermore, it must be remarked that only two weights are independent variables. For instance, $\lambda_3=1-\lambda_1-\lambda_2$. So, the graphical representation of the set $\underline{\lambda}=[\lambda_1,\lambda_2,\lambda_3]$ which gives origin to each Pareto optimal extreme point can be achieved through the decomposition of the triangle (fig. 1). In this paper we always consider the triangle of fig. 1.

In general, the triangle will be completely filled when all the Pareto optimal feasible polyhedron vertices have been found.

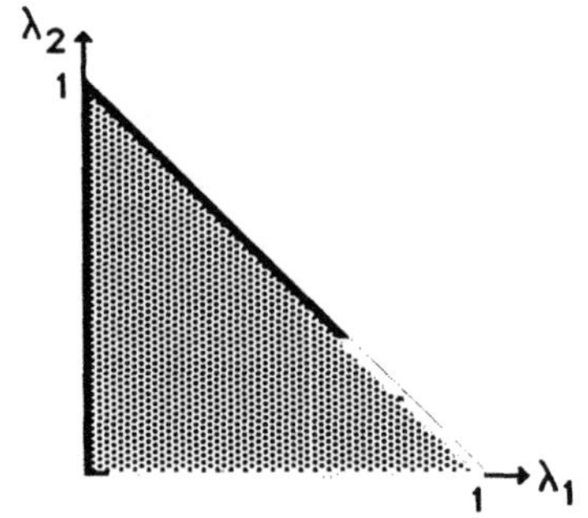

Figure 1 - The weighting space

The main purpose will be doing a progressive and selective triangle filling. In each step the DM will be called to decide whether or not the study of solutions corresponding to not yet searched triangle regions is of interest. We want to prevent the exhaustive search of regions with close objective function values (very often found in real case studies). It is also possible triangle regions be eliminated by imposing limitations on the objective functions value.

The introduction of the additional constraints can also be used to get some Pareto optimal solutions that are not vertices of the feasible polyhedron.

The selection of weights may be automatic (in this way it is possible to get a certain number of Pareto optimal extreme points, the weights that originated them being well distributed on the triangle) or manual (the selection of the weights is made by the DM's indication of the triangle zones not yet filled, for which he/she thinks important the search to be continued). Moreover, in order to get a more correct perception of the relative importance of the objective functions (if the functions are not in the same units) it is convenient to normalise them (Clímaco and Antunes, 1987 and 1989).

In each interaction two graphs will be presented to the DM.

The first one is the triangle filled with the regions corresponding to each of the already known Pareto optimal vertices. Eventual constraints on the variation of $\underline{\lambda}$ will also be presented. The second one presents a projection on the plan $f_1 f_2$ of the Pareto optimal solutions already computed. The area of each region of the triangle and the Tchebycheff distance (L_∞) to the "ideal solution" corresponding to each Pareto optimal extreme point already computed are also presented. The so-called "ideal solution" is the one that would optimise all objective functions. Of course it is not feasible in general!

1.2. The Step Method (STEM)

STEM is an interactive reduced feasible region method where, in each interaction, we seek to minimise a weighted Tchebycheff distance to the ideal solution.

In each interaction the problem to be optimised reflects the DM choices in the preceding interactions through the reduction of the feasible region due to the setting of constraints on the objective functions values. In each interaction of STEM a problem of the following type is solved:

$$\min_{\underline{x},\ \alpha} \alpha$$
$$\text{s.t.} \quad (f_j^* - \underline{c}_j' \underline{x}) b_j \le \alpha, \quad j=1,\dots,p$$
$$x \in F_k$$
$$\alpha \ge 0$$

Without loss of generality, we consider that all the objective functions are to be maximised.

This is a problem with p objective functions where f_j^* is the maximum of the objective function $\underline{c}_j' \underline{x}$ in the domain F. The elements b_j are the Tchebycheff weights computed by a specific procedure aiming at a certain normalisation of the objective functions (for details see (Steuer, 1986, chapter 13)).

The compromise solution found in each iteration is presented to the DM. If the values of the objective functions are considered satisfactory the process stops, otherwise the DM must specify those which he/she is willing to relax, and by how much, in order to improve the other functions.

The F_k domain results from the introduction of the additional constraints on the objective functions values for the feasible region F of the original problem. These constraints will have effect on the objective functions whose values were considered satisfactory, but for which the DM decides to admit a certain relaxation, in order to improve the remaining ones. For each objective function that becomes a constraint we consider $b_j = 0$.

As long as there exist any objective function values for which the DM is willing to trade-off in order to improve the others, the interactive process goes on.

1.3. The Zionts-Wallenius Method

The Zionts-Wallenius method is an interactive procedure that progressively reduces the weighting space accordingly to the responses of the DM when asked, in each interaction, about preferences on adjacent extreme points or trade-offs.

Initially a system of weights is chosen and the weighted sum of the objectives of the original problem is optimised using the simplex method, thus obtaining a nondominated extreme point of the feasible polyhedron.

From the non-basic variables of the corresponding simplex tableau for this solution, the efficient variables (those which introduction in the basis lead to a displacement along nondominated edges) are chosen. The DM is asked whether he/she prefers some nondominated vertex adjacent to the one currently considered or he/she would consider trade-offs corresponding to displacements along non-dominated edges (for details see (Steuer, 1986, chapter 13)).

The DM is asked to express his/her preferences through pairwise comparisons between the current solution and its nondominated adjacent vertices or to answer whether he/she likes the tradeoffs corresponding to displacements along non-dominated edges with origin at the current vertex. From these responses the method operates by introducing constraints in the weighting space, thereby progressively reducing the admissible domain for selecting a new set of weights. When inconsistencies occur the Zionts-Wallenius algorithm usually allows the deletion of the oldest constraints in order to continue the searching process.

The process ends when the weighting space has been reduced to a small enough region for a final solution to be identified, thus converging to the optimum of the implicit DM's utility function. The responses of the DM in the dialogue phases must be coherent with this implicit utility function. The authors admit as possible implicit utility functions, not only weighted sums of the linear objective functions but also more general pseudo-concave functions. In this case the method only guarantees the convergence to the greatest utility Pareto optimal extreme point.

2. THE CASE STUDY

The planning of new units for electrical power production is a problem whose study explicitly requires several objective functions. We have carried out a study some years ago, using a linear model which considered three objective functions: the global cost, one related to the risk of supply system, and a third one related to environmental impacts. A multiobjective linear programming algorithm by Zeleny for calculating the set of the Pareto optimal extreme points was used (Clímaco and Almeida, 1981). However, further experiments with more realistic formulations, leading to a much higher number of variables and constraints, confirmed the problems raised in these cases by the application of a method for generating all the Pareto optimal extreme points. Namely, the computational effort becomes impracticable when the Pareto optimal extreme points grow beyond certain limits; and even when it is possible to compute them, this is in most cases a worthwhile effort. In fact, the DM can not make a well-founded choice if he/she is faced to a large set of solutions, most of them with just slightly different values of the objective functions. Besides, the best trade-off solution may be a point belonging to a Pareto optimal surface which is not an extreme point. The consideration of this possibility makes even more difficult the calculations and the presentation of the results to the DM. These difficulties led us to study the possibility of applying interactive methods to this type of problem. For this purpose a survey was carried out where the advantages and shortcomings of different methods were analysed. Taking into account some general views expressed by Roy (1985) a new method named "TRIMAP" was developed (Clímaco and Antunes, 1987 and 1989).

This study is solely concerned with the planning of electricity generation in view of its dependence on load evolution. The problems of siting the new units and its effects on the transmission network expansion planning are not considered herein.

2.1. Identification of the Objective Functions

The three objective functions considered in this approach are: cost and environmental impact (to be minimised) and reliability (to be maximised). Simplifications had to be introduced because of the linear nature of the computational tools used.

2.1.1. Cost

The cost objective function is composed by the fixed charges and by the operational costs of both the new units and the units already in service at the beginning of the planning period.

No fixed charges are assumed for units already in service at the beginning of the planning period, since they are constant to every plan.

2.1.2. Environmental impact

Alternative energy generating technologies are very difficult to compare from an environmental perspective. Four indices of impact evaluation (land use, large accident/sabotage, emissions/ public health and effect on ecosystems) have been considered in order to construct an objective function quantifying the environmental impact of three generating technologies (oil, nuclear and coal). The first index has little to do with the energy output of the generating groups. The second has to do both with energy output and with the existence of the generating facilities. The remaining may be related solely to energy output.

In this context two parameters were obtained that characterise each generating alternative: one penalising the installed capacity and the other penalising the energy output. The weight assigned to the index "large accident/sabotage" has been intentionally set much higher than the others in order to penalise risk. As the nuclear option has the worst ranking of risk, the procedure intends to introduce a bias in the data in an attempt to anticipate future analysis that will take into account the Tchernobil accident.

2.1.3. Reliability

The linear objective function is an approximation since the probabilistic calculations involved in reliability assessment are intrinsically nonlinear. The fundamental parameter used to model reliability of the generating units is the load-carrying-capability (LCC) of new units. The LCC is the amount of power that the system load can increase with the addition of a new unit in order to maintain the reliability level as it was before the unit addition.

Within the set of solutions to the global optimisation problem which the DM considers acceptable, some may well not satisfy a reliability criterion if they are too biased by the other two objectives. This raises the need to have some means of evaluating a plan's performance against the risk of supply failure.

2.2. Identification of the Constraints

Two main categories of constraints have been considered. The first one basically imposes that all the generating power available at each period of time considered must be at least equal to the load in the same interval. Both the new and the previously existent generation units must be considered. The second category includes operational constraints of the generating units: the power output of a unit cannot exceed its rated power, previously affected by some availability factor. A budgetary constraint on the maximal feasible value for the cost objective function was also introduced.

3. RESULTS OF THE APPLICATION OF THE METHODS

3.1. Some TRIMAP Results

The run of the TRIMAP method began by the automatic computation of three non-dominated solutions, corresponding to the optima of the objective functions. Other solutions have been subsequently obtained through an interactive procedure by manual selection of the weights (Clímaco and Antunes, 1989). Only the solutions corresponding to the optima of the objective functions have been displayed on the projection of the objective function space on the f_1f_2 plane (fig. 2). This is done in order to avoid a multitude of hardly distinguishable points.

Table I condenses a fraction of the data obtained for 20 solutions. In table II the amounts of power generation additions in a subset of those solutions are presented with discrimination of the subperiods. Although they had been computed the energy outputs are not presented.

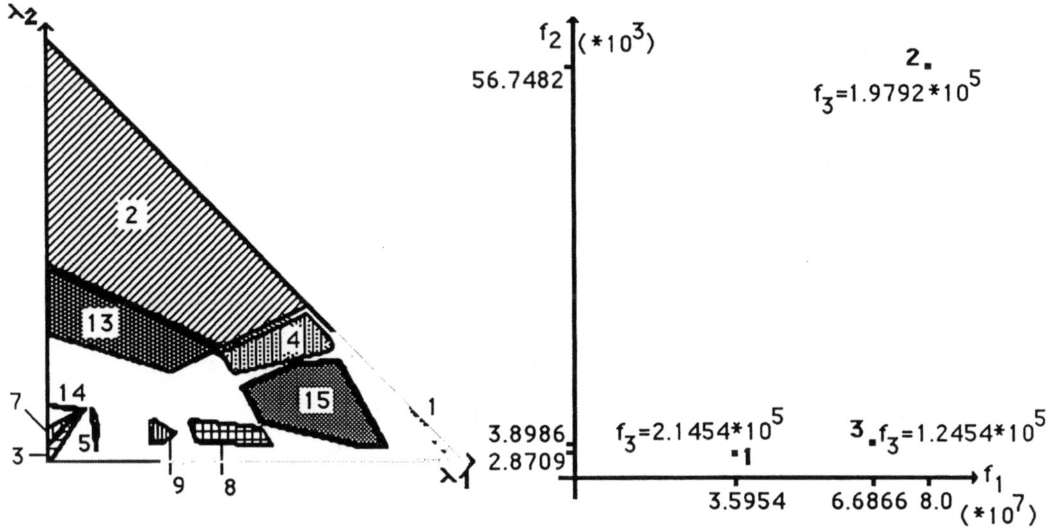

Figure 2 - The weighting space and a projection of the objective space

Minimum cost is obtained with a mix of nuclear and coal units, whereas minimum environmental impact corresponds only to addition of oil burning units. The maximum of f_2 is an unrealistic solution, where cost is sacrificed to reliability. Three distinct zones corresponding to solutions with different characteristics can be distinguished on the weighting space (figs. 2 and 3): an upper triangle sub-region where a compromise between f_2 and f_3 keeps f_1 at the allowed maximum, and two lower triangle sub-regions. In the first sub-region, nuclear units appear in just few solutions near the optimum of f_1 and a smooth cost transition occurs among solutions without nuclear technology, the majority of the lower triangle sub-region being filled by solutions corresponding to additions of only coal units. In the second sub-region near the optimum of f_3, a sharp transition occurs from all-oil plans to all-coal plans. A sharp variation of cost occurs among these three sub-regions.

Fig. 3 illustrates a feature of TRIMAP which allows bounds to be introduced on the objective function values (in this case an upper-bound $f_1 \leq 4 \times 10^7$ has been imposed), which enables a more exhaustive search of expansion plans in a restricted sub-region of the triangle, thus reducing the scope of the search.

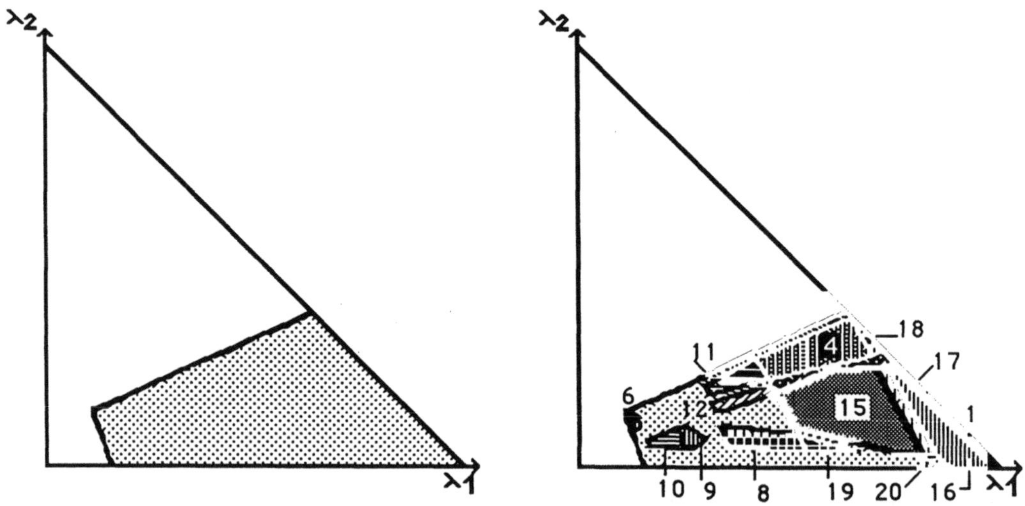

Figure 3 - The weighting space constrained with $f_1 \leq 4 \times 10^7$

At present the authors are considering the problem of discretising the results obtained from this type of studies with an adequate approach.

<u>Note</u>: The triangle decomposition is achieved using normalised objective functions. In fig. 2-b and in table I the real values are presented.

Solution	f_1 ($\times 10^7$)	f_2 ($\times 10^3$)	f_3 ($\times 10^5$)	Area (%)
1	3.5954	2.8709	2.1454	0.551
2	8.0000	56.7482	1.9792	41.439
3	6.6866	3.8986	1.2454	0.466
4	3.7902	5.7698	1.5452	3.923
5	5.7824	2.5116	1.3048	0.102
6	3.7678	4.8627	1.5328	0.076
7	6.7208	4.0949	1.2461	0.313
8	3.7330	4.4636	1.5355	1.666
9	3.7373	4.4006	1.5334	0.408
10	3.7375	4.4006	1.5333	0.508
11	3.7904	5.7698	1.5451	1.080
12	3.7702	5.3687	1.5406	0.789
13	8.0000	56.7455	1.9790	8.925
14	8.0000	19.9078	1.3825	0.033
15	3.7308	4.6535	1.5408	8.807
16	3.5968	2.8429	2.0554	3.212
17	3.6452	3.5957	1.8414	1.078
18	3.7045	4.7120	1.8459	0.304
19	3.7328	4.4636	1.5357	0.452
20	3.6451	3.5947	1.8414	0.120

Table I - Objective function values and area (%) occupied on the weighting space corresponding to solutions on figs. 2 and 3

Solution	Oil 1	Nuclear 1	Coal 1	Coal 2	Coal 3
1		3673.3			441.3
3	3994.4				
18		1826.6		2065.0	
15			1370.0	975.0	851.3
5	2573.3				
6			1222.5	1672.5	
8			1370.0	975.0	625.0

Table II - Generation additions, in MW, for some solutions (the index refers to the subperiod)

3.2. Some STEM Results

The initial solution computed by STEM presents the following objective functions values:

$f_1 = 78170369.1$

$f_2 = 54536.7$

$f_3 = 196042.0$

It is easy to verify, if we compare the results obtained with TRIMAP, that we find ourselves on a nondominated face, close to the optimum of the objective function f_2. From what is already known from the application of TRIMAP we are interested in relaxing f_2 a lot since that its value is very high as a result of building many more power plants than necessary for satisfying the demand.

It is somehow difficult for the DM to establish the values to be relaxed by doing a simple comparison of the objective function values of the initial solution computed by using STEM with the optima of the three objective functions. Nevertheless, an experiment will be presented.

Let us suppose that in the first interaction f_2 is relaxed by 51×10^3. Then we get:

$f_1 = 41491034.1$

$f_2 = 3536.7$

$f_3 = 148160.6$

At last, in the second interaction, let us suppose that the DM decides f_3 to be relaxed by 20×10^3. Then we get:

$f_1 = 36906874.6$

$f_2 = 3993.5$

$f_3 = 168160.6$

The results of the consecutive interactions are presented to the DM in bar graphs such as the ones displayed in fig. 4.

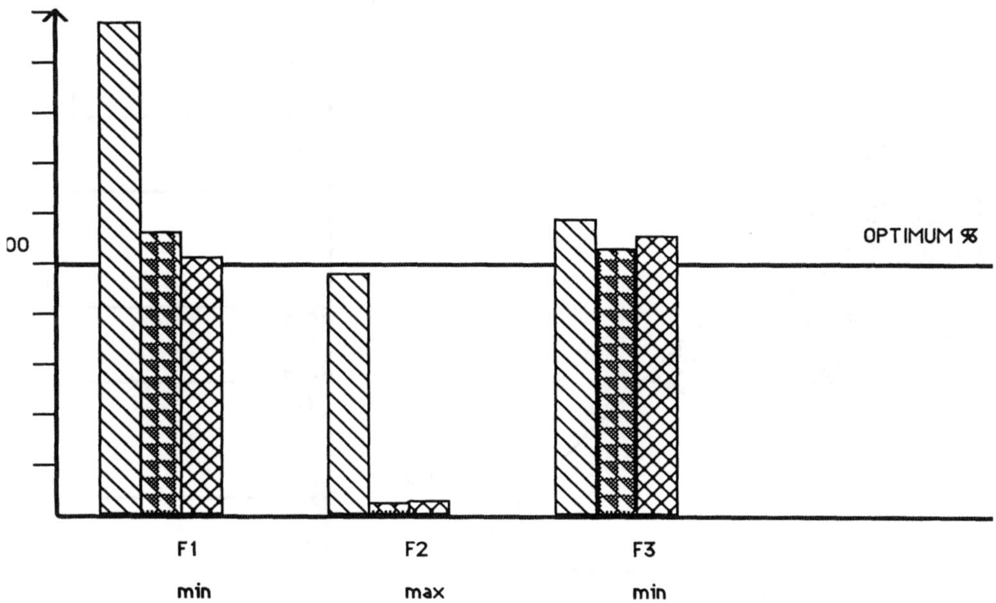

Figure 4

The corresponding expansion plan is (the index refers to the ubperiod):

Nuclear 1	Coal 1	Coal 2	Coal 3
860.6	724.6	783.8	1041.3

By comparison with the results obtained from the study using RIMAP it is possible to know to which zone of the solutions we re being driven to.

3.3. Some Zionts-Wallenius Results

In this paragraph we will present some results of the applica- ion of the Zionts-Wallenius method to our case study.

The first run of this model using the weights $\lambda_1 = \lambda_2 = \lambda_3 = 1/3$ ed to the optimal solution of f_2 (solution A) (fig. 5).

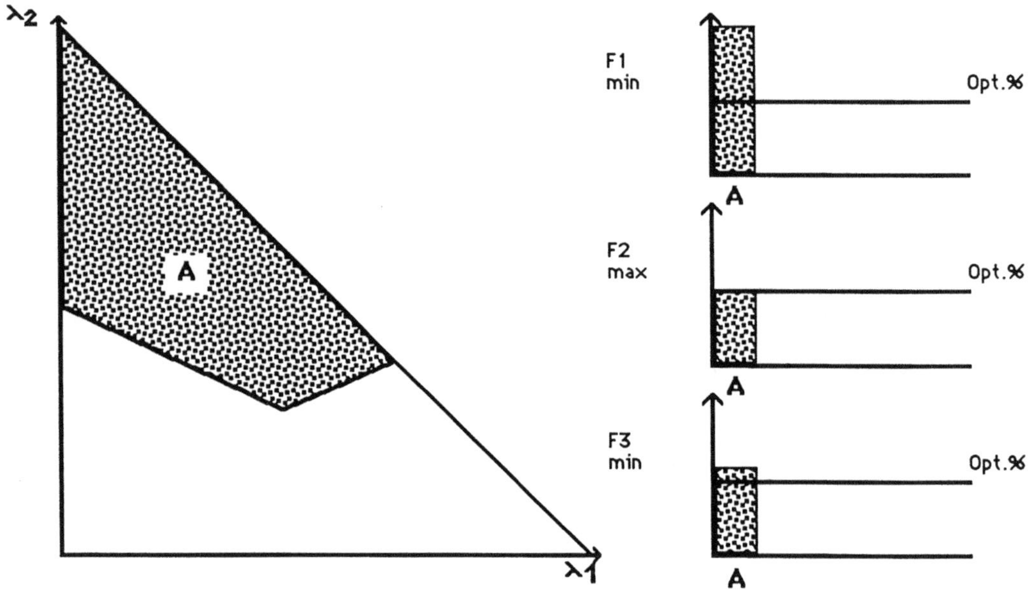

Figure 5

There are two adjacent vertices, but only the one that makes us to move to the lower right part of the triangle is distinct from the present one in the sense indicated in (Steuer, 1986, chapter 13): two adjacent vertices are distinct (meaning they are different enough for a pairwise comparison) when at least for one of the objective functions the corresponding values differ more than 10%.

That adjacent extreme point leads to

$f_1 = 40691197.5$

$f_2 = 9234.1$

$f_3 = 157474.0$

By comparing these values with the ones corresponding to the optimum of f_2 (see table I), we see that preferring this solution to the previous one is just like giving an enormous "leap on the darkness". We will admit two hypotheses:

(a) the DM does not prefer this solution to the optimum of f_2.

The constraint 1 is introduced on the weighting space. Next the DM is confronted with the trade-offs, when moving along the nondominated edges that leave the extreme point which optimises f_2.

The components of the trade-off vectors represent variations of the objective functions values per unit of a non-basic efficient variable chosen for entering the basis. The positive values indicate a decrease in the values of the objective function values.

In fig. 6 the corresponding constraint, supposing the trade-off is accepted, appears signalled by 2. The trade-off values (non-normalised) are:

f_1 : 0.0000
f_2 : 0.9530
f_3 : 8.0113

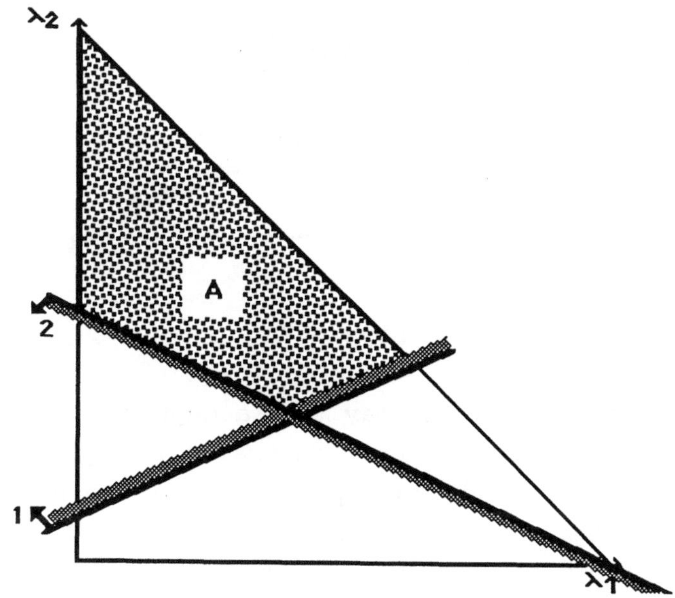

Figure 6

Supposing this trade-off is not accepted, then the constraint 2 is introduced by reversing its direction with respect to fig. 6 (see fig. 7).

Then the DM is asked if he/she is satisfied with the trade-off corresponding to the constraint 3 in fig. 7. The trade-off values (non-normalised) are:

f_1 : 1.0000
f_2 : 0.0012
f_3 : 0.0010

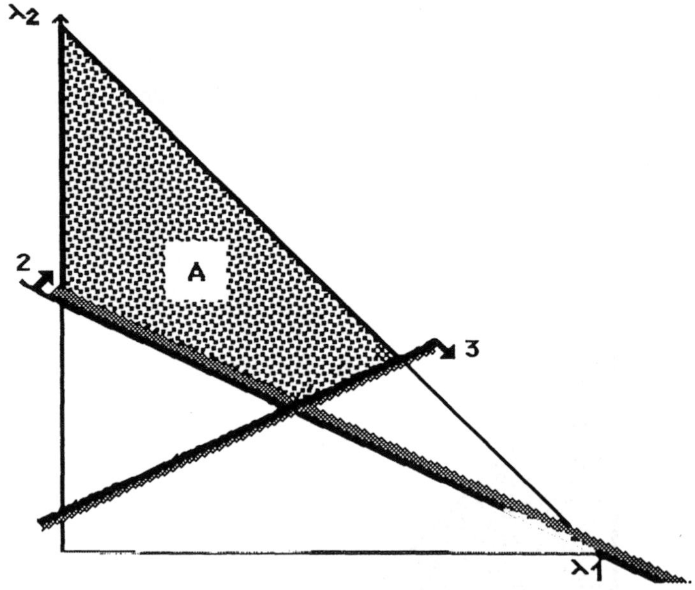

Figure 7

Note that the DM may or may not accept the trade off that leads to the elimination of the constraint of opposite direction previously introduced and its replacement by the new constraint. Let us suppose that the response is positive. Then the method selects a new weighting system which satisfies the constraints signalled in fig. 7, and calculates the corresponding non-dominated vertex (fig. 8).

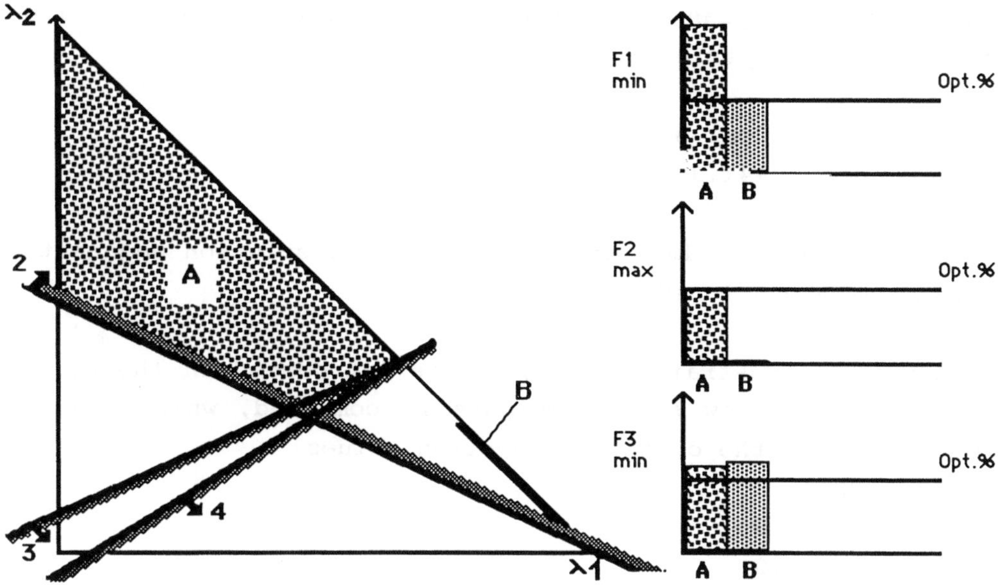

Figure 8

The vertex B (optimum of f_1) is the new candidate to the current solution and corresponds to the following values for the objective functions:

$f_1 = 35953830.8$

$f_2 = 2870.9$

$f_3 = 214535.9$

Let us suppose the DM prefers this solution to solution A. This leads to the introduction of the constraint 4 on the weighting space, reducing even more the scope of the search.

There are no distinct extreme points adjacent to the solution B, in the sense previously mentioned. Note that this will be a common situation inside the signalled area in fig. 3 (where the acceptable solutions mostly lie, as we have seen when we applied the TRIMAP method). In these circumstances, the continuation of the process will be, from this point onwards, many times based on responses regarding the possible trade-offs. So, in many situations it is difficult for the DM to make a decision.

The expansion plan corresponding to solution B is (the index refers to the subperiod):

Nuclear 1	Coal 3
3673.3	441.3

(b) We will now admit the DM prefers the solution adjacent to A, previously mentioned.

Thence the constraint 1 (fig. 9) is introduced in the weighting space and a new set of weights is computed, which leads to solution C with the objective function values:

$f_1 = 37308455.3$
$f_2 = 4653.5$
$f_3 = 154082.0$

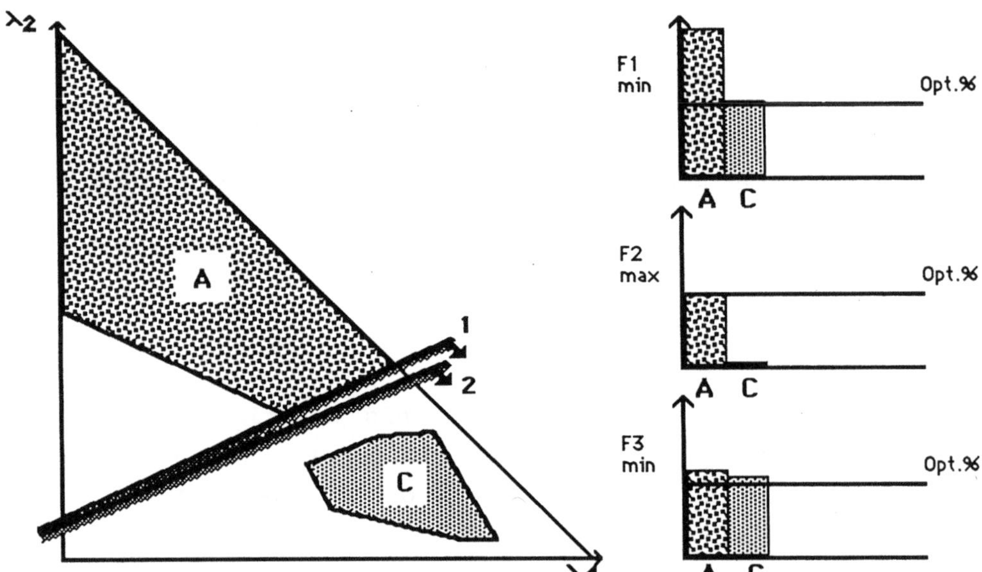

Figure 9

Let us suppose the DM prefers the solution C to the one adjacent to A. The solution C becomes the new current solution and the constraint 2 is introduced. There is only one vertex adjacent to C distinct from this one (fig. 10), whose objective function values are:

$f_1 = 37700264.5$

$f_2 = 5368.7$

$f_3 = 154082.0$

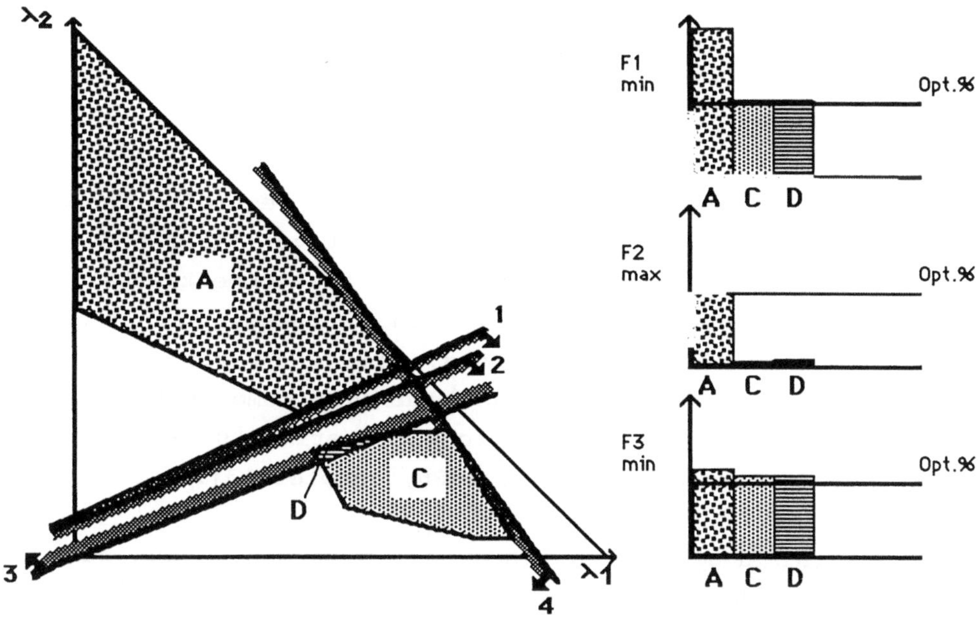

Figure 10

Note that despite these two vertices being distinct (C and D), the corresponding objective function values are not very different. Let us suppose that the DM prefers the new vertex to C. The constraint 3 (fig. 10) is generated and a new set of weights is computed which satisfies the constraints 1, 2 and 3 leading to a new candidate extreme point whose objective function values are:

$f_1 = 37045318.1$

$f_2 = 4712.0$

$f_3 = 184587.0$

If the DM prefers the vertex D to this one, it becomes the new current solution and the constraint 4 is introduced in the weighting space (fig.10). There are no new distinct solutions adjacent to D. So, the search would continue through the answers given by the DM with respect to the trade-offs corresponding to moves along the nondominated edges emanating from vertex D.

The expansion plan corresponding to solution D is (the index refers to the subperiod):

Coal 1	Coal 2
1370.0	1826.3

4. CONCLUSIONS - COMPARISON OF THE METHODS

Next we will put forward some conclusions regarding the experiments of application of the three interactive linear programming methods previously described. In general the comments that follow confirm what is stated in (clímaco and Antunes, 1987 and 1989; Steuer, 1986; Vincke, 1982; Zionts, 1989). As far as STEM and Zionts-Wallenius methods are concerned only some comments, which became evident from the computational experiments, are referred. For further details see (Steuer, 1986, chapter 13).

(a) The TRIMAP method is based on a progressive and selective "learning" of the set of Pareto optimal solutions. It is not intended to "converge" on any "best" compromise solution, but to help the DM eliminating the sub-set of Pareto optimal solutions in which he/she is not interested at all. There are no irrevocable decisions throughout the interactive process and the DM is always allowed to go "backwards" at a later interaction.

So, in each interaction (when preparing a new computation phase) the DM just gives some indications about the regions where the search for Pareto optimal solutions should be carried out, as well as eventual introduction of additional constraints. This interactive process only ends when the DM considers to have sufficient knowledge about the set of Pareto optimal solutions.

Using Roy's terminology, we may say that in TRIMAP "convergence" is replaced by "creation" and that the interactive process is a "building process" and not the discovery of any pre-existent utility function.

A problem always present in the interactive methods is the limited capacity of the human-being in processing information. The TRIMAP method is dedicated to problems with three objective functions. Although being a limitation, this allows us the use of graphical means which are specially suited for the "dialogue" with the DM. This undoubtedly enlarges the DM's capacity to process the information, by simplifying the dialogue phase.

(b) Using TRIMAP the DM considered that the search for 20 Pareto optimal extreme points gave him/her a satisfactory knowledge of the set of Pareto optimal vertices. It should be noticed that an exhaustive search would lead to much more than 100 Pareto optimal extreme points.

Note that although in the experiment described only nondominated vertices of the admissible polyhedron have been searched, TRIMAP has means to search for others solutions belonging to nondominated faces. By contrast the method of Zionts-Wallenius requires that the compromise solution will always be vertices of the feasible polyhedron.

Once a region of the weighting space we are particularly interested in exploring has been located, a procedure of the type described in (Kornbluth, 1977) could be incorporated in TRIMAP. This is an interactive procedure that starts from a solution calculated by the weighting method and then uses the information contained in the optimal simplex tableau in order to help the DM deciding about the direction (in the neighbourhood of the starting solution) where the search should carry on.

This procedure, although being less rigid than the one proposed in the Zionts-Wallenius method, contradicts the general philosophy of TRIMAP. Here we seek to widen the flexibility of the dialogue with the DM. However, the introduction of this type of procedure (or the use of the Zionts-Wallenius method) may be justified to explore some specific regions of the weighting space. It should be noticed that in many real cases (such as the power systems expansion planning case referred herein) this kind of procedure would probably not be very efficient since there are large regions of the weighting space for which the variation in the objective functions values of the corresponding solutions is very small, but steep variations among those regions occur namely as far as the cost function is concerned.

(c) The experiments carried out with the STEM method confirm its main advantages: simplicity not only computational but also from the point of view of the DM's comprehension. However, it is not always easy for the DM to state which objectives are to be relaxed and by how much.

The method becomes rather rigid if the DM is not allowed to revise decisions regarding previous relaxations (as in the standard STEM). In our case study it is obvious the difficulty for the DM to fix the relaxation quantities due to the sharp variations occurring among certain regions of the feasible polyhedron.

The method becomes more flexible if we give the freedom to relax several objective functions at once and a given function more than once (obviously increasing the computational effort).

By using STEM, the DM can obtain Pareto optimal solutions that are not extreme points of the feasible polyhedron.

The weights can be affected by values in the pay-off table that do not correspond to the maximum (or to the minimum) of the objective functions. For instance, in our case study the minimum of f_2 is 718.99 and the minimum value of f_2 in the pay-off table is 2870.9 (corresponding to the optimisation of f_1).

(d) In the experiments carried out with the Zionts-Wallenius method the DM is confronted from the beginning with a difficult decision, i.e. he/she has to decide upon passing from the maximum of f_2 to the lower right part of the triangle (the variations of some objective functions values being very high). In several runs of the program the DM ended up faced to options within the mentioned region of fig. 3. Within this region the variations of the objective functions are smooth, reason why the local information used by the Zionts-Wallenius method may turn difficult for the DM to make decisions in conformity with his/her implicit utility function.

As pointed out by Steuer (1986) the DM may sometimes feel uncomfortable responding to the trade-offs. The computational experiment with the Zionts-Wallenius method also shows the rapid reduction in the region of the weighting space where generally the search in later iterations takes place.

ACKNOWLEDGEMENTS

The authors would like to express their gratitude to their colleague J. Sá Marta who made the implementation on Macintosh II of the STEM and Zionts-Wallenius methods.

Note: All figures presenting the results are hard-copies of the graphical displays on the screen. The alphanumerical identification of the areas inside the triangles as well as the arrows denoting the direction of the constraints on the weighting space have been added for clarity of presentation.

REFERENCES

Clímaco, J. and Almeida, A. (1981), "Multiobjective power systems generation planning", *Proc. of the Third International Conference on Energy Use Management*, Pergamon Press.

Clímaco, J. and Antunes, C.H. (1987), "TRIMAP - An interactive tricriteria linear programming package", *Foundations of Control Engineering*, vol.12, 3, 101-119.

Clímaco, J., Antunes, C.H., Martins, A., Sá Marta, J. and Almeida, A. (1988), "A novel approach to power generation expansion planning using TRIMAP", paper presented at the *VIII International Conference on MCDM*, Manchester.

Clímaco, J. and Antunes, C.H. (1989), "Implementation of a user friendly software package - a guided tour of TRIMAP", *Mathematical and Computer Modelling*, vol.12, 10/11, 1299-1309.

Kornbluth, J. (1977), "The fuzzy dual: information for the multiple objective decision-Maker", *Computers and Operations Research*, vol.4, 65-72.

Reeves, G. and Franz, L. (1985), "A simplified interactive multiple objective linear procedure", *Computers and Operations Research*, vol.12, 589-601.

Roy, B. (1985), "Meaning and Validity of Interactive Procedures as Tools for Decision Making", *Cahier du Lamsade nº62*, Université de Paris-Dauphine.

Steuer, R.E. (1986), *Multiple Criteria Optimization: Theory, Computation and Application*, Wiley.

Vincke, Ph. (1982), "Présentation et Analyse de Neuf Methodes Multicritères Interactives", *Cahier du LAMSADE nº42*, Université de Paris-Dauphine.

Zionts, S. (1989), "Multicriteria mathematical programming: an update overview and several approaches", in B. Karpak and S. Zionts (eds.), *Multiple Criteria Decision Making and Risk Analysis Using Microcomputers*, NATO ASI Series F, vol.56, Springer-Verlag.

THE MULTIOBJECTIVE LINEAR PROGRAMMING DECISION SUPPORT SYSTEM VIG AND ITS APPLICATIONS

Pekka Korhonen

Helsinki School of Economics,
Runeberginkatu 14-16,
00100 Helsinki, FINLAND

ABSTRACT

We will consider a Multiple Criteria Decision Support System-VIG (**V**isual **I**nteractive **G**oal Programming) developed by Korhonen. VIG is designed to support both modelling and solving multiple objective linear programming problems. The interface is based on one main menu, spreadsheets, and interactive use of computer graphics. The corner point VIG is PARETO RACE, which enables the decision maker to freely search nondominated solutions in a dynamic way. VIG provides the decision-maker with the possibility to approach his/her decision problem in an evolutionary way. This means that the decision-maker does not have to specify the model precisely prior to solving the problem. In fact, the model evolves progressively. We also discuss applications of VIG to practical problems.

Keywords: Decision Support, Computer Graphics, Multiple Criteria.

The research is supported, in part, by grants from the Y. Jahnsson Foundation, the Foundation of the Helsinki School of Economics, and the Foundation of the Student Union of the Helsinki School of Economics.

1. INTRODUCTION

In this paper we describe the principles of VIG (**V**isual **I**nteractive **G**oal Programming), a Multiple Criteria Decision Support System developed by Korhonen (1987a). The system is user-friendly and it is designed to support both the modelling and solving of a multiple objective linear programming problem. Menus, spreadsheets, and interactive use of computer graphics play a central role. Constraints and objectives are treated uniformly; their role may be changed during a session. Such features make it possible to apply an "evolutionary approach" to solving linear decision models (Korhonen, Narula, and Wallenius, 1988). In the evolutionary approach, the decision-maker (DM) does not have to specify the role of the rows of his/her models precisely prior to solving the problem. Even the model may evolve progressively. The search for the most preferred solution takes place on the efficient frontier specified by the set of current objectives.

VIG is written in TURBO PASCAL and implemented on an IBM PC/1 microcomputer. To benefit fully from VIG, a color monitor should be used. In the system, PARETO RACE, a visual, dynamic, search procedure for exploring the efficient frontier of a multiple objective linear programming problem, plays a central role (Korhonen and Wallenius, 1988). In PARETO RACE the user sees the objective function values (flexible goals) on a display in numeric form and as bar graphs, as he travels along the efficient frontier. The keyboard controls include an accelerator, gears, brakes, and a steering mechanism.

The foundations of PARETO RACE originate in the visual interactive reference direction approach to multiple objective linear programming developed by Korhonen and Laakso (1986a and 1986b). In the original procedure, using a reference direction, a subset of efficient solutions (an efficient curve) is generated and presented for the DM's evaluation. The interface is based on a graphical representation, but it is static by nature. One picture is produced for each iteration. Therefore, the DM may partly feel oneself at the system's mercy. PARETO RACE improves upon this procedure by making it dynamic. It makes the DM feel that the system is totally under his control. The DM can move in any direction (on the efficient frontier) he likes and no unduly restrictive assumptions concerning the DM's behavior are made. Based on our experience to date, PARETO RACE is able to deal with problems involving a relatively large number of variables, constraints and objectives without any difficulty.

This paper consists of six sections. The introduction explains the underlying philosophy of VIG. The second section provides the problem formulation. The third section describes the structure of VIG. The fourth section discusses PARETO RACE. In the fifth section several applications are described. The sixth section concludes the paper.

2. A LINEAR DECISION MODEL

Let us consider the following problem, where the consequences (outcomes) y_i, i=1,...,m, of decisions (activities, actions, choices) can be stated as linear functions of the decision variables x_j, j=1,...,n:

$$y_i = y_i(x) = \sum_{j=1}^{n} a_{ij}x_j, \quad i \in M = \{1,2,\ldots,m\},$$

or equivalently in the matrix form:

$$\mathbf{y} = \mathbf{y}(\mathbf{x}) = \mathbf{A}\mathbf{x},$$

where **x** is an n-vector of decision variables, **A** is an m x n matrix of coefficients, and **y** is an m-vector of consequences or outcome variables. In the subsequent discussion, the vector **y** may include the vector **x**, if the DM imposes restrictions upon it or has preferences about the values of the decision variables. The problem is to find values for the decision variables x_j, $j \in N = \{1,2,\ldots,n\}$, such that the outcome variables, y_i, $i \in M$, would have acceptable or desirable values. If $n \geq m$, then for each desired of given value of **y**, there exist an infinite number of solutions for the model, and it can easily be solved. To avoid this trivial case, we assume that $m > n$.

Various procedures have been developed for solving the problem. These include linear programming, fuzzy linear programming, multiple objective linear programming, and "what-if"-analysis. Each of these approaches imposes certain limitations on the model that may result in misrepresenting real-world decision situations.

In linear programming, one of the outcome variables is chosen as an objective function to be maximized or minimized over a feasible set. The feasible set is defined by specifying acceptable values for the other outcome variables. As it is well known, it is possible to find an optimal solution (if one exists) for a linear programming problem. Implicitly, in linear programming we assume that the DM's preference structure (or the value function) over the feasible set can be adequately described by means of a single objective function. Furthermore, although any boundary point is acceptable, a point that lies just outside the feasible region is not. Thus, in linear programming there is a very clear conceptual difference between an objective function and the constraints. Duality theory and postoptimality analyses can be used to study the neighbourhood of the "optimal" solution, but such techniques may not provide enough flexibility for the DM.

In fuzzy linear programming, the DM is not assumed to minimize or maximize an objective function (Zimmermann, 1985). Instead, he/she is assumed to want to reach some "fuzzy" aspiration levels. In fuzzy linear programming, membership functions are used to describe the violation of constraints. Each membership function has value 1, if the constraint is fully satisfied, and 0, if the maximum tolerance is exceeded. The simplest type of membership functions are linear, but several other forms of membership functions and aggregation rules are also possible (see, for example, Sakawa, 1983). However, after one has specified such functions and aggregation rules, the solution procedure is straightforward and a unique solution is found without the DM's intervention. Some methods have recently been developed, which allow the DM to analyze the effect of "fuzziness" on the final solution (Chanas, 1983, Carlsson and Korhonen, 1986). To sum up, usually in fuzzy linear programming, the boundary of the feasible region is "soft", whereas the final solution is "hard".

In multiple objective linear programming, the DM selects a number of outcome variables as objective functions and attempts to maximize (or minimize) them, simultaneously. Since this problem has rarely, if ever, a unique solution, the DM has to choose a solution from among the set of nondominated (efficient) solutions (see, Steuer, 1986). As in linear programming, the DM has to make a sharp distinction between the constraints and the objective functions at the beginning of the solution process. Also, all preference information is assumed to be contained in the objective functions. However, sometimes seeing the solutions makes the DM realize that it is not satisfactory. Thus, despite the advanced state of research, even a multiple objective approach may not always be flexible enough in practice.

The DM may also experience difficulties in specifying what he/she wants and what it is possible to achieve. Using "what-if" -analysis, the DM can experiment with different values of the decision variables and consider the values of the outcome variables without specifying explicit bounds for them. Alternate solutions are generated until the DM is satisfied. However, "what-if"-analysis does not guarantee that the most preferred solution will be found. In fact, the final solution may be dominated.

To overcome some of the above difficulties, Korhonen et al. (1988) have recently proposed that an evolutionary approach be used to solving the problem. Briefly, in the evolutionary approach the modelling and solving phases are not separated as it is usually done, but instead are considered simultaneously.

Also, there is initially no conceptual difference between objective functions and constraints, because their roles can be changed during the solution process. The unified formulation for objective functions and constraints can be written as follows (see, Korhonen and Wallenius, 1988):

$$\begin{aligned} &\min \epsilon \\ &\text{subject to:} \\ &C\mathbf{x} + \mathbf{w}\epsilon \geq \mathbf{b}, \\ &\mathbf{x} \geq \mathbf{0}, \\ &\epsilon \text{ unrestricted}, \end{aligned} \tag{2.1}$$

where $\mathbf{w}$ is an k-vector whose components are

$$w_i \begin{cases} = 0, & \text{if } i \text{ refers to a constraint,} \\ > 0, & \text{if } i \text{ refers to an objective.} \end{cases}$$

C is a k x n matrix of coefficients, $\mathbf{b}$ is an m-vector of aspiration levels, and ϵ is a scalar variable. At the optimum of ϵ, the solution vector $\mathbf{x}$ is (at least) weakly efficient (see, e.g., Korhonen and Wallenius, 1988). Note that using the notation in section 1:

$$A = \begin{bmatrix} C \\ I \end{bmatrix} .$$

To move about on the efficient frontier, we can vary either the vector $\mathbf{w}$ or the aspiration levels $\mathbf{b}$. If only the components of $\mathbf{b}$ corresponding to the objective function rows are varied, then we move on the efficient frontier and the constraint set remains unchanged. Of course, we can also change the components of $\mathbf{b}$ corresponding to the other rows, thus changing the constraint set. Actually, then we have a classical parametric programming problem. Note that the values of the $\mathbf{w}$-vector specify, which rows are objectives and which constraints. By making changes in $\mathbf{w}$, we can easily change the role of objectives and constraints.

3. THE STRUCTURE OF THE SYSTEM

VIG is implemented in the spirit of visual interaction. Following the spirit of goal programming, constraints are regarded as the subset of goals (Ignizio, 1983). In fact, constraints are inflexible goals. VIG is described in more detail in the user's guide (Korhonen, 1987b).

To illustrate the use of the system, we consider a production planning problem, where the DM tries to find the "best" product-mix for three products: Product 1, Product 2, and Product 3. The production of these products requires the use of one machine (Mach.Hours), man-power (Man Hours), and two critical materials (Crit.Mat 1 and Crit.Mat 2). Selling the products results in profit. The A-matrix of the problem is given in Table 1:

	Product 1	Product 2	Product 3
Mach.Hours	1.5	1.0	1.6
Man Hours	1.0	2.0	1.0
Crit.Mat 1	9.0	19.5	7.5
Crit.Mat 2	7.0	20.0	9.0
Profit	4.0	5.0	3.0
Product 1	1.0	0.0	0.0
Product 2	0.0	1.0	0.0
Product 3	0.0	0.0	1.0

TABLE 1: The Coefficient Matrix of the Model

Suppose that the DM has difficulties in clearly specifying what he wants to achieve and what the bounds for the (inflexible) goals might be. For example, the DM may state:

"I would like to make as much profit as possible, but because it is difficult to obtain critical materials, I would like to use them as little as possible. One machine is used to produce the products. The machine operates without any problems for at least 9 hours, but it is very likely to break down if it is used for more than 12 hours. The length of a normal working day is 10 hours. People are willing to work overtime, but it is costly and they are tired the next day. Therefore, if possible, I would like to avoid it. Finally, product 3 is very important to a major customer, and I cannot totally exclude it from the production plan".

The preceding decision problem is typically semistructured. Although the coefficients of the model are assumed to be determined precisely, it is not quite clear what the DM would like to have. In the DSS literature (see, e.g., Sprague and Carlson, 1982), "what-if"-analysis has been proposed to solving such problems. However, the "what-if"-approach does not provide enough support for finding the best solution.

Throughout this paper we use this example to illustrate the use of our system.

VIG has four main functions:

(1) **Model Management Function**
(Allowing us to create, edit, store, and retrieve models.)

(2) **Model Development Function**
(Allowing us to provide the decision variables and outcome variables (goals) with names, specify the technology matrix describing the relationships among the variables, and determine aspiration levels for the goals.)

(3) **Problem Solving Function**
(Allowing us to tell the system, how the problem should be solved - and actually to solve it. With more than two objectives (flexible goals), PARETO RACE is invoked. With one or no objectives, no user's intervention is necessary.)

(4) **Solution Output Function**
(Allowing us to examine solutions on the screen, and store intermediate results for later consideration or for writing reports.)

Each main function consists of one or more sub-functions, which are all chosen from the main menu. At each stage, the available choices are shown using light cyan. All functions are not available all the time. For example, one cannot examine a solution before it is computed. Using only one main menu makes it easy to keep the control of the program in the user's hands. The user is welcome to choose anyone of the available functions without "having to run through the entire list". When a function is terminated (using the F10-key), the system prints the main menu. By moving the cursor up and down, we can trace all available choices.

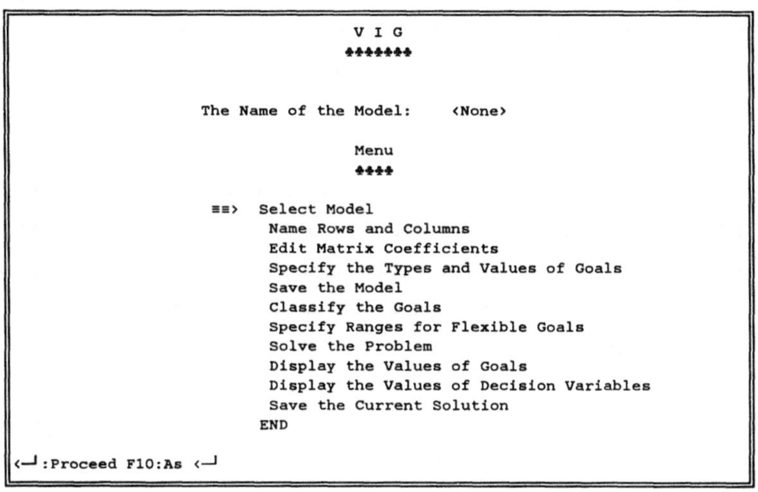

Figure 1. The Main Menu

Model Management Function

"Select Model" and "Save the Model" activate Model Management Functions. When we enter "Select Model", the system prompts us to provide the model with a name. If we are creating a model, the system uses "Temp" as a default name. If we wish to save the model, another name should be used. "Temp" will be erased at the end of the session. By entering "Save the Model" we store the current version of the model into the default directory (diskette). We can rename it or use the old name.

Model Development Function

The model is entered in three phases. Each phase has its own spreadsheet format. Using arrow keys, we can move from one field to another. The use of any other key activates the field, and makes it ready for our input.

Initially, we are asked to provide the decision variables (columns) and goals (rows) with names. At this stage, we are urged to consider the question: "What are my options and what are their consequences?", without incorporating numbers. When creating a model, only the frames of the names are provided.

Next, we enter "Edit Matrix Coefficients" and specify the technology matrix of the problem. In other words, we specify the dependencies of the goals on the decision variables. At this stage, the relevant question is: "How will my decisions affect the values of the goals?"

Finally, we define aspiration levels for the goals (the right-hand-side vector) (Figure 2: Given Values). We also define the directions of the inequalities (Types). The direction indicates, whether we wish to exceed the aspiration levels or fall below them. For example, the "greater than or equal" symbol associated with Profit indicates that we wish to maximize the increase in profit. Similarly, the "less than or equal" symbol of critical materials indicate minimization of the use of materials.

Note that up to this point we have not made any distinction between objectives and constraints. We have treated them in a uniform manner.

Editing the Types and Aspiration Levels of Goals
++

Names	Types	Given Values	Current Values
Mach.Hours	<=	9.0000000	0.0000000
Man Hours	<=	10.000000	0.0000000
Crit.Mat 1	<=	96.000000	0.0000000
Crit.Mat 2	<=	96.000000	0.0000000
Product 1	>=	0.0000000	0.0000000
Product 2	>=	0.0000000	0.0000000
Product 3	>=	0.0000000	0.0000000
Profit	>=	28.000000	0.0000000

F10:Exit to Menu

Figure 2. Aspiration Levels

Problem Solving Function

This function consists of three phases: Classifying the Goals, specifying Ranges for Flexible Goals, and Solving the Problem. Initially, all goals are assumed to be inflexible (that is, constraints). Moving the cursor to point to a row name and pressing SPACE BAR once changes the status of the goal (from inflexible to flexible, and vice versa). On the screen, brown stands for inflexible and white for flexible goals.

If none of the goals is specified to be flexible, the program will find a feasible solution, if one exists. If one of the goals is specified flexible, the program solves a linear programming problem after we have entered "Solve the Problem". This is done without user's intervention. If more than one goal is flexible, the system invokes PARETO RACE. PARETO RACE needs the user to specify approximate ranges for the flexible goals (Figure 3). In our example, initially we have three flexible goals. Mach.Hours is (initially) allowed to vary from 8 to 13, Man Hour from 9 to 14, and profit from 25 to 35.

Specifying (Roughly) the Bounds for Goals

Names	Types	Given Values	Current Values	Lower Bounds	Upper Bounds
Mach.Hours	<=	9.0000000	0.0000000	8.0000000	13.000000
Man Hours	<=	10.000000	0.0000000	9.0000000	14.000000
Profit	>=	28.000000	0.0000000	25.000000	35.000000

F10:Exit to Menu

Figure 3. Bounds for Flexible Goals

Ranges for flexible goals are needed for defining the lengths of the bars on the screen. To obtain a maximum benefit from the visual feature of PARETO RACE, we recommend careful consideration of the ranges (even though they are automatically updated by the system). Sometimes it is advisable to respecify the ranges during the race. In the use of the program, experience is a good teacher.

By entering "Solve the Problem", the PARETO RACE screen is displayed (Figure 4). Since PARETO RACE is a corner-stone of our system, it is explained more in detail in the next section.

Solution Output Function

After using PARETO RACE, the choices "Display the Values of Goals", "Display the Values of Decision Variables", and "Save the Solution" become available. They enable us to examine the values of all flexible and inflexible goals, and the values of nonzero decision variables. We can save intermediate solutions on a diskette or a hard disc. It is possible to store all solutions in the same file. Afterwards we may examine solutions using a text processing system or a text editor.

Having obtained one solution, we can go back and edit the model, change the set of flexible goals, or respecify the ranges for the flexible goals for a better visual effect. We can also return to PARETO RACE. The race starts from the current solution, if we have not edited the model in the meantime.

4. PARETO RACE

In PARETO RACE, we can freely search the efficient frontier of a multiple objective linear programming problem. Specific keys are used to control the speed and direction of motion. On a display, we see the objective function values in numeric form and as bar graphs whose lengths are dynamically changing as we move about on the efficient frontier. The keyboard controls include the following function keys (see, Figure 4 and Figure 5):

(SPACE) BAR: An "Accelerator"
We proceed in the current direction at constant speed.

F1: "Gears (Backward)"
Increase speed in the backward direction.

F2: "Gears (Forward)"
Increase speed in the forward direction.

F3: "Fix the Aspiration Level of a Specific Goal"
The current value of a specific goal is taken as an absolute lower (upper) bound.

F4: "Relax the Aspiration Level of a Fixed Goal"
The fixed goal becomes flexible again.

F5: "Brakes"
Reduce speed.

Num: "Turn"
Change the direction of motion by pressing the number key corresponding to the goal's ordinal number once or several times.

F10: "EXIT"
Exit to the main menu.

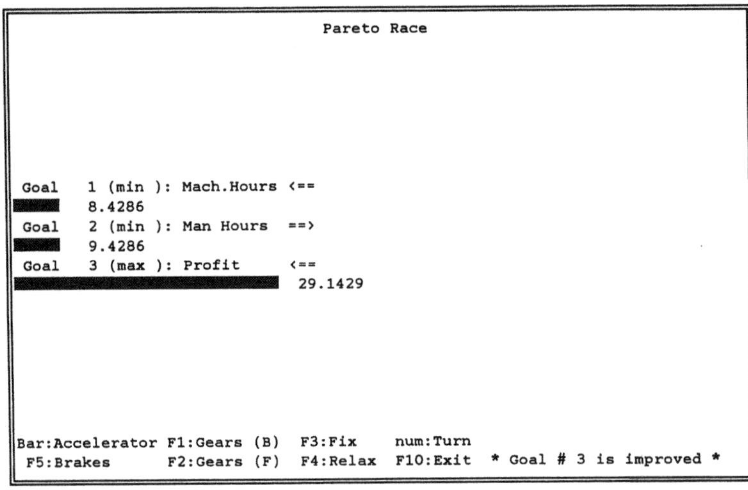

Figure 4. An Example of PARETO RACE: Initial Solution (making a Turn)

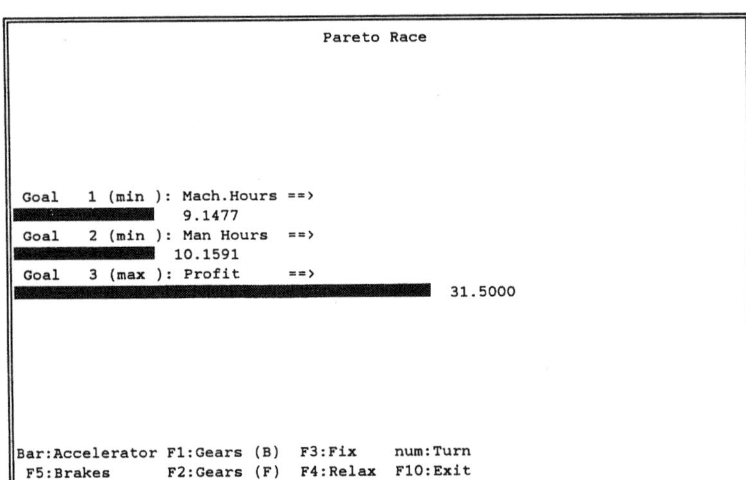

Figure 5. An Example of Another Solution of PARETO RACE

Initially, we are going to see (graphically and numerically) an efficient solution on the computer screen (Figure 4). Arrows indicate an initial direction chosen by the computer (based on our aspiration levels). If we like this initial direction, we hold SPACE BAR down and observe the solutions change. We are "travelling" at base speed. If we like to increase the speed, we press the F2-key once or several times (depending on the desired speed) and hold SPACE BAR down. If at some point the direction is no longer attractive, we initiate a turn (see, Figure 4). Assume that the user wishes to improve a certain goal. To accomplish this, the goal's corresponding number key is pressed once or several times, depending on the amount of desired improvement. Then, the program updates the direction, and so forth. To reduce speed, we use brakes (F5-key). If we wish to resume base speed, the most convenient way is to press the F1-key (gears: backward) and then the F2-key (gears: forward). We reverse direction twice and start travelling at base speed. When a flexible goal is fixed by using the F3-key, the system asks us to specify the number of the goal to be fixed; then an asterisk appears next to the name of such a goal. The goal may be relaxed by using the F4-key; the asterisk disappears.

When we terminate PARETO RACE, we are welcome to examine the values of the decision variables and goals. Just exit to the main menu. In fact, the user can examine the values of the decision variables and goals also during the race (and save such intermediate results for later use). Simply enter "Solve the Problem" in the main menu and we are back in the race. We can also change the role of flexible and rigid goals during the race. We feel that this adds flexibility to the system and extends it beyond "classical" multiple objective linear programming.

5. APPLICATIONS

VIG has been and is being applied to solving several practical problems. In this section we briefly survey a number of such applications. Additional details may be found in the original references. Up till now, most of the problems solved have been relatively small-scale (with respect to the number of rows and columns), but there clearly exists a need (and the potential) for solving large-scale problems using VIG.

One of the earliest applications deals with the determination of optimal price changes for alcoholic beverages sold by ALKO Ltd., the Finnish State Alcohol Monopoly. Pricing decisions are among the most important ones the company must make. The objective is not only to maximize profits, but to restrict sales with the intention of reducing harmful effects of alcohol consumption, and to minimize the impact of price increases on the consumer price index. A linear programming model three objective functions (reflecting the above concerns) was developed and solved using our procedure. The decision variables of the model were logarithms of relative price changes. Modifications in the input-output routines of the program enabled the user to operate with original percentages instead of their logarithms. Based on this model, the company has implemented an extended version of the model. The ultimate DM, the governmental ministers comprising the cabinet, is interested in the model. Additional details may be found in Korhonen and Soismaa (1988).

Another interesting and important application concerns the problem of stockpiling critical materials to be used in the event of a national emergency in Finland. It is essentially a bi-criteria problem. On the one hand, the purpose is to minimize the cost of stockpiling, and on the other hand to maximize the welfare of the population at large during (and after) a crisis. Using the model it is possible to make intelligent decisions regarding appropriate stockpiling levels for critical materials, and understand the implications in case of a crisis. The national Board of Economic Defense has presented a document on this project to the Ministry of Trade and Industry, and the system is being used by the board on a regular basis. because of the critical nature of this project, no scientific articles have been published on the system, but the system and its merits have been summarized in some Finnish newspapers.

In Kananen et al. (1988) we demonstrate how VIG can effectively be used for analyzing input-output models. We have applied our approach to studying the effects of economic or political crises on the Finnish economy. Examples of such crises are nuclear power plant accidents, trade embargoes, and international conflicts of various nature. An input-output model, based on the latest official input-output statistics of the Finnish economy with 17 industries (sectors) is employed. relevant objective functions include maximization of (private) consumption, employment, and total gross outlay. Our system has been implemented on a microcomputer and is currently being used on a regular basis by the National Board of Economic Defense. For large-scale problems, such as inter-temporal, regional, input-output models, a large-scale version of VIG should be used.

An interesting first application of VIG is being performed for a major Finnish wholesaler of hardware and related products. The problem concerns a division of a company that has several departments or units. The analysis starts from the income statements of each department. Each of these income statements forms a column in our model. Some of the items (rows) in the income statements (for example, profits, turnover, etc.) are defined as objective functions (flexible goals). Using VIG the company management can decide, what they want to achieve (division-wise) in terms of profits, turnover, and so forth. Then, they can find out, which departments are instrumental (critical) in achieving these overall goals, and which departments need strengthening in terms of managerial skills, capital, and labour. Interestingly, our system identified certain weak departments in the division (confirming the suspicions of management) and identified some strategic opportunities for other departments. Based on this study, the company has reached a decision to start using VIG at management level in the future. The results will be forthcoming as a future report.

We have applied VIG to selecting advertising media for a Finnish software company. The problem was to assist the management in allocating an advertising budget across various media. The purpose was to maximize audience exposure. Six different newspapers and professional magazines were initially chosen as the relevant advertising media. The relevant audience consisted of the following (target) groups: Marketing Management, Finance and personnel, ADP, Production Management, R&D, and General Management. To measure the audience exposure, readership numbers were obtained from Finnish Gallup Ltd., reflecting the current situation. The media selection problem was modelled as a linear decision problem in the spirit of VIG. The decision variables were the numbers of ads in each medium, which were also considered as consequences in our model. For additional details, an interested reader may consult Korhonen et al. (1988).

We would also like to mention the work done by Joen Forest Consulting, a Finnish consulting company specialized in forest sector management models. They are using VIG as an important part of their system for forest planning, harvesting, and replanting. The system is being used by more than half a dozen forest sector (vocational) educational institutions in Finland. It has also been used for a forest project in Zambia. (see Pukkala et al., 1989.)

The management of NumPlan, Ltd., a small Finnish software company that markets the VIG software has also used VIG in planning its marketing strategy. Alternative strategies were identified and used in a multiple objective linear programming model involving qualitative data. Based on the analysis and use of the VIG program, several strategies appeared which have substantially changed our approach to marketing VIG. See Korhonen and Wallenius (1989a).

We also cite several studies that have been made, that have not yet resulted in concrete applications.

The waters of the New York Bight are among the most intensively utilized in the world. Concern for water quality in this region is long standing. Yet, sewage sludge has been dumped at the so-called 12-mile site for more than 60 years. Now the communities using the site are shifting their operations to the more distance 106-mile site, following orders issued in 1985 by the U.S. Environmental Protection Agency. In a recent study, Wallenius et al. (1987) used VIG to re-examine the EPA decision in a way which permits simultaneous multi-site dumping. The study is based on a linear programming model, providing a framework for ocean waste disposal management. The relevant objective functions were to minimize transportation costs to New York City, and communities in New Jersey and on Long Island, and to minimize the volume of pollutants in near-shore and off-shore areas. The decision variables in the model are the number of barge trips made from source i to site j by disposal method k. The model consists of 40 decision variables and 27 constraints (inflexible goals) imposing environmental and capacity restrictions.

Duckstein et al. (1988) describe the use of VIG in forest management. More specifically, they examine the case of managing a ponderosa pine forest in the Salt-Verde river basin, situated in North-Central Arizona. The problem is to allocate to given subwatersheds six different treatment strategies corresponding to clearcutting between 0 and 100% of the area. This allocation is evaluated using multiple, conflicting objective functions, such as water yield, sediment yield, amount of recreational usage, economic benefits/costs. The problem has a number of physical constraints, some of which are fuzzy. For additional details, see Duckstein et al. (1988).

6. CONCLUSION

In this paper we have described a software package called VIG. It is essentially a Multiple Criteria Decision Support System for modelling and solving linear programming problems. The computer program is written in TURBO PASCAL and it implements PARETO RACE as an essential ingredient. The necessary hardware consists of an IBM compatible Personal Computer with at least one diskette drive and a graphics card. The minimum memory size is 256 kbytes. Preferably, a color monitor should be used. The current version of the program is capable of solving problems with a maximum of 95 variables and 100 goals, from which at most ten may be flexible, simultaneously. The size of the array may be increased, although part of the dynamic effect is then lost. The program consists of more than 3000 lines of code, most of which is used to build up an attractive interface. One can easily construct or edit one's models with computer graphics playing a central role. VIG was updated in August of 1987, with many desirable features added.

Our objective is to make the program widely available, and pursue several applications.

REFERENCES

Chanas, S. (1983), "The Use of Parametric Programming in Fuzzy Linear Programming", *Fuzzy Sets and Systems*, 11, 243-251.

Carlsson, C. and Korhonen, P. (1986), "A Parametric Approach to Fuzzy Linear Programming", *Fuzzy Sets and Systems*, 20, 17-30.

Duckstein, L., Korhonen, P. and Tecle A. (1988), "Multiobjective Forest Management: A Visual, Interactive, and Fuzzy Approach", in B.M. Kent and L.S. Davis (Eds.): *Proceedings of the 1988 Symposium on Systems Analysis in Forest Resources*, Asilomar Conference Center, Pacific Grove, California, March 29-April 1, 1988.

Ignizio, J.P. (1983), "Generalized Goal Programming", *Computers and Operations Research* 10, 277-289.

Kananen, I., Korhonen, P., Wallenius, H. and Wallenius, J. (1988), "A Multiple Objective Approach to Analyzing Input-Output Models, with an Application to Economic Defense", (unpublished manuscript).

Korhonen, P., and Laakso, J. (1986a), "A Visual Interactive Method for Solving the Multiple Criteria Problem", *European Journal of Operational Research*, 24(2), 277-287.

Korhonen, P. and Laakso, J. (1986b), "Solving Generalized Goal Programming problems Using a Visual Interactive Approach", *European Journal of Operational Research*, 26(3), 355-363.

Korhonen, P. (1987a), "VIG - A Visual Interactive Support System for Multiple Criteria Decision Making", *Belgian Journal of Operations Research, Statistics and Computer Science* 27, Nr. 1, 3-15.

Korhonen, P. (1987b), VIG (A Visual Interactive Approach to Goal programming) - user's Guide.

Korhonen, P., Narula, S., and Wallenius, J. (1988), "An Evolutionary Approach to Decision-making, with an Application to media Selection", (unpublished manuscript).

Korhonen, P.and Soismaa, M. (1988), "A Multiple Criteria Model for Pricing Alcoholic Beverages", *European Journal of Operational Research*, 37, 165-175.

Korhonen, P. and Wallenius, J. (1988), "A Pareto race", *Naval Research Logistics*, 35, 615-623.

Korhonen, P. and Wallenius, J. (1989a): "On Using Qualitative Data in Multiple Objective Linear programming", (Forthcoming in *European Journal of Operational Research*).

Pukkala, T., Saramäki, J., and Mubita, O. (1989), "Management Planning System for Tree Plantations. A Case Study for Pinus kesiya in Zambia", Working paper, *The Finnish Forest Research Institute*, Joensuu, Finland.

Sakawa, M. (1983), "Interactive Computer Programs for Fuzzy Linear Programming with Multiple Objectives", *International Journal of Man - Machine Studies*, 18, 489-503.

Sprague, R.H. and Carlson, E.C. (1982), *Building Effective Decision Support Systems*, Prentice-Hall.

Steuer, R. (1986), *Multiple Criteria Optimization: Theory, Computation, and Application*, John Wiley & Sons, New York.

Wallenius, H., Leschine, T.M. and Verdini, W. (1987), "Multiple Criteria Decision Methods in Formulating Marine Pollution Policy: A Comparative Investigation", unpublished manuscript.

Zimmermann, H.-J. (1985), *Fuzzy Set Theory and Its Applications*, Kluwer-Nijhoff Publishing, Boston.

A PERSONAL COMPUTER VERSION OF THE MCDA APPROACH STRANGE

J. Teghem Jr.(*), P. Kunsch(**), C.Delhaye(*), F.Bourgeois(*)

(*) Faculté Polytechnique de Mons,
Rue de Houdain 9, 7000 Mons - BELGIUM
(**) Belgonucléaire,
Rue du Champ de Mars, 25, 1050 Bruxelles - Belgium

1. INTRODUCTION

STRANGE as an original approach has been already described in the literature (Teghem *et al.*, 1986, Teghem and Kunsch, 1988, Slowinski and Teghem, 1988). It is a method designed to treat multiobjective and stochastic linear programming (MOSLP) problem. Some industrial applications has been solved using this method (Teghem and Kunsch, 1985, Kunsch and Teghem, 1987).

The main steps of the method are recalled in section 2 and described in the appendix. Its main features are:

(1) It is based in LP-techniques.
(2) It can manage uncertainties both in the objective functions and in the linear constraints.
(3) It is an interactive method: at each phase, the method proposes a large sample of new possible compromises from an area indicated by the decision-marker (D.M.).

The method has first been implemented on mini-computer as required by the size of the applications (Teghem and Kunsch, 1985, Kunsch and Teghem, 1987). Nevertheless, it has appeared interesting to also have a personal computer version of STRANGE to spread more largely the method. The purpose of this paper is to present this new software applied to a simple example and to demonstrate the capabilities of a friendly user-interface.

In section 3, we introduce the illustrative example, which will support the presentation of the micro-computer version in section 4.

2. SHORT DESCRIPTION OF STRANGE.

The MOSLP problem solved by STRANGE can be written as:

$$\begin{aligned} &\text{minimize} \quad z_k = c_k X \qquad k = 1, \ldots, K \\ &\text{s.t.} \qquad X \in D(X) = \{X \mid T X < b,\ X > 0\} \end{aligned} \tag{1}$$

where c_k and (T,b) are discrete random variables; more precisely:

- the k-th objective depends on S_k different scenarios; let $c_{k\,s_k}$ (s_k=1,...,S_k) be the possible values of c_k and $p_{k\,s_k}$ the corresponding subjective probabilities with

  $$\sum_{s_k=1}^{S_k} p_{k\,s_k} = 1.$$

- some elements of matrix (T,b) are imprecise; let (T_r, b_r) (r=1,...,R) be the possible outcomes of (T,b) and q_r the corresponding probabilities with

  $$\sum_{r=1}^{R} q_r = 1.$$

Three main steps can be distinguished in the method:

α) An **associated deterministic problem** is defined. In this MOLP problem, recourse variables W_r are introduced to measure the violation of the outcome r of the set of constraints and to define an additional objective expressing a global measure of violation of the constraints due to uncertainty.

β) A **first compromise** is determined using the idea of the method STEM (Benayoun *te al.*, 1971); the D.M. does not take part in this stage.

γ) Each time the D.M. is not satisfied with a new compromise, an **interactive phase** is performed. The D.M. is asked to indicate a criterion (an objective and a scenario) to be relaxed and an upper limit for the value of this criterion. STRANGE will explore the consequences of such relaxation: for each possible level of relaxation, a new compromise is proposed and the corresponding values of each criterion are given. The D.M. is provided with a complete graphical presentation of these informations so that he is able to choose the new compromise he prefers.

These steps are mathematically described in the appendix.

3. AN ILLUSTRATIVE EXAMPLE

Let us consider the following illustrative example, in a two dimensional space (other examples can be found in Teghem *et al.*, 1986 and Slowinski and Teghem, 1988).

First objective

$z_1 = 40 + c_1.(x_1,x_2)'$

This objective has two scenarios ($S_1 = 2$):

$s_1 = 1 : c_{11} = (-1,-4)$ with probability $p_{11} = 0.75$
$s_1 = 2 : c_{12} = (-2,-4)$ with probability $P_{12} = 0.25$

Second objective

$z_2 = 20 + c_2.(x_1,x_2)'$

This objective has two scenarios ($S_2 = 2$):

$s_2 = 1 : c_{21} = (-1,2)$ with probability $p_{21} = 0.50$
$s_2 = 2 : c_{22} = (-2,1)$ with probability $P_{22} = 0.50$

Constraints

There are no deterministic constraints and four uncertain constraints with three possible outcomes (R = 3):

$r = 1$ ($q_1 = 0.25$):

$$-2x_1 + 3x_2 \leq 9 \qquad (a.1)$$
$$x_1 + 3x_2 \leq 12 \qquad (b.1)$$
$$2x_1 + x_2 \leq 12 \qquad (c.1)$$
$$x_1 - x_2 \leq 5 \qquad (d.1)$$

$r = 2$ ($q_2 = 0.50$):

$$-3x_1 + 3x_2 \leq 12 \qquad (a.2)$$
$$x_1 + 3x_2 \leq 18 \qquad (b.2)$$
$$2x_1 + x_2 \leq 16 \qquad (c.2)$$
$$x_1 - 2x_2 \leq 4.5 \qquad (d.2)$$

$r = 3$ ($q_3 = 0.25$):

$$-4x_1 + 3x_2 \leq 15 \qquad (a.3)$$
$$x_1 + 3x_2 \leq 24 \qquad (b.3)$$
$$2x_1 + x_2 \leq 20 \qquad (c.3)$$
$$x_1 - 4x_2 \leq 4 \qquad (d.3)$$

The variables are non negative.

This MOSLP problem is presented graphically in Fig.1. The uncertainties in the scenarios can result in a change of the slope for the objective functions. The polyhedron of feasible solutions may also change significantly according to the different outcomes of the constraints.

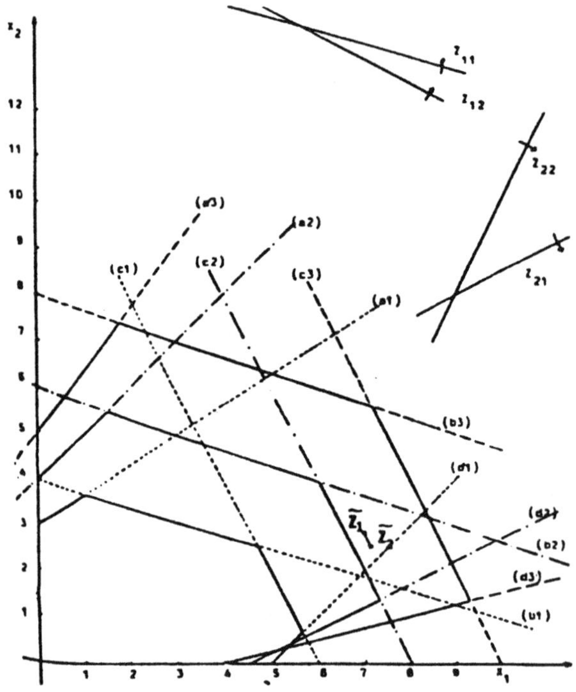

Figure 1: The illustrative example

The PC-version of STRANGE is illustrated in the next section using this example.

4. PC-VERSION OF STRANGE

The programme works on a compatible personal computer equipped with a graphic card IL. Programming language is Turbo Pascal 4.0, using the toolbox "Turbo Graphics" ([1]). Special care has been given to the user-friendliness. The programme is menu-driven and uses spreadsheet techniques. Help frames can be called before and during the execution.

---

1. Turbo Pascal is a registered product of Borland International Inc.

```
Scenarios edition                              Number of scenarios : 2
______________________________________________________________________

                          "MIN" Objective # 2
                             Scenario # 2
                        Probability : 0.500000

     20.00000 + X1  -2.00000 + X2  1.000000

______________________________________________________________________
        View          Add          Edit         Delete          Quit
Edit this scenario
```

Figure 2: Input of the scenario for an objective

```
constraints edition
______________________________________________________________________

                             Scenario # 2
                            Constraint # 3

     16.00000 ≥ X1  2.000000 + X2  1.000000

______________________________________________________________________
        View          Add          Edit         Delete          Quit
Edit this constraint
```

Figure 3: Input of the outcome for a constraint

A) Preparation of the data

The user can prepare a new problem or load an existing data set from disk. The input data of a given problem can be introduced in a sequence of menus and spreadsheets.

Fig. 2 and 3 show of the screens used for inputting the illustrative example.

B) Calculation sequence

The successive steps of the method can be easily followed. Moving forward and backward is possible at each stage. A display of solutions in the form of graphics and/or tables guides the user to a final compromise. We now describe and illustrate the example.

1. The first compromise

The user has the opportunity to analyse separately each LP problem resulting of the choice of an objective k, a scenario s_k and an outcome r. He can display the results (i.e., see appendix, the values of the variables $Y^{(r)}_{k\,s_k}$ and the optimal value the objective function $z_{k\,s_k}(Y^{(r)}_{k\,s_k})$).

The pay-off table (see appendix) can also be displayed (see fig. 4).

The user has of course the option to directly aim for the first compromise $\tilde{Y}^{(m)}$ for m = 1. The illustration is in fig 5.

PAY-OFF-TABLE

CRITERION 11	8.600	10.400	25.333	25.333	24.333
CRITERION 12	6.800	3.200	16.000	16.000	20.800
CRITERION 21	33.000	24.000	13.333	13.333	15.500
CRITERION 22	23.800	11.200	2.667	2.667	9.333
CRITERION 31	11.800	11.000	7.167	7.167	1.000

Menu Continue

Figure 4: The pay-off table

THE FIRST COMPROMISE

The values of the variables are

X1 = 6.932

X2 = 2.985

W1 = 0.000	W2 = 3.936	W3 = 4.949	W4 = 0.000
W5 = 0.000	W6 = 0.000	W7 = 0.949	W8 = 0.000
W9 = 0.000	W10 = 0.000	W11 = 0.000	W12 = 0.000

	Absolute Value	Relative Value
CRITERION 11	21.079	0.746
CRITERION 12	14.097	0.619
CRITERION 21	18.987	0.287
CRITERION 22	9.020	0.301
CRITERION 31	3.696	0.250

	Mean Value	Confidence Level
OBJECTIVE 1	19.334	0.250
OBJECTIVE 2	14.003	0.500

Menu Continue

Figure 5: The first compromise

The values of the variables X and of the recourse variables W_r (r=1,...,R) are shown, and so are, for each criterion k_{s_k}, the absolute and the relative values, the mean value and the confidence level of each objective k. If requested, the different relative values corresponding to this compromise can be displayed as chart bars (see fig. 6).

2. The interactive phases

During the interactive procedure - let us assume that phase m has been completed - all selected compromise $\tilde{Y}^{(l)}$, l=1,...,m, have been stored on disk and the user can come back at any time to each of them. In case none of these compromises is completely satisfactory, a new interactive phase can start.

The user selects on the spreadsheet the compromise to improve and then the candidate criterion for relaxation (fig.7).

Let us assume criterion z_{31} has been chosen. All desirable information regarding this criterion is displayed and the user is requested to fix an upper level for the relaxation (fig.8).

Several displays for the following parametric analysis can be selected as in the menu of fig. 9.

Either:
- the absolute values of the criteria (fig.10), or
- the relative values of the criteria (fig.11), or
- the values of the variables (fig.12);

are presented as a function of the level of relaxation (see relation (3) of the appendix; the values of parameter λ corresponding to a basis change in the dual analysis are indicated by a star on the screen).

A graphic representation of the relative variation of all the criteria is shown in fig. 13.

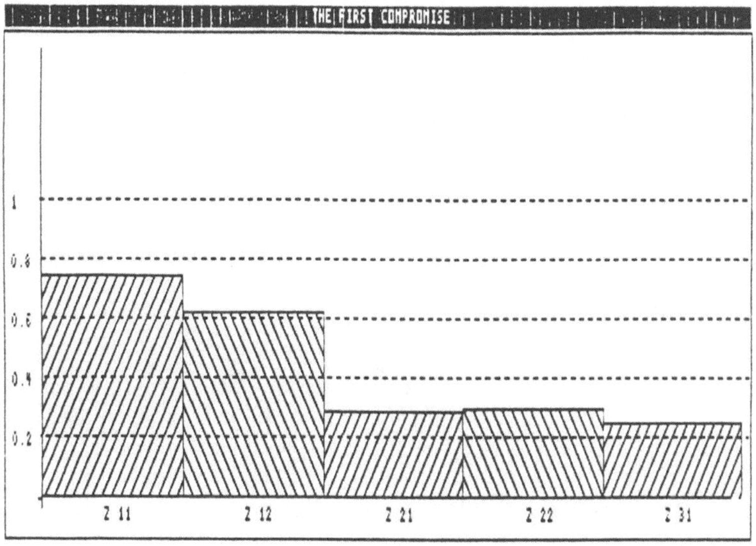

Figure 6: Display of the first compromise

Phase n° 1 Compromise n° 1

1 Relax the Criterion 11

2 Relax the Criterion 12

3 Relax the Criterion 21

4 Relax the Criterion 22

5 Relax the Criterion 31

6 Quit

Figure 7: Selection of the criterion for relaxation

CRITERION 31

Absolute Value : 3.696

Variability Range : [1.000,11.800]

Relative Value : 0.250

Fix an upper level of relaxation included in [3.696,11.800] : 11.8

Your choice in relative value : 1.000

Is it OK ?   Yes   No

Figure 8: Choice of the upper level for relaxation

Description of an Interactive Phase

1 Table of the Absolute Values of the Criteria

2 Table of the Relative Values of the Criteria

3 Table of the Values of the Variables

4 Graphics

5 Visualisation

6 Fix the Final Value of the Parameter

7 Abort this Phase

Figure 9: Menu for the parametric analysis

ABSOLUTE VALUES OF THE CRITERIA

LAMBDA	Z 11	Z 12	Z 21	Z 22	Z 31
*0.250	21.079	14.097	18.987	9.020	3.696
0.300	20.553	13.321	18.876	8.591	4.236
0.350	20.026	12.546	18.766	8.162	4.776
0.400	19.500	11.770	18.655	7.733	5.316
*0.475	18.703	10.595	18.487	7.082	6.134
*0.537	18.250	9.929	18.393	6.714	6.794
*0.553	18.147	9.776	18.371	6.629	6.974
0.603	17.849	9.338	18.308	6.387	7.514
0.653	17.552	8.899	18.246	6.144	8.054
0.703	17.254	8.461	18.183	5.902	8.594
0.753	16.957	8.023	18.121	5.659	9.134
0.803	16.659	7.584	18.058	5.417	9.674
0.853	16.362	7.146	17.996	5.174	10.214
0.903	16.064	6.708	17.933	4.932	10.754
*1.000	15.488	5.859	17.812	4.462	11.800

Figure 10: Option 1 of the menu

RELATIVE VALUES OF THE CRITERIA

LAMBDA	Z 11	Z 12	Z 21	Z 22	Z 31
*0.250	0.746	0.619	0.287	0.301	0.250
0.300	0.714	0.575	0.282	0.280	0.300
0.350	0.683	0.531	0.276	0.260	0.350
0.400	0.651	0.487	0.271	0.240	0.400
*0.475	0.604	0.420	0.262	0.209	0.475
*0.537	0.577	0.382	0.257	0.192	0.537
*0.553	0.571	0.374	0.256	0.188	0.553
0.603	0.553	0.349	0.253	0.176	0.603
0.653	0.535	0.324	0.250	0.165	0.653
0.703	0.517	0.299	0.247	0.153	0.703
0.753	0.499	0.274	0.243	0.142	0.753
0.803	0.482	0.249	0.240	0.130	0.803
0.853	0.464	0.224	0.237	0.119	0.853
0.903	0.446	0.199	0.234	0.107	0.903
*1.000	0.412	0.151	0.228	0.085	1.000

Figure 11: Option 2 of the menu

VARIABLES

LAMBDA	X 1	X 2	W 1	W 2	W 3	W 4	W 5	W 6	W 7	W 8	W 9	W 10	W 11	W 12
*0.250	6.982	2.985	0.000	3.936	4.949	0.000	0.000	0.000	0.949	0.000	0.000	0.000	0.000	0.000
0.300	7.232	3.054	0.000	4.393	5.517	0.000	0.000	0.000	1.517	0.000	0.000	0.000	0.000	0.000
0.350	7.481	3.123	0.000	4.850	6.085	0.000	0.000	0.000	2.085	0.000	0.000	0.000	0.000	0.000
0.400	7.730	3.192	0.000	5.307	6.652	0.000	0.000	0.000	2.652	0.000	0.000	0.000	0.000	0.000
*0.475	8.108	3.297	0.000	6.000	7.513	0.000	0.000	0.000	3.513	0.000	0.000	0.000	0.000	0.000
*0.537	8.321	3.357	0.000	6.393	8.000	0.000	0.000	0.393	4.000	0.000	0.000	0.000	0.000	0.000
*0.553	8.371	3.371	0.000	6.483	8.112	0.000	0.000	0.483	4.112	0.000	0.000	0.000	0.112	0.000
0.603	8.512	3.410	0.000	6.741	8.433	0.102	0.000	0.741	4.433	0.000	0.000	0.000	0.433	0.000
0.653	8.652	3.449	0.000	6.999	8.754	0.203	0.000	0.999	4.754	0.000	0.000	0.000	0.754	0.000
0.703	8.793	3.488	0.000	7.258	9.075	0.305	0.000	1.258	5.075	0.000	0.000	0.000	1.075	0.000
0.753	8.934	3.527	0.000	7.516	9.395	0.407	0.000	1.516	5.395	0.000	0.000	0.000	1.395	0.000
0.803	9.075	3.567	0.000	7.774	9.716	0.508	0.000	1.774	5.716	0.000	0.000	0.000	1.716	0.000
0.853	9.216	3.606	0.000	8.033	10.037	0.610	0.000	2.033	6.037	0.000	0.000	0.000	2.037	0.000
0.903	9.357	3.645	0.000	8.291	10.358	0.712	0.000	2.291	6.358	0.000	0.000	0.000	2.358	0.000
*1.000	9.629	3.721	0.000	8.791	10.979	0.909	0.000	2.791	6.979	0.000	0.000	0.000	2.979	0.000

Figure 12: Option 3 of the menu

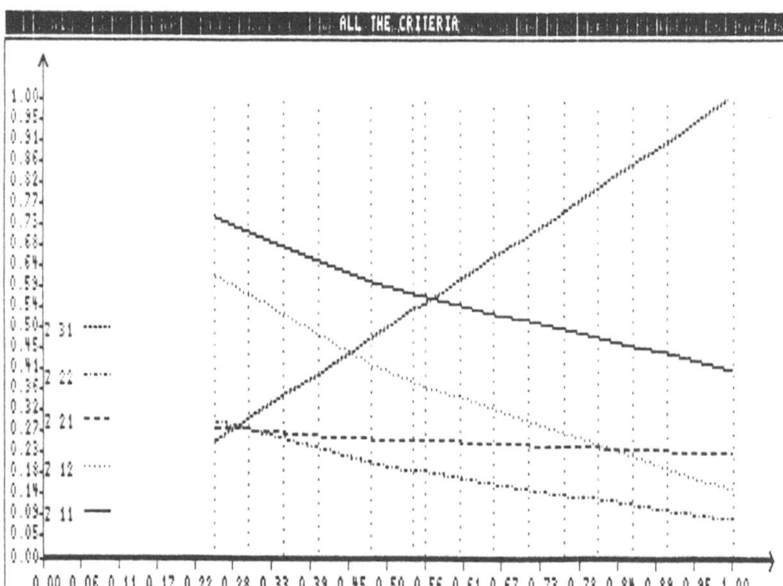

Figure 13: Option 4 of the menu

The user can select any value for the level of relaxation and call for a vizualisation of the corresponding solution. The previous compromise is displayed for comparison purposes in a separate window on the same screen (fig. 14).

The last option of the menu (fig. 9) completes the present interactive phase: the user imposes his preferred level of relaxation (for instance $\lambda = 0,6$) and the new compromise is then shown (fig. 15).

5. CONCLUSIONS

The developments of a personal computer version of STRANGE has been inspired by the two main objectives:

- First to have a demonstration module, that can be easily distributed for those potential users not aware of the approach;

- Second to demonstrate the importance of a friendly user-interface for inputting data and for reporting on results. The latter aspect is specially important for an interactive method. The user must be able to move hence and forth, while keeping track on all generated solutions as tables and graphic displays.

These objectives could not be achieved at the same cost and effort, nor with the same guarantee of portability on mainframe computers. So far, the maximum size of problems has been voluntarily limited, but the improvement in the micro-computer performances makes it possible to have both full-scale approach and comfortable user-interface.

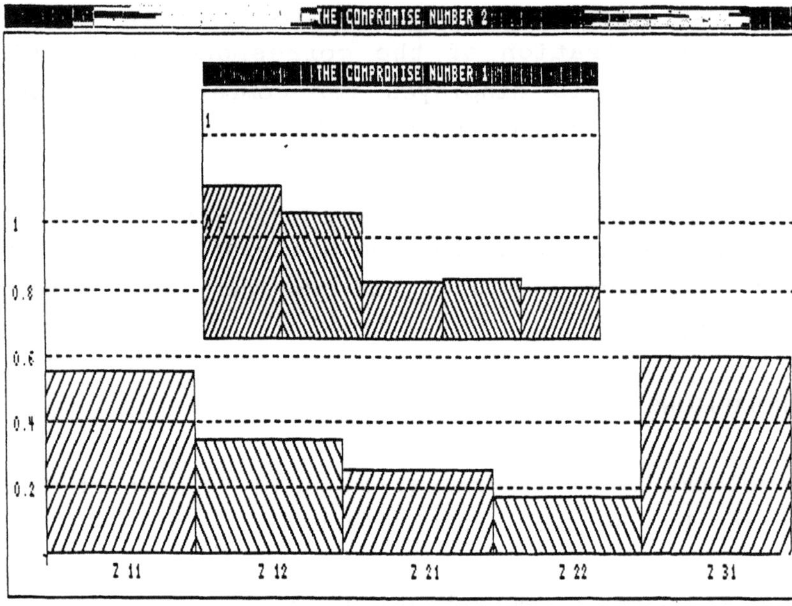

Figure 14: Option 5 of the menu for λ = 0.6

THE COMPROMISE NUMBER 2

The values of the variables are

X1 = 8.503

X2 = 3.407

W1 = 0.000	W2 = 6.725	W3 = 8.413	W4 = 0.095
W5 = 0.000	W6 = 0.725	W7 = 4.413	W8 = 0.000
W9 = 0.000	W10 = 0.000	W11 = 0.413	W12 = 0.000

	Absolute Value	Relative Value
CRITERION 11	17.868	0.554
CRITERION 12	9.365	0.350
CRITERION 21	18.312	0.253
CRITERION 22	6.402	0.177
CRITERION 31	7.480	0.600

	Mean Value	Confidence Level
OBJECTIVE 1	15.742	0.250
OBJECTIVE 2	12.357	0.500

Menu **Continue**

Figure 15: The new compromise

REFERENCES

Benayoun, R., Montgolfier, J., Tergny, J. and Larichev, O. (1971): "Linear programming with multiple objective functions: step method (STEM)", *Mathematical Programming*, 1/2, 366-375.

Kunsch, P. and Teghem, J. (1987): "Nuclear fuel cycle optimization using multi-objective linear programming", *EJOR*, 31, 2, 240-249.

Slowinski, R. and Teghem, J. (1988): "Fuzzy vs. stochastic approaches to multicriteria linear programming under uncertainty", to appear in *Naval Research Logistics*, 35, 6.

Teghem, J. and Kunsch, P. (1985): "Multiobjective decision making under uncertainty: an example for power systems", in Y. Haimes and V. Chankong (eds.) *Decision Making with Multiple Objectives*, Springer Verlag, Lecture Notes in Economics and Mathematical Systems, nº 242, 443-456.

Teghem, J., Dufrane, D., Thauvoye, M. and Kunsch, P. (1986): "STRANGE: an interactive method for multi-objective linear programming under uncertainty", *EJOR*, 26, 1, 65-82.

Teghem, J. and Kunsch, P (1988): "An interactive DSS for multiobjective investment planning, in D. Mitra et al. (eds.) *Mathematical Models for Decision Support*, Springer Verlag, 123-134.

APPENDIX

MATHEMATICAL DESCRIPTION OF STRANGE

α) Associated deterministic problem

- Each situation (k,s_k) defines one objective corresponding to scenario s_k affecting criterion k, i.e
  $z_{k\,s_k}(X) = c_{k\,s_k}.X \;(k=1,\dots,K;\; s_k=1,\dots,S_k)$.

- To deal with the uncertainty, slack variables V_r and recourse variables W_r are introduced to measure the violation of outcome r so that the set of constraints becomes
  $D_1(Y) = \{Y = (X,V_r,W_r\;(r=1,\dots,R)) \mid T_rX+V_r-W_r = b_r,\; Y \geq 0\}$.

- An additional objective denoted by $z_{K+1,s_{K+1}}$ with $S_{k+1} = 1$ expresses a global measure of violation of the constraints due to uncertainty:

$$z_{K+1,\,s_{K+1}}(Y) = \sum_{r=1}^{R} q_r(W_r)$$

- Finally, an equivalent deterministic MOLP problem is obtained:

$$\text{minimize} \quad z_{k\,s_k}(Y) \quad k = 1,\dots,K+1;\; s_k = 1,\dots,S_K$$

$$\text{s.t.} \quad Y \in D_1(Y)$$

β) The first compromise

- For each criterion $z_{k\,s_k}(Y)$ and for each outcome r, an optimal solution $Y^{(r)}_{k\,s_k}$ of the corresponding single objective LP problem is determined; this provides a best solution $\tilde{Y}_{k\,s_k}$ for criterion $k_{\,s_k}$,

$$z_{k\,s_k}(\tilde{Y}_{k\,s_k}) = \min_{r\in\{1,\dots,R\}} z_{k\,s_k}(Y^{(r)}_{k\,s_k}),$$

the ideal point $M_{k\,s_k} = z_{k\,s_k}(\tilde{Y}_{k\,s_k})$, and a pay-off table with entries $z_{(l\,t_l)(k\,s_k)} = z_{l\,t_l}(\tilde{Y}_{k\,s_k})$. If $\tilde{Y}_{k\,s_k}$ is not a unique solution, a special technique is used to avoid ambiguity in the definition of the table (see Teghem *et al.*, 1986).

- Like in the STEM method (Benayoun *et al.*, 1971), this pay-off table is used to define some normalizing weights $\pi_{k\,s_k}$ giving for each objective $z_{k\,s_k}$ the relative importance of the distances from the ideal point:

$$\pi_{k\,s_k} = \frac{\alpha_{k\,s_k}}{\sum_{k=1}^{K+1} \sum_{s_k=1}^{S_k} \alpha_{k\,s_k}} \quad \text{with} \quad \alpha_{k\,s_k} = \frac{m_{k\,s_k} - M_{k\,s_k}}{m_{k\,s_k} \cdot \| c_{k\,s_k} \|}$$

and $m_{k\,s_k} = \max_{(l\,t_l)} z_{(l\,t_l)(k\,s_k)}$

- Then, the min-max optimization is applied to determine the first compromise $\tilde{Y}^{(1)}$:

minimize δ

$$\text{s.t.} \sum_{s_k=1}^{S_k} p_{k\,s_k} \, (c_{k\,s_k} Y - M_{k\,s_k}) \, \pi_{k\,s_k} \le \delta \quad k=1,\dots,K+1 \qquad (2)$$

$Y \in D_1(Y)$

If the optimal solution of problem (2) is not unique, a special technique must be used to ensure the efficiency of the first compromise (see Teghem *et al.*, 1986).

γ) **The interactive phases**

- for each compromise $\tilde{Y}^{(m)}$ $(m \geq 1)$, the D.M. receives three pieces of information:

  - The set values $z^{(m)}_{k\,s_k} = z_{k\,s_k}(\tilde{Y}^{(m)})$ together with intervals of variations $[\,M_{k\,s_k}, m_{k\,s_k}\,]$;

  - The mean values $\bar{z}^{(m)}_k = \sum_{s_k=1}^{S_k} P_{k\,s_k}\, z^{(m)}_{k\,s_k}$

  - The confidence level of the compromise $\alpha^{(m)}_k = P(c_k.Y > \bar{z}^{(m)}_k)$.

- If the D.M. is not satisfied with the compromise $\tilde{Y}^{(m)}$, he is a asked to indicate criterion $(k\,s_k)^*$ to be relaxed and upper limit $\Lambda_{(ks_k)^*}$ for the value of this criterion.

In order to analyse the results of relaxation indicated by the D.M., a parametric LP problem is considered:

minimize δ

$$\text{s.t.} \quad \sum_{s_k=1}^{S_k} P_{ks_k}\,(c_{ks_k}\;Y - M_{ks_k})\,\pi_{ks_k} \leq \delta \quad k=1,\ldots,k+1$$

$$c_{(ks_k)^*}\,.\,Y = M_{ks_k)^*} + \lambda(m_{(ks_k)^*} - M_{ks_k)^*}) \quad \lambda_o \leq \lambda \leq \Lambda \qquad (3)$$

where λ_o and Λ are given respectively by:

$$M_{ks_k)^*} + \lambda_o(m_{(ks_k)^*} - M_{ks_k)^*}) = z^{(m)}_{ks_k)^*}$$

$$M_{ks_k)^*} + \Lambda(m_{(ks_k)^*} - M_{ks_k)^*}) = \Lambda_{ks_k)^*}$$

with $D_2(Y) \equiv D_1(Y)$

$$D_{m+2}(Y) \equiv D_{m+1}(Y) \cap \{Y \,|\, z_{(ks_k)^*}(Y) \leq z_{(ks_k)^*}(\tilde{Y}^{(m+1)})\}$$

and $\pi_{(ks_k)^*}$ set equal to 0 in the course of the procedure (see (1)).

In this way, the value accepted by the decision maker for the relaxed criterion $(k\,s_k)^*$ will not be increased during the next interactive phases. This limits the number of interactive phases.

Remark

Should the D.M. not be willing to introduce these additional constraints, an alternative way is to define

$D_{m+2}(Y) \equiv D_1(Y)$ for all m

$\pi_{(ks_k)^*}$ is set equal to zero during the next iteration, but set back to his original value during the following interactive phases. A corresponding option is provided in the software.

Using a dual simplex technique, the optimal solution $Y^{(m+1)}(\lambda)$ of this new problem is found and the sequence of bounds $\lambda_0 \le \lambda_1 \le \ldots \le \Lambda$ corresponding to the stable intervals of the optimal bases are determined; for each objective $(k\,s_k$ a piecewise linear function $z_{ks_k}(Y^{(m+1)}(\lambda))$ is displayed graphically so that the D.M. can assess the complete range of variation of all objectives when $z_{(ks_k)^*}(Y)$ is changing. The same information may also be given in relative values $(z_{(ks_k)}(Y^{(m+1)}(\lambda)-M_{ks_k})\ /\ (m_{ks_k}-M_{ks_k})$ belonging to the interval [0,1]. Basing on this information, the D.M. is able to define an acceptable level $\tilde{\lambda}$ of relaxation corresponding to a new compromise: $\tilde{Y}^{(m+1)} = Y^{(m+1)}(\tilde{\lambda})$.

INTERACTIVE MULTIFACTORIAL PLANNING: STATE OF THE ART

Jaap Spronk

Erasmus University
P.O. Box 1738,
3000 DR Rotterdam
THE NETHERLANDS

1. INTRODUCTION

Financial planning is a structured process of identification and selection of present and future capital investment projects (including disinvestments) while taking account of the financing of these projects over time.

Important elements in this process are:

a. identification of investment and financing projects;
b. description of the effects of these projects;
c. specification of criteria for project selection;
d. selection of projects.

Clearly, financial planning is a continuous process, the elements of which are linked through a series of feed-back and feed-forward loops. In this overview we will concentrate on the fourth element of the process: how to select projects and combinations of projects? Since the selection of projects requires a yardstick to compare between different (combinations of) projects, we will also pay attention to (c), the criteria for project selection and (b), the description of the projects in terms of these criteria.

As we will see below, it is reasonable to assume that the firm has to deal with a dynamic goal complex, a multiplicity of goals which vary over time. The purpose of this paper is to show how a multicriteria approach can help to choose between different (combinations of) projects. In doing so, descriptions of problem elements will play at leap-frog with descriptions of multicriteria procedures.

In the following section, some mono-criterion approaches to financial planning are discussed. Reasons to use a multiple criteria approach are given. Section 3 is devoted to multiple goal programming and its application in the area of financial planning. In section 4, we describe interactive multiple goal programming, which has been especially designed for financial planning but which is also being used in a wide variety of other applications. In section 5, we discuss the multi-factorial approach to financial planning, being a relatively new approach to deal with uncertainties in the projects' cash flows. Next, in section 6, an approach to financial planning in decentralised firms is presented, which in section 7 is combined with the multifactorial approach.

2. SINGLE MINDED OPTIMIZATION: WHY MULTIPLE GOALS?

Financial theory adopts the maximization of shareholders' wealth as the exclusive objective to evaluate alternative plans. This can be achieved by maximizing the market value of the firm's shares, which is determined on the stock market where the shares (being claims on the firm's future and uncertain cash flows) are traded. It is argued that the owners of the firm evaluate the firm in view of alternative investment opportunities. An obvious opportunity for the owners is to invest in the stock market. This gives a clear reason to use the firm's market value as a benchmark for evaluating the firm and its investments. And to close the circle: the firm's market value directly depends on the firm's cash flows and the associated risk. If the firm's cash flows decrease and/or its riskiness increases, its market value will drop. As such, the shareholders may de facto be the most powerful group of participants of the firm. Nevertheless, there are several reasons to describe the financial planning problem as one with multiple goals.

The first group of reasons, relating to the difficulty of formalising uncertain outcomes and expressing preferences with respect to uncertain outcomes, will be discussed in section 5.

Other reasons for using a multicriteria approach stem from the fact that the firm has not only to deal with the shareholders but also with a series of other participants.

Very often, it is assumed that goals other than shareholders' wealth maximisation can be translated into «cost factors» which can be adopted in the formulae for the share's value, thus re-establishing a single criterion decision problem. However, it is very hard, if not impossible to follow this procedure. Assume for instance that the firm's market value depends on such entities as the quality of its products, management's job security, the well-being of the workers, the firm's public relations and so on. In some of these cases it might be possible to specify more or less exactly how the firm's market value depends on one or more of these entities. For instance, one might try to express this market value as a function of the quality of the firm's products. More often however, it is practically or even principally impossible to specify such a function. Assume for instance that at least some of the firm's participants want to allocate the firm's means in directions other than required for value maximisation and, at the same time, have the ability to co-determine the availability and allocation of these means, the goal of value maximisation has no satisfactory mathematical definition. The goal of value maximisation then depends on instruments which are not under control of the value maximiser. Such a problem can not be described as a «maximum problem», but rather as a peculiar and disconcerting mixture of several conflicting maximum problems (Spronk, 1981).

An alternative procedure to deal with multiple goals is to formulate them as hard constraints, subject to which the shareholders' wealth can be maximised. An objection against the latter procedure is that all goals formulated as constraints are thus treated as equally important and that each of them has absolute priority over the goal variable that is to be maximised.

3. MULTIPLE GOAL PROGRAMMING

During the very early stages of the development of goal programming it was already suggested several times that this technique could be an important means of dealing with capital budgeting and financial planning involving multiple objectives (see e.g. Ijiri et al., 1963). Since then, a considerable number of authors have indeed used goal programming in financial planning models. The main advantage of goal programming is that it is able to process fairly realistic problem descriptions. In this respect, Ashton and Atkins (1981) state that «it is natural in financial planning to speak in terms of targets and goals; many of the indicators of company performance such as dividend cover, liquidity or return on capital employed have target ratios adopted by customs and practice».

The *goal programming problem* can be described as follows (for a more detailed summary we refer to Spronk, 1981; furthermore, we refer to the original publications by Charnes and Cooper, to Ijiri, 1965, and to the work of Lee (e.g. Lee, 1972) and Ignizio (e.g. Ignizio, 1976, 1985)).

Assume a number (say m) of goal variables depending on a number of instrumental variables $x_1, x_2, \ldots, x_n$ (in vector notation x). In general one may suppose

$$b = f(x) \text{ for} \quad (1)$$
$$x \in R,\ R = \{x \mid h(x) \le h\},$$

where b is the m x 1 vector of aspired levels of the m goal variables and R denotes the feasible region, being the set of admissible values of the instrumental variables x. In this paper we assume both f(x) and h(x) to be linear in x. The problem can then be formalized as finding the set of vectors x, for which

$$A.x = b \qquad (2)$$
$$x \in R,\ R = \{x \mid B.x \leq h\},$$

where A is a matrix of order m x n, B is a matrix of order k x n, b is a m x 1 vector, h is a k x 1 vector and x is a nx 1 vector.

Because of the possible incompatibility of goals, at least within the feasible region R, we have to account for deviation of the goal variables from the aspired levels $b_1, b_2, \ldots, b_m$. This can be done by introducing the (m x 1) vectors y^+ and y^-, which measure respectively the over and underachievement of the stated goals. Because over- and underachievement of a given goal cannot take place at the same time we impose the conditions $y_i^+ . y_i^- = 0$ for $i = 1, 2, \ldots, m$; $y^+ > 0$ and $y^- > 0$. Assuming that the undesirability of the deviations can be represented by some dispreference function $f(y^+, y^-)$ the problem comes to:

$$\text{Minimize } \{f(y^+, y^-)\} \qquad (3)$$

$$\text{under } A.x - y^+ + y^- = b,$$
$$\text{where } y^+, y^- \geq 0 \text{ and } y_i^+ . y_i^- = 0 \text{ for } i = 1, 2, \ldots, m$$
$$\text{and } x \in R.$$

Both to tackle the problem of incommensurable goal variables and to meet everyday practice the formulation is often extended as follows. First divide the m goals in p (p < m) classes in a way that the goals in the first class have a preemptive priority above the goals in the second class and so forth. In this in fact lexicographic approach, the goals in a certain priority class must be satisfied «as good as possible» before paying attention to the goals of subsequent classes.

By taking b_{ij} as the ith goal of the ith priority class and defining m_j as the number of goals in the jth class C_j, we can formulate this sequential problems as follows:

First: minimize $\{f_1(y^+, y^-)\}$ (4)

where $f_1(y^+, y^-) = f(y^+_{11}, y^+_{12}, \ldots, y^+_{1,m1}, y^-_{11}, y^-_{12}, \ldots, y^-_{1,m1})$

under $A.x - y^+ + y^- = b$,

where $y^+, y^- \geq 0$ and $y^+_i . y^-_i = 0$ for $i = 1, \ldots, m$

and $X \in R$

then: minimize $\{f_2(y^+, y^-)\}$

where $f_2(y^+, y^-) = f(y^+_{21}, y^+_{22}, \ldots, y^+_{2,m2}, y^-_{21}, y^-_{22}, \ldots, y^-_{2,m2})$

under $A.x - y^+ + y^- = b$,

where $y^+, y^- \geq 0$ and $y^+_i . y^-_i = 0$ for $i = 1, \ldots, m$

and $X \in R$

and under $f_1(y^+, y^-) = \min \{f_1(y^+, y^-)\}$,

then: ...,

.
.
.

then: minimize $\{f_p(y^+, y^-)\}$

where $f_p(y^+, y^-) = f(y^+_{p1}, y^+_{p2}, \ldots, y^+_{2,p2}, y^-_{p1}, y^-_{p2}, \ldots, y^-_{p,m2})$

under $A.x - y^+ + y^- = b$,

where $y^+, y^- \geq 0$ and $y^+_i . y^-_i = 0$ for $i = 1, \ldots, m$

and $X \in R$

and under $f_i(y^+, y^-) = \min \{f_i(y^+, y^-)\}$, for $i = 1, \ldots, p-1$.

Within a class of goals C_j, the function $f_j(y^+, y^-)$ is generally minimized when the undesired deviations from the goals within that class are zero, i.e. when those goals all have been satisfied completely. Because in most problems not all stated goals are attainable at the same time we accordingly will not find the minima of all functions f_j, $j = 1, 2, \ldots, p$ in the way described.

In general, each minimization in (4) reduces the feasible region left for the fulfilment of the lower priority goals. Let C_j be the first class in the priority ordering where not all goals can be fulfilled simultaneously (so for i < j all goals in C_j are attainable). Then we need to specify $f_j(y^+, y^-)$ in order to choose among the goals within $C_j(b_{j1}, b_{j2}, \ldots, b_{j,mj})$. In essence this dispreference function is a distance function, requiring some distance measure.

A general formulation of this dispreference function comes through the Minkovski metric, which is frequently used in welfare economics:

$$\min Q = \{ \sum_{i=1}^{m} |(Z_i - \bar{Z})/\bar{Z}|^p \}^{1/p} \tag{5}$$

where Z_i is the average income (or some other adequate welfare indicator) in class i (i = 1, ..., n) and $\bar{Z}$ in the overall average income. More general the term within the inner brackets can be interpreted as a standardized distance from a predetermined norm. Realizing there is not a single norm (as $\bar{Z}$ in (5)) in (3) and (4) we can adapt (5) in terms of the multiple goal problem (3) by:

$$\min Q = \{ \sum_{i=1}^{m} |(a_i.x - b_i)/b_i|^p \}^{1/p} \tag{6}$$

where a_i is the ith row of A. Unfortunately, this form does not discriminate between positive and negative deviations from the stated goal levels b_i, i = 1, ..., n. In order to bridge this gap let us define:

$$y_i^+ = \{ |a_i.x - b_i| + (a_i.x - b_i) \}/2 \quad \text{for } i = 1, \ldots, m \tag{7a}$$

to represent the positive and:

$$y_i^- = \{ |a_i.x - b_i| - (a_i.x - b_i) \}/2 \quad \text{for } i = 1, \ldots, m \tag{7b}$$

to represent the negative deviations. Then adding (7a) and (7b) gives:

$$y_i^+ + y_i^- = |a_i.x - b_i|, \tag{8}$$

which can be substituted in (6), in order to get $f(y^+, y^-)$ as a function of the stantardized deviations from the aspired goal levels:

$$\min Q = \min f(y^+, y^-) = \{\sum_{i=1}^{m} [(y_i^+ + y_i^-)/b_i]^p\}^{1/p}. \tag{9}$$

Then of course, $y_i^+ \cdot y_i^- = 0$ must hold for $i = 1, \ldots, m$ in order to exclude the possibility of y_i^+ and y_i^- being both positive at the same time.

This condition permits us to write (9) as:

$$\min f(y^+, y^-) = \sum_{i=1}^{m} [(y_i^+/b_i)^p] + \sum_{i=1}^{m} [(y_i^-/b_i)^p]\}^{1/p} \tag{10}$$

which form is preferable to (9) when not all deviations are equally valued by the decision-maker. In that case (10) permits different weights for the positive and negative deviations from the same aspiration level whereas (9) do not.

The specification of the weights in (10) may be given by:

$$\min f(y^+, y^-) = \{\sum_{i=1}^{m} \alpha_i^+ \cdot [(y_i^+/b_i)^p] + \sum_{i=1}^{m} \alpha_i^- \cdot [(y_i^-/b_i)^p]\}^{1/p} \tag{11}$$

Obviously both (9) and (10) or its weighted forms, may serve as the distance function $f(y^+, y^-)$ in (3). Moreover, subtracting (7b) from (7a) gives:

$$y_i^+ + y_i^- = a_i.x - b_i \qquad \text{for } i = 1, \ldots, m \tag{12}$$

which exactly equal the goal restrictions in (3).

Of course different metrics will usually generate different results. This means, as it was pointed out by Ijiri (1965, p. 56), that apart from the determination of the weights α_i, the prescription of the metric forms a problem of managerial choice.

There are several procedures to solve the various goal programming problems. For these we refer to the authors mentioned earlier in this section.

Despite its attractive properties, the use of goal programming is not without difficulties. Notably, its need of a considerable amount of a priori information to be given by the decision maker should be mentioned. The decision maker has to define aspiration levels, preemptive priorities and weighting factors. Once a goal obtains preemptive priority over another, trade-offs with other goals are excluded. In other words, first order priority goals determine the solution space for goals of the less important priority classes. This means that the definition of the goal levels must be completely reliable or that some kind of bargaining about these levels must remain possible. The latter is exactly what is offered by the interactive procedure described in the next section.

4. INTERACTIVE MULTIPLE GOAL PROGRAMMING

Interactive Multiple Goal Programming (IMGP) is a technique which was especially designed for financial planning but which as well has been and is being used in a variety of different, including very different, applications. IMGP is an interactive procedure which, just as goal programming, can handle a broad variety of problems. Here, we give a brief description of the procedure. For further details we refer to (Spronk, 1981).

In IMGP the decision maker has to provide information with respect to his preferences on basis of a vector of required values and a vector of ideal values presented to him (there are options to retrieve other information).

The vector of required values (lower limits for goal variables to be maximised and upper limits for goal variables to be minimised) are either defined directly by the decision maker or, when possible and accepted by the decision maker, derived from the mathematical properties of the problem. The vector of ideal values shows the optimal value, for each of the goal variables separately, given the required goal values. Note that the vectors of required and ideal values together (briefly «the potency matrix») enclose a set of solutions, every single one of which satisfies the required goal values but are never better than the ideal values. Given such a potency matrix, the decision maker gets the possibility to change one of the required values. On basis of a new vector of required values, a new potency matrix is calculated. Then, the decision maker is asked whether he accepts the «loss in potency» in exchange for the improved required value. If so, he can continue changing the required values one by one. If not, the required value which was apparently set to a too ambitious value can be reconsidered by the decision maker. Clearly, the decision maker can continue with this procedure until he ends within an epsilon environment of the (Pareto-) optimal solution. Alternatively, and in many cases more realistically, the decision maker can stop the procedure before as soon as the required value are satisfactory: there is a <u>set</u> of solutions satisfying the required goal values. The decision maker may then choose freely from this set, maybe even using viewpoints and criteria which were not (or could not be) included in the original model.

5. THE MULTI-FACTORIAL APPROACH TO UNCERTAINTY

In literature, uncertain outcomes of decision alternatives are very often modelled as probability distributions defined on the range of possible outcomes. Decision-makers are then required to assess some parameters (e.g. mean and variance) of these distributions. In addition, decision-makers often have to express their preferences (e.g. utility values) with respect to the uncertain outcomes. This information is then used in the formulation of an objective function (e.g. the expected utility of the decision alternatives) or, alternatively, chance constraints.

In our opinion, any decision supportive approach to financial planning should take account of the reasonably well established fact that people are not very good in assessing reliable probabilities to the outcomes of future events. This may be partly explained by pointing at the limitations of the human mind. Another reason is of an even more fundamental nature: many future outcomes are contingent on decisions which are still to be made. In the case of capital investment projects, for instance, the future cash flows can generally not be well represented by some probability distribution alone. In addition it is often necessary to describe as well the «rights» and «duties» connected with the project. For instance, the right to expand or to abandon the project or some legal duty to keep people on the payroll even in case it would be more economical to fire them. One might consider the use of decision trees to describe this type of situations. But in most practical cases this is impossible because of the number of possible outcomes being too large and/or because the timing of the potential outcomes (when something will happen) is largely unknown. The picture becomes even more complicated if one realises that the outcomes of a project are often not only contingent on the future decisions of the decision maker himself, but also on the future decisions of the firm's participants, which are in their turn contingent on the firm's decisions. In other words, the way the firm manages the exchange relations with its participants influences the uncertainty surrounding the firm's cash flows.

So in modelling one has to find a compromise: the modelling should be sufficiently precise whereas on the other hand the modelling should be both comprehensible and manageable by those who are intended to use the models. We take the position that, in general, not all aspects of uncertainty can be modelled. Especially when the modelling requires a high degree of precision (e.g. when utility functions and probability distributions are to be formulated), the requirements will in many cases surmount the decision-maker's capabilities. This was one of the reasons for including flexibility within the IMGP procedure described above. However, uncertain phenomenae very often show at least some structure.

Following Hallerbach and Spronk (1986) we propose the use of multi-factor models as a way of bringing some structure in uncertain situations. The results of a financial plan (for ease of exposition we take the firm's cash flow as the only relevant result variable) will depend on the one hand on the decisions made by the firm and on the other hand on the various forces and influences from its dynamic environment. We assume that, in general, it is very hard if not impossible to define a probability distribution over the value of the firm's cash flows. However, we do assume that the firm is able to define its expectations concerning these cash flows and, in addition, that it is able to assess the sensitivity of these cash flows for unexpected changes in a number of factors which may influence these cash flows. Consequently, the effect of a decision can be modelled as an expected level of the cash flows plus a series of sensitivities for unexpected changes in a number of factors influencing these cash flows. The firm does not necessarily know how these factors themselves will change in the future. Also, it may have found only some of the factors influencing its cash flows. For the factors which have been identified, the firm can estimate its firm's <u>aggregate</u> sensitivity. In this way, the firm can calculate the effect of different combinations of future factor changes.

After the identification of the most important factors and the assessment of the firm's sensitivity for these factors, a number of questions remain. A first question is whether and, if so, how the firm is prepared for unexpected changes in the factors. Obviously, both the <u>defensive</u> and the <u>offensive</u> weapons of the firm should be considered. In both cases, the firm has two possibilities to armour itself. One possibility is to neutralize or to limit a risk beforehand by «buying an insurance» with respect to this risk (where «buying an insurance» should be understood in a broad sense). For instance, firms often insure themselves against the negative consequences of unexpected changes in the factors such as fire or exchange rate fluctuations. On the other hand, the firm can assure itself of the positive effects of unexpected changes in the factors (e.g. by acquiring the exclusive selling rights of a product in development). Another possibility to face risks, instead of buying an insurance, is to create sufficient elbow-room in the firm to be able to react adequately to an unexpected change in some factor if and at the moment it occurs. An example is not to insure against fire but instead to make a large enough reservation to be able to bear the negative consequences of fire.

Creating elbow-room can be viewed as buying an option. If an unexpected change of a factor materializes, the option holder has the right to use the elbow-room to react to this change. The possessor of the elbow-room is of course obliged to bear the consequences if the elbow-room is not sufficient. Elbow-room is often labelled as *flexibility*, where a distinction is made between operational and financial flexibility (cf. Kemna and Van Vliet, 1984, and Kemna, 1988). Examples of operational flexibility are the possibility to quickly adapt production (e.g. with respect to production volume or to product specification) according to the changing needs of the product's consumers. Examples of financial flexibility are unused reserves (cash surpluses, unused credit facilities), unused debt capacity, the capability to reduce expenditures and the earlier mentioned possibility to sell assets. Not surprisingly, the creation and maintenance of elbow-room is not without costs: flexibility has its price (cf. Kemna and Van Vliet, op cit.).

For further details and examples of the use of multifactorial financial planning we refer to (Spronk, 1989) and (Goedhart et al., 1990).

6. TWO-LEVEL FINANCIAL PLANNING: AN INTERACTIVE APPROACH

To deal with financial planning within and between two hierarchically distinct decision levels, a two-level procedure has been developed (see Schaffers and Spronk, 1987 and Goedhart et al., 1988).

The upper decision level consists of one internally consistent decision unit called the principal or coordinator. The lower decision level consists of a number of mutually separate decision units (subordinates), each of which is internally consistent. The principal decides on the distribution of resources over the lower level decision units. These decisions are a.o. based on the information provided by the lower level units. Each decision unit has to deal with a complex of goal variables. The principal deals with several «central» goal variables, the subordinates have to deal both with their contribution to the central goal variables and with additional goal variables which are included in order to take account of «local circumstances» (possibly including the subordinate's own interests, thus giving rise to agency problems). Since financial planning is, per definition, dealing with an uncertain future, and since not all outcomes can be foreseen (even not in probabilistic terms) it is necessary to build a certain degree of flexibility within the planning process. Moreover, the information regarding the uncertain future is assumed to be asymmetrically distributed over the different decision units. That implies that not all flexibility in the planning process has to be concentrated at the principal decision unit, but has to be distributed over both the central and the subordinate decision units. The approach is interactive in two ways. One, each decision unit selects by means of an interactive programming approach a set of satisfactory plans. Second, a description of this set is communicated to the decision units at the other hierarchical decision level. At this level, a (new) set of plans, taking into account the desires and/or requirements of the other decision level, is determined. And so on, until the upper hierarchical level enforces its conditions on the lower level.

Both interactive procedures are structured in a systematic way, guaranteeing stepwise improvement of the resulting sets of alternative plans until a satisfactory set of plans (including sufficient flexibility for each decision unit) is reached. The proposed approach is different from other approaches to the stated problem (e.g. goal programming, decomposition techniques) in that the communication between different decision levels is structured in terms of <u>sets</u> of decision alternatives, which can be adapted in a systematic interactive way.

The central decision unit defines a series of «central» goal variables g_m, m = 1, ..., M. In principle, each of the lower level decision units contributes to each of the central goal variables; g_m^u measuring the u'th units contribution to the m'th central goal variable (u = 1, ..., U and m = 1, ..., M). The central decision level can influence the values of its central goal variables both directly and indirectly. Directly, because it controls a number of central instruments x_i, i = 1, ..., I. Examples of these instruments are the size of dividends and the amount borrowed during a certain period. The central decision unit influences the values of the central goal variables also indirectly, because it controls the distribution of a series of central resources (r_j, j = 1, ..., J) over the different lower level decision units [1]. If we define $g_m^u = g_m^u(x_1, \ldots, x_I)$ as the contribution of unit u to the m'th central goal value and if we define g_m^c has the central decision unit's direct influence on the m'th central goal variable g_m, the latter can be defined as

$$g_m = g_m^c + \sum_{u=1}^{U} g_m^u \qquad \text{for } m = 1, \ldots, M. \tag{13}$$

----

1. Here we abstract from another instrument central management has to inderectly influence the values of the central goal variables, viz. the use of incentive schemes.

The central resources can be used to support the central activities x_i or, alternatively, be distributed to the lower level decision units. Thus

$$r_j = \sum_{i=1}^{I} \theta_{ij} x_i + \sum_{u=1}^{U} r_j^u, \tag{14}$$

where θ_{ij} denotes central activity i's pro unit use of resource j and r_j^u represents the amount of resource j distributed to unit u. Obviously, θ_{ij} may be negative in the case that central activity i adds to the amount of type j resources (for example, borrowing increases the amount of funds available). In addition, it is possible to define for each central resource j, an activity variable measuring the unused amount of this resource. This type of variables can be actively manipulated in order to guarantee the required flexibility in the plans to be developed. Of course, the central decision unit is not completely free in choosing the values of the instrumental variables. We assume the choice of instrumental variables restricted to the feasible set X. Thus $x \in X$, with x the vector of instrumental variables.

Central management is assumed to be able to choose between different combinations of goal variables. However, it is assumed not to have a mathematically well-defined preference function which is «ready for an optimization algorithm». Likewise, no precise knowledge is assumed with respect to the relation between the g_m^u, the lower levels' contribution to the central goal variables, and r_j^u, the central resources distributed to these lower levels. Without the latter two assumptions, the problem would boil down to a more or less conventional optimization problem. With these assumptions included, it becomes a multicriteria problem mingled with an information problem. Next, to give a better insight into the nature of this problem, the lower level decision units' situation is sketched.

In addition to the central goal variables, each lower level decision unit has to deal with other goal variables in order to take account of «local circumstances», possibly including the subordinate's own interest (see e.g. Christensen and Obel, 1981, and Christensen, 1982). The u'th unit's contribution to the m'th central goal variable was denoted by g_m^u. The local goal variables of decision unit u are denoted by $g_{m(u)}$, with m(u) = 1, ..., M(u). So in total, decision unit u has to deal with M + M(u) goal variables. Each of these goal variables is a function of the I(u) local instrumental variables:

$$\begin{aligned} g_m^u &= g_m^u(x_{1(u)} \ldots x_{i(u)} \ldots x_{I(u)}), \text{ for } m = 1,\ldots,M; \text{ and} \\ g_{m(u)} &= g_{m(u)}(x_{1(u)} \ldots x_{i(u)} \ldots x_{I(u)}), \text{ for } m(u) = 1,\ldots,M(u). \end{aligned} \tag{15}$$

The choice of the local instrumental variables is restricted by local constraints, i.e. $x_u \in X_u$ with x_u the vector of unit u's instrumental variables and X_u the convex feasible set described by local constraints. In addition, the choice is restricted by the amounts of resources r_j^u, j = 1, ..., J, available to unit u. This leads to the constraints

$$r_j^u \geq \sum_{i(u)=1}^{I(u)} \theta_{i(u)j}\, x_{i(u)}, \quad j = 1, \ldots, J; \tag{16}$$

where $\theta_{i(u)j}$ represents decision unit's u pro unit of activity i(u)'s use of resource j. Obviously, one or more of the $\theta_{i(u)j}$ may be negative in the case that local activity i(u) adds to the type j resources. Like central management, local management is assumed to be able to choose between different combinations of goal variables. Likewise, local management is assumed not to have a mathematically well-defined preference function. With these assumptions, the problem of local management becomes a multiple criteria problem, too. This problem differs from the multiple criteria problem of top management, since given the goal variables (15), local management can determine the goal values associated with a given instrument vector x_u. If in addition, X_u and the amounts of resources r_j^u, j = 1, ..., J are known, local management can calculate, at least in principle, the best values for each of the goal variables (15) separately. Top management's problem is more difficult, since its goal variables depend among others on the results of the lower level units, which in their turn depend not only on the resources distributed to them by top management but also on local circumstances, including both local goal variables and local resources.

Basically, top management has two tools for co-ordination, one of which is the use of incentive functions. In the two-level approach, we concentrate on the second tool for coordination: the distribution of central resources. In essence, this distribution process helps top management to solve an information problem. If it had sufficient information, top management could simultaneously determine the r_j^u and the x_i values which would yield the optimal combination of gm values.

However, especially the relationship between the resources r_j^u distributed to the units, and the contribution g_m^u of these units to the central goal variables is largely unknown. Of course, each plan which is communicated from the subunits to the central authority gives some information on this relationship. If the «loop» in the described distribution process could be repeated many times, the amount of information would eventually be sufficient to find a close to optimal solution. However, it is clear that the amount of iterations (loops) should be very limited in practice. A central idea in the two-level procedure described here is to structure the communication process between top management and subunits in terms of sets of plans, thus increasing the amount of information even if the number of iterations is very small.

The two-level procedure starts with the development, by each lower level decision unit, of a set of plans. In developing these plans, lower level takes account both of the guidelines forwarded by central management and of its own interpretation of local circumstances. It is the task of lower management to find a balance between the local requirements and those formulated centrally. Obviously, this balance will depend on the amounts of resources provided by the centre. Therefore, we propose to construct, for each subunit, plans for different combinations of resources. In doing so, we propose lower management to use IMGP.

This procedure fits well to the described problem, since it generates solutions (plans) each set satisfying a combination of constraints formulated by the decision-maker. To develop a set of plans which is communicated to top management, lower management starts with the assumption that the resources are very limited. Given this limitation, the IMPG-procedure is used to derive a set of conditionally (i.e. given the limitations) satisfactory plans. In doing so, management can take account of local circumstances (both instruments and goals) while searching for a satisfactory contribution to the central goal values. The set of plans which results from this first exercise is labelled «plan one» of the decision unit concerned and is described in terms of the maximally required amounts of resources and the minimal contribution to the central goal values which can be attained given these resource levels. Given plan one, local management has to determine a second set of plans («plan two»). This is done by increasing one or more of the goal values which could be reached in plan one. Of course, this can only be done if the limitations on the resources are somewhat relaxed. In the procedure proposed here, it is local management who has to decide which of its goal values and which of the resources are to be increased. However, once plan two has been determined, it is also described in terms of the maximally required amounts of resources and the minimal contribution to the central goal values which can be attained given these resource levels. In exactly the same way, plans three, four and so on can be determined, having the property that plan k is preferred to plan (k-1), at least by local management. The plans provided by unit u can be viewed as a hierarchically ordered set of projects x_{u1}, ..., x_{uk}, k_u being the number of plans formulated by the u'th unit. To be more precise, central management has to decide whether it adopts project x_{u1} or not and, if so, whether it adopts x_{u2} and so on. For each project x_{uk}, lower level management has calculated the minimum contribution to each of the central goal variables. Thus, assuming that central management receives a series of plans from each of the lower level decision units, we can reformulate formula (13) for the m'th central goal variable as

$$g_m = g_m^c + \sum_{u=1}^{U} \sum_{k=1}^{k^u} g_m^{uk} \cdot x_{uk} \qquad \text{for } m = 1, \ldots, M \tag{17}$$

with g_m^{uk} the contribution to the m'th central goal variable in case project k of decision unit u is adopted. For each project x_{uk}, u = 1, ..., U and k = 1, ... u_k; 0 < x_{uk} <1 must hold. Also, for each decision unit u, the project x_{uk} (k = 2, ..., u_k) may only become positive if all preceding projects are equal to one. This condition can be fulfilled straightforwardly within zero-one procedures or, alternatively, by restricted entry methods.

Similarly, the central resource constraints can be written as

$$r_j = \sum_{i=1}^{I} \theta_{ij} + \sum_{u=1}^{U} \sum_{k=1}^{k^u} r_j^{uk} \cdot x_{uk} \tag{18}$$

where r_j^{uk} stands for the additional use of central resource j in case plan k of decision unit u is adopted. Relations (17) and (18), together with the constraints formulated earlier, define central management's problem as a linear programming problem with multiple goal variables. As with the local problems, we propose IMGP as a tool to solve the central problem. The main advantage of doing so is that central management can find a series of minimally required values for the central goal variables which are fulfilled by a set of «master plans». Although at this point it will not be clear yet which of these master plans will be adopted, it is clear that there is a subset of the local plans will be adopted, it is clear that there is a subset of the local plans which will be adopted in all masterplans satisfying the minimum requirements with respect to the central goal values.

This subset of projects is adopted by central management. If central management judges the associated central goal values unsatisfactory and/or if there is an undesirable pattern of local plans adopted, it can decide to give the local decision units another chance to formulate plans, in addition to those already accepted. In this case, local management has to make a new series of plans, knowing that some of its plans have already been adopted and taking account of the «signals» central management has given by adopting some of the local plans while rejecting others. Other plans than before will result because local management has got a better insight into the availability of resources (a.o. in view of the plans of other local decision units) and because its balance between central and local goal variables will probably be changed.

The whole procedure is repeated until central management is satisfied with the resulting set of solutions.

7. TWO-LEVEL FINANCIAL PLANNING AND THE MULTI-FACTORIAL APPROACH

The multifactorial approach of modelling uncertainty, as described in section 5, can be incorporated within the two-level procedure described in the preceding section.

Theoretically, each goal variable may be replaced by a set of goal variables: one to measure the expected value of the goal variable and others to measure the goal variable's influence for unexpected changes in the firm's environment.

In practice, one should treat the most important goal variable in this way: the ultimate number of goal variables would become intractable if each original goal variable would be replaced by its expected value and a series of sensitivities.

REFERENCES

Ashton, D.J. and Atkins, D.R. (1981), "Multicriteria programming for financial planning, some second thoughts", in P. Nijkamp and J. Spronk (eds.), *Multiple Criteria Analysis*, Gower, Aldershot, pp. 11-24.

Christensen, J. and Obel, B. (1981), "Incentive Issues in multilevel organisations using iterative planning", *Int. Journal of Policy Analysis and Information Systems*, Vol. 5/4, 287-304.

Christensen, J. (1982), "Communication in agencies", *Bell Journal of Economics*, 661-674.

Goedhart, M., Peters, J. and Spronk, J. (1990), "Multi-factor financial planning: an outline and illustration", to be published in the *Rivista di Matematica per le Scienze Economiche e Sociali*.

Goedhart, M., Schaffers, J. and Spronk, J. (1988), "An interactive multi-Factor procedure for two-level financial planning with conflicting goals", Report 8708/f, *Centre of Research in Business Economics*, Erasmus University, Rotterdam.

Hallerbach, W. and Spronk, J. (1986), "An interactive multi-factor portfolio model", Report 8610/F, *Centre of Research in Business Economics*, Erasmus University, Rotterdam.

Ignizio, J.P. (1976), *Goal Programming and Extensions*, Heath, Lexington.

Ignizio, J.P. (1985), *Introduction to Linear Goal Programming*, Sage, Beverly Hills.

Ijiri, Y. (1965), *Management Goals and Accounting for Control*, North-Holland, Amsterdam.

Ijiri, Y., Levy, F.K. and Lyon, R.C. (1963), "A linear programming model for budgeting and financial planning", *Journal of Accounting Research*, 198-212.

Kemna, A.G.Z. (1988), Options in Real and Financial Markets, PhD thesis, Erasmus University Rotterdam, Donner Boeken, Rotterdam.

Kemna, A.G.Z. and Van Vliet, J.K. (1984), "Onzekerheid en flexibiliteit: als heeft nu ook een prijs", in *Vermogen in Onzekerheid*, Kluwer, Deventer.

Lee, S.M. (1972), *Goal Programming for Decision Analysis*, Auerbach, Philadelphia.

Schaffers, J. and Spronk, J. (1987), "Two-level financial planning with conflicting goals, an interactive procedural approach", in Y. Sawaragi, K. Ihoue and H. Nakayama (eds.), *Towards Interactive and Intelligent Decision Support Systems, Vol. 1*, Springer-Verlag, Heidelberg, 270-279.

Spronk, J. (1981), *Interactive Multiple Goal Programming: Applications to Financial Planning*, Martinus Nijhoff, Boston.

Spronk, J. (1985), "Financial planning with conflicting objectives", in G. Fandel and J. Spronk (eds.), *Multiple Criteria Decision Methods and Applications*, Springer-Verlag, Berlin, 269-288.

Spronk, J. (1989), "Multi factorial financial planning", in: A.G. Lockett and G. Islei (eds.), *Improving Decision Making in Organisations*, Lecture Notes in Economics and Mathematical Systems, vol. 335, Springer-Verlag, Heidelberg, 380-389.

CHAPTER V

GROUP DECISION AND NEGOTIATION

AN INTRODUCTION TO GROUP DECISION AND NEGOTIATION SUPPORT

Tawfik Jelassi
Technology Management Area
INSEAD
Boulevard de Constance
77305 Fontainebleau
FRANCE

Gregory Kersten
Decision Analysis Laboratory
School of Business
Carleton University
Ottawa, Ontario KIS 5B6
CANADA

and

Stanley Zionts
Department of Management Science and Systems
School of Management
State University of New York
Buffalo, New York 14260
U.S.A.

ABSTRACT

Group decision making and negotiation are important managerial activities, yet difficult to understand and support. The associated complexity is due to the multi-person, dynamic, and ill-structured environment in which these activities take place. Recent advances in information technology create new opportunities for supporting group decision and negotiation processes.

This paper first reviews formal models for group decision making and negotiation. It then presents the different classes of multi-person decision situations. A discussion of the relationships, similarities, and differences between multi-criteria decision making and negotiation follows. The last part of the paper focuses on the notion of computer support and provides examples of conceptual frameworks and actual implementations of group decision and negotiation support systems.

KEY WORDS AND EXPRESSIONS: Group Decision Making; Negotiation; Conflict Resolution; Group Decision Support Systems; Computer-Assisted Negotiation.

CONTENTS

1. INTRODUCTION

Group decision making and negotiation are multi-person processes which are complex and ill-structured. They are dynamic in nature due to changing information, aspiration levels of the participants, preferences and tactics. Moreover, decision-makers rarely conform to classical models of rational behavior (Tversky and Kahneman, 1981; MacLean, 1985). They do not want to unveil their real interests (Fisher and Ury, 1983) and may change their strategies as the process evolves over time (Lockhart, 1979; Schaffers, 1985).

The importance of group decision making and negotiation has stimulated a large volume of research. At one end of the spectrum, behavioral scientists (e.g., Druckman, 1977; Gouran, 1979; Lewicki and Litterer, 1985a and 1985b) have examined special cases or aspects of individual behaviors in prescribed settings in an attempt to discover the importance of specific factors on the overall process. At the other end, formal attempts to understand group decision making and negotiation have utilized various forms of mathematical or logic-based representation. However, these efforts have often been at such a level of abstraction that they are rarely applied to specific real-world instances. **While the behavioral approach has attempted to discover a particular rationality which may be generalized, the formal approach has assumed a general rationality which could be applied to specific problems.**

This paper provides an overview of formal models for group decision making and negotiation with a special focus on those that can be used in developing computer-based support systems. Section 2 discusses the main features of group decision making and negotiation and introduces four classes of multi-person decision situations. Section 3 briefly describes some formal models suggested in economics and game theory. Section 4 presents the relationships, similarities and differences between multiple-criteria decision making and negotiation. Computer support for multi-person decision situations and bargaining is introduced in Section 5. Then conceptual frameworks and actual implementations of group decision and negotiation support systems are discussed. A summary of the paper and some concluding remarks are provided in Section 6.

2. GROUP DECISION MAKING AND NEGOTIATION

2.1. Classes of Multi-Person Decisions

Multi-person decision-making covers a wide range of processes. In an attempt to model such processes, one should take into account their basic features. In this section we discuss the main differences between group decision making and negotiation. We see these two types of processes as different and as members of a more general class.

We distinguish four types of multi-person decision-making situations:

(i) Individual decision-making in a group setting. The decision maker utilizes knowledge of experts, advisers or stakeholders during the process. All group members participate in the process, but only one person is responsible for the decision made. An example of such a case is that of two regional authorities, one dealing with the ecological situation and the other responsible for the industrial development (Gonçalves, 1985).

(ii) Hierarchical or bureaucratic decision-making. There are two cases to consider here: In the centralized one, it is assumed that there is one set of objectives representing the top-level decision maker, and also that he has full control over the lower-levels group members. The formal model that corresponds to this situation is Dantzig and Wolfe's (1960) decomposition model. In the decentralized case, each participant independently controls subsets of the decision variables and objectives and is responsible for his decision which serves as input to the higher-level

one. There is a coordination procedure assuring the optimality of the overall decision.

(iii) Group decision-making or one-party decision-making. Each group member participates in the process and is partly responsible for the final decision. There usually is an overall goal which is accepted by all the members, but they differ in the ways of how this goal should be achieved. The decision problem can be solved by an individual but a group possesses more resources than each of its members and the potential for making effective decision-making is greater (Gouran, 1979; Steiner, 1972).

(iv) Multi-party decision-making or negotiation. One decision maker represents one party and is responsible for the decision before this party and not before the other one(s). There is a conflict of interests because parties have separate and conflicting objectives (Lewicki and Litterer, 1985a) and they have different needs which they want to satisfy (Nierenberg, 1973). Negotiation is the chosen way to resolve a conflict out of necessity and not out of effectiveness or efficiency.

2.2. The Group Decision Problem

People make a group decision when they face a **common problem** and they all are interested in its solution. This problem may be the choice of a vacation site, the purchase of a car, or the acquisition of a house by a family; it may also be the design of a new product or a production plan for a company. An important characteristic of this situation is that all involved individuals belong to **one organization** (family, firm, or government). They may differ in their perception of the problem, and have different interests, but they are all responsible for the organization's well-being and **share responsibility** for the implemented decision.

Although the features mentioned above characterize group decision making, they do not apply to negotiation settings.

Organizations may have a different representation of the decision problem and their representatives need to negotiate in order to find a settlement. For example, a buyer's problem is different from, though complementary to, a seller's problem; the same applies to a company management and a trade union, or to a police department and a hostage-taker. The varying perceptions may be less relevant than the **different interests** and **goals** of the involved parties. The responsibility of a negotiator, who may be from outside the organization he represents, is solely to this organization and not to other negotiators. Negotiation takes place when one organization cannot make a decision without the consent of another. Because negotiation involves organizations, one can expect group decision making within the process of negotiation (e.g. when a negotiator asks for new directives).

2.3. Alternative Decisions and Compromise Solutions

In group decision making, there is **one set** of alternative solutions. During the decision process, this set may be expanded or contracted when, respectively, new options are added or non-efficient alternatives are dropped. Moreover, the considered and communicated alternatives are **feasible**, or at least subjectively feasible, i.e. perceived as such by the decision maker.

In negotiation, there may exist as many sets of alternatives as there are negotiating parties and alternatives may belong to different decision spaces. For example, in the free trade negotiations, alternatives considered by the Canadian government are different from those considered by the U.S. government. However, these different alternatives are **interdependent:** there is a transformation which converts an alternative from one set to an alternative in another one. Usually this transformation is a homomorphism or a quasi-morphism (Holland et al., 1987), i.e. an alternative corresponds to a subset of alternatives. (It seems that this is one of the reasons why negotiators try to negotiate in their opponents' decision spaces).

Negotiating parties often choose alternatives from different spaces. Then, each party presents its **compromise proposal**. The preparation of this proposal is a decision process in itself; on the basis of one alternative several proposals may be prepared differing in the consideration of the opponent's interests. In a single-issue negotiation, an alternative may be the first aspiration level (Tietz and Bartos, 1983) and the negotiator presents his first proposal higher than this level. This proposal may be unfeasible but the very nature of negotiation requires making concessions. In fact, presenting initial proposals which are feasible and close to the expected consensus is called negotiating in «bad faith» (Lewicki and Litterer, 1985a, page 13). Hence, in negotiations the initial proposals may be **unfeasible** because all parties expect **concessions**.

In group decision making, concessions are required because of **hard constraints** which determine the feasible alternatives (Kersten, 1988) and, if possible, they are avoided. For example, a marketing department expects an increase in the quality of a product but the finance department argues that funds should be channeled to paying off debts, and there is only a given amount of funds available.

Hence, in group decision making, concessions are made to improve the overall quality of a decision, while in negotiation they are part of the procedure. Moreover, in the former case, concessions change the group decision while in the latter, they affect the communicated proposal and not necessarily the settlement itself. Moreover, in group decision making, hard constraints are known, or become known during the process, to all the participants. In negotiation, each decision maker tries to hide his constraints because revealing them would weaken his bargaining position.

3. FORMAL MODELS FOR GROUP DECISION MAKING AND NEGOTIATION

There are numerous attempts to solve conflicts using theoretical frameworks. The two classical ones are economic models of bargaining and game-theoretic models and both utilize the utility concept.

3.1. Economic Models of Bargaining

Economic models are associated with Zeuthen's (1930) pioneering work and they treat bargaining as a process of convergence over time involving a sequence of offers and counter-offers. These models assume that the utility functions of the participants are fixed and known from the outset, and that a compromise zone exists, can be identified, and remains stable over time (Young, 1975). Moreover, the economic models deal with negotiations involving a single issue that is homogeneous and continuously divisible, such as money (Pen, 1952).

An important group of economic models emphasize the role of time as a bargaining factor (Cross, 1969). In the Cross model, bargaining is a routinized discovery process. The participant starts out with a set of expectations which he/she learns to correct on the basis of experience. The process produces determinate solutions given proper values governing the participants' behavior. If the participants are identical, the model yields the same result as the Nash solution (Cross, 1969, page 59). The Cross model and the one proposed by Rao and Shakun (1974) aim at providing normative recommendations for concession making. Though we are concerned here with both prescriptive and descriptive aspects of the process, with situations involving multiple criteria and «non-rational» aspects of the participants' behavior, the presented approach is partly based on the Cross model.

3.2. Game-Theoretical Models

The second type of approach is based on game theory (von Neumann and Morgerstern, 1947) and its extensions (see for example, Axelrod, 1984; Fraser and Hipel, 1984). The game models assume that the number and identity of players as well as the alternatives and utility functions are fixed and known, that players are fully rational, and that the communication takes place only within the model and it cannot affect either the form or the content of a game's payoff matrix (Luce and Raiffa, 1957).

While research built on game theory has generated important insights into the processes and outcomes of negotiation and bargaining strategies, the restrictive assumptions underlying this framework along with the computational difficulties that arise in all but the simplest cases, have meant that appropriate mathematical constructs are difficult to develop and apply. The limitations of this approach are inherent in the basic method **«... the answers provided by the theory of games are sometimes very puzzling and ambiguous»** (Simon et al., 1987, page 17).

3.3. Aggregation Models

Another often used type of approach recognizes the multiplicity of criteria underlying participants' behavior and aims at developing a decision rule(s). One assumes that the utility functions of each participant are stationary and may be first assessed separately, then aggregated by invoking the utility independence assumption. Under this assumption the bargaining process is reduced to specifying preferences and then combining them for each participant and for the whole group. The obtained group utility function (additive or multiplicative) is used to generate compromises.

The utility aggregation approach may be used in specific group decision making processes (e.g., project or scenario evaluation, joint model building) but not in typical bargaining situations where participants have conflicting objectives, display strategic behavior and withhold preferences and other information. The assumption of independent utility functions has been criticized, and an aggregate group consensus function which does not assume the existence of a von Neumann-Morgerstern utility function for each participant was proposed in the literature.

Other approaches do not require definition of a group utility or consensus function, but use decision rules defined on the alternatives themselves. An example here is ranking of alternatives by each participant and then determining compromise alternatives through expanding-contracting operations on the set of alternatives (Bui, 1985). Jarke et al. (1987) and Shakun (1988) propose another expanding-contracting procedure based on a goal/values referral process. Isermann (1985) assumes that the participants start from inferior alternatives and the negotiation process is a contracting procedure with participants determining the direction of changes. The Isermann model gives every participant the possibility to move from a worse to a better alternative so concessions are unnecessary. Another example of ranking is the use of Saaty's method in group decision making (Lockett et al., 1981).

3.4. Tactics Models

Out of analysis of labor and international negotiations, models that focus upon the tactics of bargaining have been developed (Schelling, 1966; Ikle, 1964). Tactics models lack the restrictive assumptions of the utility-based models. They do not assume that information is complete, nor that utility functions and alternatives are given from the outset and fixed. Manipulating these features of the bargaining situation is considered as one of the most important characteristics of the process.

The level of formalization in manipulative models is low. They cannot specify with determination bargaining outcomes, or attempt to represent negotiations as game-theoretic models.

They also cannot provide a consistent account of concession making as in the economic models or most multiple criteria decision models. Heckarton (1980) considers these features as deficiencies, and they are indeed if a model is to be used to replace the process, to analyze it or verify its efficiency. However, this is not the case if the model is a support tool for a participant or a mediator, if it is used to analyze the current bargaining situation and to verify and determine tactics. Since we are concerned with the latter, we find manipulative models useful.

The tactics models accept «non-rational» aspects of behavior which appear in negotiations. For example, negotiation and bargaining require communication among participants, but what is communicated is influenced by the act of communication. The sanction of communication implies that compromises may depend on the starting point rather than any underlying set of values (Raiffa, 1982, page 215). It further implies that players, while regarding their own behavior as rational, are not immune to strategies of other players (Satterthwaite, 1975). Another outcome of the importance of communication in these processes is that concessions of one player depend on concessions made by others (Komorita, 1973). There are other behaviors which may also not be seen as rational in the traditional sense because an essential part of negotiation and bargaining as a method of making decisions is the communication between players. Since players believe that such communications may be incomplete or inaccurate (at the minimum), their own behavior in response is unlikely to be rational in the traditional sense.

Tactics models assume that decisions, and thus compromises, are time, context, and strategy dependent in negotiation and bargaining settings. Even if the players' explicit utility functions are known, we may determine a compromise decision which they would not accept. This possibility arises because the decision was determined without taking into account the art of negotiation: skills of persuasion and argumentation, the ability to employ bargaining ploys, and the knowledge of how and when to use them. We have also overlooked learning effects; i.e., changes in the knowledge of the problem and of the players.

Kersten and Szapiro (1986, 1988) presented a tactics model of the negotiation process based on the information available to both players and mediators. This model assumes the dynamic nature of negotiation, varying strategies of players and the existence of secret and unveiled goals and interests. This paper is somewhat an extension of this approach and focuses on problems of supporting negotiators in their effort to determine compromise proposals.

4. MCDM AND NEGOTIATION SUPPORT

4.1. The Spaces Where Decisions Can Be Made

The models outlined in section 3 are aimed at describing and analyzing group decision and negotiation processes. Although some of them were computerized, their use was either incidental or limited to a very specific situation. There is an ongoing discussion about the usefulness and scope of use of group decision support systems (GDSS) and negotiation support systems (NSS). These systems with their underlying models and techniques are being developed, tested, and empirically validated. Therefore, generalizations cannot be drawn when answering questions such as: To what extent can GDSS be used for real-world group decision making? Is problem-oriented NSS better than process-oriented NSS? Is it possible to support ill-structured negotiation tasks? etc.

Multiple-Criteria Decision Making (MCDM) is one of the most dynamic areas of research oriented towards the understanding and support of decision making in general, and group decision and negotiation in particular. MCDM provides a framework for group decision and negotiation support. This framework may be oriented around the spaces where individuals can make decisions, in which the decisions can be evaluated and compared by an individual or by the group. More about the issues discussed here can be found in Keeney and Raiffa (1976) and Kersten et al. (1988).

4.2. The Decision Variable Space

The decision variables are the detailed actions that are taken, communicated, and when a compromise is reached, agreed upon. The level of decision variable may be coarse or fine, depending on the situation. For example, it may be the total number of sprockets produced in a month, or a total production time devoted to a class of products. The decision variable is a familiar concept in mathematical programming, where the x vector is the vector of decision variables. The decision variables together with their values which are considered at any point of time constitute a **decision** or a **compromise proposal**. Thus, group decision making may be reduced to an exchange of decisions between the group members, and negotiation - a similar exchange of compromise proposals.

The space of decision variables consists of feasible and unfeasible decisions. In the continuous case, there is a set of constraints which allow determination of the feasibility of a decision, and in the discrete case the feasible decisions can be enumerated. If all the participants in the decision process agree upon these constraints, then they are called hard constraints, in the sense that they are binding for each participant. In negotiations, often the **hard constraints** are discussed during the pre-negotiation phase when the parties determine the definitions of a particular decision variable, or its possible values.

An important feature of the decision variable space is its objectivity. The participants of the process are linked together only through a decision; the decision space provides the platform on which other spaces can be built. The participants differ in the evaluation of a decision but the requirement for the effective negotiation or group decision is the objectivity in understanding a decision.

For some problems, it may be difficult to determine decision variables or it may be convenient not to distinguish between them. A decision about the management strategy for a large corporation often involves too many decision variables to make effective choices. Even in simpler problems, it may be difficult to distinguish between a decision and an objective, as in the case of union/management salary negotiations.

4.3. The Objective Function and Weight Spaces

A second space to consider is the objective function space. The objects in this space are obtained through transformation of the decisions with the use of the objective functions. It is usually a space of lower dimension than the decision variable space which makes it easier to analyze and evaluate different alternative decisions. While the objects in the decision space are neutral, these objects when transformed into the objective function space can be evaluated. We may say, for example, that one value is better than another because it is higher and we are interested in obtaining as high a value as it is possible. Such a statement cannot be formulated about values of a decision variable.

Introduction of the objective function space makes it possible to define a **non-dominated** or efficient decision. A non-dominated decision is a feasible decision for which there is no other feasible decision which yields better values of one or more of the objective functions. Note that we say that a decision is, or is not, non-dominated, but say that we have to perform analysis in the objective function space.

The third space is the weight space, the space of linear positive (non-negative, for convenience) weights. Weights are used to aggregate the objective functions values into one value determining the goodness or the utility of alternative decisions. This aggregate is called the utility function.

There is a relation between the dimension of the weight space and the objective function space. This relation depends on the type of the utility function which is used to describe preferences of a particular decision maker. For example, this function may be linear, or multilinear with weights assigned both to individual objective functions and to their multiplications, or polynomial. In the simple, linear case the weight space is equal to the objective function space; one weight is assigned to one objective so the dimension is equal to the objective function space. (In this case the space of weights may be one dimension less than the objective function space because, without loss of generality, any set of weights may be normalized.)

Both the objective function and weight spaces are used in individual decision support systems. The support provided by, for example, VIG (Korhonen et al., 1986; Korhonen, 1990) is in the objective function space, while TRIMAP (Climaco and Antunes, 1990) and VISA (Belton and Vickers, 1990) support decision making with the use of the weight space.

4.4. The Party Utility and Weight Spaces

The three considered spaces refer to the decision problem or to the multiple criteria of one party. The party utility and weight spaces are introduced because of the multiple agents. The party space has as many dimensions as there are parties and one dimension refers to one individual in the group decision or one party involved in the negotiation. That dimension refers to the utility function of a given individual or party.

There is an analogy between the objective function space and the party space. Therefore one can introduce a concept similar to the non-dominated (efficient) solution. It is the Pareto optimality solutions in the party space. We say that a solution is Pareto optimal if and only if there does not exist another solution for which every party is at least as well off and at least one party is strictly better off. Pareto optimal solutions are considered to be desirable in negotiations. If a solution is not Pareto optimal, that means that one or more parties can negotiate gains at no cost to the other parties. This is possible if the utility functions are mutually independent and this assumption reflects rational behavior of the parties, i.e. one party's increase in its utility does not imply another party's decrease.

Continuing the analogy between the multiple objectives of an individual and multiple individuals, we may define the party weight space. Remarks given with respect to the (individual) weight space do hold here. The party weight space could be introduced, if we were to assume a societal utility function (social welfare function) or a «supra decision maker» (Keeney and Raiffa, 1976).

The party weight space may be used in the mixed «MCDM/voting» support systems. The party utility space is used in, for example MEDIATOR (Jarke et al., 1987) and it is also further discussed in Raiffa (1985). Note that instead of using the party utility space explicitly, one can also use the Cartesian product of the objective function spaces (it is called the Cartesian product). Here the difference between group decision and negotiation is that in the former case, all the dimensions are known to all the individuals, while in the latter one, a negotiating party would know only its own subspace. It would be only the computer (acting as a facilitator or mediator) which would have access to all the information. An extension of VIG from individual to group decision or negotiation would be an example of a support system in the objective function spaces. Another example is NEGO (Kersten, 1985).

The above considerations and the sequence of the geometrical spaces leading from the neutral decision space, possibly with many dimensions to one dimensional supra utility or welfare function, are general in that they do not restrict the possible approaches to NSS and GDSS design. Only one assumption should hold, which is the economic rationality of the involved individuals or parties.

Hence, one can incorporate the aspirations which may cause the objective functions or the utility functions to be defined for the aspiration levels intervals. The reservation prices (the lowest/highest acceptable values), shifts in negotiation tactics, and context dependent concessions can also be introduced. The paradigm is, however, that a Pareto optimal solution is desirable, and that the decision problem can be expressed as a problem of identifying acceptable Pareto optimal solutions.

5. COMPUTER SUPPORT FOR GROUP DECISION MAKING AND NEGOTIATION

One of the issues this paper tries to address is defining the type of computer-based information systems that has the potential to support group decision making and facilitate negotiation processes. The first part of this section describes some conceptual frameworks for GDSS and NSS that were proposed in the literature and summarizes their key features. The second part presents illustrative examples of system implementations in the area of group decision and negotiation support.

5.1. Conceptual Frameworks for GDSS and NSS

DeSanctis and Gallupe (1985) categorized GDSS technology into four separate areas: decision room, local decision network, linked decision rooms, and remote decision networks. They distinguished between the four models in terms of proximity of participants and duration of the decision making session. Their categorization helped define GDSS, their possible configurations and potential applications.

Bui and Jarke (1986) analyzed the communication requirements of various group decision settings. They also suggested an architecture for defining and enforcing dynamic application-level protocols that organize group interaction.

DeSanctis and Gallupe (1987) introduced a typology for GDSS based on three levels representing varying degrees of intervention into the decision process. Level 1 GDSS aim at facilitating communication between decision makers. Level 2 GDSS use decision modelling techniques to support involved parties. Such techniques may include multiple-criteria decision models, multi-attribute utility techniques, decision trees, etc. Level 3 GDSS have built-in knowledge and can provide advice and explanation through the use of expert systems technology. The authors also identified three environmental contingencies as critical to GDSS design: group size, member proximity, and the task confronting the group.

Jelassi and Beauclair (1987) suggested a comprehensive framework that integrates the behavioral characteristics of group decision making with the technical specifications that drive GDSS. Behavioral issues in GDSS include diffusion of responsibility, deindividuation, pressure toward group consensus, and problems of coordination. These issues should be taken into account when designing GDSS «shells». The intervention of such systems aims at reducing the negative impact of these behavioral issues.

Based on the experiments conducted at the University of Arizona, Nunamaker et al. (1989) discussed the interaction of task and technology to support decision groups. They presented several perspectives for studying group environments, including systems theory, communication, decision making, and management science. They argued that a multi-disciplinary approach is necessary for studying such environments.

Pinsonneault and Kraemer (1990) analyzed the empirical research on the impacts of electronic meetings on group processes and outcomes. This analysis is based on a framework developed from the literature on organizational behavior and group psychology. Two specific technologies that facilitate electronic meetings were studied: GDSS and Group Communication Support Systems. Issues that have affected previous research and implications for future work were discussed.

In the NSS Area, a number of conceptual frameworks have been suggested in the literature. The one proposed by Jarke et al. (1987) set the stage for several research efforts. It viewed negotiation as an evolutionary process of problem definition and solution design. In this context, information sharing and exchange coupled with concession making lead the bargaining parties, with the help of a human mediator, toward consensus.

Bui (1985) proposed a consensus-seeking algorithm called the «Negotiable Alternatives Identifier» which could be used in conjunction with preference aggregation techniques. The algorithm is characterized by a triple operation: expansion, contraction, and intersection.

Sycara (1987) developed a general approach for structuring and modelling negotiations. The main emphasis of the approach is on the use of case-based reasoning. The suggested model has been implemented in a computer program for labor-management negotiations.

Kersten (1988) suggested an interactive procedure for group decision making and negotiation. It is based on the aspiration theory and utilizes both satisficing and optimizing approaches. The outcome of the modelled process is a compromise decision which can fulfil fairness and equity criteria and may also be an efficient solution.

Jelassi and Jones (1988) addressed two general questions associated with NSS: How can computers support negotiations? What acceptable systems can we develop to help in such complicated scenarios as multi-party, multi-issue negotiations? These questions were addressed using the analytical mediation process.

Jelassi and Foroughi (1989) examined the issues involved in NSS design based on a review of previous literature. They also discussed current systems that could be used to support the negotiation process. The article laid the foundation for the development of a wide range of individual and group support tools for use in negotiation. After contrasting between «hard» (zero-sum) and «soft» (win-win) negotiations, the authors outlined five factors important to the design of NSS. These factors are: 1) Separate the people from the problem; 2) Provide communication between negotiators; 3) Help negotiators identify their real interests; 4) Generate options for mutual gain; and 5) Use objective criteria. In a subsequent paper, Foroughi and Jelassi (1990) focused on the major stumbling blocks to successful negotiation and suggested ways for alleviating or overcoming them.

In a recent article, Anson and Jelassi (1990) provided a conceptual framework for developing computer-supported conflict resolution. It addressed the following obstacles to integrative bargaining in a mediation setting: cognitive biases of negotiators, socio-emotional factors, and analytical processing difficulties. A number of practical design issues and implementation guidelines were suggested in order to build NSS capable of overcoming the above obstacles.

The conceptual frameworks outlined above and others suggested in the literature are based on different perspectives: management science, operations research, information systems, computer science, organizational behavior, communication, etc. They help advance the concepts of GDSS and NSS by providing enriching and complementary views of group decision and negotiation processes. The next step for researchers in the field is to build computer-based support systems using these frameworks and to empirically test their effectiveness.

5.2. Examples of GDSS and NSS Implementations

An increasing number of researchers are investing considerable time and effort in designing and implementing computer-based information systems for group decision and negotiation support. Anson and Jelassi (1990), Jelassi and Foroughi (1989), Jelassi and Jones (1988), DeSanctis and Gallupe (1987), Kersten (1987), Nyhart and Goeltner (1987) and Kraemer and King (1986) have identified some sixty GDSS and NSS. In the GDSS area, significant experiments took place in North American university laboratories, in particular at the University of Arizona where the PLEXSYS system was developed and at the University of Minnesota, home of the SAMM system.

The University of Arizona has two GDSS rooms in operation. The smaller system (Nunamaker et al., 1987) provides 16 personal computers whose screens are imbedded in a U-shaped table (see Figure 1). A public screen can be used to show what is on the facilitator's screen. The system's program and data reside on a data server that is also used as the facilitator's workstation. The second, larger system provides 24 workstations and two projection screens arranged in amphitheatre style (see Figure 2). This system, which became operational in November 1987, uses the same software. PLEXSYS provides a large number of tools that support brainstorming, issue analysis, voting, stakeholder identification, assumption surfacing, and recording what happened during a meeting. The facilitator's station provides access to and control over the group support tools. The system uses pop-up menus, cursor selection from menus, and keyboard instructions to communicate with the user.

The University of Minnesota GDSS uses 4 microcomputer terminals and its software consists of a public program to manage what appears on each screen and private programs for individual participants. The public program receives all communications from the participants' private programs during the meeting and acts on them. The private program initially presents each participant with a general purpose agenda that includes choices such as problem formulation, criteria/alternatives definition or selection, rating, ranking, and voting. Any participant can ask to view the current state of the decision on the screen, without having to go through the facilitator. Notice that the personal screens face toward the public screen rather than being built into the table (see Figure 3). [For more details on these and other GDSS implementations, the reader is referred to (Gray and Olfman, 1989)].

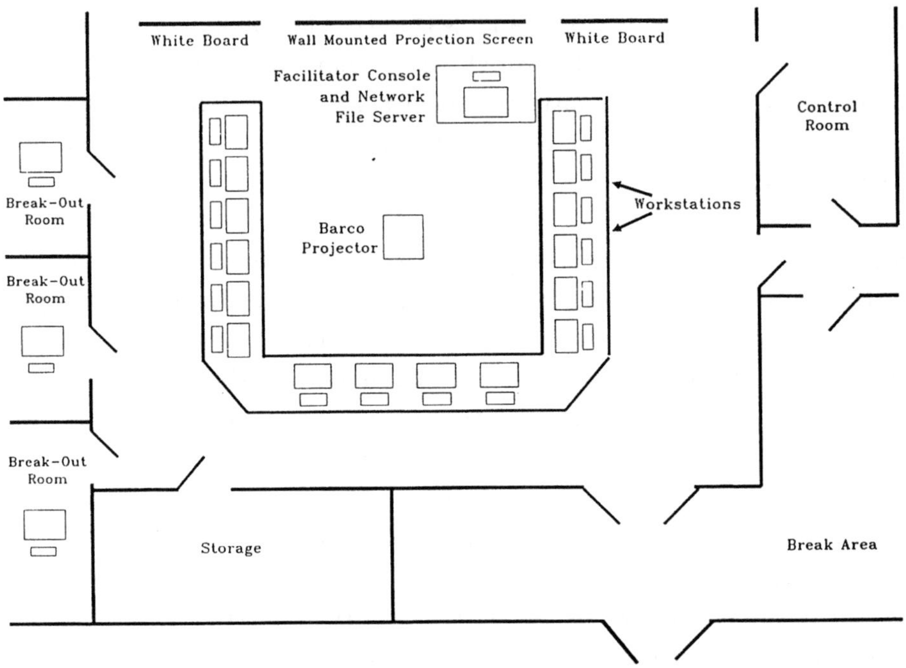

Figure 1. The University of Arizona Small GDSS Facility

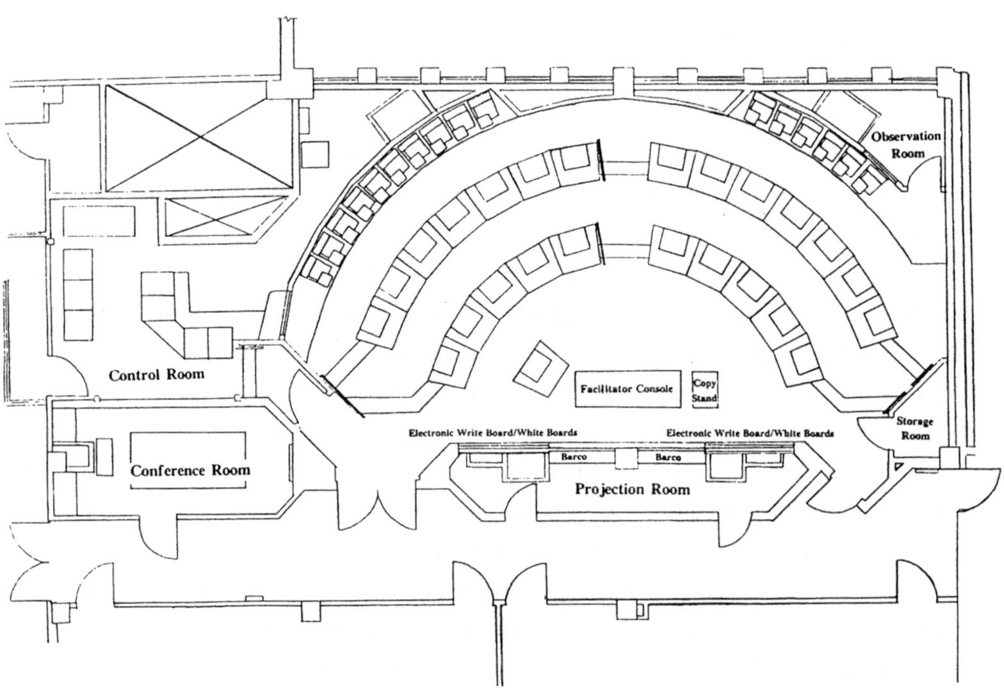

Figure 2. The University of Arizona Large GDSS Facility

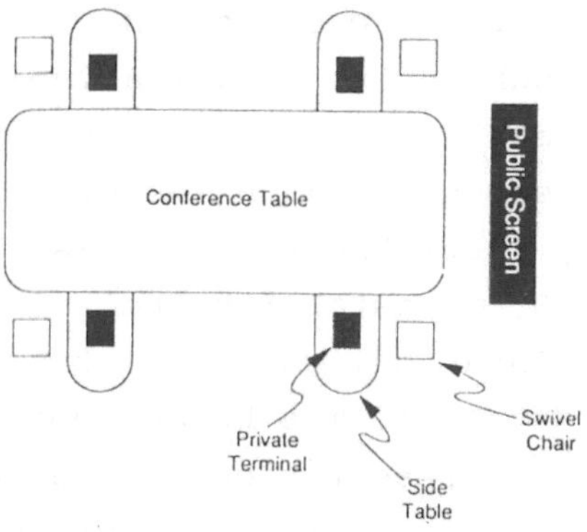

Figure 3. The University of Minnesota GDSS Facility

In the NSS area, existing implementations differ widely in the type and amount of support they can provide. At one end of the spectrum, NSS consist of a computerized model that performs computations or quantitative analysis used during the negotiation process. Such systems serve as «backroom processors» and thus play a rather passive role. Examples of these NSS include DECISION TREE (Winter, 1985) and DECISION MAKER (Fraser and Hipel, 1984). [For a detailed discussion of these systems, see Jelassi and Foroughi, 1989].

Toward the middle of the spectrum of NSS implementations are interactive systems that support a human mediator and/or the negotiating parties. Examples of such implementations include MEDIATOR (Jarke et al., 1987) and GDS1 (Kersten, 1987) which use multicriteria decision making methods, preference elicitation techniques, and electronic communication features.

MEDIATOR is a database-centered, micro-mainframe NSS used to support negotiators and human mediators in solving conflicts. It is applicable during the pre-negotiation stage where players formulate their initial bargaining position. It is also employed in the negotiation stage to help select and evaluate alternatives. MEDIATOR handles subjective (qualitative) and objective (quantitative) data and analyzes decision-maker preferences for possible solutions (agreements). Each negotiating party uses PREFCALC (Euro-Decision, 1986; Lauer and Jelassi, 1987), a single-user multicriteria DSS, to establish their individual preferences and problem representation, which are transferred then to the common (mainframe) database. The human mediator integrates these individual problem representations using relational query language capabilities to form an initial group joint problem representation (Jarke and Jelassi, 1986). Negotiations are undergone by consensus seeking through exchange of information and compromise. MEDIATOR is useful for multi-player, multi-criteria, ill- structured, dynamic problems.

Proceeding along the spectrum, we find two recently proposed systems: the KAJ NSS (Anson and Jelassi, 1990) and MEDIANSS (Carmel and Herniter, 1989). These NSS provide interactive support for the entire negotiation process through the use of GDSS features such as electronic communication and group process structuring techniques, in addition to the employment of quantitative analysis methods and game-theoretical models for conflict resolution. MEDIANSS was strongly influenced by the KAJ model. Negotiating teams are guided by the mediator through a structured set of computerized steps. These consist of the following: rule setting, role reversal, issues and reason identification, issue consolidation, ranking, package creation, proposal presentation, linking, house trading, and agreement wording.

At the extreme end of the spectrum are rule-based NSS which use expert systems techniques to play an active role in the negotiation process (Sycara, 1987). An example of such NSS is RUNE (Kersten et al., 1988) which employs an artificial intelligence approach to problem representation and solution by evaluating bargaining positions and modelling negotiating strategies. However, RUNE supports only one stage of the process, pre-negotiation planning. DeSanctis and Gallupe (1987) as well as Anson and

Jelassi (1990) suggested that rule-based intervention in negotiations could potentially include: 1) analysis of conflict contingencies, 2) suggestion of appropriate process structuring formats or analytical models, 3) monitoring of the semantic content of electronic communications in order to enforce pre-programmed, mutually agreeable interaction norms, 4) suggestion of settlements with high joint benefits, 5) automatic mediation, and 6) automated parlimentary procedure.

Recent research in applying artificial intelligence to decision making (Jelassi et al., 1987) and negotiation (Kersten et al., 1988; Kersten and Szpakowicz, 1990; Sycara, 1987) opens up new possibilities for providing effective support to decision makers as well as bargaining parties. Negotiators' knowledge is one of the forces shaping the process and influencing its outcome. This knowledge increases and dynamically evolves over time and therefore artificial intelligence may well provide the appropriate techniques and tools for its representation and manipulation. The ultimate objective here is to offer negotiating parties a means by which they can **directly** define and evaluate possible settlements. Hence the support system will allow negotiators to act as their own mediator, rather than having to go through a third party, human or machine-based, which is quite often imposed on them. Achieving this objective would be a significant step toward improving the efficiency and effectiveness of the negotiation process.

6. CONCLUDING REMARKS

The aim of this paper was to introduce some formal models of group decision making and negotiation and discuss ways in which recent advances in information technology can be applied to these managerial activities. A special focus was on the concept of computer support for such multi-person, complex, dynamic, and ill-structured processes.

As can be noticed from the sections on conceptual frameworks and actual implementations of group decision and negotiation support systems, significant developments have taken place in the field in spite of its infancy. However, although the potential of GDSS and NSS has been recognized, there is still very little empirical evidence about **how and under what circumstances computer-based tools can best assist decision makers and negotiators.** (For examples of empirical assessments of GDSS and NSS, see, respectively, Nunamaker et al., 1989 and Jones and Jelassi,1990).

There is a need for more rigorous research on the role computers can play in group decisions and conflicts and their impact on the process outcomes as well as on the participants' attitudes. Further conceptual studies are needed in the field, in conjunction with the development of more effective NSS and GDSS technologies. A necessary subsequent step in such a research agenda should be the real-world, «live» testing of these computer-based systems in order to evaluate their actual benefits as well as their shortcomings.

REFERENCES

Anson, R.G. and M.T. Jelassi (1990), "A Development Framework for Computer-Supported Conflict Resolution", *European Journal of Operational Research*, Special Issue on Group Decision and Negotiation Support Systems, forthcoming.

Axelrod, R. (1984), *The Evolution of Cooperation*, New York, N.Y.: Basic Books.

Belton, V. and S. Vickers (1990), "Use of a Simple Multi-Attribute Value Function Incorporating Visual Interactive Sensitivity Analysis for Multiple Criteria Decision Making", in this Volume.

Bui, T.X. (1985), "N.A.I.: A Consensus Seeking Algorithm for Group Decision Support Systems", *IEEE Conference on Cybernetics and Society*, Atlanta, GA., 380-384.

Bui, T.X. and M. Jarke (1986), "Communications Design for Co-oP: A Group Decision Support System", *ACM Transactions on Office Information Systems*, Vol. 4, No. 2, 81-103.

Carmel, E. and B. Herniter (1989), "MEDIANSS: Conceptual Design of a System for Negotiation Sessions", *Transactions of the 9th International Conference on Decision Support Systems*, San Diego, CA.

Clímaco, J. and C.H. Antunes (1990), "A Comparison of Microcomputer Implemented Interactive MOLP Methods Based on a Case Study", in this Volume.

Cross, J.G. (1969), *The Economics of Bargaining*, New York, N.Y.

Dantzig, G.B. and P. Wolfe (1960), "Decomposition Principle for Linear Programs", *Operations Research* Vol. 8, Nº 1, 101-111

DeSanctis, G. and B. Gallupe (1987), "A Foundation for the Study of Group Decision Support Systems", *Management Science*, Vol. 33, No. 5, 589-609.

DeSanctis, G. and B. Gallupe (1985), "Group decision support systems: a new frontier" *Data Base*.

Druckman, D., ed. (1977), *Negotiations. Social-Psychological Perspectives*. Beverly Hills, CA.: Sage Publications.

Euro-Decision, Inc. (1986), PREFCALC User Guide, France.

Fisher R. and W. Ury (1983), *Getting to Yes: Negotiating Agreement Without Giving In*, New York, N.Y.: Penguin Books.

Foroughi, A. and M.T. Jelassi (1990), "NSS Solutions to Major Negotiation Stumbling Blocks", *Proceedings of the 23rd Hawaii Conference on System Sciences*, Kailua-Kona, Hawaii.

Fraser, N.M. and K.W. Hipel (1984), *Conflict Analysis: Models and Resolutions*, New York, N.Y.: North-Holland.

Gonçalves, A.S. (1985), "Group Decision Methodology and Group Decision Support Systems", *Decision Support Systems: The International Journal*.

Gouran, D.S. (1979), *Making Decisions in Groups: Choices and Consequences*, Glenview, IL.: Scott, Foresman & Comp.

Gray, P. and L. Olfman (1989), "The User Interface in Group Decision Support Systems", *Decision Support Systems: The International Journal*, Special Issue on Group Decision Support Systems, Vol. 5, 119-137.

Heckarton D. (1980), "A Unified Model for bargaining and Conflict, *Behavioral Science*, Vol. 25, 261-284.

Holland J.H., K.J. Holyoak, R.E. Nisbett and P.R. Thagard (1987), *Induction: Process of Inference, Learning and Discovery*, Cambridge, MA.: MIT Press.

Ikle, F. (1964), *How Nations Negotiate*, New York, N.Y.: Harper.

Isermann, H. (1985), "Interactive Group Decision Making by Coalitions", in M. Grauer and A. Wierbicki (eds.): *Interactive Decision Analysis*, Berlin: Springer-Verlag.

Jarke, M., M.T. Jelassi and M.F. Shakun (1987), "MEDIATOR: Toward A Negotiation Support System", *European Journal of Operational Research*, Vol. 31, Nº 3, 314-334.

Jarke, M. and M.T. Jelassi (1986), "View Integration in Negotiation Support Systems", *Transactions of the Sixth International Conference on Decision Support Systems*, Washington, D.C., 314-334.

Jelassi, M.T. and R.A. Beauclair (1987), "An Integrated Framework for Group Decision Support Systems Design", *Information & Management*, Volume 3, 143-153.

Jelassi, M.T. and A. Foroughi (1989), "Negotiation Support Systems: An Overview of Design Issues and Existing Software", *Decision Support Systems:The International Journal*, Special Issue on Group Decision Support Systems, Vol. 5, 167-181.

Jelassi, M.T. and B.H. Jones (1988), "Getting to Yes With NSS: How Computers Can Support Negotiation", in R.M. Lee, A.M. McCosh and P. Migliarese (eds.): *Organizational Decision Support Systems*, Amsterdam: North-Holland, 75-85.

Jelassi, M.T., K. Williams and C.S. Fidler (1987), "The Emerging Role of DSS: From Passive to Active", *Decision Support Systems: The International Journal*, Vol. 3, 299-307.

Jones, B.H. and M.T. Jelassi (1990), "The Effect of Computer Intervention and Task Structure on Bargaining Outcome", *Theory and Decision*, forthcoming.

Keeney, R.L. and H. Raiffa (1976), *Decisions with Multiple Objectives: Preferences and Value Tradeoffs*, New York, N.Y.: Wiley.

Kersten, G.E. (1988), "A Procedure for Negotiating Efficient and Non-Efficient Compromises", *Decision Support Systems: The International Journal*, Vol. 4, 167-177.

Kersten, G.E. (1987), "On Two Roles Decision Support Systems Can Play in Negotiations", *Information Processing and Management*, Vol. 23, № 5, 605-614.

Kersten, G.E. (1985), "NEGO - Group Decision Support System", *Information and Management*, Vol. 8, № 5, 237-246.

Kersten, G.E., W. Michalowski, S. Matwin and S. Szpakowicz (1988), "Representing the Negotiation Process with a Rule-based Formalism", *Theory and Decision*, Vol. 25, 225-257.

Kersten, G.E. and S. Szpakowicz (1990), "Rule-based Formalism and Preference Representation: An Extension of NEGOPLAN", *European Journal of Operational Research*, Special Issue on Group Decision and Negotiation Support Systems, forthcoming.

Kersten, G.E. and T. Szapiro (1986), "Generalized Approach to Modelling Negotiations", *European Journal of Operational Research*, Vol. 26, № 1, 142-149.

Kersten, G.E. and T. Szapiro (1988), "Negotiating of Quasi-Concave Aspiration Functions", *Belgian Journal of Operational Research, Statistics and Computer Science*, Vol. 27, № 3, 17-42.

Komorita, S.S. (1973), "Concession-Making and Conflict Resolution", *Journal of Conflict Resolution*, Vol. 17, № 4, 745-763.

Korhonen, P. et al. (1986), "An Interactive Approach to Multiple Criteria Optimization with Multiple Decision-Makers", *Naval Research Logistics Quarterly*, Vol. 33, 589-602.

Korhonen, P. (1990), "The Multiobjective Linear Programming Decision Support System VIG and Its Applications", in this Volume.

Kraemer, K. and J. King (1986), "Computer-Based Systems for Group Decision Support: Status of Use and Problems in Development", *Proceedings of the First Conference on Computer-Supported Cooperative Work*, Austin, TX., 353-375.

Lauer, T.W. and M.T. Jelassi (1987), "PREFCALC - A Multi-Criteria Decision Support System: User Tutorial", Institute for Research on the Management of Information Systems, Indiana University, Working Paper #W714.

Lewicki, R.J. and J.A. Litterer (1985a), *Negotiation*. Homewood, IL.: Irwin.

Lewicki, R.J. and J.A. Litterer, (1985b), *Negotiation. Readings, Exercises, and Cases*. Homewood, IL.: Irwin.

Lockett, A.G., A.P. Muhlemann and A.E. Gear (1981), "Group Decision Making and Multiple Criteria - A Documented Application" in: *Organizations: Multiple Agents with Multiple Criteria*, J.N. Morse (ed.), New York N.Y.: Springer-Verlag.

Lockhart, C. (1979), *Bargaining in International Conflicts*, New York, N.Y.: Columbia University Press.

Luce, R.D. and H. Raiffa (1957), *Games and Decisions*, New York, N.Y.: Wiley.

MacLean, D. (1985), "Rationality and Equivalent Redescriptions", in *Plural Rationality and Interactive Decision Processes*, M. Grauer et al. (eds.), Berlin: Springer-Verlag.

Nierenberg, G.I. (1973), *Fundamentals of Negotiating* New York, N.Y.: Hawthorn Books.

Nunamaker, J.F., L.M. Applegate, and B.R. Konsynski (1987), "Facilitating Group Creativity: Experience with a Group Decision Support System", *Journal of Management Information Systems*, 5-19.

Nunamaker, J.F., D. Vogel, and B. Konsynski (1989), "Interaction of Task and Technology to Support Large Groups", *Decision Support Systems: The International Journal*, Special Issue on Group Decision Support Systems, Vol. 5, 139-152.

Nunamaker, J.F., D. Vogel, A. Heminger, B. Martz, R. Grohowski, and C. McGoff (1989), "Experiences at IBM with Group Support Systems: A Field Study", *Decision Support Systems: The International Journal*, Special Issue on Group Decision Support Systems, 183-196.

Nyhart, J.D. and Ch. Goeltner (1987), "Computer Models as Support for Complex Negotiations", *International Conference of the Society for General System Research*, Budapest, 40-48.

Pen, J. (1952), "A General Theory of Bargaining", *The American Economic Review*, Vol. 17, 24-42.

Pinsonneault, A. and K.L. Kraemer (1990), "The Effects of Electronic Meetings on Group Processes and Outcomes: An Assessment of the Empirical Research", *European Journal of Operational Research*, Special Issue on Group Decision and Negotiation Support Systems, forthcoming.

Raiffa, H. (1985), "Back from Prospect Theory to Utility Theory", in Grauer, M. et al. (eds.): *Plural Rationality and Interactive Decision Processes*, Berlin: Springer-Verlag.

Raiffa, H. (1982), *The Art and Science of Negotiations*, Cambridge, MA.: Harvard University Press.

Rao, A.G. and M.F. Shakun (1974), "A Normative Model for Negotiations", *Management Science*, Vol. 20, Nº 10.

Satterthwaite, M.A. (1975), "Strategy-Proofness and Arrow's Conditions: Existence and Correspondence Theorems for Voting and Social Welfare Functions", *Journal of Economic Theory*, Nº 10, 187-217.

Schaffers, H. (1985), "Design of Computer Support for Multicriteria and Multiperson Decisions in Water Resource Planning", in G. Fandel and J. Spronk (eds.): *Multiple Criteria Decision Methods and Applications*, Berlin: Springer-Verlag.

Schelling, T.C. (1966), *Arms and Influence*, New Haven, CT.: Yale University Press.

Shakun, M.F. (1988), *Evolutionary Systems Design: Policy Making Under Complexity and Group Decision Support systems*. Oakland, CA.: Holden-Day.

Simon, H.A. et al. (1987), "Decision making and Problem Solving", *Interfaces*, Vol. 17, Nº 5, 11-31.

Steiner, I.D. (1972), *Group Process and Productivity* New York, N.Y.: Academic Press.

Sycara, K. (1987), "Resolving Adversarial Conflicts: An Approach Integrating Case-based Reasoning and Analytic Methods", Ph.D Thesis, School of Information and Computer Science, Georgia Institute of Technology.

Tietz, R. and O.J.Bartos (1983), "Balancing of Aspiration Levels as Fairness Principle in Negotiations" in R. Tietz (ed.) *Aspiration Levels in Bargaining and Economic Decision Making*, Berlin: Springer-Verlag.

Tversky, A. and D. Kahneman (1981), "The Framing of Decisions and the Psychology of Choice", *Science*, Vol. 211.

Von Neumann, J. and O. Morgerstern (1947), *Theory of Games and Economic Behavior*, Princeton, N.J.: Princeton University Press.

Winter, F. (1985), "An Application of Computerized Decision Tree Models in Management-Union Bargaining", *Interfaces*, No. 15, 74-80.

Young, O.R. (1975), "Strategic Interaction and Bargaining", in O.R. Young (ed.), *Bargaining: Formal Theories of Negotiations*, Urbana, IL.: University of Illinois Press.

Zeuthen, F. (1930), *Problems of Monopoly and Economic Warfare*, London: Routledge & Kegan Paul.

GROUP DECISION MAKING: METHODOLOGY AND APPLICATIONS

Günter Fandel
Fernuniversität Hagen
D - 5800 Hagen 1 - WEST GERMANY

Group decision problems are decision problems with several decision makers and different utility functions. Game and bargaining approaches can be taken into account as solution methods. They are characterized by the actual decision rule which describes, or rather determines, the decision behaviour of the group members.

1. DESCRIPTION OF THE DECISION SITUATIONS IN GROUPS

For the formal description of group problems let

- $\mathbb{N}$ be the set of the natural numbers,
- $\mathbb{R}$ be the set of the real numbers,
- $n \in [N] = \{1,...,N\}$ be the decision makers - units or persons - in the group, $N \in \mathbb{N}$ and $N \geq 2$,
- $A \subset \mathbb{R}^N$ be the set of decision alternatives $a = (a_1,...,a_N)$ of the group, and
- $U \subset \mathbb{R}^N$ be the set of utility vectors $u = (u_1,...,u_N)$ of the group which develop as a mapping of A under the individual utility functions $u_n = u_n(a)$, $n \in [N]$, of the decision makers, that is to say $U = u(A)$.

In order to obtain a reasonable economic and mathematical formulation of the probelm let us further assume that the set of decision alternatives A is convex, bounded and closed, and that the utility functions u_n, $n \in [N]$, are concave, continuous and different from each other, that is to say, $u_n \neq u_{n'}$ is true for n, $n' \in [N]$ and $n \neq n'$ in particular. Without loss of generality we further suppose that every decision maker n controls one and only one decision component a_n of a vector a, where $a_n \in A_N$ and $a \in A = A_1 \times \ldots \times A_N$; A_n designates the set of decision alternatives of the n-th decision maker.

Then the group decision problem consists of choosing alternatives $a \in A$ or, equivalently, in determining utility vectors $u \in U$ that the decision makers will regard as solutions to their decision process. The common decision rule of the group members determines the choice of such $a \in A$ or $u \in U$. This rule can at the same time serve to characterize the solution approach used. In this connection, the notion of decision rule means an operation $Q: \mathbb{R}^N \to \mathbb{R}^N$, which for each utility set $U \subset \mathbb{R}^N$ chooses a subset $L_Q \subseteq U$, and thus for each decision set $A \subset \mathbb{R}^N$ a subset of decision alternatives A_Q with $u(A_Q) = L_Q$. $L_Q = \{u' \in U \mid u' = Q[u(a)],\ a \in A\}$ can be designated as set of the Q-optimal solutions to the group decision problem.

For practical reasons the solution L_Q of the decision problem is to fulfil the following requirements:

$$L_Q \neq \phi, \tag{A1}$$

that is to say, there must exist at least one solution to each decision problem.

$$L_Q \subset U, \tag{A2}$$

that is to say, only such utility vectors will be suitable for solutions which can be obtained by corresponding feasible decision alternatives.

Designate $M(U) := \{u \in U \mid u \geq \bar{\bar{u}}\}$ where the utility vector u indicates the utility level $\bar{\bar{u}}_n = \max_{a_n} \min_{\bar{a}_n} u_n(a_n, \bar{a}_n)$, $\bar{a}_n = (a_1, \ldots, a_{n-1}, a_{n+1}, \ldots, a_N)$, $(a_n, \bar{a}_n) \in A$ and $n \in [N]$ which the individual decision makers can at least obtain within the group, then let

$$L_Q \subset M(U). \tag{A3}$$

Thus postulate (A3) requires the solution to be individually rational (Luce and Raiffa, 1958, pp. 192/193).

Let $P(U) = \{u \in U \mid w \geq u \wedge w \in U \Rightarrow w = u\}$ be the set of all Pareto-optimal utility vectors of U. Then, the solution is to satisfy the condition

$$L_Q \subset P(U), \tag{A4}$$

that is to say, consider efficient results only.

L_Q is a one-element set (A5)

insures the uniqueness of the solution.

The conditions (A1)-(A5) allow a comment on the quality of the solution proposals still to be presented, that means, a comment with respect to their contribution to the optimal decision in groups. In this connection, existence, feasibility and individual rationality of the solution are quite obvious postulates derived from plausibility assumptions. From the economic point of view the requirement of Pareto-optimality corresponds to the use of synergic effects which can emerge in groups due to the joint effort of several members. Under the conditions (A1)-(A4) the necessity of a unique solution results from the following fact: If there exist several Pareto-optimal solutions then some decision makers will profit more from one result vector than from another, and vice versa, so that with interests conflicting, a final solution to the decision problem has not yet been found. Furthermore, uniqueness is necessary for a stable or equilibrated decision behaviour of the persons involved (Harsanyi, 1963, p. 219; Friedman, 1971, p. 7).

2. GAME- AND BARGAINING-THEORETIC CONTRIBUTIONS TO THE SOLVING OF GROUPS DECISION PROBLEMS

2.1. Game-Theoretic Solution Approaches

a) For two-person cooperative games without side payments Nash (1953, p. 136ff.) has indicated an axiomatically founded solution which can be extended to N persons and is then characterized as follows:

$$L_Q = \{u^* \in U \mid \prod_{n=1}^{N} (u_n^* - t_n) = \max_{a \in A} \prod_{n=1}^{N} [u_n(a) - t_n],\ t \in U\}.$$

In this expression $t \in U$ designates a disagreement vector - which is not Pareto-optimal - from the interior of U; t may be given definitely by the rules of the game, or be determinable by threat strategies (Harsanyi, 1963, p. 195ff.) of the players. The optimal solutions u^* are then characterized by the fact that they maximize the product of all utility increases with respect to the disagreement vector $t \in U$ for the decision makers involved. Obviously, this so-called cooperative Nash solution satisfies the requirements (A1) and (A2), as well as postulate (A3) after the construction of t. On account of the strictly monotonically increasing and strictly convex goal precept which follows from the underlying axioms, the u^* are Pareto-optimal and with the possible unique choice of the disagreement vector $t \in U$ also unique, that is to say, they satisfy the conditions (A4) and (A5). Thus, as a whole, the cooperative Nash concept is well suitable for solving group decision problems. Furthermore, it has the properties which are desirable for utility-theoretic considerations, that the solution u^* is invariant with respect to linear utility transformations, symmetric with respect to the decision makers and independent of irrelevant decision alternatives.

b) For solving of N-Person cooperative games with side payments and transferable utilities Shapley (1953) has formulated the value $\gamma(v)$ of a game. Being defined on the characteristic function v which describes the game it assigns the payoff $\gamma_n(v)$ to each of the decision makers $n \in [N]$ at the end of the game. In this connection $v: P[N] \rightarrow \mathbb{R}$ is a mapping of the power set of [N] into the real numbers, and for each coalition $\bar{S}$, $\bar{S} \in P[N]$ or $\bar{S} \subseteq \bar{N}$, $v(\bar{S})$ indicates the common payoff under transferable utilities, which it can obtain by the maximin strategy at the expense of the coalition $\bar{N}-\bar{S}$ of the other players:

$$v(\bar{S}) = \max_{a^{\bar{S}}} \min_{a^{\bar{N}-\bar{S}}} \sum_{n \in \bar{S}} u_n(a^{\bar{S}}, a^{\bar{N}-\bar{S}}) \text{ for all } \bar{S} \subseteq \bar{N},\ a=(a^{\bar{S}}, a^{\bar{N}-\bar{S}}) \in A; \quad (1)$$

$\bar{N}$ designates the coalition consisting of all decision makers. Because of the axiomatic requirements which, according to Shapley, the mathematical structure of γ has to satisfy it can be shown that there exists a unique function γ describing the value of the game for each player $n \in [N]$ and reading as follows:

$$\gamma_n(v) = \sum_{\bar{S} \subseteq \bar{N}} \frac{(S-1)!\ (N-S)!}{N!} [v(\bar{S}) - v(\bar{S} - \{n\})],$$
$$n \in [N], \quad S = \| \bar{S} \| \text{ and } N = \| \bar{N} \|. \quad (2)$$

The solution corresponding to the Shapley value can be formally represented as follows:

$$L_Q = \{u^* \in U \mid u^* = \gamma(v)\}.$$

With regard to expression (1) existence and uniqueness of the solution u* follow directly from formula (2). Pareto-optimality of u* is guaranteed by the axiomatic construction of γ, since the maximal payoff which can jointly be obtained for all players will be distributed fully to them due to the solution vector u*, that is to say $\sum_{n=1}^{N} u_n^* = v(\bar{N})$ holds. u* continues to be individually rational since the function γ satisfies the conditions $\gamma_n(v) \geq v(\{n\})$ for all $n \in [N]$. The feasibility of u* is ascribable to the classification of the considered game by admitted side payments and transferable utilities, since the utility set is then characterized by

$$U = \{u \mid \sum_{n=1}^{N} u_n \leq v(\bar{N})\} \tag{3}$$

and $u^* \in U$ holds because of Pareto-optimality. Thus, in case of a possible equivalent mapping of the group decision problem by the game situation discussed here, all requirements (A1)-(A5) for its solution are satisfied, so that in this sense the Shapley value can serve as a concept of solution. Its workability in real cases, however, is questionable due to the fact that it can only be used for solving games with transferable utilities in which side payments take place. Under practical aspects these assumptions are critical and clearly limit the efficiency of the Shapley value with respect to the solution of decision problems in organizations.

c) Shapley himself has indicated a way which allows the extension of his solution idea to cooperative games without side payments; thus it is made more attractive for the application to organizational decision problems. As opposed to the concept of value of the game this proposal is referred to as evaluation of a game (Shapley, 1964; Shapley and Shubik, 1969). The starting point for deriving the Shapley evaluation once more is the characteristic function v which, however, is now given by the mapping $v: P([N]) \rightarrow \mathbb{R}^S$, $S \in [N]$, since there are no side payments. It assigns to each coalition $\bar{S} \subset \bar{N}$ a subset $v(\bar{S}) \subset \mathbb{R}^S$, $S = \|\bar{S}\|$, of feasible payoff vectors. The sets $v(\bar{S})$ are subsets of the utility set $U = v(\bar{N}) \subset \mathbb{R}^N$, that is to say $v(\bar{S}) \subseteq v(\bar{N})$ for all $\bar{S} \subseteq \bar{N}$; they are assumed to be convex, closed and nonempty. Its vectors $u^N = (u_n)_{n \in \bar{S}} \in v(\bar{S})$ result from the projection of corresponding $u \in U$ into $\mathbb{R}^S$, and consequently contain just as many components as there are members in the coalition $\bar{S}$. In order to extend the Shapley approach represented in b) to game situations of this kind the following procedure is taken for determining the Shapley evaluation.

1. By rescaling the utility function of the decision makers by a vector

$$\lambda = (\lambda_1, \ldots, \lambda_N) \geq 0, \; \lambda \neq 0, \tag{4}$$

the utility set $U=v[\bar{N}]$ of the cooperative game without side payments will be transformed into the utility set

$U' = v'(N) = (u' \mid u' = (\lambda_1 u_1, \ldots, \lambda_N u_N) =: >\lambda, u<, \; u \in U)$ of another cooperative game without side payments, where

$$v'(\bar{S}) = \{u'^{\bar{S}} \mid u'^{\bar{S}} \Rightarrow \lambda^{\bar{S}}, u^{\bar{S}}<, \; u^{\bar{S}} \in v(\bar{S})\}, \bar{S} \subseteq \bar{N} \text{ and } \lambda^{\bar{S}} = (\lambda_n)_{n \in \bar{S}} \tag{5}$$

holds.

2. The cooperative game without side payments with the transformed characteristic function v' is now treated as a corresponding cooperative game with side payments and transferable utilities. The latter then possesses the characteristic functions $v''(S)$.

3. On the basis of v'' compute the Shapley value

$$\gamma(v'') = \tilde{u} \tag{6}$$

according to (2).

4. If now $\tilde{u} \in U' = v'(\bar{N})$ holds, that is to say, if there exists a $u^* \in U = v(\bar{N})$ with the property

$$\gamma(v'') = \tilde{u} \Rightarrow \lambda, \; u^*<, \tag{7}$$

then $\tilde{u}$ - and consequently u^* - can be obtained also without drawing on side payments. $u^* \in U$ which, after rescaling the utility functions, corresponds to $\tilde{u} \in U'$ on returning to the original cooperative game without side payments is to be regarded as the solution of this game, that is to say, let

$$L_Q = \{u^* \in U \mid >\lambda,\ u^*< \ = \ \gamma(v'')\}.$$

$\lambda.u^*$ is referred to by Shapley as the evaluation of a game if and only if λ and u^* satisfy the requirements (4) and (7).

The Shapley evaluation $\lambda.u^*$ fulfils the existence postulate (A1) for any finite N-person cooperative game without side payments (Shapley, 1964); according to its construction the appropriate solution vector u^* will then also obey the requirements (A2)-(A4). As opposed to the Shapley value, in this case the uniqueness is dependent on the possible unique choice of the scaling vector λ, so that the fulfilment of (A5) cannot generally be insured. Precisely this deficiency, however, gives rise to the strongest objections as regards the practical use of the Shapley evaluation for solving organizational decision problems. An additional difficulty is the fact that the relative utility positions of the decision makers shift in the solution vector u^* when the relative utility weights are changed by the choice of λ (Fandel 1979, p. 52). Therefore, in higher-dimensional problems it is hardly any longer possible to predict in which way solution u^* will behave in case of variation of λ if there exist several evaluations $\lambda.u^*$ for a cooperative game without side payments.

d) Extending the minimax criterion developed for two-person zerosum games Nash (1951) designates the set of equilibrium decisions a^* in common N-person non-cooperative games as their solution. This so-called non-cooperative Nash solution formally reads as follows for the considered group decision situation

$$L_Q = \{u^* \in U \mid u^*=u(a^*),\ a^* \in A, \text{ and } u_n(a^*) = \max_{a_n \in A_n} u_n(a_n, \bar{a}^*_n) \text{ for all } n \in [N]\}.$$

Nash has shown that each non-cooperative game of this kind possesses at least one equilibrium vector $a^* \in A$. Such equilibria are at the same time feasible and individually rational. As opposed to these positive statements with respect to the postulates (A1)-(A3) the requirements (A4) of Pareto-optimality and (A5) of uniqueness cannot normally be insured for the non-cooperative equilibria (Luce and Raiffa, 1958, p. 106ff.; Shubik, 1960). Therefore, the non-cooperative Nash concept cannot generally be considered to be a satisfactory approach to the solution of group decision problems.

e) In order to come from non-efficient equilibria in non-cooperative games to such equilibria with Pareto-optimal utility vectors Friedman (1971) starts from the formulation of a super-game consisting in the infinite periodical repetition of a given normal game. To deal with it, a new class of non-cooperative supergame equilibria is introduced by definition, first assuming that the normal game possesses only a (non-efficient) equilibrium $c \in A$. In this connection for each decision maker $n \in [N]$ a super-game strategy σ'_n on the basis of a decision vector $a' \in A$ which strictly dominates the normal equilibrium c with respect to the utility - that is to say, for which $u(a') >> u(c)$ holds - may be constructed as follows:

$$\sigma'_n = (a_{n1}, a_{n2}, \ldots, a_{nt}, \ldots) \text{ with}$$
$$a_{n1} = a'_n,$$

$$a_{nt} = \begin{cases} a'_n, \text{ if } a_{n'\tau} = a'_n, \ n \neq n', \ \tau=1, \ldots, t-1, \ t > 1, \\ c_n \text{ otherwise.} \end{cases} \tag{8}$$

Allowing (8), the supergame strategy $\sigma^{a'} = (a', a', \ldots)$ represents a non-cooperative equilibrium if it fulfils the condition

$$\sum_{t=1}^{\infty} \alpha_n^{t-1} u_n(a') > u_n(\bar{a}'_n, b_n) + \sum_{t=2}^{\infty} \alpha_n^{t-1} u_n(c) \text{ for all } n \in [N]. \tag{9}$$

α_n designates the discount rate of the decision maker n; it is constant for all periods t. $u_n(\bar{a}'_n, b_n) = \max\{u_n(\bar{a}'_n, a_n) \mid a_n \in A_n\}$, $n \in [N]$, indicates the maximal yield which he can achieve at the expense of all other players by deviating once from a'. Since condition (9) after splitting, applying the sum formula and regrouping, is equivalent to

$$u_n(a') > u_n(c)+(1-\alpha_n)[u_n(\bar{a}'_n,b_n)-u_n(c)] \equiv u^+_n(a'), \; n \in [N], \tag{10}$$

the class of the non-cooperative supergame equilibria can now be described in the utility space by the following set $U_G^{\sigma'} \subset U$:

$$U_G^{\sigma'} = \{u \in U \,|\, u(a') >> u^+(a') \text{ and } u(a') >> u(c), \; a' \in A\}. \tag{11}$$

If (10) is transformed into (12)

$$\frac{\alpha_n}{1-\alpha_n}[u_n(a')-u_n(c)] > u_n(\bar{a}'_n,b_n)-u_n(a'), \; n \in [N], \tag{12}$$

it can be seen the supergame strategy $\sigma^a=(a, a, \ldots)$, resulting from the infinite repetition of a strategy $a \in A$ of the normal game dominating the equilibrium c belongs to $U_G^{\sigma'}$, if for each decision maker the single net gain obtainable by deviating from a - right hand side of the inequation (12) - is smaller than the cash value of the permanent utility losses to be expected on account of the reaction of the partners - left hand side of the inequation (12). According to (8) any other supergame strategy σ^n for player n is weakly dominated either by $\sigma'_n = (a_{n1}, a_{n2}, \ldots)$ or $\sigma''_n=(b_n, c_n, c_n, \ldots)$.

Now, in order to particularly mark a Pareto-optimal σ^{a*} in the sense of (8) as solution among the equilibria of the supergame, Friedman proposes that the temptation to deviate from a* is to be equally large for all players, that means

$$L_Q=\{u^* \in U \;|\; u^*=u(a^*) \in P(U_G^{\sigma'}) \text{ and } v_n(a^*)=v_{n'}(a^*) \text{ for all } n, n' \in [N]\}$$

The temptation to deviate $v_n(a^*)$ is defined by:

$$v_n(a^*) = \frac{u_n(\bar{a}^*_n, b_n) - u_n(a^*)}{u_n(a^*) - u_n(c)}, \; n \in [N]. \tag{13}$$

The existence of such a decision alternatives a* is not generally guaranteed since the set $U_G^{\sigma'}$ - and thus also L_Q - can become empty accordding to condition (10) in case of small discount rates (that is high time preferences) of the decision makers. For sufficiently high discount rates, however, the fulfilment of postulate (A1) can be insured. In this case a* will also satisfy the requirements (A2)-(A4). The uniqueness of the solution, however, cannot be guaranteed at the same time (Friedman, 1971, p. 8ff.), which means that on account of a possible violante of (A5) the Friedman concept can be used for solving the group decision problem in special cases only.

2.2. Bargaining-Theoretic Solution Approaches

a) The N-person bargaining model developed by Harsanyi (1963) is based on the idea to generalize the cooperative Nash solution (Nash, 1950 and 1953) which has been conceived for two-person decisions. As opposed to the determination of solutions described in 1.a) the N-person decision on the Harsanyi model, however, must first be decomposed into a set of two-person subgames between all possible pairs n and n' from [N] according to the mathematical concept of the theory of bargaining. Allowing for their interdependence the resulting subgame will then have to be formulated so mutually consistently with respect to the partial conditions of solutions that subsequently, the total solution of the group decision problem can be composed of them in the form of an equilibrium strictly taking two-person subgames as a basis.

Let the function f(u)=0 describe the efficient border P(U) of the utility set U in parametric form, and let it be differentiable. With the analytic properties of the cooperative Nash solution for the two-person decision problem the optimal total solution to the general N-person bargaining problem

$$L_Q=\{u^* \in U \mid u^* = \bar{u}^N\}$$

which in the Harsanyi model has successively been composed of the solutions to two-party subgames can then be characterized by the following system of necessary conditions (Harsanyi, 1963, pp. 214/215):

$$f(u^{\bar{N}}) = 0 \tag{14.1}$$

$$c_n = f_n(u^{\bar{N}}) = \partial f/\partial u_n \mid u^{\bar{N}};\ n \in \bar{N}; \tag{14.2}$$

$$u_n^{\bar{S}} = u_n(a^{\bar{S}^o}, a^{\bar{S}'^o});\ n \in \bar{S},\ \bar{S} \subset \bar{N},\ \bar{S}' = \bar{N} - \bar{S} \tag{14.3}$$

$$t_n^{\bar{S}} = \sum_{\substack{\bar{R} \subset \bar{S} \\ n \in \bar{R}}} (-1)^{s-r+1} u_n^{\bar{R}};\ n \in \bar{S};\ \bar{S} \subseteq \bar{N},\ s = \|\bar{S}\| > 1,\ r = \|\bar{R}\|; \tag{14.4}$$

$$c_n(u_n^{\bar{N}} - t_n^{\bar{N}}) = c_{n'}(u_{n'}^{\bar{N}} - t_{n'}^{\bar{N}}) \text{ for all } n, n' \in \bar{N}; \tag{14.5}$$

$$\sum_{n \in \bar{S}} c_n u_n^{\bar{S}} - \sum_{n' \in \bar{S}'} c_{n'} u_{n'}^{\bar{S}'}$$

$$= \sum_{n \in \bar{S}} c_n u_n(a^{\bar{S}^o}, a^{\bar{S}'^o}) - \sum_{n' \in \bar{S}'} c_{n'} u_{n'}(a^{\bar{S}^o}, a^{\bar{S}'^o}) \tag{14.6}$$

$$= \max_{a^{\bar{S}} \in A^{\bar{S}}} \ \min_{a^{\bar{S}'} \in A^{\bar{S}'}} \left[\sum_{n \in \bar{S}} c_n u_n(a^{\bar{S}}, a^{\bar{S}'}) - \sum_{n' \in \bar{S}'} c_{n'} u_{n'}(a^{\bar{S}}, a^{\bar{S}'}) \right],$$

$$\bar{S},\ \bar{S}' \subset \bar{N},\ A^{\bar{S}} = \mathop{\mathsf{X}}_{n \in \bar{S}} A_n;$$

$$\text{subject to: } c_n(u_n^{\bar{S}} - t_n^{\bar{S}}) = c_k(u_k^{\bar{S}} - t_k^{\bar{S}}),\ n, k \in \bar{S};$$

$$c_{n'}(u_{n'}^{\bar{S}'} - t_{n'}^{\bar{S}'}) = c_l(u_l^{\bar{S}'} - t_l^{\bar{S}'}),\ n', l \in \bar{S}'.$$

Conditions (14.1), (14.2), (14.5) and (14.6) are expressive of the fact that for a consistent construction of the bargaining result according to the cooperative Nash concept, the criteria of optimality of the total solution must be of prime importance for all two-person or two-party subgames, too. According to (14.3) the utility of the decision maker n in the coalition $\bar{S}$ is dependent on the choice of optimal threat strategies $a^{\bar{S}^o}$ and $a^{\bar{S}'^o}$ by the two coalitions $\bar{S}$ and $\bar{S}'$. When he joins this coalition his disagreement payoff, however, according to (14.4) will consist of the cumulated utility increases achieved by him in all subcoalitions $\bar{R} \subset \bar{C}$ of which he was a member previously.

The optimal solution $u^{\bar{N}}$ of the problem (14.1) - (14.6) which, in its formulation, tries to make the most perfect use of the Nash axioms for two-person cooperative games with respect to the theory of bargaining in general, fulfils the requirements (A1)-(A4). Difficulties, however, in unrestrictedly accepting $u^*=u^{\bar{N}}$ as optimal solution to the group decision problem may arise from the fact that the disagreement payoffs are variable on account of threat strategies, so that the solution $u^{\bar{N}}$ need not necessarily be unique. This violation of (A5) is, however, avoidable by uniquely presetting $t_n^{\bar{S}}$.

b) As an alternative to such game-theoretically founded approaches of the theory of bargaining Contini and Zionts (1968) have conceived a concessive bargaining model in which the solution to the group decision problem is simultaneously determined by agreement of all group members. This agreement is reached under threat of an imposed solution $y \in U$ by means of a process of concession which is continuous with respect to time and leads to the solution

$$L_Q = \{u^* \in U \mid u^* \geq z(t) \geq y\}.$$

Here z(t) designates the decision makers' aspiration levels at time t, declining in the course of the process of concession. At the beginning let $z(0)=\hat{u}$, that is to say, let the aspiration levels tally with the individual utility maxima of the group at time t=0. The concession behaviour of the decision makers which is achieved by the threat of an imposed solution is described by the following system of conditions:

$$dz_n(t)/dt = \begin{cases} k_n(t)<0, \text{ if } L_Q = \phi \text{ and } z_n(t)>y_n \text{ ,} \\ \\ 0 \text{ otherwise,} \end{cases} \qquad n \in [N]. \qquad (15)$$

Consequently, every decision maker is ready to make concessions only as long as no feasible solution has been found and his aspiration level remains above the payoff which would be yielded to him in case of the imposed solution. If one of the conditions does no longer apply, then the concession rate $k_n(t)$ will become equal to zero.

In order to insure according to the postulate (A1) the existence of a solution, that is to say $L_Q \neq \phi$, the functions $k_n(t)$ are subjected to the following additional sufficient conditions that for every $k_n(t)$ there exists a t'_n, $0 \leq t'_n < \infty$, with

$$- \int_0^{t'_n} k_n(\tau)\ d(\tau) = \hat{u}_n - y_n, \qquad n \in [N]. \qquad (16)$$

Thus, even if the condition $L_Q = \phi$ in (15) is neglected the concession path $z(t)$ for a finite $t' = \max \{t'_n \mid n \in [N]\}$ would at nay rate have at least to lead to the imposed solution $y \in U$ as stationary solution. With the existence the feasibility requirement (A2) is at the same time fulfilled by the solution u^*. The individual rationality (postulate (A3)) of u^*, however, is not automatically given; it is rather dependent on the clever choice of the imposed solution y. Sufficient for the guarantee of individually rational solutions is $y \geq \bar{u}$. Pareto-optimality and uniqueness of the solution u^*, as well, cannot generally be insured without additional conditions. Contini and Zionts have shown that (A4) and (A5) are fulfilled if the utility set is strictly convex. It is also sufficient if for the location of the imposed solution $y \in U$ $\eta(y) \cap R(U) \subseteq P(U)$ holds with $\eta(y) = \{u \in \mathbb{R}^N \mid u \geq y\}$ and and $R(U)$ as border of the utility set U (Fandel 1979, p. 115).

It is evident that the concessive bargaining model developed by Contini and Zionts can be used for solving the group decision problem on certain additional conditions only, the existence of which must always be separately examined. Nevertheless, this concept with its spontaneous elements of behaviour - expressed in concession rates - may by all means be regarded as a practice-oriented alternative to the axiomatically founded solution approaches. The question, however, in how far the concession rates of the decision makers can rationally be accounted for, remains still to be answered. This problem will be treated in the next section.

2.3. On the Rationalizing of the Concession Behaviour

It is unsatisfactory, that so far there are only two-person concepts available for rationally elaborating the concession behaviour characterized according to (15) (Fandel 1979, p. 120 ff.). Their ability to give information about the solution of the group decision problem is limited correspondingly; yet, they indicate possibilities of development with respect to elaborating the rational foundations of the concessive bargaining models under the aspect of their-usefulness in practice.

Most of the approaches can be reduced to the idea of the bargaining theorem developed by Zeuthen (1930, chapter 4). According to this theorem the concession behaviour for jointly determining a Pareto-optimal solution vector $u^* \in P(U)$ depends on the subjective probabilities of conflict which can maximally be accepted by the two bargaining partners; the former result from the Pareto-optimal proposals and counterproposals which are made by the two players in the course of the process of concession. At the beginning let u^1 and u^2, $u^1 \neq u^2$, be the proposal and couter-proposal of the first and the second decision maker, respectively. Let $\bar{u} \in S$ represent the disagreement or conflict point, and let $U' = \{u \in U \mid u \geq \bar{u}\}$. Then the following conclusions can be drawn with respect to the first decision maker:

1. He can definitely achieve the utility u_1^2 if he accepts the counterproposal u^2 of the second decision maker (action 1).

2. If, however, he rejects u^2 persisting in his proposal u^1 (action 2) the expected value of his utility will be $\bar{u}_1 \pi + u_1^1(1-\pi)$ where π indicates the probability that the decision maker 2 will risk a conflict.

The subjective conflict probability p_1 which is maximally acceptable for decision maker 1 and with which he is just able to keep up his proposal then results from the indifference between the two actions indicated, that is to say, if one has

$$u_1^2 = \bar{u}_1 \, p_1 + u_1^1 \, (1-p_1) \text{ or } p_1 = (u_1^1 - u_1^2) \,/\, (u_1^1 - \bar{u}_1). \tag{17}$$

Correspondingly, for the second decision maker, one obtains:

$$p_2=(u_2^2 - u_2^1)/(u_2^2 - \bar{u}_2). \tag{18}$$

According to Zeuthen now a concession is made by the decision maker $n \in \{1,2\}$ for whom

$$p_n \le p_{n'}; \; p_n, \, p_{n'} > 0 \text{ and } n, \, n' \in \{1,2\}, \; n \neq n' \tag{19}$$

holds, that is to say, by that decision maker who is not able to cope with a greater probability of conflict than his bargaining partner. The concession consists of a new proposal (e.g. $\hat{u}^2$ instead of u^2) which causes the probability of conflict maximally endurable to become greater again than that of the contrahent, thus forcing the latter to make concessions for his part. This process is iteratively continued until a common agreement solution $u^* \in P(U')$ is obtained. If, allowing for the expressions (17) and (18) condition (19) is transformed into the equivalent inequation

$$(u_n^n-\bar{u}_n)(u_{n'}^n-\bar{u}_{n'}) \le (u_n^{n'}-\bar{u}_n)(u_{n'}^{n'}-\bar{u}_{n'}); \quad n, n' \in \{1,2\}, \; n\neq n' \tag{20}$$

for N=2, it can be seen that the rationalizing of the concession behaviour in the bargaining theorem by Zeuthen is identical with the axioms of the cooperative Nash solution (see section 2.1.a)). The embedding of a concession behaviour thus rationalized in the system (15) of the bargaining model by Contini and Zionts is quite obvious now if there the process is discretized by proposals and counterproposals $u^n(t) \in P(U')$, $n \in \{1,2\}$, at times t, t=0,1,2,... Using the old symbols we then have:

$$y=\bar{u},\; z_n(0)= \hat{u}_n = \max\{u_n \mid u \in U'\} = u_n^n(0)$$

and

$$z_n(t+1)-z_n(t) = \begin{cases} k_n(t)<0, \text{ if } L_Q=\phi \text{ and } \prod_{r=1}^{2} [u_r^n(t)-\bar{u}_r] \le \prod_{r=1}^{2} [u_r^{n'}(t)-\bar{u}_r] \\ 0 \text{ otherwise}, \\ n,n' \in \{1,2\} \text{ and } n \neq n' \end{cases} \qquad (15')$$

with

$$\begin{cases} k_n \; u_n^n(t+1)-u_n^n(t) \text{ and} \\ u^n(t+1) \text{ with the property that } n \; u \in P(U') \\ \text{exists with } u_n > u_n^n(t+1) \text{ and} \\ \prod_{r=1}^{2} (u_r-\bar{u}_r) > \prod_{r=1}^{2} [u_r^n(t+1)-\bar{u}_r], \\ n \in \{1,2\} \text{ and } t \in \{0, 1, 2,...\}. \end{cases}$$

The extension of the thus described rational concession behaviour according to (20) or (15′) to N persons, N>2, is obvious, but it can no longer formally be concluded from the bargaining theorem by Zeuthen (condition 19). The above remarks show, however, that the combination of game-theoretic axioms and concessive bargaining models which are based on spontaneous elements of bahaviour may yield valuable suggestions for the solving of group decision problems.

3. APPLICATION OF GROUP DECISION MAKING CONCEPTS TO WAGE BARGAINING

3.1. Preliminary Remarks

Wage bargaining may formally be treated as decision problems among two parties with different utility functions. The decision consists of determining the increase of the wage rate for the time of the tariff contract commonly accepted by the both tariff parties representing the employers and the employees respectively. So far, on principle, the same solution methods of the game and the bargaining theory can be applied to such decision situations as they come into question for multi-person-decision problems in groups. However because of the special problem structure in the case of wage bargaining one may restrict mainly to cooperative solution conccepts without side payments.

In the following especially it will be analysed, how far the methodical approaches of Nash (1953), Zeuthen (1930) and Contini and Zionts (1968) can be used in practice in order to explain and reconstruct the courses of and the agreements within wage bargaining processes. Thereby the investigations refer to wage bargaining in the metal industry of the Federal Republic of Germany between 1961 and 1979.

The subject of the analysis will be the wage disputes between "Gesamt-metall" as organization representing the employers and "IG Metall" (Metalworkers Union) as the employees' represention in the Federal Republic of Germany in 1961/62 to 1979. The years 1964, 1967 and 1972/73, however, had later to be excluded from the study, since in these years agreements were either reached unusually quickly on account of political or economic events or were simply taken over from other areas; in these cases there were not any signs of bargaining processes.

3.2. The Naive Solution and the Two-Thirds Hypothesis

Introducing the application-oriented considerations let us first of all deal with an assumption which is often expressed in connection with the wage disputes in the metal-processing industry with respect to the contract eventually signed by workers and employers. This assumption is based on a general rule of the economic practice and says that, usually, the later agreement with respect to the wage increase rate lies halfway between the union's initial claim and the employers' initial ofter, that is to say approximately corresponds to the arithmetic mean of these two quantities (naive solution). The initial claim, the initial ofter and the agreement are to be expressed in percentages of the last basic wage.

In order to be able to verify the validity of this assumption by means of the regression analysis Table 1 has been made up; it specifies the initial claims of IG Metall $\bar{w}_t$, the initial offers made by Gesamt-metall $\underline{w}_t$, the actual contracts w_t^*, the naive solution $\tilde{w}_t$ as well as the two-thirds values w_t' for the years t from 1961/62 to 1979.

Setting up the homogeneous linear regression equation

$$w_t^* = \beta \tilde{w}_t + \epsilon_t \tag{21}$$

for the connection between the actual contract and the naive solution where the perturbation variable ϵ_t is to be normally distributed with the expected value zero, one obtains the regression line

$$\hat{w}_t = 0{,}96 \tilde{w}_t \tag{21'}$$

with the correlation coefficients r = 0,97. On account of this correlation coefficient, the statistically founded correlation, and the fact that the regression coefficient β is near one it may reasonably be said that the connection expressed by this assumption cannot simply be brushed aside.

Table 1: Survey of the wage negotiations 1961/62 to 1979

Year t	Initial claim of IG Metall $\bar{w}_t$	Initial offer made by Gesamtmetall $\underline{w}_t$	Actual contract achieved w_t^*	Naive solution $\tilde{w}_t = \frac{1}{2}(\bar{w}_t + \underline{w}_t)$
1961-62	10.0	1.5	6.0	5.75
1963	8.0	3.0	5.67	5.5
1965-66	9.0	2.4	6.63	5.7
1968	7.0	4.0	5.5	5.5
1969	12.0	5.0	8.7	8.5
1970	15.0	7.0	11.0	11.0
1971	11.0	4.5	7.0	7.75
1974	18.0	8.5	12.25	13.25
1975	11.0	6.0	6.8	8.5
1976	8.0	4.0	5.4	6.0
1977	9.5	4.75	6.9	7.125
1978	8.0	3.0	5.4	5.5
1979	6.0	3.0	4.3	4.5

3.3. Application of the Cooperative Nash Solution

The close connection between the naive solution and the actual agreement which has been described in the last section calls for an enquiry of the question to what extent the agreement points of the wage negotiations are interpretable in the sense of the cooperative Nash solution, since in case of linearly transferable utility this game-theoretic concept divides the bargaining cake available between two parties into equal shares. For an analytic reconstruction of this possibility of interpretation, let us assume in the following that the cake to be divided between IG Metall and Gesamtmetall is each time defined by the difference of the changes of the wage sums resulting from the initial claim of the union and the initial offer made by the employers, thus reads

$$K_t = (\bar{w}_t - \underline{w}_t) \cdot L_t \qquad (22)$$

and is consequently fixed, L_t denoting the wage sum before the wage dispute in the year t, and workers and employers possessing linearly homogeneous utility functions with respect to the shares which they eventually obtain of this bargaining cake. If IG

Metall and Gesamt-metall are assigned the indices n=1 and n=2 respectively, the utility functions remaining constant over the years with respect to the shares can be written as follows:

$$u_{1t} = a. \frac{w_t - \underline{w}_t}{\bar{w}_t - \underline{w}_t} \tag{23}$$

and

$$u_{2t} = b. \frac{\bar{w}_t - w_t}{\bar{w}_t - \underline{w}_t} \tag{24}$$

respectively, the quantities a and b indicating the constant marginal utility. It is easy to see that the utility of IG Metall (Gesamtmetall) grows (declines) linearly with the rising wage increase rate w_t which the two parties have to agree upon in year t and for which $\underline{w}_t \leq w_t \leq \bar{w}_t$ holds in general. For $w_t = \bar{w}_t$ IG Metall reaches the highest utility, that of Gesamtmetall becoming equal to zero; correspondingly the ratio is inverse for $w_t = \underline{w}_t$. Moreover, let $\bar{w}_t > \underline{w}_t$ be presupposed. The cooperative Nash solution (Nash 1953) is characterized by the fact that the two negotiating parties involved in the case considered here agree upon a wage increase rate w_t^* or, which is equivalent, upon shares of the bargaining cake by which the product of their utility increases with respect to a disagreement vector $\bar{u} = (\bar{u}_1, \bar{u}_2)$ is maximized. To simplify matters the disagreement vector can be fixed by the zero utility levels of the negotiation parties. The following assumption which is fairly plausible is to justify this understanding: the bargaining cake corresponds to that part of the return of production for which in future they will have to work in common and the distribution of which to the factors labour and capital must be agreed upon within the framework of the wage disputes; if one of the parties claims the total share this will be met by strike and lockout measures respectively by the other party; in such a case, the burden for the fighting fund on the union's part on the one side is opposite to the capital expenditure for the plant facilities on the employer's part on the other.

Assuming a disagreement vector $\bar{u} = (\bar{u}_1, \bar{u}_2) = (0, 0)$ in this sense, and considering (23) and (24) the cooperative Nash solution can be determined as follows:

$$\max u_t = (u_{1t} - 0)(u_{2t} - 0) = u_{1t} \cdot u_{2t} = a.\frac{w_t - \underline{w}_t}{\bar{w}_t - \underline{w}_t} \, . \, b.\frac{\bar{w}_t - w_t}{\bar{w}_t - \underline{w}_t} \tag{25}$$

$$= \frac{a.b}{(\bar{w}_t - \underline{w}_t)^2} (w_t - \underline{w}_t)(\bar{w}_t - w_t) = c(w_t - \underline{w}_t)(\bar{w}_t - w_t).$$

This expression is exclusively dependent on the wage increase rate w_t as variable which has to be determined optimally by both parties in the form of an agreement. As necessary condition for determining such an optimal w_t^o one obtains from (25)

$$\frac{du_t}{dw_t} = c[-(w_t - \underline{w}_t) + (\bar{w}_t - w_t)] = c(-2w_t + \bar{w}_t + \underline{w}_t) = 0 \tag{26}$$

and from this, because of $c \neq 0$,

$$w_t^o = \frac{\bar{w}_t + \underline{w}_t}{2}. \tag{27}$$

Consequently, assuming the utility functions in (23) and (24) to apply, and taking the assumptions concerning the bargaining of the two parties and their disagreement vector as a basis, the cooperative Nash solution w_t^o tallies with the naive solution $\tilde{w}_t$, and its explanatory value with respect to the actual agreements reached w_t^*, as may be seen from Table 1, can be estimated accordingly. In this connection, the cooperative Nash solution has been derived from the initial claim $\bar{w}_t$ of the union and the initial offer $\underline{w}_t$ made by employers; no statement, however, has thus been made about how these two initial values were obtained. As far as this is concerned, it may be enough to say that both employers and employees probably take the data of the past or the future economic trend as a basis; due to its limited methodical performance the Nash concept does certainly not allow these data to be elucidated and verified. Similarly, the deviations of the actual contracts w_t^* from the analytically derived values w_t^o cannot be explained on the basis of the Nash approach. Since these deviations are not too important, as a rule, they could be ascribed to the differences which are usually to be found between the rationally postulated and the empirically observable decision behaviour. Assuming constant marginal utilities for both parties, however, seems to be unproblematic in view of the fact that wage movements are generally of special economic importance, and IG Metall as well as Gesamtmetall represent a very large number of persons interested.

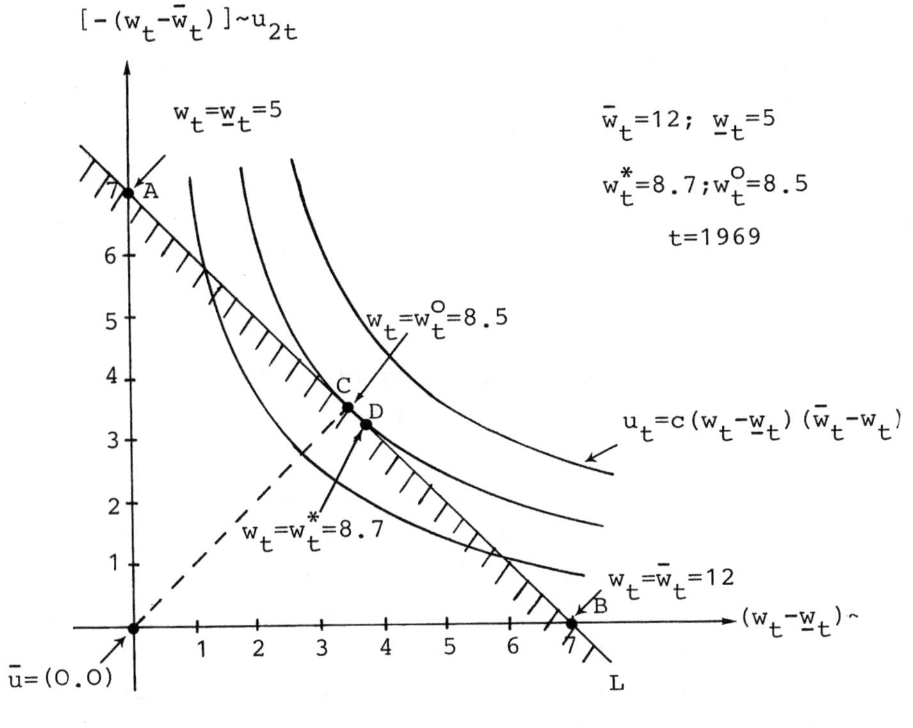

Figure 1. Nash solution w_t^o and actual agreement reached w_t^* for the wage negotiations in 1969.

For the wage negotiations of 1969 the above-mentioned analytic solution according to Nash is graphically represented in Figure 1. Here the coordinate axes are denoted by the utility arguments $(w_t-\underline{w}_t)$ and $(\bar{w}_t-w_t)$ or $[-(w_t-\bar{w}_t)]$ from the expression in (25), so that the utility values u_{1t} and u_{2t} of both parties increase positively with the direction of the coordinates. Between the points A and B line L marks all contracts w_t for which $\underline{w}_t=5 \leq w_t \leq 12=\bar{w}_t$ holds.

While any points above L are not feasible since for them the union's claims are always higher than the offers made by the employers, which means that no agrements can be reached there, the points below L represent a waste of the cake to be divided. From the condition of optimality $w_t^o - \underline{w}_t = \bar{w}_t - w_t^o$ according to (26) follows the Nash solution $w_t^o = 8.5$, with $8.5 - 5 = 12 - 8{,}5 = 3.5$, which in Figure 1 is near point C; the actual contract signed $w_t^* = 8.7$, however, corresponds to point D on L. As can be seen from (26) and (27) w_t^o is independent of parameter c in (25) due to the invariance of the cooperative Nash solution as to linear utility transformations; this parameter only represents a level constant with respect to the product of the utility increases which has to be maximized in common by both contrahents in comparison with the disagreement vector.

3.4. Verification of the Wage Bargaining Processes with the Aid of the Theorem by Zeuthen

The bargaining theorem by Zeuthen (1930), which serves for rationalizing the concession behaviour of decision makers in conflicting decision situations is identical with the axioms of the cooperative Nash solution, so that both concepts are equivalent with respect to determining an optimal agreement solution between the bargaining partners. But in comparison with the cooperative Nash solution, the bargaining theorem by Zeuthen represents a much more efficient instrument for verifying the behaviour-theoretic consistency of the decisions made by the bargaining partners. While according to Nash the optimal solution is determined statically, the bargaining theorem by Zeuthen, in order to reach this aim, requires a dynamic process consisting of proposals and counterproposals which consider the concessions made by the parties; in this way, the optimal solution is interactively approximated step by step and eventually reached.

Table 2

ʼey of the claims and offers $w_t^n(r_t)$ (in terms of bargaining rounds) submitted by IG Mettal (n=1)
Gesamtmettal (n=2) during the bargaining processes 1961/62 to 1979.

t	1961/62		1963		1965/66		1968		1969		1970		1971		1974		1975		1976		1977		1978		1979	
n	1	2	1	2	1	2	1	2	1	2	1	2	1	2	1	2	1	2	1	2	1	2	1	2	1	2
	10		8		9		7		12		15		11		18		11		8		9.5		8		6	
		1.5		3		2.4		4		5		7		4.5		8.5		6		4		4.75		3		3
	10		7		9		6.5		11		15		7.9		14		7		8		9.5		8		6	
		3		3.5		5.8		4.57		5		10		6		11.33		6.5		4.5		5.25		4.6		4
	8.5		6		5.4		6.5		10		11		7.5		12.25		6.8		5.4		7		5.7		4.3	
		4		4.5		5.8		4.67		7		11		6.5		12.25		6.8		5.4		6.5		5		4.3
	6		5.5		5		5		8.7				7								6.9		5.4			
		6		5.5		6.63		4.83		8.7				7								6.9		5.4		
			5.67		6.63		5.5																			
				5.67				5.5																		
			1)		1)																					

ʼhese values are not considered in the verification according to the theorem by Zeuthen

In order to be able to properly check the rationality of the concessions made by IG Metall and Gesamtmetall during the pay talks of 1961/62 to 1979 according to Zeuthen while strictly maintaining the utility functions introduced in (23) to (25) the respective claims and offers, which were submitted by the bargaining partners until the agreement was reached, have been compiled in Table 2 for these different wage disputes in terms of bargaining rounds (Krelle, 1976, p. 617ff.). The claims and offers are given in wage increase rates and according to (25) can immediately be converted into the utility values that are necessary for applying the theorem of Zeuthen; to simplify matters, the constant c can be neglected here, i.e. set equal to one. Simultaneously, Table 2 shows that the wage negotiations are always to begin with an initial claim by the unions in the first round, the following claims are always to be made in the respective odd rounds, and the offers by Gesamtmetall are to be made only in the respective even rounds. Moreover, as far as those cases are concerned in which between two different claims and/or offers made by the one side there was no reaction by the other side, it has been presupposed that the other side has maintained its last claim and/or its last ofter in the intermediate round. In 1963 the last two rounds and in 1965/66 the last four rounds are not taken into account in the following considerations, since in these two years agreements could have been reached earlier, but the negotiations were continued for the time being due to the fact that additional claims were dropped and finally a higher wage increase rate was agreed upon. Accordingly, it should be noted that in 1968 the wage claim of the union increases again in round 9; consequently, no concession is made.

Let $R_t = \{r_t \mid r_t=1,2,\ldots,\bar{r}_t\}$ denote the set of round indices in year t, where $\bar{r}_t$ indicates the number of bargaining rounds in this year required for a (possible) agreement. Let $w_t^n(r_t)$ characterize the proposal made by partner n, $n \in \{1,2\}$ in the round $r_t \in R_t$ of year t; in this connection it should be pointed out that according to the arrangement described in the last paragraph index n=1 (IG Metall) can appear only in the case of odd and index n=2 (Gesamtmetall) only in the case of even round indices. Moreover, let $\hat{R}_t = \{\hat{r}_t \mid \hat{r}_t \in R_t\}$ be the set of round indices of year t in a ascending order, for which one of the bargaining partners makes a concession, thus $w_t^1(\hat{r}_t) < w_t^1(\hat{r}_t - 2)$ holds for the union, and $w_t^2(\hat{r}_t) > w_t^2(\hat{r}_t - 2)$ holds for Gesamtmetall with $\hat{r}_t \in \hat{R}_t$ and $(\hat{r}_t - 2) \in R_t$. Then let the mapping $\rho : \hat{R}_t \rightarrow \{1,2\}$ be defined by the fact that it indicates for each concession round $\hat{r}_t \in \hat{R}_t$ that bargaining partner $\rho(\hat{r}_t) \in \{1,2\}$ who made the concession in this round. The new suggestion $w_t^n(\hat{r}_t)$, $n \in \{1,2\}$ of that bargaining partner which results from this concession leads - according to (25) - to the product of the utility increases

$$u_t^n(\hat{r}) = c[w_t^n(\hat{r}_t) - \underline{w}_t][\bar{w}_t - w_t^n(\hat{r}_t)]. \tag{28}$$

Moreover, in order that this concession may be rational according to the theorem by Zeuthen

$$u_t^n(\hat{r}_t - 2) \leq u_t^{n'}(\hat{r}_t - 1);\ n, n' \in \{1,2\},\ n \neq n';\ (\hat{r}_t - 2), (\hat{r}_t - 1) \in R_t \tag{29}$$

must hold.

If the fact whether such a concession was rational or not is then mapped by the binary attribute function $\gamma(\hat{r}_t)$ with

$$\gamma(\hat{r}_t) = \begin{cases} 1, \text{ if } \rho(\hat{r}_t) = n \text{ and } u_t^n(\hat{r}_t - 2) \leq u_t^{n'}(\hat{r}_t - 1) \\ 0 \text{ otherwise} \end{cases} \tag{30}$$

then the result of the analysis can be illustrated in a simple form by Table 3. Here index i indicates the concession steps of the years 1961/62 to 1979. In order to be able to partly reconstruct the evaluation of the results of Table 3, by way of explanation the derivation of the results for the year t = 1969 is demonstrated in Table 4 using the symbols introduced. The common utility value $u_t^n(r_t)$ resulting from the proposals and counter-proposals of the partners can be marked as points on Line L in Figure 1; for the sake of clarity, however, this has not been done in this case.

Table 3

Result of the verification with respect to Zeuthen rationality of the concessions made by IG Metall and Gesamtmetall during the wage negotiations 1961-62 to 1979

t	1961-62					1963						1965-1966		1968					
r_t	4	5	6	7	8	3	4	5	6	7	8	4	5	3	4	6	7	8	10
$\rho(r_t)$	2	1	2	1	2	1	2	1	2	1	2	2	1	1	2	2	1	2	2
$\gamma(r_t)$	1	1	1	1	1	1	1	0	1	0	1	1	1	1	1	0	1	1	1
i	1	2	3	4	5	6	7	8	9	10	11	12	13	14	15	16	17	18	19

t	1969					1970			1971						1974			
r_t	3	5	6	7	8	4	5	6	3	4	5	6	7	8	3	4	5	6
$\rho(r_t)$	1	1	2	1	2	2	1	2	1	2	1	2	1	2	1	2	1	2
$\gamma(r_t)$	1	0	1	1	1	1	1	1	1	1	0	1	0	1	1	1	0	1
i	20	21	22	23	24	25	26	27	28	29	30	31	32	33	34	35	36	37

t	1975				1976			1977					1978					1979		
r_t	3	4	5	6	4	5	6	4	5	6	7	8	4	5	6	7	8	4	5	6
$\rho(r_t)$	1	2	1	2	2	1	2	2	1	2	1	2	1	1	2	1	2	2	1	2
$\gamma(r_t)$	1	1	0	1	1	1	1	1	1	1	0	1	1	1	1	0	1	1	1	1
i	38	39	40	41	42	43	44	45	46	47	48	49	50	51	52	53	54	55	56	57

As can be seen from Table 3, 47 out of 57 considered concession steps by the bargaining partners in the years 1961/62 to 1979 are rational according to Zeuthen; thus for the proportional value f of the Zeuthen steps

$$\hat{f}(\gamma=1) = \frac{47}{57} = 0.82 \tag{31}$$

holds.

Table 4

Derivation of the results in table 3 for the year t=1969

r_t	$w_t^1(r_t)$	$w_t^2(r_t)$	$\hat{r}_t$	$\rho(\hat{r}_t)$	$u_t^n(r_t)$	$\gamma(\hat{r}_t)$	i
1	12		-	-	0	-	-
2		5	-	-	0	-	-
3	11		3	1	6	1	20
4		5	-	-	0	-	-
5	10		5	1	10	0	21
6		7	6	2	10	1	22
7	8.7		7	1	12.21	1	23
8		8.7	8	2	12.21	1	24

If one starts from the assumption that γ is a random variable of quantitative binary quality and the concession steps of the years 1961/62 to 1979 represent a sample of the size m=57 from an infinite dichotomic parent entirety consisting of the set of all concession steps made by the bargaining partners IG Metall and Gesamtmetall in the Federal Republic of Germany and for whose elements only the property $\gamma=1$ or $\gamma=0$ is interesting, then the number of Zeuthen steps in such a sample of the size m is binomially distributed with the parameters m and f, where f may designate the percentage of Zeuthen steps of the finite parent entirety. If on this basis the zero hypothesis is tested, that is to say the percentage of Zeuthen steps in the parent entirety is at the most 70%, i.e.

$$H_0\colon\ f \le f_0 = 0.70 \tag{32}$$

and if the sample proportion $\hat{f} = 0.82$ is chosen as suitable test variable which, due to $m.f_0.(1-f_0) = 57 \,.\, 0.7 \,.\, 0.3 = 11.97 > 9$ can be assumed to be approximately normally distributed, then the zero hypothesis is rejected with an error probability $\alpha = 0.05$ since the test variable $\hat{f}$ exceeds the upper limit c_0 of the zero hypothesis (Wetzel, 1973, p. 195 ff.), i.e.

$$\hat{f} = 0.82 > 0.81 = c_0 \tag{33}$$

Thus by inverse conclusion, it may be stated with 95% certainty that the percentage of Zeuthen steps in the parent entirety is higher than 70%. Moreover, if on account of the sample one determines the confidence interval for the parameter f on the level significance of $\alpha = 0.1$, where because of $m.\hat{f}(1-\hat{f}) = 57 \,.\, 0.82 \,.\, 0.18 = 8.41 < 9$ an aproximation of the binominal distribuition by the normal distribution is not possible, but the F distribution must be used, then one obtains the interval

$$f \in [0.72;\ 0.90], \qquad (34)$$

that is to say, with 90% certainty the percentage of the Zeuthen steps in the parent entirety is between 72 and 90%. The latter two statements which have been derived statistically on the basis of the empirical analysis suggest that the theorem by Zeuthen, together with the utility functions (23) and (24) assumed when applying the cooperative Nash solution in the last section, is by all means acceptable as a methodical basis for substantiating the wage bargaining processes in the metal-processing industry under decision-theoretic aspects. Thus simultaneously, the explanatory value of the cooperative Nash solution and the naive solution is persistently backed.

3.5. Discussion of the Concept of Concession Functions Developed by Contini and Zionts

In this last section, eventually the question is to be examined to what extent the data available for the wage negotiation processes of 1961/62 to 1979 can be regarded as a basis for proving the existence of concession functions such as they have been assumed in the spontaneous bargaining model by Contini and Zionts (1968), and whether such concession functions actually follow decreasing convex courses, which is stated on account of empirical investigations. Convex decreasing concessions - counted in positive quantities - however, call for like-wise convex decreasing utility values of that bargaining partner who makes the concessions. But on the other hand, according to (23) and

(24) convex decreasing utility values are identical to the fact that the wage claims by IG Metall decrease convexly and/or the wage offers made by Gesamtmetall increase concavely. In order to simplify matters the discussion about concession functions can so far be reduced to the following two partial aspects:

1) Are there certain rules applying to the annual wage claim and/or wage offer curves of the bargaining partners which find expression in uniform function parameters, and

2) do the wage claim curves of the union and/or the wage offer curves of the employers take decreasing convex and/or increasing concave courses?

In order to clarify these two partial aspects a series of methodical restrictions and simplifying additional conditions are required on account of the nature of the object to be investigated. For one thing, the data consisting of three to five wage claims and/or offers per year by the bargaining partners is rather meager, so that, when deriving annual wage claim and/or wage offer functions for IG Metall and Gesamtmetall with the aid of the regression analysis, sometimes a statistical proof of the correlation for time dependent wage proposals cannot be established. As far as the time variable is concerned the wage bargaining processes can be considered in terms of days or bargaining rounds. While generally the scantiness of the data has to be put up with, thus not representing a specific objection, the preliminary investigations have revealed that with respect to the regressions for the description of time dependent wage proposals for both bargaining partners, the analysis in terms of rounds is to be preferred to the analysis in terms of days, since in the latter case no usable results are obtained. Under the first partial aspect of uniform function parameters, this is quite obvious, since the duration of the wage negotiations - between 10 days in 1969 and 86 days in 1978 - was subject to substancial variations, a consideration in terms of bargaining rounds, however, leads to something like a standardization with respect to time. In view of these preliminary reflections, the regression analysis has eventually been based on the data prepared in terms of bargaining rounds in Table 2.

With respect to the clarification of the second partial aspect a further problem consists in choosing such regression approaches which due to their mathematical structure do not ex ante determine the individual course of the curve, but which rather leave it open for the time being, admitting as results convex as well as concave structures. In view of the limited quantity of data available and the number of function parameters to be estimated, it seems reasonable here to confine oneself to such regression approaches of the logarithmic, exponential and potential forms which are linearizable, thus representing no computational difficulties. As things are, of these regression approaches only the potential approach is eligible for deriving wage offer functions for Gesamtmetall if for positively increasing wage offers the question concerning concavity or convexity of the course of the curves is to remain freely answerable. Circumstances are less favourable with respect to IG Metall, since there in all regression approaches mentioned positively decreasing wage claims would mean that simultaneously the course of the curve of the wage claim function is definitely fixed, this course being for example ex ante convex in the potential approach. In order to overcome this difficulty, a non-linearizable exponential approach has been used which showed that the estimated wage claim functions of IG Metall follow concave courses for the four years 1961/62, 1965/66, 1970 and 1979, but in fact take convex courses for the other nine years. On account of these additional considerations, eventually a potential approach has been chosen for the wage claim curves of IG Metall too. Moreover, with respect to both bargaining partners in an ex-post analysis, the potential approach has proved quite successful in comparison with the other linearizable approaches without being dominant in each individual case.

Finally, another critical point to be dealt with in this connection is the fact that in the regression approach continuos courses of the annual wage claim and/or wage ofter functions of the two bargaining partners are assumed which could contradict the discrete round-oriented expressions of wage proposals made by the two bargaining partners. This discrepancy might perhaps be modified by making the following helpful assumption: All rounds are considered to be of equal lenght, at the end of the individual rounds the bargaining partners are asked by turns to submit their new wage proposals, the wage claims of IG Metall and the wage offers by Gesamtmetall are subject to a continuous change between these discrete points of information, and this behaviour is approximated by the regression approaches.

In view of these methodical preliminary remarks, now the round-oriented potential regression approaches for the annual wage claim curves of IG Metall and/or the annual wage ofter curves of Gesamtmetall can be formulated as follows:

$$w_t^1(r_t) = a_t^1 . r_t^{b_t^1} + \epsilon_t \qquad (35)$$

and/or

$$w_t^2(r_t) = a_t^2 . r_t^{b_t^2} + \epsilon_t \qquad (36)$$

where the exponents b_t^1 and b_t^2 can be interpreted as a measure of the two partners' readiness to make concessions. The regression results obtained on the basis of the data in Table 2 and by using (35) and (36) are specified in Table 5 where also the individual correlation coefficients r are given. For IG Metall the correlation coefficients r are naturally bad in those years in which the preliminary reflections on the basis of a non-linearizable exponential approach called for concave wage claim functions. The wage claim curves according to (35) are decreasing and convex on account of $a_t^1 > 0$ and $b_t^1 < 0$. As far as Gesamtmetall is concerned, however, the potential approach according to (36) left ex ante open the possibilities of convex or concave courses of the curves of the wage ofter functions. Strikingly enough, the regression has shown here all wage ofter functions of the years under review increase and follow concave courses on account of $a_t^2 > 0$ and $0 < b_t^2 < 1$, thus the wage ofter behaviour of Gesamtmetall can be approximated fairly well by a function whose curves follow these courses. While consequently the empirically founded proposition concerning the existence of decreasing convex concession functions is supported by the regression results for Gesamtmetall, the same cannot be said with respect to the case of IG Metall, since expression (35) implies convex courses of the curves with wage claims decreasing, but the preliminary reflections mentioned did not lead to such unique results in this sense.

Table 5

Regression results for the annual wage claim and wage ofter functions of IG Metall and Gesamtmetall.

t	IG Metall			Gesamtmetall		
	a_t^1	b_t^1	r	a_t^2	b_t^2	r
1961-62	10.89	-0.22	-0.77	0.76	0.97	0.99
1963	8.18	-0.19	-0.97	2.12	0.43	0.96
1965-66	9.39	-0.24	-0.74	1.65	0.72	0.93
1968	7.23	-0.13	-0.93	3.54	0.17	0.95
1969	12.34	-0.15	-0.81	3.46	0.40	0.88
1970	15.66	-0.16	-0.74	5.31	0.42	0.98
1971	10.76	-0.23	-0.98	3.69	0.32	0.98
1974	18.04	-0.24	-0.99	6.80	0.34	0.98
1975	10.73	-0.32	-0.96	5.55	0.11	0.99
1976	8.45	-0.21	-0.74	3.28	0.26	0.96
1977	10.00	-0.18	-0.84	3.80	0.28	0.96
1978	8.52	-0.21	-0.85	2.34	0.42	0.97
1979	6.28	-0.17	-0.74	2.41	0.34	0.98

When comparing the regression coefficients of Table 5, the fact strikes us that for the estimated IG Metall wage claim curves of the 13 wage negotiations the parameter b_t^1 is confined to the relatively small range [0.32; -0.13]. This impression could further be improved by excluding the year 1975 which is not typical to a certain extent, since the agreement then reached exceeded the initial ofter by Gesamtmetall only by 0.8% which means that IG Metall lost comparatively many feathers. A correspondingly uniform picture for Gesamtmetall can be obtained only from 1969, however, if here, too, the year 1975 is excluded on account of the reasons mentioned, the exponent b_t^2 of the estimated wage offer curves for Gesamtmetall is then in the interval [0.26; 0.42] over these eight years. Moreover, it is interesting that the regression coefficients a_t^1 and a_t^2 of Table 5 are quite well accounted for by the initial claims $\bar{w}_t$ and initial offers $\underline{w}_t$ of the bargaining partners (see Table 1). Setting, as in (21), homogeneous linear regression of the form

$$a_t^1 = \beta_2 . \bar{w}_t + \epsilon_t \quad \text{and/or} \quad a_t^2 = \beta_3 . \bar{w}_t + \epsilon_t \qquad (37), (38)$$

one obtains the results

$$\hat{a}_t^1 = 1.03\,\bar{w}_t \qquad \text{and/or} \quad a_t^2 = 0.78\,\bar{w}_t \qquad (37'), (38')$$

with the correlation coefficients r_2=0,99 and r_3=0,98 respectively. Consequently, while in the case of IG Metall the coefficient a_t^1 is approximately idential with the individual initial claim, in the case of Gesamtmetall a_t^2 amounts to about 78% of the individual initial offer. This difference is due to the functions assumed in (35) and (36) respectively, and to the representation in terms of rounds of the wage bargaining processes. a_t^1 a_t^2 represent the estimated values for the wage proposals by the bargaining partners in the first round r_t=1 of each year of wage bargaining. By representing the wage bargaining processes in terms of rounds it had, however, been laid down that in each first round r_t=1 only the initial claim of IG Metall is known, whereas the initial offer by Gesamtmetall will be submitted only in each second round r_t=2. As an example, for 1971 the wage proposal curves of the bargaining partners which have been estimated according to (35) and (36) are illustrated in Figure 2. Doubtless there is no definite answer to the question to what extent the concession behaviours of IG Metall and Gesamtmetall can actually be rationally accounted for in the sense of Contini and Zionts using such wage proposal curves. Yet, due to the results of the regression analysis it can reasonably be assumed that certain rules do exist. This should encourage further investigations with a view to clarifying the rationality of wage bargaining processes. For, even irrespective of the methodical reflections made here, the naive solution and the data compiled in Table 2 allow the assumption that the decision making of the two bargaining partners shows certain features of a heuristic procedure: On an average the partners expect about four double rounds of wage bargaining until an agreement is reached, and knowing the initial proposals they adapt their concessions accordingly, so that eventually the commonly defined bargaining cake is divided almost into halves.

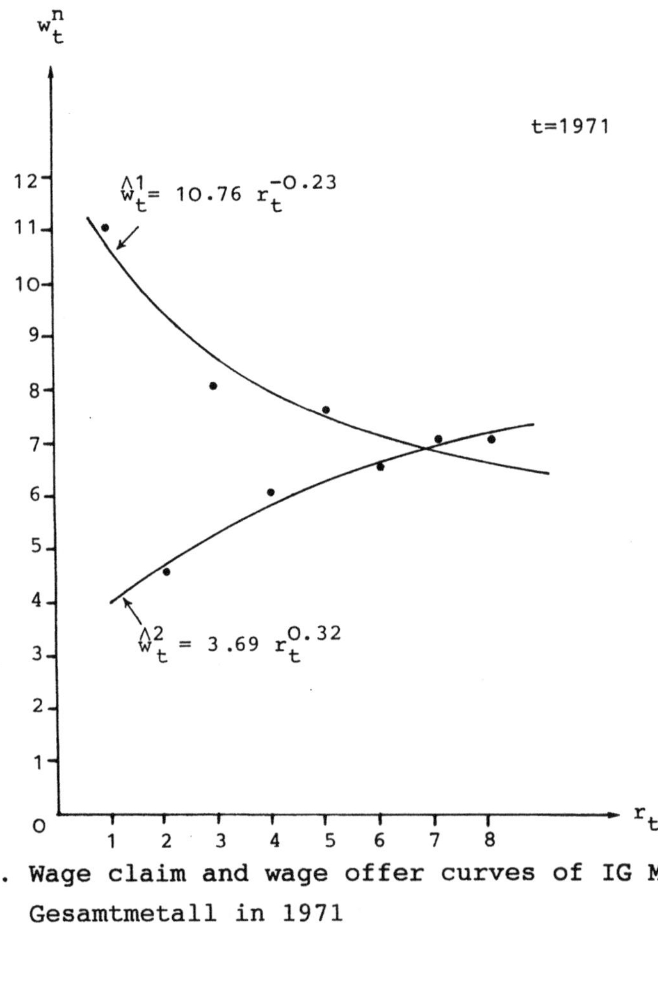

Figure 2. Wage claim and wage offer curves of IG Metall and Gesamtmetall in 1971

REFERENCES

Contini, B. and Zionts, S. (1968): "Restricted bargaining for organizations with multiple objectives", *Econometrica*, 397-414.

Fandel, G. (1979): *Optimale Entscheidungen in Organizationen*, Berlin-Heidelberg-New York.

Friedman, J.W. (1971): "A non-cooperative equilibrium for supergames, *Review of Economic Studies*, 1-12.

Harsanyi, J.C. (1963): "A Simplified bargaining model for the n-person cooperative game", *International Economic Review*, 194-200.

Krelle, W. (1976): *Preistheorie Teil II*, 2nd edition, Tübingen.

Luce, R.D. and Raiffa, H. (1958): *Games and Decisions*, New York.

Nash, J.F. (1950): "The bargaining problem", *Econometrica*, 155-162.

Nash, J.F. (1951): "Noncooperative games", *Annals of Mathematics*, 286-295.

Nash, J.F. (1953): "Two person cooperative games", *Econometrica*, 128-140.

Shapley, L.S. (1953): "A Value for n-person games", in H.W. Kuhn and A.W. Tucker (eds.), *Contributions to the Theory of Games, Volume II*, Princeton/New Jersey, 307-317.

Shapley, L.S. (1964): "Values of Large Market Games: Status of the Problem", Memorandum RM-3957-PR, *The Rand Corporation*, Santa Monica/California.

Shapley, L.S. and Shubik, M. (1969): "Pure competition, coalitional power and fair division", *International Economic Review*, 337-362.

Shubik, M. (1960): "Games decisions and industrial organization", *Management Science*, 455-474.

Wetzel, W. (1973): *Statistische Grundausbildung für Wirtschaftswissenschaftler, Volume 2: Schiebende Statistik*, Berlin, 195-220.

Zeuthen, F. (1930): *Problems of Nomopoly and Economic Warfare*, London.

Zionts, S. (ed.) (1978): *Multiple Criteria Problem Solving*, Berlin-Heidelberg-New York.

SUPPORTING DECISION PROCESSES: AN APPROACH AND TWO EXAMPLES [1]

Gregory E. Kersten and Wojtek Michalowski

Decision Analysis Laboratory
School of Business
Carleton University
Ottawa, Ontario, CANADA

ABSTRACT

In this paper problems of support of discrete decision processes are discussed. A rule-based formalism is used to represent the decision problem, to model interactions with the environment, and to modify problem representation. A three-valued valuation function is used to give the flexibility in selection of a particular solution. The modelling approach and the solutions are illustrated with simple yet compelling example. An expert system shell NEGOPLAN is discussed and examples of its use to support negotiations with a hostage-taker are given. The negotiator is a police commander who should have an ability to modify a problem representation as a hostage-taking incident evolves. The possibility of implementing the proposed approach to decision problems in a dynamic environment is also discussed.

----

1. The authors would like to thank Mike Connolly of the Canadian Police College and Peter Mcnaughton of the Royal Canadian Mounted Police for help in learning about negotiation with terrorists. The participation of Zbig Koperczak of University of Ottawa, Department of Computer Science in the development of the knowledge base described in this paper is also acknowledged. This work has been partially supported by grants from the Natural Sciences and Engineering Research Council of Canada.

1. INTRODUCTION

Many decision processes comprise of a sequence of decisions leading to achievement of one or more of the established goals. Such processes have several common features, namely:

- any decision from a sequence contributes to the achievement of the established goal,
- decisions are interdependent, i.e. making one of them influences subsequent decisions,
- decisions are context-dependent as they are determined by previous decisions and by the environment,
- the environment changes independently from the decisions, but decisions also shape the environment,
- there is strategic interaction between the decision-maker and the environment; decision-maker in his choices takes into account possible states of the environment,
- the decision process is discrete and consists of repetitive cycles of problem identification, formulation, decision selection, and the environment evaluation.

This paper deals with a support for strategic decision processes where the decision-maker does not possess all the necessary information, but during the decision process new information may be obtained (Mintzberg, 1976). This new information may change his perception of the problem, and his preferences. Decision-maker displays strategic behavior which involves learning about his understanding of the problem and his preferences, and learning about actions and preferences of others (Kersten, 1989). The decision problem which initially may be unstructured becomes progressively structured during the decision process.

There are several approaches to model decisions with strategic interactions. A number of game-theoretical models have been developed (Harsanyi, 1977; Colman, 1982), and, to overcome some of their shortcomings, meta-games and hyper-games models have been proposed (Fraser and Hipel, 1984). Out of the analysis of labor and international negotiations the tactics models have been developed (Young, 1975). The multicriteria decision analysis is another approach to model decisions with strategic interaction (Shakun, 1988; Yu, 1985) and recently, attempts are being made to use expert system technologies, and, among them, a rule-based formalism (Sycara, 1987). Most of the research, however, assumes that both the decision problem and the environment do not change and, therefore, there is no need to modify description of the decision problem, change preferences of the decision-maker, or redefine the relationships between the problem and the environment. As Hammond (1988, p. 3) stated:

> *judgement and decision-making in dynamic tasks has been hardly touched; virtually all the research has focusedon static tasks.*

It is proposed to use a rule-based formalism to model time - and context - dependent decision processes. The prototype of the decision support shell NEGOPLAN uses such formalism to support decision-makers in resolving ill-structured problems (Kersten *et al.*, 1988; Matwin *et al.*, 1987). The main difference between proposed approach and other rule-based systems is the use of a three-valued valuation function, and the ability to adapt the knowledge base to the current situation as it is perceived by the decision-maker.

The concepts discussed in the paper are illustrated with a simple example, and an application of NEGOPLAN to model and support the negotiations with a hostage-taker (Kersten and Michalowski, 1988) is presented. There are several attempts to represent and support this type of negotiation. Atkinson and Sandler (1987) present a regression model to compare and analyze a number of negotiation processes, and to verify the factors important in dealing with a hostage taker. Faure and Shakun (Shakun, 1988) present an approach to support negotiation with a hostage-taker based on evolutionary system design framework. Vedder and Mason (1987) present a rule-based model which is designed to recommend what actions to take in a given hostage-taking incident.

These approaches, however, do not aim at supporting decision-makers during the process (simulated or real) because they lack dynamic modelling capabilities.

In Section 2 of the paper we present the modelling framework and we introduce basic concepts used in rule-based representation. Discussion of the theoretical concepts is illustrated with a simple example. In Section 3 we describe some basic features of the negotiations with a hostage-taker. This particular form of decision making was used to illustrate operation of NEGOPLAN. In Section 4 we describe a computer-based modelling tool NEGOPLAN. This tool is an implementation of our ideas presented in the previous section. The particular issues of hostage-taker negotiations as modelled in NEGOPLAN are also presented in this section. Remarks concerning the flexibility of the general approach and the utility of the particular system conclude the paper. Formal presentation of the discussed concepts is given in Appendix.

2. MODELLING FRAMEWORK

The rule-based formalization of decision processes is first introduced using the "Get rich" example which is adapted from Jackson (1986, pp. 35-37):

To be rich you have to have money and there are three ways to get money: earn <u>or</u> steal <u>or</u> borrow money. To earn money you have to find a job <u>and</u> keep it, and to find a job you have to prepare your CV <u>and</u> look respectable <u>and</u> visit an employment agency. To keep the job you have to work hard <u>and</u> be cooperative. If you steal money you have to find an appropriate place <u>and</u> you must not be caught. If you decide to borrow money you can ask your friend, <u>or</u> you can see a bank manager <u>and</u> find security <u>and</u> look respectable.

The above statements about ways of getting rich describe a decision problem with the following properties:

- a problem is decomposable into smaller, logically interconnected subproblems/statements which we later represent with predicates. The relations between the predicates are expressed as connectives *and*, *or* and *not* (in the example these connectives are underlined);
- some of the predicates are used to explain meaning of others. This is achieved by using a connective *If...then*. The predicate which is explained is called *consequent* and the predicates which explain the consequent are *antecedents*, e.g. "be rich" is a consequent and "earn money" is one of its antecedents;
- there are predicates which cannot be further explained. They are called *facts*, e.g. "prepare cv" is a fact because it is not a consequent in any sentence;
- there is one predicate which does not explain any other. It is called the *principal goal*; "be rich" is the principal goal - it is not an antecedent in any sentence.

2.1. Problem Representation

Using the terminology of rule-based formalism one may rewrite the problem of becoming rich using a predicate calculus. The activity of being rich is a predicate BE_RICH which is explained by three other predicates EARN_MONEY, BORROW_MONEY and STEAL_MONEY. Relationship between these predicates is captured using the connective *or*. Thus, BE_RICH is a consequent of three antecedents and the corresponding production rule describing this situation is:

BE_RICH ⇐ EARN_MONEY *OR* BORROW_MONEY *OR* STEAL_MONEY.

where "⇐" represents "*If...then...*" connective. Similarly, one can write more rules and obtain a model of the "Get rich" problem which was described previously using English statements. This model is as follows:

BE_RICH	⇐	EARN_MONEY *OR* STEAL_MONEY *OR* BORROW_MONEY.	
EARN_MONEY	⇐	FIND_JOB *AND* KEEP_JOB.	
FIND_JOB	⇐	PREPARE_CV *AND* LOOK_RESPECTABLE *AND* VISIT_AGENCY.	
KEEP_JOB	⇐	WORK_HARD *AND* BE_COOPERATIVE.	(1)
STEAL_MONEY	⇐	FIND_PLACE *AND NOT* GET_CAUGHT.	
BORROW_MONEY	⇐	ASK_FRIEND *OR* (VISIT_BANK *AND* FIND_SECURITY *AND* LOOK_RESPECTABLE).	

An important feature of the rule-based formalism is the ability to infer the values assigned to a predicate when values assigned to other predicates are known. Usually, two values can be assigned: *true* and *false*. In the approach taken in NEGOPLAN also value *any* may be assigned (Kersten *et al.*, 1988). Value *any* means that a particular predicate may be either *true* or *false* and neither value changes the truth value of the principal goal which in this example is a predicate BE_RICH.

A value is assigned to a predicate through a valuation function γ (see Appendix for an explanation). For the sake of simplicity, its valuation is omitted and directly predicate's truth value is written (e.g. we write BE_RICH ::= *true* instead of γ(BE_RICH) = *true*; sign "::=" means "is").

It is sometimes convenient to represent predicates and the relationships among them in a form of an and/or tree (Charniak and McDermott, 1985). Such a tree for model (1) is presented in Fig. 1.

In Fig. 1 an unnamed predicate is added - an antecedent of the consequent BORROW_MONEY. This is to indicate that one can borrow money asking a friend or, by making sure that the three facts describing successful borrowing from a bank are *true*. Such a predicate is called an *artificial* one and it is introduced when ambiguity may arise as to how a value of a consequent may be inferred. Note that for this reason the brackets in model (1) are introduced in the production rule describing BORROW_MONEY.

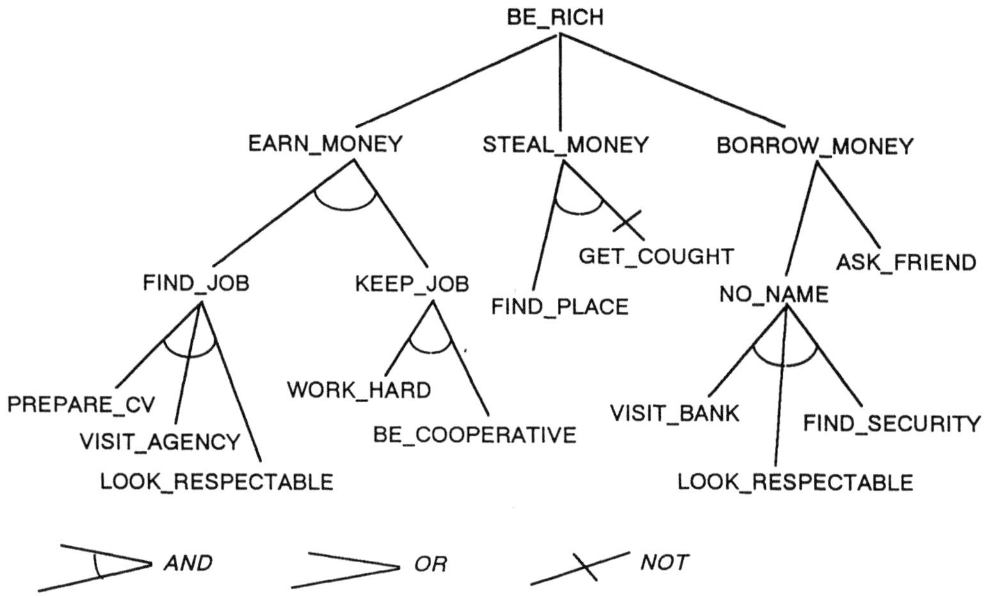

Figure 1. AND/OR tree for the "Get rich" problem

2.2. Problem Solutions

A solution of a problem "Get rich" is an advice what one should do to achieve its principal goal. We assume that the requested value of the principal goal is *true*. A *feasible solution* is an assignment of truth values to the sequence of all the facts such that one may infer the value *true* of the principal goal.

A feasible solution of (1) can be obtained by application of truth tables (Enderton, 1972). The systematic procedures for truth assignments to predicates usually employ one of the following techniques: forward chaining or backward chaining (Winston, 1986). The reasoning in forward chaining starts from facts and a value of the principal goal is deduced. The backward chaining assigns *true* value to the principal goal, and going "top-down" deduces values of the antecedents down to the facts level.

Note that if one borrows money from a friend (BORROW_MONEY is *true*) then, according to the model (1), one is rich (BE_RICH is *true*). No matter if the remaining facts are *true* or *false* the principal goal BE_RICH is *true*. It means that the remaining facts may be assigned any value (*true* or *false*). Valuation of such facts is considered to be *any*. In other words, a fact has valuation *any* if it may be assigned value *true* or *false* without changing the valuation of a principal goal. Such facts are called *flexible* ones. The *inflexible* fact has assigned value *true* or value *false* and changing this value results in the principal goal becoming *false*.

Kersten and Szpakowicz (1988) propose procedures for solving rule-based models with three-valued valuation function. The basic procedure is used to obtain an *induced solution*, i.e. a feasible solution in which only the inflexible facts have *true* or *false* values, and flexible facts have value *any* (see Appendix for further discussion). Once an induced solution is determined one can obtain a number of feasible solutions. To do so, value *any* in the induced solution is replaced with *true* or *false* for one or more flexible facts. An example of an induced solution and two feasible solutions (Feasible 1 and Feasible 2) is given in the Table 1.

Table 1

Induced and feasible solutions for the "Get rich" problem

Fact	Induced	Feasible 1	Feasible 2	Infeasible	Most Flexible
PREPARE_CV	*true*	*true*	*true*	*true*	*any*
VISIT_AGENCY	*true*	*true*	*true*	*true*	*any*
LOOK_RESPECTABLE	*true*	*true*	*true*	*any*	*any*
WORK_HARD	*any*	*any*	*false*	*true*	*any*
BE_COOPERATIVE	*any*	*any*	*any*	*any*	*any*
FIND_PLACE	*any*	*false*	*true*	*true*	*any*
GET_CAUGHT	*any*	*true*	*any*	*true*	*any*
VISIT_BANK	*any*	*false*	*false*	*false*	*any*
FIND_SECURITY	*any*	*false*	*false*	*any*	*any*
ASK_FRIEND	*any*	*any*	*false*	*any*	*true*

There are two more solutions given in Table 1. An *infeasible solution* is such for which the valuation of the principal goal is *false*. Note, that in the infeasible solution valuation of the first two facts is *true*, but the valuation of LOOK_RESPECTABLE is *any*. Thus, the valuation of goal FIND_JOB is *false*, and therefore, BE_RICH is also *false*.

There are situations when one wants to find such a solution that has as many open options as possible. This is interpreted as the ability to obtain the principal goal with the minimal number of inflexible facts and a solution which has this feature is called the *most flexible solution*. In our example there is only one most flexible solution and it is given in Table 1. If one asks a friend, then one is rich no matter whether one looks respectful or not, has cv or not, works hard or not, etc.

There are many strategies to determine a solution of a rule-based model (see Winston, 1986; Holland *et al.*, 1987) but in NEGOPLAN the user is provided with the ability to determine lower and upper bounds on most flexible solutions, to obtain solutions within these bounds, and to introduce preferences with regard to facts (Kersten and Szpakowicz, 1988). The notion of flexibility and flexible solutions is considered to be one of the most powerful features of NEGOPLAN.

An additional feature of NEGOPLAN is the *set-option* (see Section 4.2 for an illustration). Let us assume that one does not want or cannot look respectable. Then, the initial value of LOOK_RESPECTABLE is set to *false*, and the system tries to determine an induced solution with this value set *a priori*. It is possible to set values for many facts, however the result may be that no feasible solution can be found.

2.3. Response and classification rules

There are other features of a decision problem which may be modelled within the rule-based formalism. One of them is a possibility to select an action to be taken following information which is not included in the model. This *external* information can be taken into account with the use of *response rules* - rules which determine values of some predicates on the basis of values of the other predicates which are not necessarily included in the model (see Appendix for formal definition).

Consider a situation when one wants to be rich but does not want to work or to borrow money. The expected advice is "find a place to steal money and do not be caught" (see Fig. 1). We know, however, that one would probably change his mind under certain circumstances, e.g. when police announce a crackdown on thieves. Such a situation is represented as the following response rule:

$$\text{STEAL\_MONEY} ::= \textit{false} \quad \Leftarrow \quad \text{POLICE\_CRACKDOWN} ::= \textit{true}. \tag{2}$$

Until now only variable-free predicates were considered. It is possible to introduce numerical, logical or symbolic variables into predicates. To illustrate this let us assume that there are three types of cv: cv of a university graduate (good cv), cv of a person with elementary education and long work experience (average cv), and cv of a high-school graduate (basic cv). To include this in model (1) variable cv_type may be introduced:

$$\text{cv\_type} \in \{\ \textit{good\_cv},\ \textit{average\_cv},\ \textit{basic\_cv}\ \}, \tag{3}$$

and the predicate PREPARE_CV is replaced with PREPARE_CV (*cv_type*).

Let us also assume that cv has an impact on the salary and to include this in model (1) variable job_type is introduced in FIND_JOB predicate:

$$\text{job\_type} \in \{\ \textit{good\_salary},\ \textit{average\_salary},\ \textit{basic\_salary}\ \}. \tag{4}$$

Using (3) and (4) the production rule in model (1)

$$\text{FIND\_JOB} \quad \Leftarrow \quad \text{PREPARE\_CV} \ \textit{AND} \ \text{VISIT\_AGENCY} \ \textit{AND} \ \text{LOOK\_RESPECTABLE}.$$

is replaced with three following production rules:

FIND_JOB (*good_salary*) ⇐ PREPARE_CV (*good_CV*) *AND* VISIT_AGENCY
AND LOOK_RESPECTABLE.
FIND_JOB (*average_salary*) ⇐ PREPARE_CV (*average_CV*) *AND* VISIT_AGENCY (5)
AND LOOK_RESPECTABLE.
FIND_JOB (*basic_salary*) ⇐ PREPARE_CV (*basic_CV*) *AND* VISIT_AGENCY
AND LOOK_RESPECTABLE.

It is possible to introduce variables into other predicates and obtain more flexible notation.

Variables are required when one wants to use classification rules because their role is to infer a value of a variable. To illustrate an application of the classification rules let us use a variable person_type. Then, for example, a variable-free predicate EARN_MONEY becomes EARN_MONEY(*person_type*). It is also assumed that a type of a person may be determined by checking what activity this person is willing to undertake. So, for example, if it is preparing a cv, then this person is classified as a hard worker; otherwise it is easy going (looking for a loan), or a thief (looking for place to steal food). Hence, the following three classification rules may be defined:

hard_worker /person_type ⇐ PREPARE_CV(cv_type)::= *true AND*
WORK_HARD ::= *true.*
easy_going /person_type ⇐ VISIT_BANK ::= *true OR* ASK_FRIEND ::= *true.* (6)
thief /person_type ⇐ FIND_PLACE ::= *true.*

Observe, that the information used in (6) may also be used to infer values of facts. For example, if one knows that a person is a hard worker, then one may conditionally bound values of predicates PREPARE_CV and WORK_HARD to *true* and then solve the problem (1). The rule used in this conditional bounding is as follows:

2.4 Problem modification and restructure rules

The inherent feature of a strategic decision process is the impact of the environment on the problem representation. Usually, decision is made, then reaction of the environment is observed, corrected or a new decision is made, the reaction observed, etc. The reactions yield new information which may require changes in the initial problem representation. These anticipated reactions are considered with a help of *restructure rules* which modify the model according to the dynamically changing decision context (see Appendix for a efinition).

The environment reacts to the decisions made and these reactions may be anticipated. If someone decides to steal money one may anticipate that he may be caught. How a model should change to reflect this new situation? The changes are introduced with the following restructure rules:

GET_CAUGHT ::= *true*

⇒

[STEAL_MONEY ⇐ *NOT* GET_CAUGHT *AND* FIND_PLACE] (8)

—>

[GET_LOW_SENTENCE ⇐ GET_LAWYER *AND* LOOK_RESPECTABLE].

GET_CAUGHT ::= *true*

⇒

[BE_RICH ⇐ EARN_MONEY *OR* STEAL_MONEY *OR* BORROW_MONEY] (9)

—>

[BE_RICH ⇐ GET_LOW_SENTENCE *AND* BE_PREPARED;

BE_PREPARED ⇐ STUDY_WHILE_IN_PRISON *AND*

GET_RID_OF_BAD_HABITS] .

where "⇒" is "*If... then...*" connective and " → " is the replacement operator: it replaces left-hand side with the right hand side in model (1). Thus, restructure rule (8) means that if someone is caught then the statement "In order to steal money you find place and you are not caught" is replaced with new statement "In order to get low sentence you get a lawyer and you look respectable". Note that one restructure rule may introduce several new rules as in (9), and it also may introduce new predicates.

Restructure rules (8)-(9) drastically change the representation of the "Get rich" problem. Applying these rules to model (1) a new model is obtained:

$$\begin{array}{lll} \text{BE\_RICH} & \Leftarrow & \text{GET\_LOW\_SENTENCE } \textit{AND} \text{ BE\_PREPARED.} \\ \text{GET\_LOW\_SENTENCE} & \Leftarrow & \text{GET\_LAWYER } \textit{AND} \text{ LOOK\_RESPECTABLE.} \\ \text{BE\_PREPARED} & \Leftarrow & \text{STUDY\_WHILE\_IN\_PRISON } \textit{AND} \\ & & \quad \text{GET\_RID\_OF\_BAD\_HABITS.} \end{array} \tag{10}$$

The and/or tree representation of model (10) is given in Fig. 2.

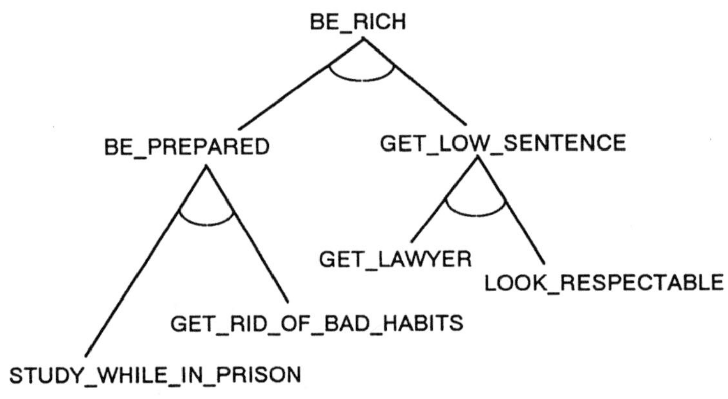

Figure 2. AND/OR tree for the modified "Get rich" problem

3. NEGOTIATIONS WITH A HOSTAGE-TAKER

NEGOPLAN was used to model a negotiation process between the police commander and a hostage-taker. This type of application was selected because it has several features which are difficult to model with "classical" support tools. These features will be outlined in this section.

Negotiation with a hostage-taker is an example of distributive bargaining, where the closer a solution is to one side, the greater must be the sacrifice of the other side (Walton and McKersie, 1965). At the same time, each side must resist pressures which come from the opposite side (Schellenberg, 1982). This is one of the reasons why a police negotiator does not make decisions but only conveys decisions made by the police commander. Tactics of commitment are also an important feature of a distributive bargaining. The commitment of the police to law and order may effectively eliminate some elements of the terrorist's position from further discussion (Harris and Hautzinger, 1979).

Unlike most negotiation situations, negotiation with a hostage-taker is initiated by one side at an unexpected time and place. Hence, the initiating side is in a semi-dictatorial position, trying to impose its most preferred outcome on the other side. The goal of a police is to change this situation in its favour (Mirabella and Trudeau, 1981). The police commander makes decisions and prepares responses to hostage-taker demands. These responses are then conveyed to a terrorist by a police negotiator. This separation of functions (decision-making and direct negotiating) minimizes, both psychological influence of the situation and of terrorist's behavior, on the decision-making process.

The approach discussed here is designed to support the police commander. Because of the strategic nature of the process and the expertise of the user, the intelligent support system has to be interactive and cooperative, in a sense that the user (police commander) has to have the ability to control the system's reasoning, to change its operation and to introduce his new elements.

3.1 The knowledge base

Implementation of our earlier research clarified a role NEGOPLAN can play in supporting negotiation with a hostage-taker (Michalowski *et al.*, 1988). Initially, it was assumed that the person to be supported would be the police negotiator. However, after discussions with members of a police force, it was concluded that the police commander himself should be an ultimate user of the system. This information also influenced the understanding of how NEGOPLAN should be used in practice. We assumed the system to be used to support training activities only. The experts that we consulted suggested, however, that NEGOPLAN might be used during actual hostage taking incidents. The framework presented in this paper is designed for the latter, more active and challenging option.

The knowledge about police/hostage-taker negotiations was initially obtained from cases prepared by police negotiators for the purpose of educating the police force in general. We also held some consultations with police experts who instructed us on the aspects of the available literature that they considered to be of prime importance.

An example of a negotiation model created by NEGOPLAN from its knowledge base is given below:

GOAL (*police*) ⇐ SHORT_TERM_GOAL (*police*) *AND*
LONG_TERM_GOAL (*police*).

SHORT_TERM_GOAL (*police*) ⇐ *NOT* LOSE_LIVES (*POLICE*).

LOSE_LIVES (*police*) ⇐ *NOT* RELEASE_WOMEN(*terrorist*) *OR*
NOT RELEASE_MEN (*terrorist*).

LONG_TERM_GOAL (*police*) ⇐ [*NOT* LOSE_IMAGE (*police*) *AND*
ERADICATE_TERRORISM (*police*)] *OR*
INTRODUCE_TOUGHER_LAWS (*police*).

LOSE_IMAGE (*police*) ⇐ CASUALTIES (*police*) *OR*
YIELD_WITHOUT_FIGHT(*police, terrorist*).

ERADICATE_TERRORISM (*police*) ⇐ *NOT* RELEASE_PRISONERS (*terrorist*) *OR*
MAKE_RECRUITING_DIFFICULT (*police*).

MAKE_RECRUITING_DIFFICULT (*police*) ⇐ KEEP_LOW_PROFILE (*police*) *OR*
[DIE_IN_AMBUSH (*terrorist*) *AND*
NOT TV_COVERAGE (*terrorist, police*)] *OR*
NOT [OPEN_PRESS_BUREAU (*terrorist*) *AND*
RECOGNIZE_MOVEMENT (*police*)].

KEEP_LOW_PROFILE (*police*) ⇐ RANSOM_DEMAND (*terrorist*) *AND*
NOT LOSE_LIVES (*police*).

YIELD_WITHOUT_FIGHT(*police, terrorist*) ⇐ [RELEASE_PRISONERS (*terrorist*) *AND*
NOT GET_INFO (*terrorist*)] *OR*
[RANSOM_DEMAND (*terrorist*) *AND*
LOSE_LIVES (*police*)] *OR*
[RELEASE_PRISONERS (*terrorist*) *AND*
LOSE_LIVES (*police*)].

GET_INFO (*terrorist*) ⇐ STAND_TRIAL (*terrorist*) *OR*
OPEN_PRESS_BUREAU (*terrorist*).

3.2 The meta-knowledge base

Negotiations with a hostage-taker depend to a great extend on the profile of a terrorist. It is possible to select appropriate profile on the basis of the analysis of terrorist's demands. NEGOPLAN has the ability to classify a hostage-taker into one of the following classes (Maksymchuk, 1982):

- *psychological* : a terrorist who is very unpredictable, self-centered, sacrificial, and who may resort to extreme violence;
- *criminal* : a terrorist who is task-oriented and rational, and who will generally concede to negotiation if there is no other way out;
- *radical* : a terrorist who is well-trained, disciplined, and militant, and who, in the case of an ideological zealot, may become extremely violent.

Classifications, responses and restructuring is inherent for negotiation dynamic. All the knowledge about these dynamics is encoded in the meta-knowledge base. The classification rules are used to determine a profile of a hostage-taker. Examples of these rules are as follows:

criminal /type ⇐ RANSOM_DEMAND (*terrorist*) ::= *true AND*
FIREARMS_ON_SCENE (*terrorist*) ::= *true.*
radical /type ⇐ RELEASE_PRISONERS (*terrorist*) ::= *true AND*
OPEN_PRESS_BUREAU (*terrorist*).
psycho /type ⇐ RANSOM_DEMAND (*terrorist*) ::= *false AND*
RELEASE_PRISONERS (*terrorist*) ::= *false AND*
TV_COVERAGE (*terrorist*) ::= *true.*

The classification rules can be used when hostage-taker's demands are known. If, because of vague demands, no classification rule can be applied, police have to respond to the demands in order to obtain information which may be later used to determine the profile of a hostage-taker. In either case the very nature of the negotiation process requires responses on both sides. These responses are obtained from the application of the response rules. NEGOPLAN has 23 such rules, and some of them are given below:

RELEASE_WOMEN(*terrorist, criminal*) ::= *true* *AND* KILL_MEN (*terrorist, criminal*) ::= *false*

⇐

OBTAIN_RANSOM (*police*) ::= *true* *AND* STAND_TRIAL (*police*) ::= *false* *AND*

FOOD_SUPPLY (*police*)) ::= *any* .

RELEASE_MEN (*terrorist, criminal*) ::= *true*

⇐

RELEASE_WOMEN(*terrorist, criminal*) ::= *true* *AND* STAND_TRIAL (*police*) ::= *false* .

RELEASE_WOMEN(*terrorist, psycho*) ::= *true*

⇐

STAND_TRIAL (*police*) ::= *any* *AND* FOOD_SUPPLY (*police*)::= *any* .

KILL_MEN (*terrorist, radical*) ::= *true*

⇐

INFLICT_CASUALTIES(*attack, police*) ::= *true* .

NEGOPLAN generates responses for either side by forward chaining of the response rules. The values of facts constituting police response are introduced as bounds. Applying NEGOPLAN's procedures, an induced solution is sought for which the bounds are satisfied. When a feasible solution of a problem cannot be inferred through the application of the response rules, NEGOPLAN consults the restructure rules. Forward chaining of these rules introduces applicable modifications to a problem. These modifications depend on the actual positions of both sides, and on terrorist's profile and negotiation's phase. The latter factors are introduced through variables in a definition of appropriate predicates.

The modifications in problem representation resulting from the evolution of negotiations are dealt with the restructure rules. Restructure rules may be introduced by the user of NEGOPLAN or they can be built into the knowledge base and applied automatically. There are 18 restructure rules in the meta-knowledge base. Below are three examples of them:

KILL_MEN (*terrorist*) ::= *true*

⇒

[SHORT_TERM_GOAL (*police*) ⇐ *NOT* LOSE_LIVES (*police*)]

—>

[SHORT_TERM_GOAL (*police*) ⇐ INFLICT_CASUALTIES(*deception*, *police*) *OR*

INFLICT_CASUALTIES(*attack*, *police*)].

FOOD_SUPPLY(*police*) ::= *true*

⇒

[PSYCHO_PRESSURE (*police*) ⇐ *NOT* FOOD_SUPPLY (*police*) *AND*

STAND_TRIAL (*terrorist*)]

—>

[PSYCHO_PRESSURE (*police*) ⇐ *NOT* TV_COVERAGE (*terrorist,police*) *AND*

STAND_TRIAL(*terrorist*)].

RELEASE_MEN (*terrorist,impasse*) ::= *false*

⇒

[PSYCHO_PRESSURE(*police*) ⇐ *NOT* FOOD_SUPPLY(*police*) *AND*

STAND_TRIAL(*terrorist*)

—>

[PSYCHO_PRESSURE(*police*) ⇐ SHOW_OF_FORCE(*police*) *AND*

POSTPONE_NEGO(*police*)].

Police commander may also change his perception of a negotiation problem as negotiations enter different phases. Identification of the hostage-taker's profile brings to a close the first, *initiation*, phase of negotiation (Abbott, 1986) where the situation is dominated by the hostage-taker's dictatorial position. Abbott distinguishes three subsequent phases in typical hostage negotiation. The initiation phase is followed by the demands phase in which the possible negotiation issues are explored. In this phase, the police try to establish which of the demands are important.

The third phase is the *impasse/suicide* phase characterized by a stalemate and the evaluation of previously unexplored issues. During this phase the risk of failure increases. The hostage-taker, whatever his type, becomes tired and frustrated, and hence more prone to use violence (Stanley, 1984). When the stalemate of this phase is broken (through persuasion or force), the negotiations proceed to their final phase. The final phase following impasse/suicide is the *surrender* phase, in which successful negotiation tactics should result in the achievement of the negotiation goals. However, in the case of a deadlock, which would become evident in the impasse/suicide phase, the surrender phase involves a final assault by a Tactical Rescue Unit team.

The transition between phases depends on the progression of terrorist's demands. Identification of the appropriate negotiation's phase is obtained through the assignment of values to a variable phase which is similar to one used in the classification of terrorist's profile. In the meta-knowledge base there are 28 rules used to infer values of the variable phase and three examples are given:

demands/phase ⇐ RELEASE_WOMEN(*terrorist, criminal*) ::= *false AND*
RELEASE_MEN (*terrorist, criminal*) ::= *false AND*
RANSOM_DEMAND (*terrorist, criminal*) ::= *true.*

impasse/phase ⇐ FOOD_SUPPLY (*police, psycho*) ::= *false AND*
STAND_TRIAL (*terrorist, psycho*) ::= *true.*

surrender/phase ⇐ RELEASE_PRISONERS (*terrorist, radical*) ::= *true AND*
STAND_TRIAL (*police, radical*) ::= *any.*

Observe that phases were already introduced to the definitions of the restructure rules. This makes it possible to make changes in the model depending on the phase of negotiation, what is accomplish by assigning variable phase to appropriate predicates in a restructure rule.

The application of the classification rules, the response rules, and the restructure rules is well defined and focuses on the specific negotiation situation. However, the context-limiting approach used to control the transformation between phases is also general enough to allow for:

- restructuring a problem representation regardless of the phase of negotiations;
- restructuring a problem representation locally, within one phase, without necessarily transforming it in other phases;
- transition from the state t directly to the state $t + k$ ($k > 1$).

4. NEGOPLAN: A COMPUTER-BASED MODELLING TOOL

4.1 NEGOPLAN architecture

NEGOPLAN is an expert system shell designed to support a decision making process. The system has been written in Quintus Prolog 2.2 to run on a Sun Microsystems Sun 3.2 workstation, and the other version is written in Arity Prolog to run on IBM PS/2 Model 60 and higher computers.

The architecture of the NEGOPLAN is determined by the representation of knowledge about the decision problem. It requires that knowledge is stored in the form of production rules. There are five modules which operate on this knowledge base: (i) tree builder, (ii) *tree solver*, (iii) forward chainer, (iv) graphics output, and (v) editors. The tree builder allows the user to construct a negotiation problem representation as a variant of an and/or tree. The goal tree created by the tree builder represents the user's knowledge base. If the user decides to preset the truth values of the selected facts, this action can be also interpreted by the tree builder. Presetting the value of a fact means assigning a particular value (*true*, *false* or *any*) to this fact regardless of the results of inference on the and/or tree. Thus, it is possible that for a given value, a feasible solution of a decision problem does not exist.

Once the and/or tree is created, the *tree solver* solves this tree by applying three procedures (Kersten and Szpakowicz, 1988). The first procedure uses DeMorgan's laws to "push" negations down to the fact level. The second one reduces the tree obtained from the first procedure to a subtree with only inflexible facts. The third procedure determines an induced solution. Each induced solution defines a *position*, from which other feasible solutions may be obtained.

Decision making involves changes in the position, or redefinition of the problem representation. These two types of changes require the use of additional knowledge regarding relationships between responsive environment and the way the decision problem is perceived by the party involved. In NEGOPLAN we represent this meta-knowledge using classification, response and restructure rules.

The *forward chainer* operating on the meta-knowledge base allows for changes in a knowledge base according to a predefined scenario. The only information available to the user is related to generated positions, and proposed changes in a problem representation. Therefore both the tree solver and the forward chainer operate in the background.

The *graphics output* of NEGOPLAN enables the user to browse through the goal tree, and if greater detail of the analysis is required, a particular part of this tree can be exploded into a full window.

The *editors allow* the user to enter and edit working environment (knowledge base, meta-knowledge base). They work in both a terse and a normal mode, accepting either complete items (fact, rule, restructure rule) to be manipulated, or allowing for the construction of an item with expanded description in a natural language. This description is later saved in a lexicon. The other function of the editors is related to the possibility of setting the elements of the position under consideration. If the user wants a predicate to have a particular value, then he may signal it by using the set option of the editor.

4.2 NEGOPLAN session

Delivery of the demands of a hostage-taker starts the session with NEGOPLAN and initiates the application of the classification rules. Prior to responding to these demands, a police commander can preset the truth values for selected facts. Observe that the setting of a value *any* should be interpreted in this case as a police commander being indifferent between either *true* or *false* values for a given fact. The appropriate dialog window for a set option which is used to preset the truth values is presented in Fig. 3.

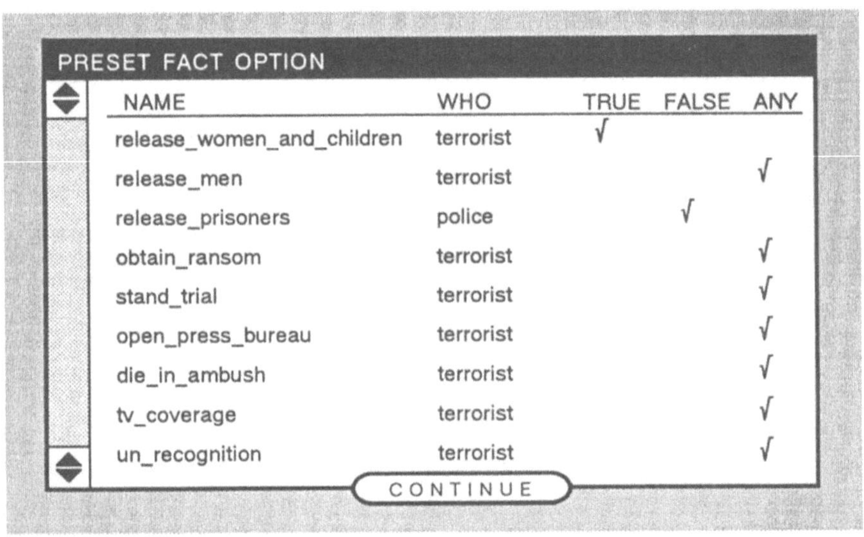

Figure 3

One of the roles NEGOPLAN can play is to help a user learn more about the problem at hand. This can be accomplished by, for example, searching through a goal tree created by the tree builder. Part of such a tree is presented in Fig. 4.

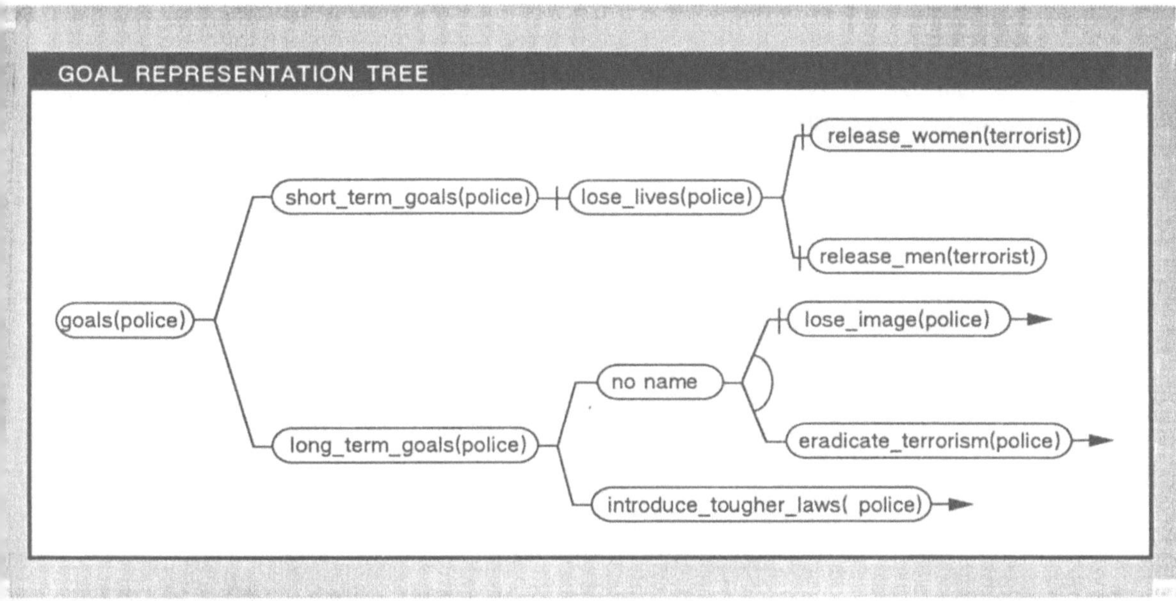

Figure 4.

The tree solver operates on a goal tree and generates an induced solution. Elements of a feasible solution obtained from an induced one may be presented in a graphical form or in a text window. Fig. 5 depicts the later option.

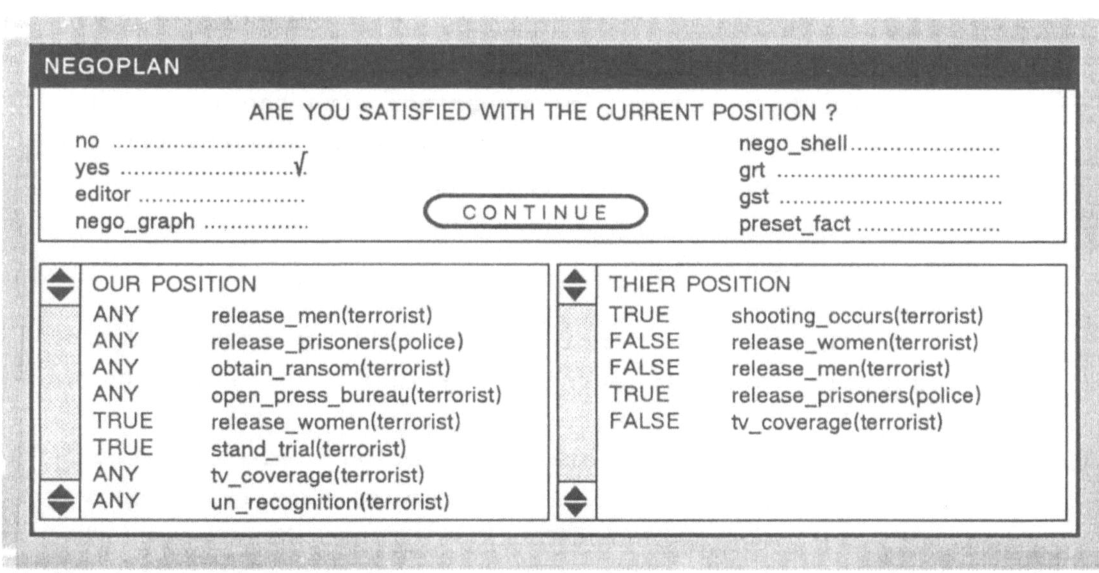

Figure 5.

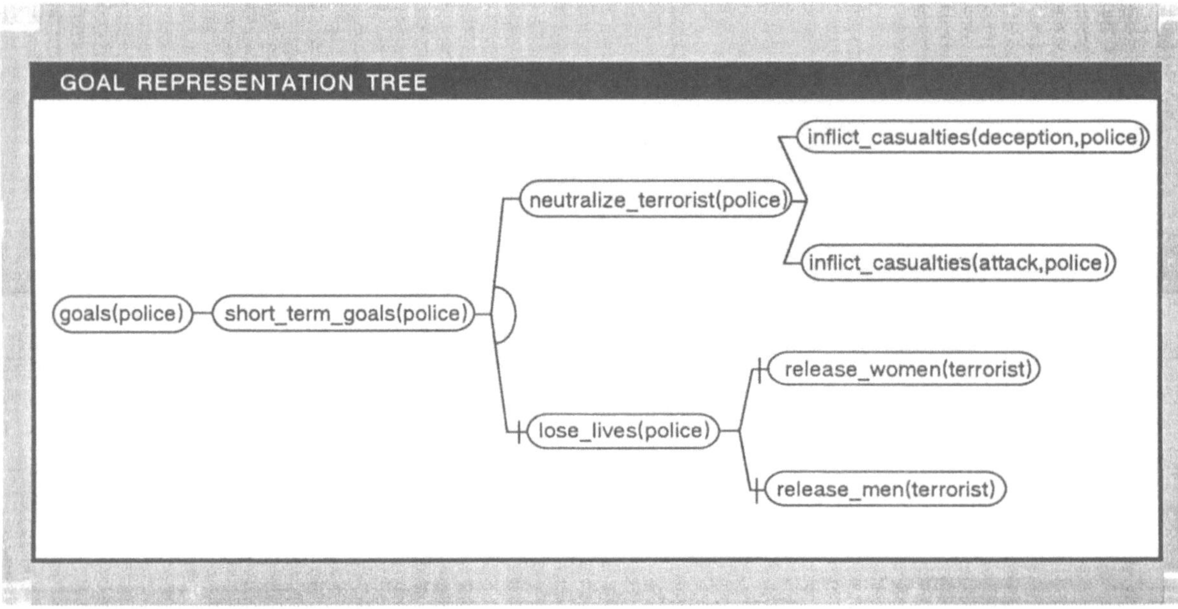

Figure 6.

Through the chaining of the response rules, the new position of a terrorist is inferred. This is followed by chaining the restructure rules, which modifies the problem representation. The modified representation, as communicated by a NEGOPLAN, is presented as an and/or tree in Fig. 6. Observe that the goal tree presented in Fig. 6 is reduced in comparison with the initial one (compare with Fig. 4) to two general options: neutralizing a hostage-taker and preserving the lives of captives.

5. SUMMARY AND CONCLUSIONS

The paper discusses a particular modelling framework for decision support, and illustrates its applicability with an example of police/hostage-taker negotiations. The framework itself is applicable to any decision situation where the assumptions outlined in the paper hold. For example, this framework was applied to model management/union bargaining (Matwin *et al.*, 1989).

Decision making requires flexibility in modelling and support. We believe that it is possible to provide this flexibility with a modelling framework which accommodates changes in problem understanding, and changes in problem perception. A clear distinction between the knowledge base and the meta-knowledge base provides the ability to modify a knowledge base according to decision dynamics. The knowledge base changes along with incremental learning about the decision problem. These changes are controlled within a certain time and knowledge frame. This, combined with the standard explanatory capabilities of an expert system, makes the NEGOPLAN implementation of a rule-based framework attractive for potential users. Observe also that actual use of the system does not follow traditional lines of expert systems applications: NEGOPLAN does not give advice, instead it supports the simulation of reasoning which allows for better understanding of a decision problem.

Appendix

A predicate is denoted as $\alpha(\beta,\ldots,\beta_{nii})$ where $i \in I = \{0,1,\ldots,n\}$ is the predicate index, β_{ji} is a variable, and $j \in J_i = \{1,\ldots,n_i\}$ is the variable index. To simplify the notation a variable-free predicate a_i is often used, assuming that whenever necessary the predicate may include one or more variables.

P is the decision problem. The entity $\boldsymbol{\alpha} = [\alpha_0, \alpha_1, \ldots, \alpha_n]$ is a sequence of n+1 predicates about the problem P. α_0 is the principal goal. The index set of facts is I_f, $(I_f \subset I)$, and the index set of goals is I_g $(I_g = N_f)$.

A value is assigned to a predicate through a valuation function $\subset(\alpha_i) = v$, where $v \in V = \{\mathit{true}, \mathit{false}, \mathit{any}\}$, $(i \in I)$. The required value of the principal goal α_0 is *true*, and there are at least two predicates describing problem P.

Definition 1. Predicate α_i, $(i \in I_g)$ is a goal and the consequent of antecedents $[\alpha_{i1}, \ldots, \alpha_{i\,mi}]$, if there is a well-formed formula r_i, called rule, such that:

$$\alpha_i \Leftarrow r_i([\alpha_{ij}]_{j \in Ji}), \tag{1}$$

where $\Leftarrow$ is the connective *If...then...*, the permissible connectives in r_i are: and, or and not, and $J_i = \{1, \ldots, m_i\}$ is the index set of antecedents of α_i.

Definition 2. A rule-based representation of the problem **P** is a model $P=\{\boldsymbol{\alpha},R\}$, where $R= [r_i]_{i \in Ig}$.

Model **P** must have the following properties:

1.For any fact α_i, $I \in I_f$ there is at least one sequence of goals $\alpha_{i1}, \alpha_{i2}, \ldots, \alpha_{ip}, \alpha_0$, such that α_i is the antecedent of α_{i1}, α_{i1} is the antecedent of α_{i2}, and ... α_{ip} is the antecedent of α_0. Moreover, any goal α_i, $i \in I_g$, belongs to at least one sequence.

2.Circular relations are not permitted; goal a_i, may appear only once in a sequence $\alpha_{i1}, \alpha_{i2}, \ldots, \alpha_{ip}, \alpha_0$.

A solution of problem P is a sequence of values assigned to all its facts. It can be obtained with a user's arbitrary valuation of facts, or by inferring their valuations through chaining the rules of model P.

Definition 3. $v^* = [\gamma(\alpha_i)]_{i \in If}$ is a feasible solution of problem **P** if $\gamma(\alpha_0) = \mathit{true}$.

In a given feasible solution v^* its elements have values *true*, *false* and *any*. In brief, a $v\cdot$ is an *induced solution* if for every $v_i (I \in I_f)$ equal *true* or *false* a change in the value of v_i (e.g. from *false* to *true*) results in infeasible solution.

An induced solution $v\cdot$ defines two sequences of facts $\alpha_{inflex, v\cdot}$ and $\alpha_{flex, v\cdot}$ such that

$$\alpha_{inflex, v\cdot} = [\ \alpha_i,\ i \in I_f:\ \gamma(\alpha_i) = \mathit{true} \text{ or } \gamma(\alpha_i) = \mathit{false}\ ], \quad (2)$$
$$\alpha_{flex, v\cdot} = [\ \alpha_i,\ i \in I_f:\ \gamma(\alpha_i) = \mathit{any}\ ]. \quad (3)$$

Classification of predicates' variables is performed through the analysis of valuations of facts. A classification rule infers a value of a variable used in a definition of appropriate predicates. In order to write a classification rule one needs to introduce a well-formed formula described on valuations:

$$s([\ \gamma(\alpha_j)\ ]_{j\in J}) = [\ \gamma(\alpha_1) = v' \ \mathit{and/or/not}, \ldots, \mathit{and/or/not}\ \gamma(\alpha_K) = v''\ ]. \quad (4)$$

Using (4) a classification rule is written as follows:

$$\mathit{value}/\beta \Leftarrow s([\ \gamma(\alpha_j)\ ]_{j\in If}), \quad (5)$$

where value is an element of the set of values for a variable β.

$P(t)$ is a model of a decision problem $P(t)$ at the moment $t (0 \le t \le T)$. The transition function T, such that $T:\ P(t-1) \Rightarrow P(t),\ (t = 1, \ldots, M)$ describes possible changes of the decision problem during period $[t-1, t]$. The transition is modelled by a set of restructure rules (Holland *et al.*, 1987), which applied to the model P(t-1) transform it into model P(t). Application of a restructure rule directly depends on the current situatión which is described by the valuations of facts.

A restructure rule is defined on the collection of rules and its only connective is *and*.:

$$s(\ \gamma([\alpha_i]_{i \in If})$$
$$\Rightarrow \quad (7)$$
$$p([\alpha_i \Leftarrow r_i([\alpha_{ij}]_{j \in Ji})]_{i \in Ig}) \text{ -> } p([\alpha_l <= r_l([\alpha_k]_{k \in I^*})]_{l \in L}),$$

where: "$\Leftarrow$" indicates the replacement of the left-hand side with the right-hand side in the problem representation **P**(t) : on the left-hand side we have rules affected by the current situation described by $s(\gamma([\alpha_i]_{i \in I})$, and on the right-hand side there are the rules which replace the affected rules; L is the index set of replacement rules, and I* is the index set of new predicates introduced by the restructure rule.

References

Abbott, T.E. (1986) "Time-Phase Model for Hostage Negotiation", *The Police Chief*. No. 4.

Atkinson, S.E. and T. Sandler (1987) "Terrorism in a Bargaining Framework", *Journal of Law and Economics*. Vol. XXX.

Charniak, E. and D. McDermott (1985) *Introduction to Artificial Intelligence*. Reading: Addison-Wesley.

Colman, A. (1982) *Game Theory and Experimental Games. The Study of Strategic Interaction*. New York: Pergamon Press.

Enderton, H.B. (1972) *A Mathematical Introduction to Logic*. San Diego: Academic Press.

Fraser, N.M. and K.W. Hipel (1984) *Conflict Analysis. Models and Resolutions*. New York: North-Holland.

Hammond, K.R. (1988) "Judgement and Decision Making in Dynamic Tasks", *Information and Decision Technologies*. Vol 14, No. 1 13-14.

Harris, R.C., and J. R. Hautzinger (1979) "Hostage Negotiation - The Ultimate Game", *R.C.M.P. Gazette*. Vol. 41, No. 9.

Harsanyi, J.C. (1977) *Rational Behavior and Bargaining Equilibrium in Games and Social Situations*. Cambridge: Cambridge University Press.

Hogger, C.J. (1984) *Introduction to Logic Programming*. London: Academic Press.

Holland, J.H., K.J. Holoyoak, R.E. Nisbett and P.R. Thagard (1987) *Induction: Process of Inference, Learning and Discovery*. Cambridge: MIT Press.

Jackson, P. (1986) *Introduction to Expert Systems*. Wokingham: Addison-Wesley.

Kersten, G.E. (1989) "Expert System Technology and Strategic Decision Support", *Mathematical Modelling*. (forthcoming).

Kersten, G.E., W. Michalowski, S. Matwin and S. Szpakowicz (1988) "Representing the Negotiation Process with a Rule-based Formalism", Theory and Decision. Vol. 25, 225-257.

Kersten, G.E. and W. Michalowski (1988) "A Cooperative Expert System for Negotiation with a Hostage-taker", School of Business, Carleton University, WP 88-9.

Kersten, G.E. and S. Szpakowicz (1988) "Rule-based Formalism and Preference Representation: An Extension of NEGOPLAN", Department of Computer Science, University of Ottawa, WP 88-14.

Maksymchuk, A.F. (1982) "Strategies for Hostage-Taking Incidents", *The Police Chief*. No. 4.

Matwin, S., S. Szpakowicz, Z. Koperczak, G.E. Kersten and W. Michalowski (1989) "NEGOPLAN: An Expert System Shell for Negotiation Support", IEEE Expert. (forthcoming).

Matwin, S. , S. Szpakowicz, G.E. Kersten and W. Michalowski (1987) "Logic-based System for Negotiation Support", *Proceedings of the 1987 Symposium on Logic Programming*. San Francisco, IEEE Computer Society Press.

Michalowski, W. , G.E. Kersten, Z. Koperczak, S. Matwin and S. Szpakowicz (1988) "Negotiation with a Terrorist: Can an Expert System Help?", in M.G. Singh *et al*. (eds), *Managerial Decision Support Systems*. Amsterdam: North-Holland, 193-200.

Mintzberg, H. *et al*. (1976) "The Structure of "Unstructured" Decision Processes", *Administrative Science Quaterly*. Vol. 21, 246-275.

Mirabella, R.W. and J. Trudeau (1981) "Managing Hostage Negotiations", *The Police Chief*. No. 5.

Schellenberg, J.A. (1982) *The Science of Conflict*. New York: Oxford University Press.

Shakun, M.F. (1988) *Evolutionary Systems Design:* Policy Making Under Complexity and Group Decision Support Systems. Oakland: Holden-Day.

Stanley, R. L. (1984) "Hostage Negotiations and the Stockholm Syndrome", *Military Police Journal*. Vol. 11, No. 2.

Sycara, E.P. (1987) "Resolving Adversarial Conflict: An Approach Integrating Case-based Reasoning and Analytic Methods". Ph.D. Thesis, School of Information and Computer Science, Georgia Institute of Technology.

Vedder, R. G. and R.O. Mason (1987) "An Expert System Application for Decision Support in Law Enforcement", *Decision Sciences*. Vol. 18, No. 3.

Walton, R.E. and R.B. McKersie (1965) *A Behavioral Theory of Labor Negotiations*. New York: McGraw-Hill.

Winston, P.H. (1986) *Introduction to Artificial Intelligence*. Reading: Addison-Wesley.

Young, O.R. (1975) *Bargaining: Formal Theories of Negotiations*. Urbane: University of Illinois Press.

Yu, P.L. (1985) *Multiple-Criteria Decision Making. Concepts, Techniques, and Extensions*. New York: Plenum Press.

CHAPTER VI

THE SCHOOL CASE-STUDY

PRESENTATION OF THE SCHOOL CASE-STUDY: EVALUATION OF PERSONNEL HOW TO RANK OFFICERS FOR PROMOTION ?

Carlos A. Bana e Costa
IST - CESUR
Av. Rovisco Pais 1000 Lisbon
PORTUGAL

José Cervaens Rodrigues
Centre for Naval Operations Research (C.I.O.A.)
Portuguese Navy - PORTUGAL

1. SHORT DESCRIPTION

For promotion and selection for certain posts, members of the Armed Forces, Civil Servants and others, who will be refered henceforth as *candidates*, are regularly evaluated by a certain number of *judges* according to a set of predefined *attributes*.

For the purpose of this case-study and in accordance with current practice in the Portuguese Armed Forces, it will be assumed that each candidate has been assessed twice yearly. The judge is normally his Commanding Officer or the person to whom he is directly responsible.

It will also be assumed that the evaluation concerns 7 sets of attributes:

A_1, intelectual capacity;
A_2, character;
A_3, social and moral attributes;
A_4, military attitude;
A_5, leadership capacity;
A_6, professional attributes;
A_7, physical fitness.

A five point ordinal scale, *common to the seven attributes*, is used by each judge in the evaluation of the candidates (see a detailed description in section 2).

The aim of the study is to develop and discuss approaches to the problem of how to rank 12 candidates (C_1 ... C_{12}) for promotion or selection, based on the multi-attribute and multi-judge information given in Annex A.

2. SCALE OF MEASUREMENT

To evaluate the aptitudes of a candidate, the evaluator uses a 5 point ordinal scale:

- Insufficient (S)
- Level I
- Level II
- Level III
- level IV.

Insufficient - should be used whenever the individual has a performance which is below the minimum standards required for his rank and job. The individual should be made aware of the fact.

Levels I to IV - this degrees of the ordinal scale should be used whenever the individual reaches or exceeds the minimum standard of performance required for his rank and job.

It must be remembered that this scale has no natural zero *and gives no information about the difference between two grades in sequence*. It only establishes that IV ≻ III ≻ II ≻ I (where "≻" designates the strict preference relation).

Level I - is to be used when the individual reaches minimally the standard of comparison:

a. The performance is acceptable but not more than that, and
b. he displays lack of experience or minor deficiencies which can be corrected.

Level II - is to be used when the individual reaches consistently the degree of comparison but does not overpass it.

Level III - is to be used when the individual exceeds frequently the standards.

Level IV - Is to be used when the individual exceeds consistently the standards or frequently shows an exceptional level of the aptitude.

SUMMARY

LEVELS	THE INDIVIDUAL	
INSUFFICIENT	... DOES NOT REACH (SHOWS SHORTCOMING)...	...THE
I	... REACHES MINIMALLY (OR CAN CORRECT MINOR DEFECTS) ...	STANDARDS
II	... REACHES CONSISTENTLY...	FOR HIS
III	... EXCEEDS FREQUENTLY...	RANK
IV	...EXCEEDS VERY FREQUENTLY... (OR EXHIBITS AN EXCEPTIONAL LEVEL)...	AND JOB

ANNEX A

CANDIDATE C_1

ATTRIBUTES	FREQUENCIES									
	ABSOLUTE					RELATIVE (%)				
	S	1	2	3	4	S	1	2	3	4
A_1	0	0	31	83	13	0	0	38	46	16
A_2	0	0	6	24	12	0	0	14	57	29
A_3	0	0	3	44	9	0	0	5	79	16
A_4	0	0	1	20	7	0	0	4	71	25
A_5	0	0	4	27	9	0	0	10	68	22
A_6	0	0	1	8	1	0	0	10	80	10
A_7	0	0	0	28	0	0	0	0	100	0
TOTALS	0	0	46	189	41	***				

CANDIDATE C_2

ATTRIBUTES	FREQUENCIES									
	ABSOLUTE					RELATIVE (%)				
	S	1	2	3	4	S	1	2	3	4
A_1	0	0	9	57	26	0	0	10	62	28
A_2	0	0	9	25	16	0	0	18	50	32
A_3	0	0	4	37	20	0	0	6	61	33
A_4	0	0	2	26	8	0	0	6	72	22
A_5	0	0	0	26	10	0	0	0	72	28
A_6	0	0	2	6	0	0	0	25	75	0
A_7	0	0	0	8	10	0	0	0	44	66
TOTALS	0	0	26	182	90	***				

CANDIDATE C_3

ATTRIBUTES	FREQUENCIES ABSOLUTE S	1	2	3	4	RELATIVE (%) S	1	2	3	4
A_1	0	2	26	74	6	0	2	24	68	6
A_2	0	0	7	37	13	0	0	12	65	23
A_3	0	0	1	29	44	0	0	1	39	60
A_4	0	0	0	23	13	0	0	0	64	36
A_5	0	1	12	31	11	0	2	22	56	20
A_6	0	0	2	7	4	0	0	15	54	31
A_7	0	0	2	27	4	0	0	6	82	12
TOTALS	0	3	50	228	95	***				

CANDIDATE C_4

ATTRIBUTES	FREQUENCIES ABSOLUTE S	1	2	3	4	RELATIVE (%) S	1	2	3	4
A_1	0	1	33	58	28	0	1	28	48	23
A_2	0	0	13	26	21	0	0	22	43	35
A_3	0	0	10	63	7	0	0	12	79	9
A_4	0	0	5	34	1	0	0	13	85	2
A_5	0	0	27	31	0	0	0	47	53	0
A_6	0	0	2	27	11	0	0	5	68	27
A_7	0	0	1	27	10	0	0	3	71	26
TOTALS	0	2	91	266	78	***				

CANDIDATE C_5

ATTRIBUTES	FREQUENCIES ABSOLUTE					RELATIVE (%)				
	S	1	2	3	4	S	1	2	3	4
A_1	0	2	27	32	17	0	2	35	41	22
A_2	0	2	16	26	10	0	4	30	48	18
A_3	0	5	14	28	12	0	9	24	47	20
A_4	0	0	3	17	6	0	0	12	65	23
A_5	0	4	13	31	5	0	8	25	58	9
A_6	0	0	6	19	3	0	0	21	68	11
A_7	0	0	4	9	0	0	0	31	69	0
TOTALS	0	13	83	162	53	***				

CANDIDATE C_6

ATTRIBUTES	FREQUENCIES ABSOLUTE					RELATIVE (%)				
	S	1	2	3	4	S	1	2	3	4
A_1	0	2	29	45	15	0	2	32	49	17
A_2	0	3	23	21	8	0	5	42	38	15
A_3	0	2	15	29	22	0	3	22	43	32
A_4	0	0	8	21	3	0	0	25	66	9
A_5	0	0	17	22	9	0	8	35	46	19
A_6	0	0	0	9	1	0	0	0	90	10
A_7	0	0	3	15	4	0	0	14	68	18
TOTALS	0	7	95	162	62	***				

CANDIDATE C_7

ATTRIBUTES	FREQUENCIES ABSOLUTE S	1	2	3	4	RELATIVE (%) S	1	2	3	4
A_1	0	0	18	77	7	0	0	18	75	7
A_2	0	0	11	35	5	0	0	21	69	10
A_3	0	0	9	38	20	0	0	13	57	30
A_4	0	0	0	28	8	0	0	0	78	22
A_5	0	0	4	26	10	0	0	10	65	25
A_6	0	0	0	15	1	0	0	0	94	6
A_7	0	0	7	17	2	0	0	27	65	8
TOTALS	0	0	49	226	53	***				

CANDIDATE C_8

ATTRIBUTES	FREQUENCIES ABSOLUTE S	1	2	3	4	RELATIVE (%) S	1	2	3	4
A_1	0	0	30	85	2	0	0	25	73	2
A_2	0	6	32	13	9	0	10	53	22	15
A_3	0	0	25	50	4	0	0	32	63	5
A_4	0	0	3	36	1	0	0	8	90	2
A_5	0	6	32	19	1	0	10	55	33	2
A_6	0	0	8	27	1	0	0	22	75	3
A_7	0	0	10	23	2	0	0	28	66	6
TOTALS	0	12	140	253	20	***				

CANDIDATE C_9

ATTRIBUTES	FREQUENCIES									
	ABSOLUTE					RELATIVE (%)				
	S	1	2	3	4	S	1	2	3	4
A_1	0	0	2	75	43	0	0	2	62	36
A_2	0	0	13	7	40	0	0	22	11	67
A_3	0	0	21	47	11	0	0	27	59	14
A_4	0	0	5	26	9	0	0	12	65	23
A_5	0	0	1	24	35	0	0	2	40	58
A_6	0	0	7	6	23	0	0	19	17	64
A_7	0	0	0	34	3	0	0	0	92	8
TOTALS	0	0	49	219	164	***				

CANDIDATE C_{10}

ATTRIBUTES	FREQUENCIES									
	ABSOLUTE					RELATIVE (%)				
	S	1	2	3	4	S	1	2	3	4
A_1	0	0	46	64	1	0	0	41	58	1
A_2	0	0	13	41	1	0	0	24	74	2
A_3	0	0	16	46	13	0	0	21	62	17
A_4	0	0	2	26	10	0	0	5	69	26
A_5	0	0	7	38	3	0	0	15	79	6
A_6	0	0	7	8	1	0	0	44	50	6
A_7	0	0	12	25	0	0	0	32	68	0
TOTALS	0	0	103	248	29	***				

CANDIDATE C_{11}

ATTRIBUTES	FREQUENCIES ABSOLUTE					FREQUENCIES RELATIVE (%)				
	S	1	2	3	4	S	1	2	3	4
A_1	0	0	57	26	1	0	0	89	9	2
A_2	0	2	23	19	4	0	4	48	40	8
A_3	0	1	18	40	1	0	2	30	66	2
A_4	0	0	8	19	12	0	0	28	65	7
A_5	0	1	23	15	1	0	2	58	38	2
A_6	0	0	10	6	0	0	0	63	37	0
A_7	0	0	14	10	0	0	0	58	42	0
TOTALS	0	4	153	135	9	***				

CANDIDATE C_{12}

ATTRIBUTES	FREQUENCIES ABSOLUTE					FREQUENCIES RELATIVE (%)				
	S	1	2	3	4	S	1	2	3	4
A_1	0	1	25	38	19	0	1	30	46	23
A_2	0	3	21	15	4	0	7	49	35	9
A_3	1	2	28	18	7	2	4	50	32	12
A_4	0	1	5	21	1	0	4	17	75	4
A_5	0	4	15	15	3	0	10	41	41	8
A_6	0	1	3	2	0	0	17	50	33	0
A_7	0	1	10	11	0	0	5	45	50	0
TOTALS	1	13	107	120	34	***				

A REPORT ON THE STUDY OF THE PORTUGUESE NAVY CASE

Marc Pirlot

Université Libre de Bruxelles
BELGIUM

FOREWORD. As author of this report and participant to the Summer School, I have to make a preliminary remark. In writing the present report, I tried to restate the atmosphere and outline the main trends in the work done on the Navy case rather than provide an exhaustive account. This can maybe result in a subjective presentation, but I have paid attention to using the «I» pronoun as soon as I was aware of expressing my own views on some point. Moreover, as written records of about half of the working groups activity were not available, I had to invoke my personal memories to fill in the gaps. I hope I did not forget too many important things nor I did (too much) distort the rest. Anyway, I am asking for indulgence both from the participants and the reader.

The problem of Portuguese Navy Officers evaluation and promotion was presented to the participants by Carlos Bana e Costa at the very beginning of the School (see the presentation of the case by Bana e Costa and Rodrigues in this book). It was asked to the participants to constitute a number of working groups of about eight people each, with at least one permanent advisor (in general, a lecturer who would stay during the whole School) and at least one Navy Officer who would answer the questions concerned with the concrete situation. Hervé Raynaud was in charge of the coordination of the groups and of the organisation of the two meetings where each group's delegate presented an account of the group activity.

1. GENERAL ATMOSPHERE

At the first meeting, on Thursday in the first week of the School, it was planned that each group would send a delegate to inform the audience of his group's first impressions and program of action. Before the meeting, one could see rather excited people meet in the corridors and explain each other the ranking they had obtained with one or another method.

Very quickly after the meeting, people realised this unexpected thing: nearly everybody had obtained about the same ranking. At first glance, all methods from the most sophisticated to the «quick and dirty» weighted sum yielded about the same results. The enthusiasm of the first week was followed by a rather general disappointment. A more in-depth analysis of the data seemed impossible and the apparently incontestable first ranking appeared difficult to justify. To this situation, the groups reacted in various ways that we shall try to summarise in the sequel.

2. FIRST LOOK AT THE DATA

The best way for the reader to get a feeling of the case is probably to have a look at the histograms reproduced in Appendix 1 (for the original data, the reader is referred to the presentation of the case in this book). Drawing histograms was actually the first approach of many participants. For each candidate, C_1 to C_{12}, an histogram is drawn representing the relative frequencies of the levels S, I, II, III, IV, obtained by counting the assessments in each class without distinguishing between the attributes.

As a general remark, the levels S and I are very seldom observed (53 occurrences of «I» and only one of «S» in a set of 4,164 assessments, i.e. a relative frequency of 1.3%). For almost all candidates (with the exception of C_{11}, C_{12} and C_9) class III is a very strong mode so that what seems essential at first glance is the balance between the frequency of «IV»'s and the frequency of «II»'s. This could suggest the following purely ordinal ranking method. First, rank the candidates according to their modal class (i.e. the most frequent assessments), then break the (many) ties by ranking them in decreasing order of the difference:

(% of assessments above the mode)
- (% of assessments below the mode).

For all candidates but C_{11}, the modal class is III. One obtains the following ranking:

$$C_9 > C_2 > C_3 > C_7 > C_1 > C_4 > C_6 > C_5 > C_{10} > C_8 > C_{12} > C_{11}.$$

Were there no «S» and no «I» at all, the above criterion for breaking ties would yield exactly the same ranking as with giving equidistant valuations to II, III and IV (the same for all attributes) and calculating frequency-weighted means: the only difference with the former ranking is an inversion of the positions of C_7 and C_1. The mild dispersion of the assessments can certainly be considered as one of the main stability factors.

After the exploratory phase, many people tried to go beyond the empirical methods and made attempts to «solve» the problem on the basis of general rational methods, but this raised so many questions about the data that people got discouraged. Close examination of the data and discussions with each group's Navy Officer showed that the information in its present state was hardly interpretable. Among the problems mentioned at the first meeting of the groups or in informal discussions held during the first week are the following:

a) What is the meaning of the «scale» {S, I, II, III, IV} ? Is it an ordinal scale, an interval scale, a ratio scale ? (For a discussion of these notions, see J.-C. Vansnick's paper in this book). For instance, is it meaningful to ask whether the difference between level «S» and level «I» is the same as the difference between levels «III» and «IV»?

(It is indeed meaningful if the scale is an interval scale). In the case it is meaningful, one has no possibility of asking the judge who made the assessment to compare those differences. Similarly for ratio scales. In a less formal way, how much importance shall we give to exceptional assessments like «S» or «I» ? Moreover, one should point out that the above questions can have different answers depending on the judge and also the attribute.

b) What is the relative importance of the attributes? (For the board who makes the decisions? For the judges?) Can this importance be reflected in weights ? Has level «III» for instance the same meaning for each judge on each attribute?

c) It is impossible to identify the judges who made the assessments while this could be an important information as:

* some of the candidates have been assessed many times by the same judge, others by many different judges;
* all candidates did not receive the same number of assessments: this number ranges from 275 to 436;
* the assessments were made at different epochs; do the ancient ones have the same relevance as the most recent ones?

d) Some of the attributes are aggregates of sub-attributes and the detailed assessments on the sub-attributes have been aggregated in some way.

3. ATTITUDES

In front of such a situation, there was a wide variety of reactions among the groups. In this report, I did not try to give a complete description neither of the groups activity during the second week nor of the presentations that were made by the end of the School: the reader can get an overview of the whole work by looking at Appendix 2.

In this section I present a classification of the main activities in different trends.

3.1. Those who wanted to "solve" the case...
... by constructing a rigorous scientific method adapted to the case

This was an ambition of a few of the most theoretically inclined participants (mainly lecturers). But, during the School, nobody could conceive an idea of a rigorous method developed on the basis of explicit assumptions that the case could reasonably fulfil.

3.2. Simple methods vs sophisticated methods

The first attitude can be summarised by a sentence like : «How can I help the decision maker ?». Some of the teams went on working with the essentially descriptive elementary methods used during the preliminary phase. For instance Group 4 worked with histograms using the Lotus 123 spreadsheet. In his presentation, the group made it clear that the results obtained through a clever use of Lotus are rather similar to those obtained with more specific tools like VISA (Group 3) (see Belton and Vickers' paper in this volume). Anyway, in the presence of badly conditioned and little reliable data, one attitude is to use as simple tools as possible in order to examine the data, present them in an informative way to the decision maker(s), helping him (them) to make up his (their) mind(s) and allowing him (them) to manipulate the data. In the case, graphics have proved of invaluable help: just a look at such simple tools as the histograms in Appendix 1 shows a lot of things as already mentioned.

The «sophisticated» attitude, at the opposite, can be characterised in our case by the point of view reflected in the following sentence: «It is a good exercise for the participants to try a method on the data». This resulted sometimes in the group trying his advisor's favourite method.

The methods used were (in alphabetic order) ELECTRE, EXPERT CHOICE, MAPPAC-PRAGMA, ORESTE, PREFCALC, REGIME, VISA (see papers by B. Roy, D. Vanderpooten, E. Forman, B. Matarazzo, E. Jacquet-Lagrèze and R. Janssen et al. in this book) and probably others that I cannot remember. I shall not enter into the detail of the results, but they very often confirmed the preliminary observations; in particular, C_9, C_2 and C_3 are nearly always the leading candidates. A lot of sensitivity analysis was done on the parameters of the models (which were in general determined by questioning each group's Navy Officer): this work confirmed the great stability of the ranking.

3.3. A particularly sophisticated method

An extreme variant of the sophisticated attitude is illustrated by the work of Group 2 who applied the ORESTE method (Roubens, 1982; Pastijn and Leysen, 1989). To obtain the rankings (weak orders) of the candidates on each point of view, the group invokes some pieces of statistical theory. Taking argument of the large number of assessments, they consider the «evaluation» of each candidate on each point of view as a frequency distribution on the set {S, I, II, III, IV}. With each candidate C_i is associated the set $\{F_{i,k}(x)\}$ of <u>cumulative</u> distribution functions, one for each attribute A_k. The difference in favour of candidate C_i w.r.t. candidate C_j on attribute A_k is evaluated by:

$$\Delta_k(i,j) = \max_{x} \{F_{i,k}(x) - F_{j,k}(x)\}$$

This evaluation, once renormalised to take into account the different numbers of assessments made about C_i and C_j on A_k, is used to define the ranking of the candidates on each point of view, by means of pairwise comparisons. The use of cumulative distribution functions and the distance Δ_k above are inspired by the Kolmogorov-Smirnov test in Statistics and the group invoked the statistical theory in justification of their approach.

In my view, this is an interesting example of a common attitude which can maybe make impression on a decision maker if he is sensible to the prestige of Theory but can deter some other DM's from using scientific methods they do not understand. This approach can also be a tactics for imposing the solution of the specialist, but it will probably not help a decision maker to understand the problem. I do not pretend however that simplicity is in general a good criterion to choose a decision aid method: a sophisticated method can be used if its hypotheses are fulfilled. Here, I think that invoking Kolmogorov - Smirnov is merely reasoning by analogy but cannot be considered as a scientifically valid justification. Indeed, the sets of assessments associated to each candidate on each point of view cannot be considered as samples of a random variable.

3.4. Experiments with the methods

Another trend was to use the data for testing the methods rather than solving the problem. Group 5 experimented with the valuation of the ordinal scale (S, I, II, III, IV) on an interval scale (see the paper by J.-C. Vansnick in this book). Using the two degrees of freedom of the scale, they attributed value 0 to «II» and value 10 to «III». Afterwards they asked eleven Navy Officers to locate the other «levels» w.r.t. to 0 and 10, for each attribute. For attribute A_7 e.g., this resulted in Figure 1 which shows the relative position of level «IV» for the different officers. Similar experiments showed that the same person gives very different valuations to the «same» level IV depending on the attribute: e.g. «IV» is 22 on A_2 and 40 on A_1. From the great variability in the valuations the group concluded in the necessity of using an ordinal method in this case.

Group 7 also experimented with scalings and, in addition, with «weights» on the points of view. The objective was to compare the results provided by a software like EXPERT CHOICE (see the paper by E. Forman in this book) to a direct intuitive evaluation.

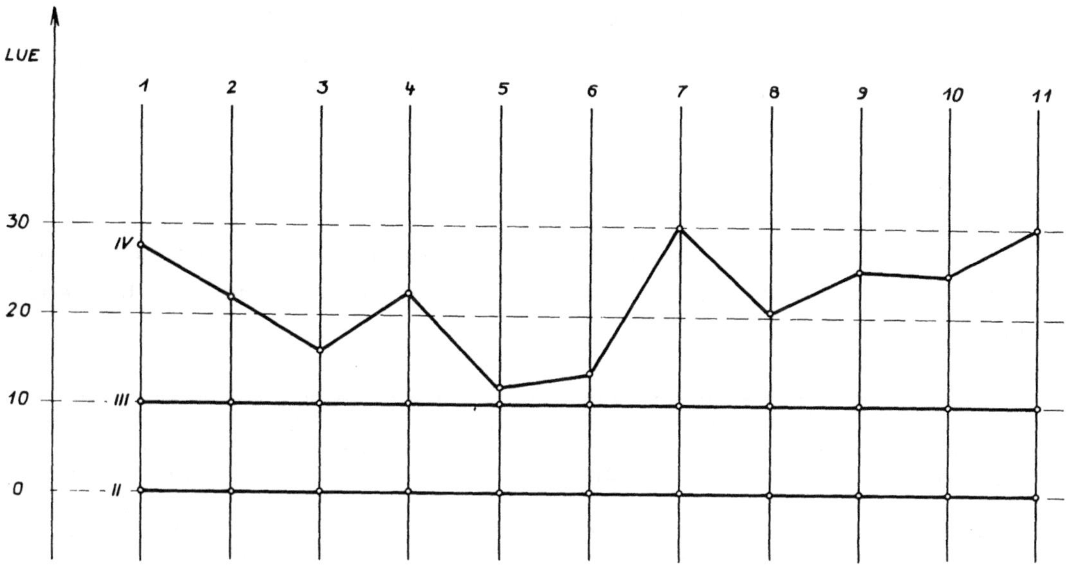

Figure 1. Observed variations in the valuation of the level IV on attribute A_7 by 11 officers

The group begun with asking his Navy Officer for intuitive valuation of the levels {S, I, II, III, IV} which resulted in the first row of Table 1. Then they asked the same Navy Officer to use the EXPERT CHOICE software to obtain the pairwise comparison matrix on the basis of which EXPERT CHOICE operates. This yielded the valuations (as a ratio scale) given in the second line of Table 1.

	S	I	II	III	IV
Intuitive	-40	0	10	20	50
EXPERT CHOICE	.028	.050	.113	.237	.572

TABLE 1. Comparison of two scalings of the levels

While the Navy Officer was answering the questions of EXPERT CHOICE, it was evident that he was reasoning in terms of <u>differences</u> between levels, but EXPERT CHOICE mapped his comparisons into a ratio scale.

The same phenomenon occurred in the assessment of weights to the attributes (see Table 2). The ordinal (intuitive) ranking by the Navy Officer was:

$$A_5 > A_2 \sim A_6 > A_1 \sim A_3 \sim A_4 > A_7.$$

In this case, the ordinal ranking is not even respected by EXPERT CHOICE as A_3, for instance, receives the second highest mark.

	A_1	A_2	A_3	A_4	A_5	A_6	A_7
Intuitive	25	40	25	25	50	40	50
EXPERT CHOICE	.074	.107	.224	.115	.295	.116	.068

TABLE 2. Comparison of sets of weights

This experiment shows that one should be extremely careful when using a decision aid software whatever it is: the decision maker can feel comfortable in answering the questions raised in an interactive software without agreeing with the fundamental hypothesis whereupon the method relies.

4. CONCLUSIONS

Though the case was not «solved» during the School, it has been a particularly important element in the success of the School. First of all it has been highly profitable for the participants to think about a real situation and become aware of all kinds of difficulties that can arise in concrete cases. Second, the case was a good exercise for familiarising the participants with the methods taught during the School and this motivated them and maintained their attention at a high level during the whole course of this intensive series of conferences.

In conclusion, one could not put enough emphasis in recommending to all future organisers of such schools to consider the

APPENDIX 1. Relative frequencies of assessment levels for candidates C_1 and C_{12}

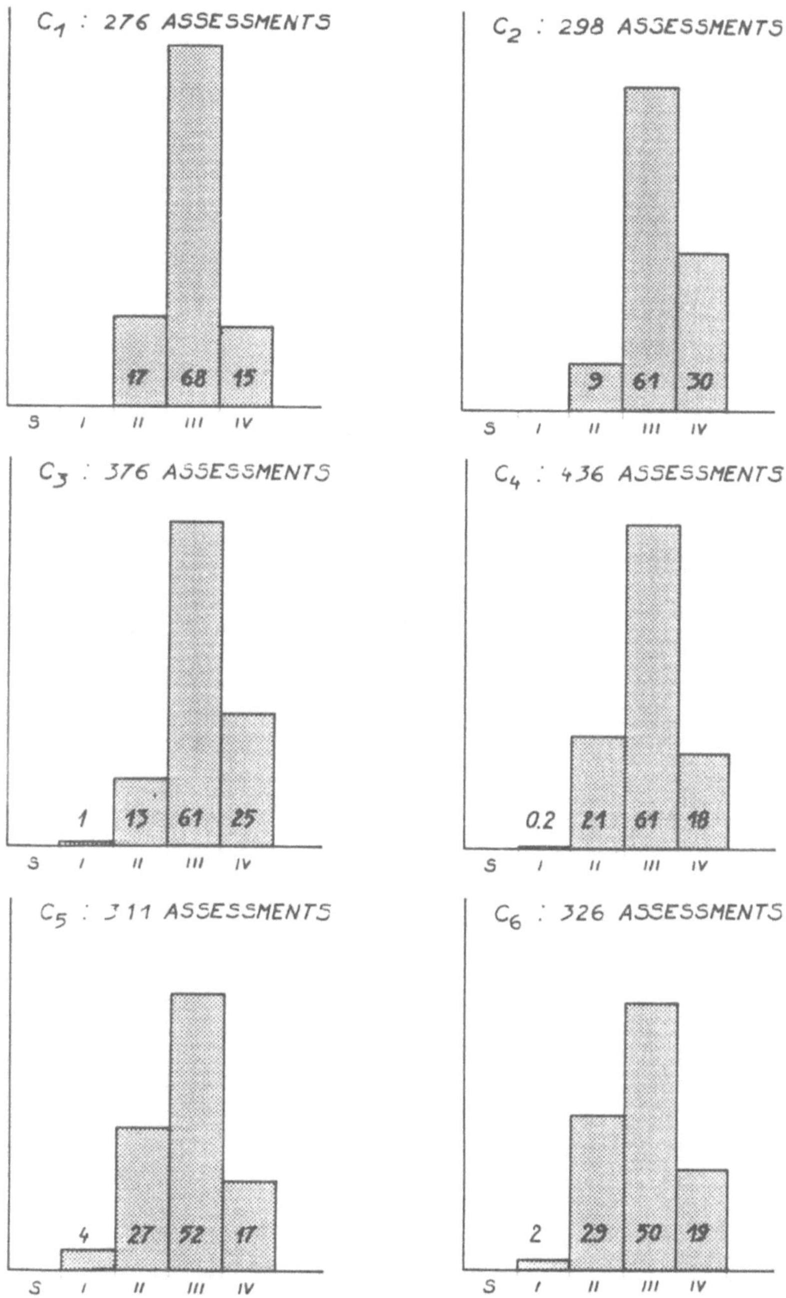

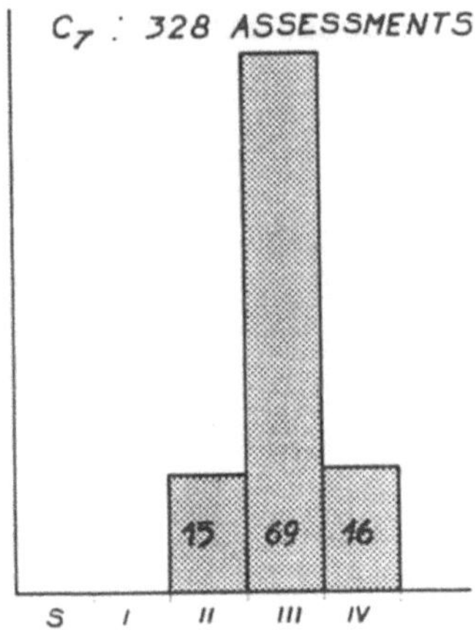
C_7 : 328 ASSESSMENTS
15
69
16
S I II III IV

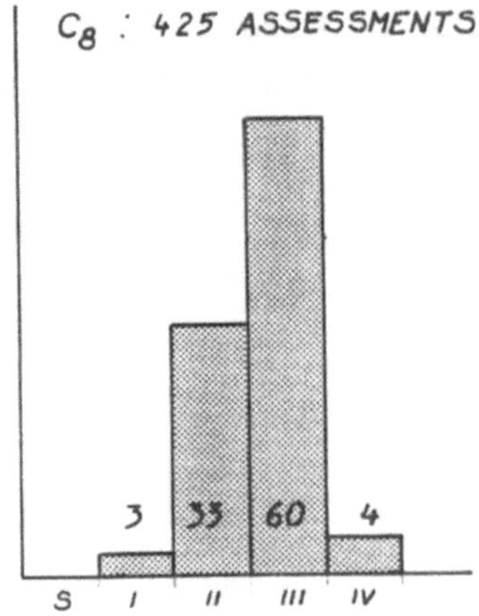
C_8 : 425 ASSESSMENTS
3
33
60
4
S I II III IV

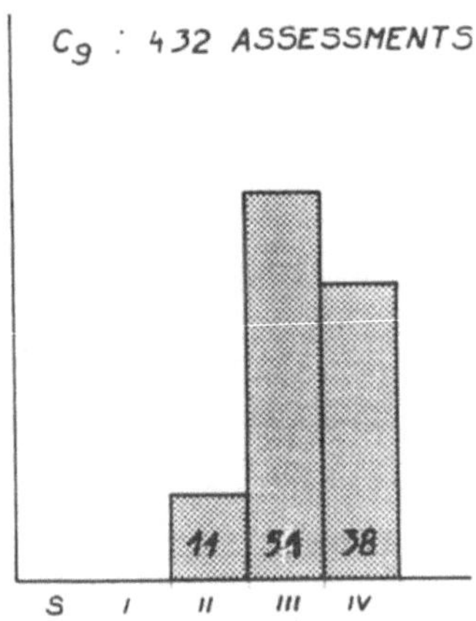
C_9 : 432 ASSESSMENTS
11
51
38
S I II III IV

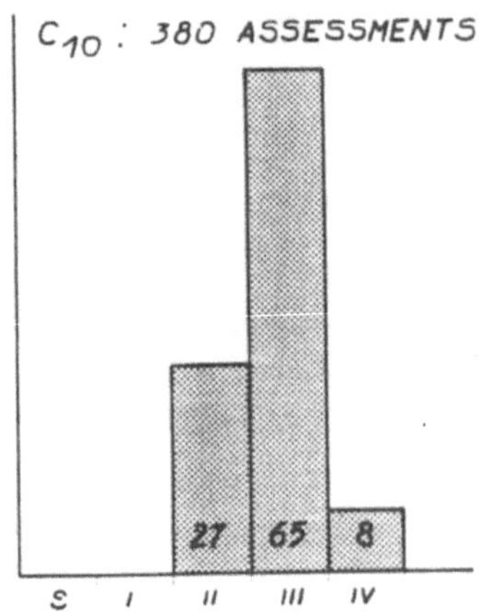
C_{10} : 380 ASSESSMENTS
27
65
8
S I II III IV

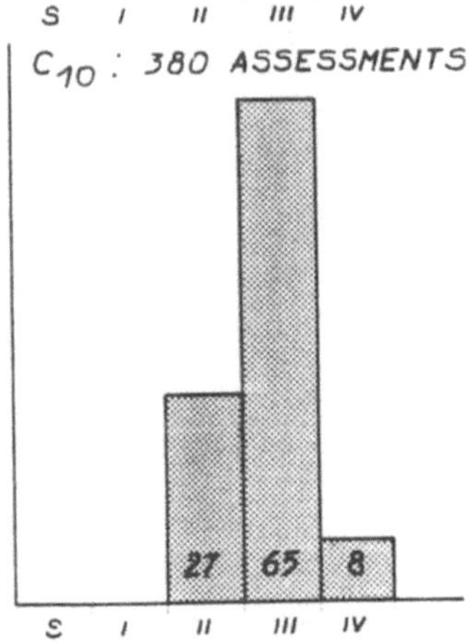
C_{10} : 380 ASSESSMENTS
27
65
8
S I II III IV

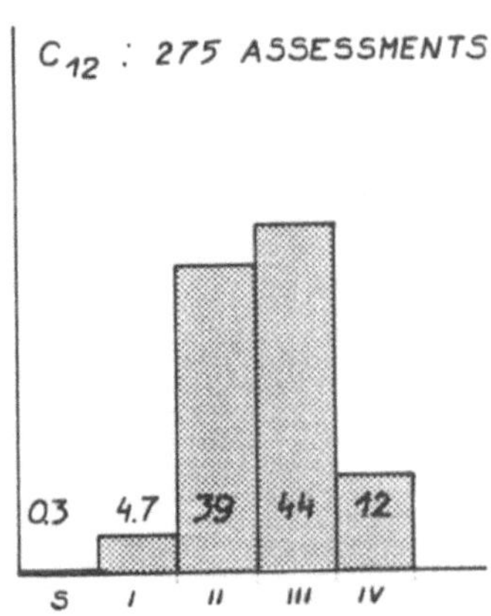
C_{12} : 275 ASSESSMENTS
0.3
4.7
39
44
12
S I II III IV

APPENDIX 2. List of the groups and main activity

GROUP 1. Advisors: B. Matarazzo, J. Spronk.
Report presented by M. Leotta and G. Munda.
Main topics: application of MAPPAC-PRAGMA method as an illustration of the conference of B. Matarazzo.

GROUP 2. Advisors: H. Pastijn, M. Roubens.
report presented by L. Lillich.
Main topics: application of the ORESTE method.

GROUP 3. Advisor: D. Bouyssou.
Report presented by D. Bouyssou.
Main topics: use of LOTUS 123 spreadsheet and its graphic facilities.

GROUP 4. Advisors: V. Belton, E. Forman.
Report presented by V. Belton.
Main topics: use of VISA and EXPERT CHOICE softwares.

GROUP 5. Advisor: J.-C. Vansnick.
Report presented by J.-C. Vansnick.
Main topics: experiments with scales of measurement.

GROUP 6. Advisor: D. Vanderpooten.
Report presented by D. Vanderpooten and R. Janssen.
Main topics: application of ELECTRE and REGIME methods.

GROUP 7. Advisor: M. Pirlot.
Report presented by J. Faria.
Main Topics: experiments with weights and scales; stability of the rankings with different methods.

REFERENCES

Bana e Costa, C.A. and Rodrigues, J.C. (1990), "Presentation of the school case-study: evaluation of personnel - how to rank officers for promotion?", in this volume.

Belton, V. and Vickers, S. (1990), "Use of a simple multi-attribute value function incorporating visual interactive sensitivity analysis", in this volume.

Forman, E.H. (1990), "Multi criteria decision making and the analytic hierarchy process", in this volume.

Jacquet-Lagrèze, E. (1990), "Interactive assessment of preferences using holistic judgments - the PREFCALC system", in this volume.

Janssen, R., Nijkamp, P. and Rietveld, P. (1990), "Qualitative multicriteria methods in the Netherlands", in this volume.

Matarazzo, B. (1990), "A pairwise criterion comparison approach: the MAPPAC and PRAGMA methods", in this volume.

Pastijn, H. and Leysen, J. (1989), "Constructing an outranking relation with ORESTE", *Mathematical and Computer Modelling*, Vol. 12 (10/11), 1255-1268.

Roubens, M. (1982), "Preference relations on actions and criteria in multi-criteria decision making", *European Journal of Operational Research*, Vol. 10 (1), 51-55.

Roy, B. (1990), "The outranking approach and the foundations of ELECTRE methods", in this volume.

Vanderpooten, D. (1990), "The construction of prescriptions in outranking methods", in this volume.

Vansnick, J.-C. (1990), "Measurement theory and decision aid", in this volume.

MIX
Papier aus verantwortungsvollen Quellen
Paper from responsible sources
FSC® C105338

FSC
www.fsc.org

If you have any concerns about our products,
you can contact us on
ProductSafety@springernature.com

In case Publisher is established outside the EU,
the EU authorized representative is:
Springer Nature Customer Service Center GmbH
Europaplatz 3, 69115 Heidelberg, Germany

Printed by Libri Plureos GmbH
in Hamburg, Germany